EDUCATIONAL INNOVATION IN SOCIETY 5.0 ERA: CHALLENGES AND OPPORTUNITIES

PROCEEDINGS OF THE 4TH INTERNATIONAL CONFERENCE ON CURRENT ISSUES IN EDUCATION (ICCIE 2020), YOGYAKARTA, INDONESIA, 3–4 OCTOBER 2020

Educational Innovation in Society 5.0 Era: Challenges and Opportunities

Edited by

Yoppy Wahyu Purnomo & Herwin
Faculty of Education, Universitas Negeri Yogyakarta, Indonesia

CRC Press/Balkema is an imprint of the Taylor & Francis Group, an informa business

© 2021 selection and editorial matter, the Editors; individual chapters, the contributors.

Typeset by MPS Limited, Chennai, India

All rights reserved. No part of this publication or the information contained herein may be reproduced, stored in a retrieval system, or transmitted in any form or by any means, electronic, mechanical, by photocopying, recording or otherwise, without written prior permission from the publisher.

Although all care is taken to ensure integrity and the quality of this publication and the information herein, no responsibility is assumed by the publishers nor the author for any damage to the property or persons as a result of operation or use of this publication and/or the information contained herein.

Library of Congress Cataloging-in-Publication Data

A catalog record has been requested for this book

Published by: CRC Press/Balkema
                Schipholweg 107C, 2316 XC Leiden, The Netherlands
                e-mail: enquiries@taylorandfrancis.com
                www.routledge.com – www.taylorandfrancis.com

ISBN: 978-1-032-05392-9 (hbk)
ISBN: 978-1-032-07231-9 (pbk)
ISBN: 978-1-003-20601-9 (ebk)
DOI: 10.1201/9781003206019

Educational Innovation in Society 5.0 Era: Challenges and Opportunities – Purnomo & Herwin (Eds)
© 2021 the authors, ISBN 978-1-032-05392-9

Table of contents

Preface   ix
Committees   xi

Society 5.0 and education in Japan   1
H. Masami

Innovation in early childhood and primary education   7
G. Fragkiadaki, M. Fleer & P. Rai

Strengthening resilience for learning transformation and anticipatory education in the era
of society 5.0   11
S.I.A. Dwiningrum

The roles of social capital to promote soft skills among university students in facing the challenge
of society 5.0 transformation   18
A.R. Ahmad, M.M. Awang & N.A. Mohamad

Perception and attitude of student to character education towards society 5.0 era   27
K. Fajriatin & A. Gafur

Implementation of civic virtue in character education in the era of Society 5.0   31
Y. Mahendra, Nasiwan & S.H. Rahmia

The importance of using e-learning in teaching social science during and after the Covid-19 pandemic   36
E. Suprayitno, Aman & J. Budiman

The development of instrument analysis for elementary school children's social interaction patterns
in the era of revolution 4.0   40
Firmansyah, A. Senen, Mujinem, Hidayati & S.P. Kawuryan

The effectiveness of batik learning activities on the improvement of character values in grade VII
students of SMP PGRI 8 Denpasar   45
I.K. Mahendra & N.M.M. Minarsih

The need fulfillment of assistive technology for students with intellectual disabilities in Indonesia   50
Ishartiwi, E. Purwandari, R.R. Handoyo & A. Damayanto

Learning mathematics online during the Covid-19 pandemic: Is it without problems?   54
Desmaiyanti & Sugiman

Assessing the discriminant validity of the curiosity scale using confirmatory factor analysis   60
H. Sujati

Collaboration practices between educators in inclusive education before and during Covid-19   64
W. Hardiani & Hermanto

Investigating students' self-regulated learning and academic procrastination on primary school during
distance learning   69
T. Nugraha & S. Prabawanto

Exploring e-learning platforms used by students in Indonesia during the Covid-19 pandemic   74
S. Rahayu & Supardi

Pre-service teacher education reform in Indonesia: Traditional and contemporary paradigms   80
A. Mustadi, P. Surya & M.-Y. Chen

Envisaging Montessori visions on a K-2 learning environment in a digital form 91
M.H. Ismail

Online learning feedback for elementary school during the Covid-19 pandemic 96
F.N. Ismiyasari, E. Rahmawati, W. Kurniawan, Sutama, C. Widyasari, Z. Abidin & Z. Arifin

Multiliteracy education challenges: Will narrative texts used in textbooks open students' minds to critical literacy? 101
H.N. Fadhlia & W. Purbani

Adaptive learning 4C skills during and after Covid-19 in elementary school 107
A. Yatini & B.E. Mulyatiningsih

Multiliteracy pedagogy challenges: EFL teachers' multicultural attitudes in the literacy classroom practices 113
U. Sholihah & W. Purbani

Implementation of wooden craft vocational learning to improve life skills in students with disabilities 119
A. Sulistyo & Kasiyan

Support of parents and schools for online learning during the Covid-19 pandemic 125
I.W. Liasari & A. Syamsudin

The influence of a gamification platform and learning styles toward student scores in online learning during the Covid-19 pandemic using split-plot design 132
W.P. Hapsari, Haryanto & U.A. Labib

Strengthening beginners character education in facing the 2020 general election 138
D. Purba, B. Juantara & I. Bulan

Revitalization of local wisdom values in strengthening cooperation character towards community civilization 5.0 142
S.F. Shodiq, D. Budimansyah, E. Suresman & M. Hidayat

The implementation of project based learning through mind mapping to increased student creativity 147
U.M. Sadjim & R. Jusuf

The dilemma of Timorese education in the COVID-19 pandemic 151
Syahrul, Arifin & A. Datuk

Model of project based learning in online learning during and after the Covid-19 pandemic 157
E. Wijaya, Nopriansah & M. Susanti

Communal learning model as blended learning strategy in primary school during the Covid-19 pandemic 163
K.I. Sujati, A. Syamsudin, Haryanto & W.P. Hapsari

Online learning in the medicinal education during the pandemic era: How effective are the platforms? 169
Y. Febriani, H. Haritani, P. Hariadi, T.P. Yuliana, A. Rafsanjani, M. Azim & E.E. Oktresia

The effectiveness of the storybooks on the love of the homeland character trait for elementary schools 175
W. Wuryandani, Fathurrohman, E.K.E. Sartono, Suparlan & H. Prasetia

Strengthening a student's character in the era of society 5.0 in primary school 178
Jamilah, T. Sukitman & M. Ridwan

Healthy school behaviour in public elementary school in Sanden Bantul 182
S.N. Isvandari & C.S.A. Jabar

Cooperative learning model talking stick type: To improve speaking skills? 186
A.J. Verrawati, A. Mustadi & W. Wuryandani

Building the critical thinking skills of elementary students through science thematic learning using a guided inquiry model 192
P. Pujiastuti & D. Rahmawati

Analysis of improving student's statistics thinking mathematic education 196
T.H. Nio, B. Manullang, H. Suyitno, Kartono & Sc. Maryani

The effect of porpe strategy to improve the understanding of the concept of social sciences in online learning 199
L.R. Hidayah, M.N. Wangid & S.P. Kawuryan

A case study on the implementation of marketing competencies for deaf children in entrepreneurship education 204
E. Bunyanuddin & N. Azizah

The strategy of the principal of the elementary school of Jogja Green School in facing the Covid-19 pandemic 209
B.D. Jaswanti & E. Purwanta

Development of Star Book Media to influence writing skills and carrying attitude in grade IV elementary school students 214
E. Zubaidah, S. Sugiarsih & A. Mustadi

Study motivation and students' participation in distance learning during the Covid-19 pandemic 220
H. Nuryani & Haryanto

Increasing ecological intelligence through strengthening social studies education 226
R.A. Basit, A. Yuliyanto, B. Maftuh & H.E. Putri

Teacher's quality of pedagogical influence on a student's character in the society 5.0 era 232
N.K. Suarni, G.N. Sudarsana & M.N.M.I.Y. Rosita

The effect of a simplified integrated learning environment on plagiarism behavior 238
G.N. Sudarsana, N.K. Suarni & I.K. Dharsana

The profile of pre-service elementary school teachers in developing lesson plan for science instruction 244
W.S. Hastuti, P. Pujiastuti, Pujianto & Purwono

Documenting factors influencing children like learning English as a foreign language 249
R. Rintaningrum & S. Iffat Rahmatullah

The content validity analysis of the elementary school students' tolerance character measurement instrument 255
H. Sujati, Haryani, B.S. Adi, Kurniawati & T. Aprilia

Impoliteness language on social media: A descriptive review of PGSD UNY students 260
O.M. Sayekti, A. Mustadi, E. Zubaidah, S. Sugiarsih & E.N. Rochmah

Analysis of students' historical empathy in history education 266
S. Dahalan & A.R. Ahmad

Analysis of the implementation of primary school teachers' professional duties during the Covid-19 pandemic in Sleman Yogyakarta 270
A. Hastomo, B. Saptono, S.D. Kusrahmadi, F.M. Firdaus & A.R. Ardiansyah

Fun and interesting learning to improve students' creativity and self-confidence in the 5.0 social era 275
M. Susanti, Y.P. Sari, K. Karim & Sabri

Goal-orientation measurement model: A study of psychometric properties using a ranking scale 281
F.A. Setiawati & T. Widyastuti

Multiliteracy education in 5.0 era on learning entrepreneurship with the project based learning model 286
B. Afriansyah & N.V. Yustanti

Author index 291

*Educational Innovation in Society 5.0 Era: Challenges and
Opportunities – Purnomo & Herwin (Eds)
© 2021 the authors, ISBN 978-1-032-05392-9*

Preface

Educational innovation in the era of Society 5.0 is directed to resolve various social challenges, issues, and problems relating to educators, students, the dynamics of the education system, and social dynamics. Era Society 5.0 is an answer to the challenges that arose due to problems resulting from the Industrial Revolution 4.0 era by utilizing innovations in technology that integrate cyberspace and the physical world. This is expected to balance economic development and solve social problems. Based on the background of the situation, there was a need for a forum that was able to explore and publish various results of studies and research related to educational innovations in the era of Society 5.0. The 4th International Conference on Current Issues in Education (ICCIE) 2020 took place in Yogyakarta on October 3-4, 2020. The conference was organized by Yogyakarta State University (UNY) in collaboration with Universiti Kebangsaan Malaysia (UKM).

There were 226 participants from countries all over the world attending the conference. The scientific program consisted of in total 92 talks, a big part of them presented in 10 mini-symposia. Five talks were invited plenary lectures given by Assoc. Prof. Dr. Hayashi Masami (Japan), Prof. Dr. Glykeria Fragkiadaki (Australia), Prof. Dr. Juliane Stude (Germany), Prof. Dato' Dr. Abdul Razak Ahmad (Malaysia), and Prof. Dr. Siti Irene Astuti Dwiningrum (Indonesia).

We would like to express our appreciation to the many people who contributed to the success of the conference: the plenary and keynote speakers, the authors, the participants, the session chairs, and the members of the Committees who nominated plenary and keynote speakers. The editors are especially grateful to those who reviewed the manuscripts included in this book.

The Editors

Yoppy Wahyu Purnomo
Herwin

Committees

Steering Committee
Dr. Sujarwo, M. Pd.
Dr. Cepi Safruddin Abd Jabar, M.Pd.
Drs. Bambang Saptono, M.Si.
Joko Pamungkas, S.Pd., M.Pd.
Dr. Anwar Senen, S.Pd., M. Pd.

Chair
Dr. Ali Mustadi S.Pd., M. Pd.

Vice Chair
Dr. Fery Muhamad Firdaus, S.Pd., M.Pd.

Secretary
Vinta Angela Tiarani, M.Si, M.Ed., Ph.D.
Amalia Rizki Ardiansyah, S.Pd., M. Pd.

Scientific Committee
Prof. Juliane Stude, Ph.D. (*Münster University, Germany*)
Prof. Dr. Maryani (*Universiti Pendidikan Sultan Idris, Malaysia*)
Prof. Poornsook Tantrarungroj, Ph.D. (*Chulalongkorn University, Thailand*)
Chanita Rukspollmuang (*Chulalongkorn University, Thailand*)
Asc. Prof. David Evans, Ph.D. (*University of Sydney, Australia*)
Prof. Dato' Dr. Abdul Razak Ahmad (*Universiti Kebangsaan Malaysia, Malaysia*)
Dr. Mahzan Awang (*Universiti Kebangsaan Malaysia, Malaysia*)
Salleh Amat (*Universiti Kebangsaan Malaysia, Malaysia*)
Ming-Huan Lin (*National Chiayi University, Taiwan*)
Ruyu Hung (*National Chiayi University, Taiwan*)
Mei-ying Chen (*National Chiayi University, Taiwan*)
Laura Apol (*Michigan State University, United States*)
Vijay Kumar Mallan (*University of Otago, New Zealand*)
Maria I. E. Manzon (*Nanyang Technological University, Singapore*)
Sharon Russo (*The University of South Australia, Australia*)
Harri Lappalainen (*Turku University of Applied Sciences, Finland*)
Mark Bray (*The University of Hongkong, Hongkong*)
Prof. Tan Oon Seng (*National Institute of Education, Singapore*)
Prof. Aurora Adina, Ph.D. (*Stefan cel Mare University of Suceava, Romania*)
Dr. Wuri Wuryandani, S.Pd, M.Pd. (*Universitas Negeri Yogyakarta, Indonesia*)
Dr. Sigit Sanyata, M.Pd. (*Universitas Negeri Yogyakarta, Indonesia*)
Dr. Ali Mustadi, S.Pd., M.Pd. (*Universitas Negeri Yogyakarta, Indonesia*)
Dr. Yoppy Wahyu Purnomo, S.Pd., M.Pd. (*Universitas Negeri Yogyakarta, Indonesia*)
Vinta Angela Tiarani, M.Si, M.Ed., Ph.D. (*Universitas Negeri Yogyakarta, Indonesia*)
Nu Azizah, S.Pd., M.Ed. Ph.D. (*Universitas Negeri Yogyakarta, Indonesia*)
Ika Budi Maryatun, M.Pd. (*Universitas Negeri Yogyakarta, Indonesia*)
Pujaningsih, S.Pd., M.Pd. Ed.D. (*Universitas Negeri Yogyakarta, Indonesia*)
Siti Rohmah Nurhayati, S.Psi ., M.Si. (*Universitas Negeri Yogyakarta, Indonesia*)

Organizing Committee:
Dr. Wuri Wuryandani, S.Pd., M. Pd.
Octavian Muning Sayekti, S.Pd., M. Pd.
Widayanti
Sekar Purbarini Kawuryan, S.I.P., M.Pd.
Yunita Fitriatun, S.Pd.
Tika Aprilia, S.Pd., M. Pd.
Dr. Amir Syamsudin, M.Ag.
Rahmat Fadhli, S.IIP., M.A.
Irfan Wahyu Prananto, S.Pd., M. Pd.
Rendy Roos Handoyo, S.Pd., M. Pd.
Tria Widyastuti, S.Psi., M.A.
Agung Hastomo, S.Pd., M. Pd.
Budiono
Supartinah, S.Pd., M.Hum
Dr. Yoppy Wahyu Purnomo, S.Pd., M. Pd.
Dr. Herwin, S.Pd., M. Pd.
Ernisa Purwandari, M.Pd.
Septinda Rima Dewanti, M.Pd.

Educational Innovation in Society 5.0 Era: Challenges and
Opportunities – Purnomo & Herwin (Eds)
© 2021 the authors, ISBN 978-1-032-05392-9

Society 5.0 and education in Japan

H. Masami
Tokyo Gakugei University, Koganei, Japan

ABSTRACT: Japan calls the future, which is impossible to predict with precision, Society 5.0. Current occupations will be replaced by machines and human jobs will decrease. However, humans will have to do new jobs to make a living. Japan examined what kind of policies are needed for peace and sustainability for humanity. As a result, we decided that we need to master artificial intelligence (AI) and train people who cannot be replaced by AI. However, specific education is still being explored in all countries. In this paper, I have first identified the characteristics of Japan's educational reform. Then, the future of education in Japan was discussed based on the OECD's Learning Compass 2030.

1 INTRODUCTION

1.1 *The purpose of the investigation*

The purpose of this presentation is to introduce Japan's Society 5.0 and relativize its advantages and challenges. In times of slow change, the future can be predicted by analyzing past data. However, at times, there is a significant disconnect between the past and the future. For example, the present and the future are not continuously connected, as exemplified by the pandemic of the new coronavirus disease. We live in these unpredictable times. First of all, we need to be able to analyze big data and choose the best option. On the other hand, we also need to be able to deal with unpredictable situations that differ from past trends and how best to deal with them. Society 5.0, as proposed by Japan, is the inevitable future. The decision on what kind of education is needed in Society 5.0 era will differ from country to country. This decision will probably change the future of each country.

1.2 *The problem being investigated*

I will clarify what kind of Society 5.0 is proposed by Japan, its advantages, and the direction in which it should be incorporated into education. It should be noted that Society 5.0 is not an ideal future, but a reality brought about by technological innovation. Technological innovation allows machines to fill in many parts of human activities. It could be said that machines will take away human jobs. This includes the question of how humans will live in a future where machines will replace many parts of human work. Education, as proposed by the OECD, may provide a hint of a solution. I believe that one of the answers is an education that allows students to exercise competency and agency towards well-being.

1.3 *The background*

In today's world of the Internet, new future strategies are becoming an important issue for the industry as well. In conjunction with this, education strategies need to be reoriented for optimization. Governments need to examine their education policies for the sake of peace and a sustainable future for humanity. Here, I would like to show how Society 5.0, which is being considered and implemented by the Japanese government, is changing the educational sector in Japan. Society 5.0 will serve as a reference point for education policy in other countries and for teacher training in universities in other countries, and this will contribute to educational reform.

1.4 *General approach*

This report describes how the Japanese Ministry of Education, Culture, Sports, Science, and Technology (MEXT) and the educational community in Japan have responded to the Japanese government's policy of Society 5.0 and how they are putting it into practice. One common approach is to develop indicators and case studies to test their effectiveness. Since we have yet to measure effectiveness, I will adopt the case study method of the MEXT. I believe that many countries can take a number of cues from Japan's education reforms, which may or may not be positive.

1.5 *The criteria for my study's success*

The goal of the study is to be able to interpret and explain the development of Japan's Society 5.0 strategy in the education world based on OECD's Education 2030 competencies and agencies. Japanese education up to now has tried to work on the individual to make him or her into the desired condition. As

DOI 10.1201/9781003206019-1

a result, students have become more knowledgeable, skilled, and nurtured in their humanity. The Japanese educational community emphasizes the merits of individual education. However, less attention has been paid to how students live in a world of technological innovation and globalization. This is a point of reflection. Therefore, I decided to use the OECD project, which is examining educational models that respond to technological innovation and globalization, to clarify the characteristics of Japan. Tokyo Gakugei University, to which I belong, is conducting a project to provide evidence for educational reform in collaboration with the OECD and MEXT.

2 RESULTS

For predicting the future of employment, a study by Michael A. Osborne of the University of Oxford was the catalyst. There, it was published that many of the jobs that currently exist will be replaced by machines (AI) in the near future (Freya & Osborn 2017).

Influenced by research in this direction, Japan's Nomura Research Institute released a report that said 49% of current jobs in Japan will disappear in the next 15 years. In 2015, Japan's Nomura Research Institute estimated the establishment of computer technology replacements in the next 10-20 years for each of Japan's 601 occupations. It found that about 49% of Japan's working population could be technologically replaced by artificial intelligence (AI). On the other hand, occupations that require knowledge to organize and create abstract concepts, such as art, history, archaeology, philosophy, and theology, were found to be difficult to replace with AI. It was also found that professions that require cooperation with others, understanding of others, persuasion, negotiation, and service orientation are also difficult to replace with AI. Comparing Japan, the United Kingdom, and the United States, it was also found that Japan's workforce is more replaceable by robots and other technologies (Nomura Research Institute 2015). The Nomura Research Institute's comparison of the UK and the US is based on the work of Dr. Osborne and Dr. Frey, among others (Figure 1).

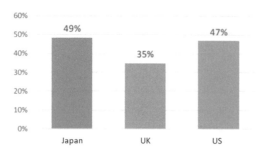

Figure 1. Percentage of the workforce likely to be replaced by AI and robots, compared to Japan, the UK, and the US (Nomura Research Institute 2015).

In the case of Japan, education policy has been developed around the concept of Society 5.0. Society 5.0 is defined by the Japanese government as follows. "A human-centered society that balances economic advancement with the resolution of social problems by a system that highly integrates cyberspace and physical space" It follows the hunting society (Society 1.0), agricultural society (Society 2.0), industrial society (Society 3.0), and information society (Society 4.0). Today's complex society is Society 5.0 (see Figure 2).

The difference between Society 4.0 and Society 5.0 is as follows: in Society 4.0, the Cloud is used, but not yet as big data. This is a society where AI does the analysis (Figure 3).

However, we are not yet at the stage where AI can analyze big data and use it consciously and creatively. Therefore, the MEXT, with an eye on society 5.0 vision of a society, is trying to increase the number of people who can engage in research and development related to AI, and to train people with human strengths that cannot be replaced by AI.

The Ministry of Education, Culture, Sports, Science, and Technology (MEXT) is currently working on three leading projects: Leading Project 1, which aims to optimize learning through the accumulation of student study logs. At present, the results are still to be determined, but support strategies have been initiated for the GIGA (Global and Innovation Gateway for All) school initiative.

For the GIGA school concept, a package of measures for the realization of "one computer per student" was created by MEXT. (December 2020) MEXT's roadmap for the realization of the GIGA school concept calls for organizing digital devices for all classes, from elementary school to high school, by the end of 2022 as a start. And it is planned to implement online education through the use of digital textbooks. The plan is to install wireless LANs in all public high schools by the end of the 2020 school year. The plan is to install wireless LANs in 80% of public schools in elementary and junior high schools by the end of the 2020 school year. The government's budget for this is 231.8 billion for the fiscal year 2019. The purpose of this project is to achieve individualized learning without leaving any of the diverse students behind by using a learning log. In the past, Japanese education has offered a choice of subjects, but the degree of individual optimization was not sufficient. From now on, we will be able to provide more customized learning in the future.

Leading project 2 is a reform of the high school entrance examination. In Japan, high school entrance examinations and university entrance examinations are regarded as important, and the subjects on the Common Entrance Examination will be especially focused on. It has been decided to add "information" to the list of subjects to be included in the university entrance examination from 2024. Related to this, data science and statistics education are also being emphasized in the elementary, middle, and high school curricula.

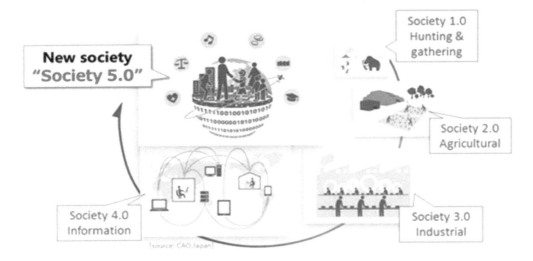

Figure 2. What is Society 5.0? https://www8.cao.go.jp/cstp/english/society5_0/index.html (July 24, 2020)

Figure 3. How Society 5.0 works, https://www8.cao.go.jp/cstp/english/society5_0/index.html (July 24, 2020)

MEXT emphasizes the promotion of information utilization skills and the enhancement of information morality education as a way to promote the informatization of education. A 2016 survey on information utilization skills is the basis of the policy. It had the following results (Table 1).

The results of the survey show that there are challenges in the ability to use information. In particular, collecting the necessary data (10) and doing numerical processing (11) has a very low percentage of correct answers. Besides, MEXT also gives students a message about moral education: do not give out passwords, do not open email links or files immediately, always keep your computer updated, and talk to adults if you need help. There have also been changes in university education, with the creation of the faculty of data science in Japan's national universities (Figure 4).

Along with the emphasis on data science in Japan's high schools and universities, a different kind of teaching is being sought at the compulsory education level as well. At Tokyo Gakugei University, we have been

Table 1. Implementing entity MEXT.

	Practical skills in the use of information	Correct answer rate
1	Reading the current situation from the text including figures and tables	77.7%
2	Question about organizing based on web pages	73.6%
3	Question about solutions based on evidence	37.2%.
4	Suggesting reasons from multiple charts of variance	Perfectly correct answer 9.8%. Proper Answer 32.1% *There is a problem in organizing and expressing multiple statistical information. Scientific understanding of information and attitude to participate in society
5	The question to be able to complete a flowchart	46.2%
6	Question about being against information morality	80.0%
7	This is a question about processing with attention to portrait rights	40.6%.
8	Questions about sources and citations	Perfectly correct answer 3.8%. Proper Answer 54.4%
9	Problems related to fraudulent billing on web pages	54.7%. *There is a problem in dealing with sources, citation, and portrait rights. Newly identified issues
10	What kind of data should I obtain?	14.9%
11	A question about calculating with a spreadsheet software	16.3%. *There is a problem in determining what information is needed and processing the data numerically.

Note:
Subjects 10th grade, 135 schools, 4,552 students surveyed.
Time period 2015-2016
Number of characters entered per minute (24.7)
Source MEXT, 2018, Information Use Capacity Survey (High School) Summary, https://www.mext.go.jp/a_menu/shotou/zyouhou/detail/__icsFiles/afieldfile/2017/01/17/1381046_01_1_1.pdf (September 25, 2020)

Figure 4. An example from the Faculty of Data Science at Shiga University. https://www.ds.shiga-u.ac.jp/en/ (July 24, 2020)

conducting research on competency and agency in the classroom, as proposed by the OECD.

In this study, we analyzed the classes at Oizumi Elementary School affiliated with Tokyo Gakugei University. A professor at Tokyo Gakugei University visited the elementary school as a guest teacher and gave a lesson using a robot. The content of the class was to think about the future of Japan and humanity as students interacted with the robot. The students thought about the alternative possibilities for robots and the unique missions of humans (Figure 5).

Figure 5. The robot class at Oizumi Elementary School, Tokyo Gakugei University.

Leading project 3 is the reform of upper secondary schools. The World Wide Learning Consortium was established in the high school to promote research on global social issues, and so forth. Specifically, they plan to set up one administrative school per 60,000

Figure 6. WWL's high school exchange with Indonesia. http://www.sakado-s.tsukuba.ac.jp/test201907/wp-content/themes/sakado/pdf/Indonesia_EN.pdf (August 6, 2020).

high school students. At present, they plan to establish one administrative school and about 10 additional schools in total.

The WWL has a new budget of 167 million yen planned for FY 2019. And it is trying to train global innovators. The WWL will break away from the traditional and poorly functioning high school system, which is divided between the humanities and the sciences and will be characterized by a high school-university connection. There are about 10 designated schools, and the maximum amount of annual financial support per school is 15 million yen. WWL has been actively promoting cooperation with other countries, including fieldwork in a rural village and a national park in Indonesia (Figure 6).

3 DISCUSSION

What AI can do and how it can be used is currently being explored. However, in Japan, we are trying to get students to think about the professions that AI will replace and equip them with the ability to use AI. It is not clear how and what these skills can be measured for non-AI alternative creative occupations. It is a challenge that we face. For the part about what forces cannot be replaced by AI, I believe that the OECD's Learning Compass 2030 (Figure 6) can be used. This is the same direction as the OECD competencies and agencies.

The OECD's Education 2030 project Learning Compass 2030 aims to be a state of well-being in 2030. Well-being here includes both personal well-being and social well-being. It is very good to set a period time, for example, 2030 as a goal in education reform. It makes the timeline easier to create.

The OECD's Learning Compass 2030 has a three-layered structure. The core competencies are knowledge, skills, attitudes and values, which have been consistently emphasized in the past. And the second layer is the core foundations, which include data literacy, and so on. The transformative competencies are located in the outer layer. It is Creating New Value, Taking Responsibility and Reconciling Tensions & Dilemmas.

In Japan, curricula are often considered in parallel with other subjects, so the OECD model of a three-layered approach is refreshing. The need to Reconciling Tensions & Dilemmas has not been deliberately incorporated into education in Japan, and this is a new need that has been focused on. In Japan, reconciling tensions and dilemmas have been taught unintentionally as part of special activities such as classroom activities, club activities, student council activities, and school events. From now on, this is the part that needs to be taught more intentionally. And this part is also a human-specific activity that is difficult for AI to replace.

The OECD's Learning Compass also modeled the learning cycle. It is an AAR cycle model of anticipation, action and reflection. It is similar to the three-step model of introduction, expansion and summation that often appears in the Japanese lesson plan for school education. In Japan, the teacher has the students go through the whole lesson at the introduction and the students are made to reflect on the whole lesson at the summary.

Traditionally, the OECD has taken the stance of making the OECD key competencies clear. However, the OECD's Learning Compass 2030 focuses on not just having a compass of competencies, but using the OECD's Learning Compass. It is called student agency and co-agency; the OECD learning model expects students to gain competency and move proactively toward 2030 well-being. It is not just an individual activity. Students are also intended to work collaboratively with peers, teachers, parents, and community members to increase their agency (Figure 7).

Figure 7. The same direction as the OECD competencies and agencies. https://www.oecd.org/education/2030-project/teaching-and-learning/learning/ (July 24, 2020).

4 CONCLUSION

In Japan, as in the United States and the United Kingdom, about half the jobs will be replaceable by AI and robots by 2030. For this reason, Japan is emphasizing both education to develop the ability to use AI and robots and education to develop competencies and agencies that cannot be replaced by AI and robots.

The Cabinet Office of Japan aims to train leaders who are capable of technological innovation by using– AI. The Japanese Ministry of Education is looking for ways to educate all students, including those from the local community to support society 5.0. I hope that each student will be able to achieve well-being both personally and socially.

ACKNOWLEDGMENTS

I would like to thank the Research Organization for the Next Generation Education, Tokyo Gakugei University, for their cooperation in this study.

REFERENCES

Cabinet Office in Japan, What is Society 5.0? https://www8.cao.go.jp/cstp/english/society5_0/index.html (July 24, 2020)

Faculty of Data Science at Shiga University, https://www.ds.shiga-u.ac.jp/en/ (July 24, 2020)

Freya, C., & Osborn, M. 2017. The future of employment: How susceptible are jobs to computerisation? *Technological Forecasting and Social Change*, 114: 254–280, https://doi.org/10.1016/j.techfore.2016.08.019.

Ministry of Education, Culture, Sports, Science and Technology in Japan, "Society, Human resource development for Society 5.0 (Society 5.0に向けた人材育成について)", https://www.mext.go.jp/component/a_menu/education/detail/__icsFiles/afieldfile/2018/11/19/1411060_02_1.pdf (July 24, 2020)

Ministry of Education, Culture, Sports, Science and Technology in Japan. 2018. Information Use Capacity Survey (High School) Summary (情報活用能力調査（高等学校）概), https://www.mext.go.jp/a_menu/shotou/zyouhou/detail/__icsFiles/afieldfile/2017/01/17/1381046_01_1_1.pdf (September 25, 2020)

Nomura Research Institute. 2015. 49% of Japan's workforce to be replaced by artificial intelligence, robots, etc. - Estimated replacement probability by computer technology for each of 601 occupations (要日本の労働人口の49%が人エボット等で代替可能に種の職～ 601業ごとに、コンピューター技術による代替確率を試算 ～)", https://www.nri.com/-/media/Corporate/jp/Files/PDF/news/newsrelease/cc/2015/151202_1.pdf (September 26, 2020)

OECD. 2019. Learning Compass 2030, https://www.oecd.org/education/2030-project/teaching-and-learning/learning/ (July 24, 2020)

Sakado High School, University of Tsukuba, Let's travel to Sarongge village, http://www.sakado-s.tsukuba.ac.jp/test201907/wp-content/themes/sakado/pdf/Indonesia_EN.pdf (August 6, 2020)

Educational Innovation in Society 5.0 Era: Challenges and
Opportunities – Purnomo & Herwin (Eds)
© 2021 the authors, ISBN 978-1-032-05392-9

Innovation in early childhood and primary education

G. Fragkiadaki, M. Fleer, & P. Rai
Monash University, Australia

ABSTRACT: Early childhood and primary education are in a continuous search for innovation for the quality improvement of learning experiences during everyday educational reality. A wide range of important ideas, views, and suggestions coming from different disciplines, several theoretical standpoints, and diverse perspectives have been introduced over time. However, does innovation need to be a complex newcomer? This paper suggests that innovation lies behind the essentials of childhood, such as play and imagination. The paper seeks to explore play and imagination as critical aspects of a child's learning and development. *Conceptual PlayWorlds* as a collective model of practice for the development of play and imagination are introduced. The study suggests that quality learning experiences with advanced learning outcomes can emerge through children's imaginary play. An indicative case example of learning and development in Science, Technology, Engineering, and Mathematics is presented. The paper concludes with an overview of the essence and the qualities of a play-based pedagogy that can support everyday educational practice and inform policy.

1 INTRODUCTION

The core idea and the major anticipation that lies behind almost every initiative in early childhood and primary education is the discovery of innovation for the quality improvement of learning experiences during everyday educational realities. At the same time that complex and usually hard to follow practices are introduced in the field, the child's leading activity and the teachers' major strength is neglected; as this is play and play practices. Despite being "the elephant in the room," play is often underestimated as a teaching and learning practice and this is realized more as the child makes transitions within school life, beginning from the early years and moving towards primary education. The present paper presents the argument that innovation does not need to be a complex newcomer. The paper suggests that innovation lies behind the essentials of childhood, such as play and imagination. Play and imagination as critical aspects of a child's learning and development are explored. *Conceptual PlayWorlds* (Fleer, 2017, 2018, 2019) as a collective model of practice for the development of play and imagination are introduced. *Conceptual PlayWorlds* attempt to create developmental conditions that support children's faculties for wondering, imagination and conceptualizing formation all while playing. The study suggests that quality learning experiences with advanced learning outcomes can emerge through children's imaginary play. An indicative case example of learning and development in Science, Technology, Engineering, and Mathematics (STEM) is presented. We know from previous research (https://www.monash.

edu/education/research/projects / conceptual-playlab/ publications) that when STEM problems are introduced during children's play it results in children's sustained engagement in the activity setting and hence creates the possibility to study children's conceptualization in a compressed manner. The example presented in this paper shows how play and imaginary play can create developmental opportunities for concept formation and conceptual learning and development. The paper concludes with an overview of the essence and the qualities of a play-based pedagogy that can support everyday educational practice and inform policy.

2 INNOVATION IS THE ESSENTIALS: PLAY AND IMAGINATION

In Vygotsky's writings (1966; 2004), play is conceptualized as the experience of an imaginary situation that a child creates and lives through. Thus, play and imagination are understood as a unity. Children's imaginary play stretches, transforms, and expands the child's overall reality in multiple ways. In line with Vygotsky's conceptualization, during their imaginary play, young children:

a) change the meaning of objects by giving imaginary characteristics and attributes to the objects that surround them (e.g., a carton may act in place of a spacecraft),
b) change the meaning of an action by pretending they are acting in a different way they actually do (e.g., they move their hands up and down pretending they are flying),

DOI 10.1201/9781003206019-2

c) change the meaning of space by giving imaginary dimensions, characteristics, and spatial relations to their surrounding space (e.g., a part of the room may be considered the sea),

d) change the meaning of time by experiencing the past again (e.g., pretending they are infants), by living through the present (e.g., undoing a decision and resetting a story), and by imagining the future (e.g., pretending that they are grown-up),

e) use diverse types of tools, signs, and artifacts such as drawings, gestures, technological equipment as well as particular forms of language,

f) make multiple connections with everyday life using and expanding their everyday knowledge and understandings, and

g) share abstract intellectual spaces, creating and building on collective imaginary experiences both with peers and adults (e.g., educators, parents).

Taken together, imagination and imaginary play are critical psychological functions and complex processes in the early years. Entering and living through an imaginary situation, young children can expand their experience (Zittoun & Cerchia, 2013) and their learning. They can shift from an actual, present, and concrete points towards an abstract, flexible, and more developed one. Through their imaginative play, young children construct, deconstruct, and reconstruct their understanding of their social and cultural environment as well as the surrounding natural, technical, and technological world.

Although play-based pedagogies are the pivot in several curriculums worldwide, a lack of systematic and efficient pedagogical models are available to step the children and the teachers through playful learning. The following subsection introduces *Conceptual PlayWorlds* as an innovating model for learning and development through play.

3 CONCEPTUAL PLAYWORLDS: AN INNOVATING MODEL FOR LEARNING AND DEVELOPMENT THROUGH PLAY

Conceptual PlayWorld is a model of practice created to support teaching of STEM concepts through play in early childhood settings (Fleer 2017; 2018; 2019). A *Conceptual PlayWorld* can be inspired by a children's book or a fairy tale story. The drama enacted through the story helps children develop empathy with characters and be motivated to learn and solve problems. Through play-based experiences within imaginary situations, young children form and understand abstract concepts.

There are five key pedagogical characteristics of a *Conceptual PlayWorld* (https://www.monash.edu/conceptual-playworld):

1. The first characteristic is selecting a story that engages children through emotionally charged scenarios.

2. The second characteristic is designing a space that becomes the imaginary space of the chosen story.

3. The third characteristic is planning the entry and exit into that space - children and adults decide upon the role they will take in the role-play of the story and join the imaginary situation.

4. The fourth characteristic is introducing a problem situation that needs a STEM concept for solving the problem.

5. The fifth characteristic is planning how the teachers will support learning in the imaginary situation based on the pedagogical positioning they will take, such as being with the children investigating, or leading an inquiry, or asking children for help.

The five characteristics of a Conceptual PlayWorld act as a flexible pedagogical model that can be used in early childhood settings, primary school settings as well as family and community settings (see research and practice examples here: https://www.monash.edu/education/research/projects/conceptual-playlab). The opportunities and the possibilities of adjusting the model in diverse social realities and children's interests and needs are unlimited. A case example of the implementation of a *Conceptual PlayWorld* is illustrated in the following section.

4 AN INDICATIVE CASE EXAMPLE: FORMING STEM CONCEPTS THROUGH PLAY

This section illustrates how play can be dialectically interrelated with learning within everyday educational reality. A case example of young children's engagement with STEM is presented. Children were engaged with a wide range of STEM concepts as part of their participation in a Conceptual PlayWorld based on the story "We're going on a bear hunt" written by Michael Rosen. The story describes the experiences of a family as they go through several sites and locations such as a river or a forest in search of a bear. The *Conceptual PlayWorld* was organized as an educational experiment (Hedegaard, 2008); that is, as a planned intervention designed to create condensed forms for the children's conceptual development. Following the methodology of the educational experiment, the learning environment was designed by the collaboration between the early childhood teachers of an early childhood center in Australia and the research team. Following the methodology of the *Conceptual PlayWorld*, the early childhood teachers along with the children created an indoor space to facilitate children's imaginary play (Figure 1). The figure has been taken from Fleer, Fragkiadaki, and Rai (2020b) https://www.monash.edu/education/research/projects/conceptual-playlab/publications. The space was created to allow, inspire, and support children's explorations and interactions.

The children and the early childhood teachers were in the role of the family members and experienced

Figure 1. Creating the indoor space for children's imaginary play

Table 1. STEM concepts that can be formed through children's imaginary play inspired by the story

The episode of the story	STEM Concept
grass	- Biological external characteristics of the plant (e.g., observing, describing, and documenting the figures of the plants in the indoors and outdoors spaces of the classroom) - The circle of life of plants (e.g., planting, observing, describing, and documenting the process) - The ecosystem (e.g., observing, describing, and documenting the small ecosystem that can be created on a plant)
river	- Floating and sinking (e.g., using a boat to cross a big bowl of water) - Design process (e.g., designing a boat to cross the bowl of water) - The kinetic energy of water (e.g., a watermill in a big bowl of water)
mud	- Dissolution (e.g., creating mud) - Spatial relations, counting, and measurement (e.g., playing with mud to follow a food recipe for a pretend lunch)
forest	- Light and shadow (e.g., playing with natural (sun) & artificial (torch) created light to create shadows)
snow	- Rendering and melting (e.g. creating and melting ice-cubes)
cave	- Statics, dynamics, and controlling material (e.g., building a cave with diverse materials)
house	- Biological external characteristic of the bear (e.g., observing, describing and documenting characteristics) - Hibernation (e.g., searching on the internet for information about the concept of hibernation) - Habitat and habits of bears (e.g., using a diverse resource such as the internet to learn more about what the bear likes)

the same imaginary situation of searching for a bear together. Being within the *Conceptual PlayWorld together,* they planned to visit several imaginary spaces such as an area with grass, a river, a muddy area, a forest, an area with snow, a cave, and a house. Each of these imaginary areas could create an opportunity for the children to explore diverse STEM concepts. The set of the STEM concepts available for exploration in line with each episode of the story is presented in the following table (Table 1). An extended version of the table is published in Fleer, Fragkiadaki, and Rai (2020b) https://www.monash.edu/education/research/projects/conceptual-playlab/publications.

The concepts presented above are indicative. Early childhood teachers had the opportunity to choose to work with one or more concepts during the period the *Conceptual PlayWorld* was implemented. They were also encouraged to add to this list in a way that aligned with children's interests and the elaboration of the imaginary play. Early childhood teachers were encouraged to use a scientific language to step the children through STEM-oriented experiences. The scientific narrative crafted by the early childhood teachers focused more on describing and explaining what was happening and how during the STEM experiences rather than using advanced terminology related to the content knowledge of each concept explored.

What is important here is that a rich learning environment was created for the children allowing deep and extensive engagement with the STEM concepts through play. The way the concepts are introduced here reflects meaningful as well as enjoyable experiences for young children. Being in an imaginary play situation, children have the opportunity to wonder, explore, learn, and reflect on the natural, technical, and technological world that surrounds them. The concepts can come in the service of play and respond to real-life needs for the children as expressed through their everyday play. A continuum and a balance between advanced learning goals and play are introduced.

5 TOWARDS PLAY-BASED PEDAGOGY

As we have argued elsewhere, imagination and concept formation starts early in children (Fleer et al, 2020a). Infants start to see the distinction between rolling, sliding, pushing, and pulling even within their first 1000 days. These distinctions hint at children's capacity to imagine and distinguish between various actions from very early years. Play emerges as a result of the dialectics between children's psychological functioning and the social and material conditions of their environment. Children's imaginary play developed through *Conceptual PlayWorlds* shows their developmental tendencies in a condensed form. The example presented in this paper showed how sustained engagement in these imaginative play situations can help in the learning of STEM concepts. Play needs to be seen as something beyond a mere pleasure-giving activity for children. It is, as Vygotsky argued, the leading factor for children's development in the early years and a critical factor as the child enters primary education. Being in the imaginary situation as a play partner, the teacher engages children in the collective act of imagining and learning. Taking a step further, the *Conceptual PlayWorld* described in this paper helped in creating a collective imaginary situation that invited children to explore their thinking and hence also learn robust STEM concepts. Following a cultural-historical line the effort has also been not to see these imaginary engagements as one-off moments of imagination but a sustained historical engagement of children in developing their social situations of development. Children's concept learning is thus employed for supporting their imagination and future exploration. The transformative character of imagination is worth highlighting here as it creates the possibility for the children to be more aware of their actions and develop empathy in the play setting. The study concludes by suggesting that a play-based pedagogy as illustrated through the Conceptual PlayWorld model can provide an innovative approach for teaching, learning, and development over the school life of children beginning from the early years and continuing during primary education.

REFERENCES

Fleer, M., Fragkiadaki, G., & Rai, P. 2020a. STEM begins in infancy: Conceptual PlayWorlds to support new practices for professionals and families. *International Journal of Birth and Parent Education*, 7(4): 29–33.

Fleer, M., Fragkiadaki, G., & Rai, P. 2020b. Exploring *STEM Concepts in Early Years*. Conceptual PlayLab, Faculty of Education, Monash University. https://www.monash.edu/education/research/projects / conceptual-playlab/publications

Fleer, M. 2017. Scientific playworlds: A model of teaching science in play-based settings. *Research in Science Education*, 49: 1257–1278. https://doi.org/10.1007/s11165-017-9653-z

Fleer, M. 2018. Conceptual Playworlds: the role of imagination in play and learning. *Early Years*, 1–12. https://doi.org/10.1080/09575146.2018.1549024

Fleer, M. 2019. Conceptual PlayWorlds as a pedagogical intervention: Supporting the learning and development of the preschool child in play-based setting. *Obutchénie*, 3(3): 1–22. https://doi.org/10.14393/OBv3n3.a2019-51704

Hedegaard, M. 2008. The educational experiment, In M. Hedegaard, and M. Fleer (eds.). *Studying children. A cultural-historical approach*, (pp. 101–201). England, Open University Press.

Vygotsky, L.S. 1966. Play and its role in the mental development of the child. *Voprosy psikhologii*, 12(6): 62–76. https://doi.org/10.2753/RPO1061-040505036

Vygotsky, L.S. 2004. Imagination and Creativity in Childhood. *Journal of Russian and East European Psychology*, 42(1): 7–97. https://doi.org/10.2753/RPO1061-0405280184

Zittoun, T., & Cerchia, F. 2013. Imagination as expansion of experience. *Integrative Psychological and Behavioral Science*, 47(3): 305–324.

Educational Innovation in Society 5.0 Era: Challenges and
Opportunities – Purnomo & Herwin (Eds)
© 2021 the authors, ISBN 978-1-032-05392-9

Strengthening resilience for learning transformation and anticipatory education in the era of society 5.0

S.I.A. Dwiningrum
Universitas Negeri Yogyakarta, Indonesia

ABSTRACT: Society 5.0, which aims at building a quality and prosperous life, requires education as a key component. Education towards society 5.0 requires an effective learning transformation and holistic approach based on culture. The principle of developing continuous education is the basis for improving the curriculum to adjust to the relevant socio-economic and cultural changes. The aim of education in society 5.0 is to be anticipatory, which means to be able to direct knowledge and age-based skills in the right places in order to provide the required skills consisting of three skill groups, namely learning and innovation skills, information, media and technology, and life and career skills. Learning transformation in society 5.0 requires resilience at the community, school, family, and individual levels, which is a system built with strong synergy between all elements of society and the government all working to achieve the same goals. This goal is to build a society which has a better quality of life in this increasingly accelerated technological development era without having to reduce the role of human beings.

1 INTRODUCTION

The increasingly accelerated technological development has been reducing the roles of human beings. The Japanese concept of society 5.0 is actually an effort to put back the foundations of humanity in the development era of industrial revolution 4.0. This concept has been discussed by Japanese researchers, namely Hayashi, Sasajima, Takayanagi, and Kanamaru at the Annual Conference of the Society of Instrument and Control Engineers of Japan (SICE) in 2017. The result of the discussion stated an intention "[to] create new values by collaborating and cooperating with several different systems, and plans, standardization of data formats, models, system architecture, etc. and development of necessary human resources. In addition, it is expected that enhancements of intellectual property development, international standardization, IoT system construction technologies, big data analysis technologies, artificial intelligence technologies and so on encourage Japan's competitiveness in a super smart society." All in all, the era of industrial revolution 4.0 is known as the cyber- physical system (CPS) which is an integration between physical systems, computing, and communication, while society 5.0 is a refinement of CPS, turning cyber-physical-human systems into a central issue in the transitional era. In this case, humans do not only play the role of objects (passive elements), but also play an active role as subjects (active players) who work with the physical system in achieving certain goals.

Dehumanization and devaluation of education are social phenomena which occur in people's lives after the industrial revolution (Dwiningrum, 2017). Likewise, the disruption of education continuously occurs in the life of the global community (Yoga 2017, 2018), which is related to the aspect of the neo-millennial generation of students or the generation which was just born world, creating a "generation gap" with the previous generation. The second related aspect is hyper-demanding parents, which is a condition of an increase in the number of "middle-class" families causing an increase in needs, including needs in education such as homeschooling. The third aspect is disruptive technology, which is the condition of extremely fast technological development changing fundamentally the methods of learning. The fourth one is namely irrelevant skills obtained at school (memorization, multi-choice, etc.) which are "not in accordance" with newly required skills (being creative, innovative, etc.). Disruption is innovation which replaces the entire old system with new ways related to digital technology that will produce something completely new, more efficient, and more useful (Kasali, 2017). In this context, the act of self-disruption is more important than allowing oneself to be disrupted by newcomers. The concept of disruption is needed in order to better understand related social phenomena in the disruption era.

The role of education is increasingly reduced in the era of industrial revolution 4.0 if its social role is not revitalized. Furthermore, it is dealing with the era

DOI 10.1201/9781003206019-3

of the development of Massive Open Online Courses (MOOCs) which are said to obliterate whatever is in front of them (like an avalanche)(Dhakidae, 2017). Educational practice has become a continuously debatable issue because most of the implementation of education has not generated maximum results. The educational process requires proper formulation so that educational goals can be optimally achieved. Criticisms about educational practice so far have proven two important things, firstly that the ideal model for educational development in the industrial revolution 4.0 era has not been well established, and second, the failure of the educational goals designed by the government. The failure in the education development process is supported by the persistence of two major problems of education development in Indonesia, namely the imbalance in the quality of education at all educational levels as well as the unequal educational distribution and opportunities for all Indonesian people.

Education experts state that the education process has not yet optimally produced well-characterized individuals in facing unpredictable changes and adapting to the challenges of the industrial revolution 4.0. In fact, education has an important role in building and shaping individuals with good characters. The power of educators in mastering the basic concepts of the educational sciences is an important basis for equipping students to have a more comprehensive frames of thinking. Educators are not only required to teach subject materials, but the most important thing is also ensuring that students can utilize the materials being taught in their social lives. Therefore, studies on educational theories and teaching methods should be designed appropriately and applicably according to education in the era of industrial revolution 4.0. In addition, educators in the era of the industrial revolution 4.0 must be able to apply the basic concepts of anticipatory education according to the needs of society; educating is not just a teaching process, it is also a process to make every human being an independent learner in responding to highly accelerated developments in science and technology. There will be two major aspects discussed in this paper, namely the importance of social transformation and learning to achieve anticipatory educational goals, and the importance of strengthening school resilience in facing the challenges of life in society 5.0.

2 SOCIAL AND LEARNING TRANSFORMATION IN SOCIETY 5.0

Education will lose its meaning when it is unable to carry out its social role optimally. Educational institutions have a role to socialize new values in human life by collaborating and working with various technological results produced in the era of the industrial revolution 4.0, with artificial intelligence, The Internet of Things (IoT), and Big Data. The transformation of character values in millions of people through the internet requires innovative learning strategies in order to achieve anticipatory educational goals. In this context, traditional approaches and teaching methodologies tend to be questioned. Therefore, social transformation using a participatory approach must to be considered (Disterheft et.al., 2015). Likewise, learning organizations which aim at improving students' understanding of the consequences of technological developments in the era of industrial revolution 4.0 need to be reconstructed. Efforts to apply a holistic, culture-based approach are needed to build an understanding of the ease gained from technological developments, such as human works, and to reduce social problems caused by the inability of humans to adapt to accelerated technological changes (Ramos et. 2015).

The goal of the new type of education needed in society 5.0 is anticipatory education. Educational activities are expected to be able to lead students to live their lives. Therefore, education aimed at building society 5.0 must still be built on its main principles which are three types of balance: 1. balance between spiritual and physical education, 2. balance between natural knowledge and social and cultural knowledge, and 3. balance between current and contemporary knowledge to come. This is very important, so that students can prepare for not only their current but also future lives. Currently, Indonesia is facing "the world system" which means students are accustomed to seeing the existence of a nation in relation to other nations and world problems and instilling national and regional awareness. To achieve anticipatory education requires a learning process that teaches students: 1. to make living, 2. to lead a meaningful life, and 3. to enable life (Buchori, 2001).

Learning transformation requires a high commitment. An important step that must be taken is to motivate the emergence of innovative ideas in changing contents and methods to be realized. Examples of transformation in the curriculum field can be made towards continuous learning. (Filho et al., 2018). This new perspective reflects the wider community debate, namely the emergence of public awareness of the importance of developing sustainable education that can be integrated into the curriculum (Ramos et al, 2015). This proves that maintaining the continuous education process still needs to be seriously studied, because there is a tendency that the education process has not been carried out with the principles of continuous education, so that educational development has not been optimally successful.

The principle of developing continuous education is the basis for improving the curriculum to adjust to the relevant socio-economic and cultural changes. The gap which occurs between what is learned in college and what college graduates must know in the era of the industrial revolution 4.0 still occurs in Indonesia. Currently there is a tendency for the learning process to focus more on developing technical discipline knowledge while ignoring the fundamental and social dynamics to be integrated into the academic curriculum. Creativity becomes an important aspect needed to adapt to changes that are more complex and dynamic than those in the past decade because it raises complex problems, so it is important to rethink educational practices to be able to respond to new social structures and their challenges better. The main problem is the extent to which the educational institutions have adaptive capabilities. Along with some academic realities, there comes an inability to adapt quickly, which raises the question of whether this is an ideal environment for creative people.

Technological and scientific developments require all the parties to be flexible and proactive. In particular, education must instill a high adaptive attitude to its students. Today's world moves at lightning speed, requiring curious, flexible and proactive people. This is an educational concern, not only for teachers and students, but also for institutions. The ability of educational institutions to adapt to changes is an important aspect of maintaining the existence of their social role. The various strategies which must be developed form the basis for the development of continuous education, namely a) the ability to develop various skills which are relevant to required needs, b). the application of multidisciplinary learning, c) the development of digital innovation within a partnership system, d) the increase of the partnership network between the world of education and industry, and e) the production of competitive graduates. These five aspects need to be developed by educational institutions, especially universities, so that they are able to produce professional graduates who are aware of their social roles, able to contribute to solving complex problems, and strong enough to be agents in the transition to sustainability and social change. For schools, this principle should be strengthened so that the inputs which go to the higher educational level have the same spirits. This is because there is a strong tendency, in the case that the education process at the basic educational level to higher education does not maintain its quality values, for students to experience less qualified learning process which will then have an impact on their personal qualities. Therefore, the problem of education in Indonesia is still very complex when the learning experience is not supported by the same quality of education between levels of education. Likewise, the issue of the quality of education is still a big problem in education development in Indonesia as quality gaps and knowledge gaps continue to occur in the midst of changes in society that continue to move in the era of the industrial revolution 4.0. To get to society 5.0 still requires strong synergy between all elements of society and the government to achieve the same goal. That is, to build a society which has a better quality of life in this increasingly accelerated technological development era without having to reduce the role of human beings.

The basis for learning transformation in the era of the industrial revolution 4.0 was built on three systemic and synergistic aspects, namely foundational knowledge (to know), humanistic knowledge (to value), and meta knowledge (to act). Learning transformation must be developed in a curriculum modification that is able to integrate these three aspects in all subjects taught in schools. The learning strategies developed by teachers will vary according to the substance of the knowledge being taught. Currently, teachers must be creative and innovative, as well as multitalented in facing accelerated social changes due to the increasing development of science in the era of the industrial revolution 4.0. This learning transformation needs to be directed at developing knowledge age-based skills in order to provide the required skills consisting of three skill groups, namely learning and innovation skills, information, media and technology, and life and career skills (Trilling & Fadel, 2009). More specifically, learning and innovation are the main focus of skills in the era of industrial revolution 4.0 related to critical thinking and problem solving, communication, and collaboration, as well as creativity and innovation. Information, media, and technology skills are the skills which students need to access, evaluate, use, organize, and add to the wealth of information and media. Life and career skills are needed in working life in a fast-changing world. Flexibility and adaptability are fundamental skills for learning, working and living in society in the 21st century. Likewise, initiative skills, social and cross culture skills, productivity and accountability, and leadership and responsibility are very much needed in life and work in the era of society 5.0.

Support for building a transformative society 5.0 requires strengthening social capital as social energy which can be explored in people's lives. Social capital will facilitate the process of learning transformation towards society 5.0 because all elements in social capital can be used to help overcome various obstacles faced by society 5.0. Why is social capital important to consider in education in Indonesia? There are two main reasons. The first is that social capital is starting to weaken in Indonesian society. Second, social capital has not been considered important in supporting education development. Even though, conceptually, social capital has an important role in the development process and it has been widely studied by social scientists (Dwinigrum, 2014).

Social capital plays a really important role in learning transformation. As explained by Bourdieu (Field, 2010), social capital is a number of resources, actual and virtual, which gather in an individual or in groups because they have durable networks in the form of a reciprocal relationship of introductions and recognitions that is slightly institutionalized. The concept of the social capital that is needed to form a social agent in constructing the world around it. The concept is not something that stands alone but is related to various other capitals. In people's lives, there are three types of capital, namely economic capital, cultural capital, and social capital (Hauberer, 2011). The problem is that schools are not aware enough of the role of social capital in dealing with society 5.0. Educational institutions as agents of change are required to strengthen all elements of social capital towards society 5.0. By strengthening social capital, it is expected that disparities in structure and culture can be overcome more effectively. Many research results prove that social capital has an important role in improving the quality of education (Dwiningrum, 2016; Zamroni, 2017).

There are two main reasons why social capital is important to face society 5.0. The first reason is that social capital is starting to weaken in the life of Indonesian society. Secondly, the social capital has not been considered important in supporting education development. However, in fact, social capital conceptually has an important role in the development process, and this has been widely studied by social scientists. Several books which examine social capital have been widely read and criticized by various disciplines, such as economics, management, politics, education and social work (Dwinigrum, 2014).

The data presented in Table 1 illustrates that schools can develop the elements of social capital needed for society 5.0. This agrees with Putnam (Suharjo, 2014) who explains that "social capital refers to connections among individual-social networks and norms of reciprocity and trustworthiness that arise from them. In that sense, social capital is closely related to what some have called "civic virtue." Putnam provides a brief definition of social capital: "by 'Social capital' I mean features of social life - networks, norms, and trust - that enable participants to act together more effectively to pursue shared objectives" (Field, & Schuller, 2000; Suharjo, 2014). The three elements include networks, norms, and trust. Even from the results of research in schools, the elements of social capital are broader, namely: trust, participation in social networks, mutual exchanging of kindness, social norms, social values, and proactive actions.

The success of social transformation into society 5.0 is determined by transformational leadership. The concept of transformational leadership is defined by Burns (2003): "when leaders and followers achieve high levels of motivation and morality, we have transformational leadership." The stance that there is a commitment from the followers in which they are led to go beyond their own interests by commitment, driving change and high performance, with the goals of the organization, is also an effect of transformational leadership. Transformational leadership theory is a useful tool in a competitive business environment (Bass et al., 2003). This theory seeks to study how the leader behaves during the organizational transition phase and how he or she develops ways to achieve the desired future. In transformational leadership, the leader cares and shows respect for employees; he is aware of the individuality of each person. He focuses on developing employee loyalty, trust, and trust in fairness and works to improve employee self-esteem, self-confidence and effectiveness (Rego & Cunha, 2007). Therefore, to prepare for community 5.0, the most important thing is to have a transformative leader at every level of society and every level of education. With transformative leadership, it is expected that learning transformation will be more effective. The process towards society 5.0 is a social process that requires adaptability from all Indonesian people. Particularly for the educational environment, the social transformation process will be more effective if schools have strong resilience.

Table 1. Social capital and community learning transformation 5.0

Elements	Indicators
Trust	1. Designing schools with superior IT-based programs
	2. Developing professional teachers
	3. Preparing graduates which meet the competence requirements of the 4.0 industrial revolution
Participation in social networks	1. Strengthening community participation in school programs
	2. Strengthening family participation in school programs
	3. Designing IT access for online-based learning processes
	4. Designing IT-based prime services
	5. Building partnerships with industry
Mutual exchange of kindness	1. Strengthening social relations with schools which have the same goal
	2. Using experts to provide constructive input
	3. Respecting differences and avoiding discrimination
	4. Having social concern for the change process
Social norms and social values	1. Having clear rules governing society 5.0
	2. Building strong character values in the community.
	3. Creating a conducive learning environment to develop competencies in the community 5.0.
Proactive action	1. Using the mass media as a source of information
	2. Producing innovative products needed by community
	3. Becoming an associative member in accordance to the goals of the school

3 RESILIENCE OF SCHOOLS IN SOCIETY 5.0

The construction of the resilience concept continues to be reviewed from various disciplines from time to time. (Day 2013). There is an agreement that resilience is a "complex, dynamic and multi-dimensional phenomenon" (Mansfield.et al. 2012) or a "composite construction" (Gun & Li, 2013). Research on resilience continues to develop in accordance with the objectives and research approach. Resilience related to the educational process is a very interesting topic, because the results can contribute to strengthening personal resilience. Resilience becomes an important asset to build a strong personality as a personal resource needed to face increasingly complex community life. As the process of resilience is the focus, researchers "focus not on the key attributes of teachers or resources in the environment, but on the strategies that teachers use" (Castro et al. 2010). This fact proves that ecologically and contextually, lecturers play an important role in building experiences in broader social, cultural and political arenas (Johnson et al. 2014). This resilience is built as a collective construction, because resilience is the culmination of collective and collaborative efforts (Gu & Li 2013).

Resilience is determined by ecological conditions socially and physically. If the environment is conducive and safe enough to develop student potential, it will produce a resilient personality. If it is related to the school environment, the resilience of teachers and students have different forming factors, so if it is applied in the university environment, it is possible that the resilience of lecturers and students might be different. In this context, research is still needed specifically to reveal the dominant forming factors of the resilience between teachers and students, and/or lecturers and students. To provide an initial description that there are differences in forming resilient qualities of teachers and students, the first state is that the differences which occur in the formation of teacher resilience are related to the capacity to maintain educational goals, and the ability of teachers to manage uncertainty and carry out their social roles. Whereas in the context of students, it emphasizes the aspect of ability in carrying out role values based on everyday experiences in their childhood. (Wosnitza et al., 2014)

The response of schools in building resilience in the industrial revolution era is determined by the conditions of the community environment (Dwiningrum, 2015). It is in this context that the mapping of school capacity in developing school resilience is determined by the aspects needed to build school resilience. Building resilience is not an easy job. Resilience and vulnerability become materials for the decision-making process by taking into account various aspects of life such as human and social capital, infrastructure, economic capital, and institutional organizations, or according to the stages of a distracting event, for example, preparing, absorbing, restoring and adapting (Cutter, 2016). Resilience requires the right strategy, model, and evaluation to make society more resilient. In addition, social capital is needed to build community resilience for disaster mitigation. Lucini (2013) states in a sociological perspective that to create social resilience and make efforts to reduce social vulnerability requires social cohesion, by strengthening social relations and the role of social capital as in prevention and preparedness, planning, warning communication, physical and psychological impacts, emergencies and disaster response, recovery and reconstruction with the specific aim of increasing social resilience and seeking to reduce social vulnerability.

In addition, building school resilience requires strong social relations with the community. Likewise Kumaraswamy, Zou, and Zhang (2015) concluded the importance of synergy between, public, private, and community partnerships in development. In line with this, Shiwaku (2011) explains that the key action to increase resilience is to strengthen the relationship between schools and communities. The three ways for schools to develop relationships with the community are as follows: a) involving the community in school activities, b) being involved in activities managed by community associations, and c) collaborating with the community to carry out joint activities (Shiwaku, 2011). The suitable type for building resilience should be chosen depending on the local context. However, schools need to make efforts to involve the community or parents in school activities and make connections with key people in the community (Siwakhu, 2011). All in all, a socio-cultural approach is also needed to build school resilience (Caroline & Clauss, 2010).

School resilience is largely determined by the contribution of teachers and students as well as school residents. As a result, the contribution of teachers and students is important in building school resilience, especially for the schools located in disaster-prone areas. This is also supported by the results of the research conducted by Shiwaku (2011) which concluded that students will have toughness and independence and be more resilient due to these following eight main attributes, namely: 1) having stable relationships with peers, 2) having the skills to develop problem solving strategies, 3) designing realistic future plans, 4) striving positively to achieve and handle all tasks effectively, 5) experiencing success in one or more areas of life, 6) being able to communicate effectively, 7) having a strong attachment to at least one adult, and 8) showing acceptance of responsibility for themselves and their behavior (Siwakhu, 2011).

The personal resilience of teachers and students determines the success of a resilient school. Resilient teachers and students have adaptive abilities and are responsive to the changes needed in the era of society 5.0. Personal resilience that needs to be developed in teachers and students is related to 7 aspects. For teachers, having a very strong resilience is very important in helping students who experience unpleasant events because of the disaster to stay enthusiastic about learning. Teachers who are not resilient will have a bad impact in building a learning atmosphere in the classroom. Personal or individual resilience as described by Reivich K. and Shatte A. (1999) states that resilience includes 7 abilities, namely: emotional regulation, impulse control, empathy, optimism, causal analysis, self-efficacy, and reaching out. The seven abilities are also referred to as 7 factors of resilience. The teacher's strategy in strengthening school resilience can be done by: 1) having a stable relationship with peers, (2) having problem solving skills, (3) designing realistic future plans, (4) striving positively to achieve and handle all tasks effectively, (5) experiencing success in one or more areas of life, (6) having the ability to communicate effectively, (7) having strong bonds with at least one adult, and (8) being responsible for themselves and their behavior (Shiwaku et al., 2016). With resilience, schools will be able to make learning transformations more effectively in developing antipatory education towards society 5.0.

4 CONCLUSION

Society 5.0 is a social structure which will be formed ideally if all elements of society have the knowledge and literacy needed to lead to a prosperous society. To get to society 5.0 requires a social transformation which progresses in accordance with the stages of society from the industrial revolution 1.0 to society 5.0. The ability of society to proceed towards society 5.0 is determined by the transformation of learning. Social capital will support the learning transformation process by building three systemic and synergistic aspects, namely foundational knowledge (to know), humanistic knowledge (to value), and meta knowledge (to act). Learning transformation is developed through a curriculum that integrates all subjects in the learning process which is important for the efforts to achieve meaningful education. The aim of education in society 5.0 should lead to the development of age-based knowledge skills in order to provide the required skills consisting of three groups of skills, namely learning and innovation skills, information skills, media and technology skills, and life and career skills. To be successful in the learning transformation of society 5.0, it is necessary to build resilience at the community, school, family, and individual levels.

REFERENCES

Bass, B. M., Avoilio, B. J., Jung, D. I., & Berson, Y. 2003. Predicting unit performance by assessing transformational and transactional leadership. *Journal of Applied Psychology*, 88(2): 207–218.

Burns, J. M. 2003. *Transforming leadership: A new pursuit of happiness*. New York: Atlantic Monthly Press.

Buchori, M. 2001. *Pendidikan antisipatoris*. Yogyakarta; Penerbit Kanisius.

Castro, A. J., Kelly, J., & Shih, M. 2010. Resilience strategies for new teachers in high-needs areas. *Teaching and Teacher Education, 26*(3): 622–629. doi:10.1016/j.tate.2009.09.010

Caetano, N., Lopez, D., & Cabre, J. 2015. *Learning Sustainability and Social Compromise Skills: a New Track Is Born*. New York: ACM Press.

Cutter, S. L. 2016. Resileince to what? Resilience for whom?. *The Geografical Journal*, 182(2): 110–113.

Clauss-Ehlers, C. S. 2010. Cultural Resilience. *Encyclopedia of Cross-Cultural School Psychology*, 324–326.

Day, C., & Gu, Q. 2014. *Resilient teachers, resilient schools*. London & New York: Routledge Taylor & Francis Group.

Gu, Q., & Day, C. 2013. Challenges to teacher resilience: Conditions count. *British Educational Research Journal*, 39(1): 22–44.

Dhakidae, D. 2017. *Era disrupsi: Peluang dan tantangan pendidikan tinggi Indonesia*. Jakarta: Akademi Ilmu Pengetahuan Indonesia.

Disterheft, A., Caeiro, S., Azeiteiro, U. M., & Leal, W. F. 2015. Sustainable universities – A study of critical success factors for participatory approaches. *Journal Cleaner Production, 106*(1): 11-21.

Dwiningrum, S. I. A. 2018. *The role of social capital in developing effective and creative schools in primary schools*. Makalah dipresentasikan pada ISCEI 2018. Nayang Universitas Singapura, Singapura.

Dwiningrum, S. I. A. 2017. Developing school resilience for disaster mitigation: A confirmatory factor analysis. *Disaster Prevention and Management: An International Journal*, 26(4): 437-451.

Dwiningrum, S. I. A. 2017b. *Role of high school on building academic resilience: Comparative study in high school student in Indonesia and Japan*. Makalah dipresentasikan pada 3rd International Conference on Education. Mudzaffar Hotel, Ayer Keroh, Malaka, Malaysia.

Dwiningrum, S. I. A. 2016. *Teori persekolahan*. Pascasarjana Universitas Negeri Yogyakarta.

Dwiningrum, S. I. A. 2016. *Building social harmony: Reinforce the foundation of reseaching multicultural education practices in Indonesia and New Zealand*. Makalah dipresentasikan pada 41th Pasific Circle Consortium Jepang. JMS Aster Plaza, Hiroshima Japan.

Dwiningrum, S. I. A. 2014. *Modal sosial: Dalam pengembangan pendidikan perspektif teori dan praktik*. Yogyakarta: UNY Press.

Dwiningrum, S. I. A. 2014. Schools in education and media hegemony in the perspective of multicultural education. Dalam *Proceeding International Conference on Fundamentals and Implementation of Education (ICFIE)*. Universitas Negeri Yogyakarta, Yogyakarta.

Field, J. 2010. *Modal Sosial*. Medan : Bina Medai Perintis

Gu, Q., & Li, Q. 2013. Sustaining resilience in times of change: Stories from Chinese teachers. *Asia-Pacific Journal of Teacher Education, 41*(3): 288–303.

Henderson, N. (Ed.) 2007. *Resilience in action: practical ideas for overcoming risks and building strengths in youth, families, and communities*. Paso Robles, CA: Resiliency in Action, Inc.

Hauberer, J. 2011. *Social capital theory*. VS Research.

Johnson, D., Johnson, R., Roseth, C., & Shin, T. 2014. The relationship between motivation and achievement in interdependent situations. *Journal of Applied Social Psychology*. http://dx.doi.org/10.1111/jasp.12280.

Johnson, B., Down, B., Le Cornu, R., Peters, J., Sullivan, A., Pearce, J., & Hunter, J. 2014. Promoting early career teacher resilience: A framework for understanding and acting. *Teachers and Teaching: Theory and Practice*, 20(5): 530–546.

Kasali, R. 2017. *Disruption*. Jakarta : PT Gramedia.

Kumaraswamy, M., Zou, W. & Zhang, J. 2015. Reinforcing relationships for resilience by embedding end-user 'people' in public-private partnerships. *Civil Engineering & Environmental Systems*, 32(1–2): 119–129.

Lucini, B. 2013. Social capital and sociological resilience in megacities context. *International Journal of Disaster Resilience in the Built Environment*, 4(1): 58–71.

Mansfield, C., Beltman, S., Weatherby-Fell, N., & Broadley, T. 2016. *Classroom Ready? Building Resilience in Teacher Education. Teacher Education*, 211–229.

Ramos, T. B., et al. 2015. Experiences from the implementation of sustainable development in higher education institutions: Environmental Management for Sustainable Universities. *Journal of Cleaner Production, 106*(1): 3–10.

Reivich,K. & Shatte, A. 2002. *The Resilience Factor.* New York: Broadway Books.

Rocha, H. 2019. Paradigma sfift in design education. Retrieved from https://medium.com/age-of-awareness/paradigm-shift-in-design-education-ef02769bec93

Rego, A., Pina E. Cunha, M., & Souto, S. 2007. Workplace Spirituality, Commitment, and Self-Reported Individual Performance: An Empirical Study. *Management Research*, 5(3): 163–183.

Shiwaku, K., Ueda, Y., Oikawa, Y., & Shaw, R. 2016. School disaster resilience assessment in the affected areas of 2011 East Japan earthquake and tsunami. *Natural Hazards*, 82(1): 333–365.

Suharjo 2014. *Peranan modal sosial dalam perbaikan mutu sekolah dasar di Kota Malang* (Disertasi). Progam Pascasarjana Universitas Negeri Yogyakarta, Yogyakarta.

Trilling, B., & Fadel, C. 2009. *21st Century Skills: Learning for Life in Our Times.* San Francisco, CA: John Wiley & Sons.

Wosnitza, M., et al. 2014. *Teachers resilience in Europe. A theoretical framework.* Aachen: ENTREE. Retrieved from http://entree- project.eu/wp content/uploads/2014/11/ENTREE1-new-v2_EN-1.pdf

Yoga, D. 2017. *Membangun budaya inovasi di perguruan tinggi.* Materi disampaikan pada Stadium General UNY.

Zamroni. 2017. *Pendidikan multikultural sebagai upaya untuk mengurangi ketimpangan prestasi.* Yogyakarta: UNY.

Educational Innovation in Society 5.0 Era: Challenges and
Opportunities – Purnomo & Herwin (Eds)
© 2021 the authors, ISBN 978-1-032-05392-9

The roles of social capital to promote soft skills among university students in facing the challenge of society 5.0 transformation

A.R. Ahmad, M.M. Awang, & N.A. Mohamad
The National University of Malaysia, Malaysia

ABSTRACT: This paper discusses the roles of social capital as used to promote soft skills among university students in order to prepare them for the society 5.0 transformation. Society 5.0 aims at tackling challenges the digitalization across all levels of society and how society can be transformed to fit the needs of the new world. One of the important elements for the new world is to have a great social capital network which has a close link with the quality of human soft skills. Therefore, this study surveyed 264 university students in Malaysia, where results demonstrated that the majority of them had a high social capital level and level of soft skills. Results revealed that there is no significant difference between genders with regards to social capital. This means that both genders had participated in various activities actively. However, there is a significant relationship between gender and level of soft skills. It has been found that male students have higher soft skills if compared to female students. This study found that there is a positive significant relationship between social capital and soft skills which indicates that both elements are very important in facing challenges in the new world. By having a great social network, they would have better opportunities in developing their careers. Therefore, this study recommends an extraordinary curriculum in higher education as it will benefit students in facing the challenges of society 5.0.

1 INTRODUCTION

Society 5.0 is described by the Japanese government as the "Super Smart Society" which aims at creating a society where people can resolve various social challenges by incorporating innovations such as robots and big data into society. Society 5.0 is a social state in which material and information are highly integrated. This means that today's graduates need a lot of modern skills. And to enable them to master various skills, graduates need to master social networks across geographical boundaries. Soft skills are among the elements that have been identified as very critical in the global world of work, not to mention the rapid technological change and development. Therefore, graduates should wisely assess these needs to meet the needs of the increasingly challenging job market and daily life. Academic achievement alone does not promise bright job opportunities to graduate graduates but soft skills are among the skills that are a guideline for employers during the interview process. This means that graduates are not only focused on excellent academic achievement at the university but should also take the opportunity in every activity and effort carried out by the university to develop soft skills in each student.

1.1 *Issue and challenges in society 5.0*

Based on the 2011 graduate workforce statistics report, a total of 33,800 graduates and 31,700 of diploma graduates were reported to be unemployed and did not get a place in the public or private fields. This figure is a large and alarming number as it indicates the level of marketability of graduates is still low in the country. Studies show that unemployment is influenced by various factors, among them is the attitude of choosing a job and failure to meet the demands of the industry and the job market, Nurita Juhdi (in Huzili, Azman & Shukri, 2008).

Lack of industrial training, poor communication skills and English proficiency, problem solving skills, the inclination to frequently change workplaces and a lack of self-confidence are among the five factors that cause graduates to fail to get a job after graduating from university. The failure of graduates to get a job is because they do not have the soft skills needed by employers. Political media is also associated with English language weakness, narrow-mindedness, lack of leadership characteristics, and lack of communication skills.

Academic achievement alone is not enough to hold positions in the public or private sectors. Employers are beginning to view employability as one of the

18

DOI 10.1201/9781003206019-4

necessary skills to enable graduates to get a place in the field of employment. Employability skills include personality, attitudes, habits and behaviors, ways of communicating, problem solving and decision-making skills as well as organizational management processes (Madar et al., 2008). There are many studies on soft skills have been conducted by researchers in and outside the country that focus more on the application and importance of soft skills in the marketability of graduates in the field of employment. However, studies on the role of social capital in shaping the soft skills of students at the university have not been conducted. Therefore, this study was conducted to see one of the angles of social capital in playing a role in shaping soft skills among students at the university.

1.2 Social capital concept

There are various terms that have been interpreted by scholars on social capital according to their respective roles. However, the main concept of social capital, which is an interaction involving a society that is interdependent in order to achieve the same goal is an important element in interpreting the meaning of this social capital. According to Putnam, social capital is the appearance of social organizations, such as trust, norms (or reciprocity), and networks (of community bonds), which can improve community efficiency by facilitating coordination and cooperation for mutual benefit. The two main constructs in social capital are structural constraints that involve relationships and interactions within the community and cognitive constructs that include perceptions in support and beliefs that are more conducive to attitudes and perceptions (Hashim & Ali, 2005).

Sociologist Pierre Bourdieu was among the first researchers to expand the use of the concept of economic capital to other fields such as cultural capital and social capital. According to Bourdieu, social capital consists of a network of institutions such as families, classes or political parties that are interdependent and involve cultural exchange within their group only. The development of an individual's social capital depends on the extent to which the network of relationships can be built and mobilized and the amount of capital in the form of economic, cultural and symbolic belonging to community members (Bourdieu, 1986).

In Hashim and Ali (2005), Putnam defines social capital as a value of mutualtrust (trust) between members of society and society towards its leaders. Social capital is defined as a social institution that involves networks, norms, and social trust that encourages social collaboration (coordination and cooperation) for the common good. Social capital is seen as a network of interaction networks that involves high trust in its community members to achieve common aspirations. Cohen and Prusak (in Inayah, 2012) state that social capital is the stock of active relationships between communities. Each pattern of relationships that occur is bound by trust, mutual understanding, and shard values, which binds group members to make it

possible for joint action to be carried out efficiently and effectively.

Coleman (1988), on the other hand, is more inclined to discuss social capital in the areas of academic achievement and adult behavior that focuses more on the mechanisms and role of social capital in the family structure. Coleman defines social capital as a variety of entities that have two common elements that they consist of several aspects of social structure and interaction in social capital facilitates the solution of a particular action that cannot be solved individually.

A key element in the formation of social capital is involvement in a community or group. The ability of a community to participate in social networking that is done on a voluntary basis (volunteer), equality, freedom and civilization. The commitment of community members to attend community gatherings is a benchmark for the formation of social capital in the community (Inayah, 2012). The second element is trust where trust is a key condition to the formation of strong social capital in society. Trust can be defined as a sense of mutual trust between individuals and between communities. Next is reciprocity i.e. exchanging kindness. Social capital involves interactions that often benefit both parties over a long and continuous period of time (Hasbullah, 2006). To form strong social capital, community members must have the same norms and social values that is rules that are expected to be followed by community members in order to prevent individuals from doing things that are contrary to the values of the local community (Agus et al., 2009).

Overall, it can be concluded that social capital as a resource generated through interaction between society and institutions that give birth to a sense of trust, reciprocity, social networks, norms and values that shape the structure of society to achieve the same goal. The loss of social capital in a society can be seen through the divisions and social problems that result due to the decline of the value of trust, the emergence of suspicion and individualism between communities in society. Weak social capital will dampen the spirit of gotong royong, increase poverty and unemployment as well as hinder every effort to improve the well-being of the population and further slow down the country's development process.

1.3 Interaction in the community in society 5.0

The term community refers to a grassroots local community that inhabits a particular area and has relationships with each other by interacting daily between them. Communities are not only characterized by physical characteristics such as residence, occupants and economic patterns but also include socio-cultural aspects such as values, norms, customs, beliefs, helpfulness, identity, and common religious practices (Wibowo, 2007). According to Fukuyama (2000) a true community is bound together by the values, norms and experiences shared by its members. Community relationships will become closer if the values and norms they hold are held stronger.

Interaction in the community means the interaction that exists in a communication that involves communication and bilateral relationships between members in the community. Interactions can be built as a result of communication or even involvement with community members in an activity carried out by the community. In addition, interaction is also established when members establish relationships and depend on each other to defend the idea of sharing and helping each other. The community serves as a gathering place for community members, a place to interact and communicate, a place to study, and even a place to shop.

An organization is a social unit of people that is structured and manages to meet needs or to achieve common goals. All organizations have a management structure that determines the relationship between different activities and members, and divides and assigns roles, responsibilities and powers to carry out different tasks. Organizations are open systems - they are influenced by their environment. Interaction within the organization is a network of social relationships that occur between members of the body of the organization. Organizational bodies can consist of associations, game clubs, NGOs and even co-curricular activities.

Through co-curricular activities such as associations, students are able to develop their potential in terms of leadership skills, communication, interpersonal, working in groups and managing. In addition, aspects of self-skills development can be strengthened through co-curriculum such as identity, discipline, patience, resilience and motivation (Esa, Md Yunos, & Kaprawi, 2007). It clearly shows that, through the interactions that occur through involvement in such activities, the skills required by students referred to as soft skills can be applied more comprehensively. The importance of co-curricular subjects can be seen through the formation of students' personalities in co-curricular activities such as respecting others and the spirit of cooperation with teammates (Som, 2001).

Madar et al. (2008) in a study of team skills mastery through participation in co-curriculum has shown that the ability to interact with group members, the ability to work with group members to achieve the same objectives and the ability to respect the opinions of group members are among three the highest element of team skills mastered by the respondents. These three abilities are among the important characteristics and elements that students need to master before they enter the career field in the future. Someone who works in a team should be able to interact and communicate well.

Programs or activities in the form of additional co-curriculum activities are non-academic support programs. The main objective of these non-academic focused support activities or programs is to help students improve their mastery of soft skills that are not directly related to academic affairs but are very helpful in shaping the personality and professionalism of students (Kementerian Pengajian Tinggi, 2006).

1.4 *Soft skills society 5.0*

Soft skills are generic skills that are also referred to as employability skills, soft skills, demonstrated skills and various other terms. However, the meaning of soft skills leads to the mastery of a student in skills that are non-academic but more focused on the development of personal skills, such as personality and humanity. Soft skills include aspects of communication skills, critical thinking and problem solving skills, teamwork skills, continuous learning and information management, entrepreneurial skills, professional ethics and morals, and leadership skills (Kementerian Pengajian Tinggi, 2006).

1.5 *Research objectives*

This study aims to identify the level of student involvement in community, organizational and political activities and the level of soft skills of students. In addition, this study also looks at the differences in the level of student involvement in community, organizational and political activities in terms of gender, race and field of study and the difference in the level of soft skills in terms of gender, race and field of study. This study also identifies the relationship between social capital and soft skills among students.

2 METHOD

This study uses respondents consisting of university students in Malaysia. A total of 90 male students and 174 female students were randomly selected consisting of students with different fields of study. Respondents consisted of 91 students from the field of science studies while 149 students from the field of social science studies. To find out the formation of social capital in university students, the instrument was built by modifying existing instruments on social capital that have been built by foreign researchers. The instrument was based on a questionnaire prepared by Grootaert, Narayan, Veronica and Woolcock (2004) in Measuring Social Capital: Integrated Questionnare and a set of questionnaires by P. Bullen and J. Onyx (1998) in Measuring Social Capital in Five Communities. Pilot tests were conducted on 20 university students and found that the Chronbach Alpha value of all items for the two constructs was high (0.967-0.993).

3 RESEARCH FINDINGS AND DISCUSSIONS

Respondents for this study involved 264 University students of which 34.1% (n = 9) were male students and 59.1% (n = 174 people) were female students. 91.7% (n = 242) of respondents are students from Indonesia. As many as 4.5% (n = 12) of Chinese students, 2.35 (n = 6) of Indian students and 1.5% (n = 4) of other races.

3.1 Student social capital level

To answer the first research question, namely the level of social capital among students, there are 2 constructs that have been built using 21 items to measure the involvement of respondents in the local community and the involvement of respondents in clubs or associations. Low scores indicate high levels of social capital among students, while high scores indicate low levels of social capital among students. Overall, based on Table 1, it was found that student involvement in community and club activities showed a high level with a community mean = 2.1340 while the club mean = 2.0283.

Table 1. Social capital.

Social Capital	Mean	SD	Level
Community involvement	2.1340	.71923	High
Club involvement	2.0283	.75522	High
Average	2.0913	.70528	High

Table 2 is the mean details for respondents' involvement in local community activities and club activities according to item breakdown.

To measure student involvement in community activities, a total of 12 items of questionnaires were constructed based on a questionnaire by Ollen (1997). Overall, the mean for all activities involving respondents' involvement in the community indicates a high level with items "I will visit sick neighbors/relatives, visiting neighbors when sick" (mean = 1.94) shows the highest mean while items "I trust the people who are around me," (mean = 2.43) shows the lowest mean.

Table 3 shows the second construct in measuring social capital. A total of 9 items were built to test the respondents' involvement in club/association activities. The results of the study found that overall, the involvement of respondents was at a high level with a mean of 2.028. Item "I engage in associations that can benefit myself and the local community, involvement in activities that benefit myself shows the highest score" had a mean of 1.96, while item "I often engage in association activities at universities or NGOs, involvement in activities and clubs" showed the lowest score (mean = 2.17).

Based on the findings of the study, it was found that the level of social capital among students in the context of this study is high. For the first construct, which is involvement in the community, visiting activities during illness are the highest elements that contribute to the formation of social capital. Nevertheless, the element of trust is seen to have the lowest mean in this construct. These findings contradict previous studies because the value of trust should be the most important element in shaping social capital. High trust among members will create a good bond and cooperation between members in a community that leads to honesty with each other (Prusak & Cohen, 2001). In addition, Inayah (2012) also agrees that social capital exists as a result of attitudes of trust in a community.

Table 2. Involvement in the local community.

Item	Mean	SD	Level
I think I am valued in society	2.10	.898	High
I will visit sick neighbours/relatives	1.94	.921	High
My friends and I often eat and go out together	2.01	.975	High
My family and friends always refer to me when they have personal problems	2.21	.882	High
My friends and I have the same interests	2.32	.917	High
When I am in trouble my friends will help me	2.08	.927	High
I always feel comfortable with my friends	2.07	1.026	High
I feel safe when in my neighborhood/college	2.20	1.039	High
I trust the people around me	2.43	.988	High
When experiencing financial difficulties, my friends/family will help me	2.14	1.069	High
The community in my residence is from a variety of different backgrounds but we respect each other and get along	2.05	.873	High
Have you ever picked up another person's rubbish in a public place?	2.02	.764	High
Average	2.134	.719	High

As for involvement in association club activities, the item of self-involvement in activities that benefit oneself is seen to have been the highest score. This is because association clubs are social agents that make it easier for an individual to obtain information. The diversity of members with different backgrounds and skills allows a variety of information to be channeled into the organization in a short time. For both of these contractors, it was found that involvement in association club activities gave a higher score in the formation of students' social capital.

Table 4 shows the soft skills construct constructed using 7 soft skill elements with a total of 49 items. Overall, the level of insnaiah skills among the respondents is high with an average mean of 2.134 and

Table 3. Participation in club/association activities.

Item	Mean	SD	Level
I often involved in uninterested activities in universities or NGOs	2.17	1.007	High
I like to engage with charity/volunteer work	2.08	.884	High
I involved in associations that can benefit myself and the local community	1.96	.837	High
I often mingle with members who have similar interests with me	2.02	.834	High
I often attend meetings and activities carried out by association/clubs	2.01	.902	High
It is easy for me to get information when joining associations/clubs	1.98	.958	High
We often work together in association activities	1.99	.856	High
My relationship with the club members is very good	2.03	.878	High
I love working and carrying out activities in associations/clubs that I joined	2.03	.928	High
Average	2.028	.755	High

Table 4. Soft skills level.

Soft Skills	Mean	SD	Level
Communication	2.1692	.67324	High
Leadership	2.1111	.71157	High
Team work	2.0019	.80342	High
Continuous learning	2.1084	.76190	High
Critical Thinking	2.1495	.71490	High
Ethics and Professional	2.1751	.72386	High
Entrepreneurship	2.2230	.76334	High
Average	2.1344	.66704	High

a standard deviation of 0.667. the group work element gives the highest score (mean = 2.0019) and the entrepreneurial skills element gives the lowest score (mean = 2.223) in the formation of students' soft skills.

3.2 Soft skills level

The following are the mean details according to the breakdown of the elements that make up soft skills among respondents.

Table 5 shows the overall mean of the 7 items found in communication skills. Item "I can make power point slides and use an LCD projector in presentation, use of projector slides" (mean = 2.02) shows the highest mean score while items "I often use email to communicate with friends/lecturers" and "use of email in communication" (mean = 2.47) show the lowest score in shaping communication skills among respondents.

Table 6 shows the overall mean for the 7 items found in leadership skills. With item "When making a decision in a group, I will listen to the opinions of each member of my group, taking into account the opinions of others" (mean = 1.87) showing the highest mean score while item "I am often chosen to be the leader in a group, group leader" (mean = 2.58) shows a score lowest in shaping leadership skills among respondents.

Table 7 shows the overall mean for the 7 items found in group work skills. Item "I do not look down on other friends and always respect them, respect others" (mean = 1.87) shows the highest mean score while item "I like to do group work, like to work in groups" (mean = 2.58) shows the lowest score in developing group work skills among the respondents.

Table 8 shows the overall mean for the 7 items found in continuous learning skills. Item "I am willing to

Table 5. Communication skills level

Item	Mean	SD	Level
I can speak English and Malay fluently	2.20	.905	High
I can create power point slides and use an LCD projector in presentations	2.02	.883	High
I use pictures and charts to convey my ideas	2.10	.883	High
I can complete a written task well	2.12	.867	High
Friends often ask me when they are unclear with what the lecturers convey	2.17	.829	High
I often use emails to communicate with friends/ lecturers	2.47	.921	High
I often use appropriate body language when chatting with friends,	2.08	.843	High
Average	2.169	.673	High

Table 6. Leadership skills

Item	Mean	SD	Level
I will hear the opinions of each member of my group when making decision	1.87	.816	High
I accept the responsibility given by lecturers and friends to me	1.96	.816	High
I will distribute the assignments in the group to members according to the ability of each member	1.99	.826	High
I always confident when communicating in a group	2.21	.898	High
I often chosen to be the leader in a group	2.58	.988	High
I will help other members of the group in completing the assignment	2.16	.900	High
I learned from mistakes and am always willing to accept comments	2.01	.832	High
Average	2.1111	.712	High

Table 7. Group work skills

Item	Mean	SD	Level
I love group work	2.27	1.003	High
Each member of the group has the right to speak up	1.90	.865	High
I often complete assignments in groups	2.06	.864	High
I produced good practical work through teamwork	2.02	.859	High
I am a flexible person and willing to hear the opinions of each member of the group	1.92	.877	High
I don't underestimate other friends and always respect them	1.88	.956	High
I always support friends to complete their tasks well	1.92	.964	High
Average	2.002	.803	High

Table 8. Continuous learning skills

Item	Mean	SD	Level
I can find relevant information from a variety of sources	2.02	.893	High
I am willing to learn new technologies	1.98	.854	High
I can identify problems in complex situations	2.23	.903	High
I can find another alternative to solve a problem	2.11	.805	High
I can adapt myself in a new work environment	2.24	.983	High
I like to find opportunities to learn something new	2.08	.881	High
I am ready to attend seminars and external courses that can improve my knowledge	2.09	.930	High
Average	2.108	.762	High

learn new technology, ready to learn new technology" (mean = 1.98) shows the highest mean score while item "I can identify problems in complex situations, identify problems" (mean = 2.23) shows the lowest score in shaping continuous learning skills among the respondents.

Table 9 shows the overall mean of the 7 items found in critical thinking and problem solving skills. The item "I often give other alternatives in solving a problem, find alternatives in solving problems" (mean = 2.08) shows the highest mean score while item "I often reflect after solving a problem, make reflection" (mean = 2.37) shows the lowest score in shaping thinking skills critical and problem solving among respondents.

Table 10 shows the overall mean of the 7 items found in professional ethics and morals. Item "Receiving intentionally excess balance money is wrong, the offense of receiving excess balance money" (mean = 1.88) shows the highest mean score while item "I am willing to report to the lecturer if my friend commits fraud in completing assignments, reporting imitation and fraud of other friends" (mean = 2.58) shows the lowest score in shaping continuous learning skills among respondents.

Table 11 shows the overall mean of the 7 items found in entrepreneurial skills. My items are highly motivated because I know what can make me successful, optimistic and find new opportunities (mean = 1.88) show the highest mean score while items I can receive and share about new ideas, share new

Table 9. Critical thinking and problem-solving skills

Item	Mean	SD	Level
I look at issue from a variety of different angles and evaluated it based on clear information	2.11	.883	High
I look for the main cause of the problem before trying to fix it	2.13	.876	High
I often give other alternatives in solving a problem	2.08	.854	High
I often contribute ideas in discussions	2.11	.839	High
When dealing with a problem, I will find information in advance regarding the problem	2.16	.921	High
I use a variety of skills in looking at a problem	2.11	.915	High
I often make a reflection after solving a problem	2.37	.903	High
Critical Thinking	2.150	.715	High

Table 10. Professional ethics and morals

Item	Mean	SD	Level
I always dress neatly while attending classes	2.19	1.007	High
I am willing to report to the lecturer if my friend commits fraud in completing the task	2.58	1.121	High
I completed the task within the stipulated time	2.14	.989	High
I do not plagiarize/imitate while completing my assignment	2.30	.916	High
I am always willing to accept comments to fix my weaknesses	1.95	.895	High
Receiving excessive balance of money intentionally is a wrong thing	1.88	1.081	High
Average	2.1751	.72386	High

ideas (mean = 2.58) show score lowest in developing continuous learning skills among respondents.

Based on the findings of the study, it was found that the level of soft skills of students is at a high level with the elements of group work skills show

Table 11. Entrepreneurial skills

Item	Mean	SD	Level
Prioritize families in filling vacancies is the wrong thing	2.31	.997	High
I enjoyed attending a reunion to get a new friend	2.41	.942	High
I am willing to take the risk of planning and starting a business	2.07	.850	High
I can accept and share about new ideas	2.38	.997	High
I am sure I can draft the business well	2.27	1.064	High
I am interested in venturing into entrepreneurship/opening a business if given the opportunity	2.13	.896	High
I was highly motivated because I knew what could make me successful	1.99	.931	High

the highest score while entrepreneurship skills show the lowest score in shaping soft skills among students. Leadership skills can be developed in students through co-curricular activities such as clubs and association activities because such activities are seen to develop the potential of the individual self as described by Walker (2002). Student involvement in as a committee member and implementation in an activity can improve the student's skills from the aspect of leadership. For entrepreneurial skills, the findings of the study are contrary to the findings of the study by Malek (2009) who found that students are ready and have the necessary entrepreneurial characteristics.

3.3 Difference analysis

Based on Table 12, it is found that the t-value for the comparison of the level of involvement in club activities for male and female students is t = −0.138 and the significance level p = 0.052. This significance

Table 12. Gender differences in community engagement and clubs/societies

	Gender	N	Mean	SD	t-value	p
Club	Male	88	2.019	.866	−.138	0.052
	Female	174	2.033	.695		
Community	Male	90	2.208	.804	1.209	0.357
	Female	174	2.095	.670		

Table 13. Gender differences in soft skills

	Gender	N	Mean	SD	t-Value	p
Soft Skills	Male	9	2.247	.770	1.987	0.028
	Female	174	2.076	.601		

Table 14. Relationship between social capital and soft skills

		Social Capital	Soft Skills
Social Capital	Pearson Correlation	1	.845**
	Sig. (2-tailed)		.000
	N	264	264
Soft Skills	Pearson Correlation	.845**	1
	Sig. (2-tailed)	.000	
	N	264	264

**. *Correlation significant at levels 0.01 (2-tailed).*

level is greater than 0.05 ($p > 0.05$). Thus, there was no significant difference in the level of involvement in club activities for male and female students. The level of involvement of male students in club activities (mean = 2.0193) is the same as the involvement of female students (mean = 2.0329).

The t-value for the comparison of the level of involvement in community activities for male and female students is t = 1.209 and the significance level $p = 0.357$. This significance level is greater than 0.05 ($p > 0.05$). Therefore, the null hypothesis (Ho1) is accepted. Thus, there was no significant difference in the level of involvement in community activities between male and female students. The level of involvement of male students in community activities (mean = 2.2083) is the same as the involvement of female students (mean = 2.0955).

Based on Table 13, it is found that the t-value for the comparison of Soft Skills level for male and female students is t = 1.987 and significant level $p = 0.028$. This significance level is smaller than 0.05 ($p < 0.05$). There is a significant difference in the level of soft skills for male and female students. The level of soft skills of male students (mean = 2.019) is higher than the level of soft skills of female students (mean = 2.033).

3.4 Relationship analysis

Table 14 shows all items in the Social Capital and Soft Skills constructs positively correlated with a significant 0.845. This indicates that social capital has a significant relationship with soft skills. The higher the level of social capital, the higher the soft skills of a student.

There is a positive relationship between social capital and soft skills. Increasing the level of social capital is able to increase the level of soft skills in students. Through involvement in co-curricular activities such as associations and clubs, various skills can be learned by students indirectly such as group work skills, communication skills and leadership skills (Walker, 2002). In addition, the relationship with the environment such as the relationship between students and the institute was found to have a positive relationship with communication skills, critical and problem solving skills, teamwork skills, continuous learning and information management, and entrepreneurial skills (Kee, et al., 2011).

4 CONCLUSION

Social capital is the interaction that exists as a result of the togetherness and trust of members in a community that involves the active involvement of individuals in a social group such as NGOs, associations, organizations, clubs, surau committees and so on. Soft skills are non-academic skills but are more focused on the development of personal skills, personality and humanity. A total of seven elements have been identified and selected by HEIs to determine soft skills among university students. Soft skills include aspects of communication skills, critical thinking and problem-solving skills, teamwork skills, continuous learning and information management, entrepreneurial skills, professional ethics and morals, and leadership skills. In the context of this study, it was found that the respondents have a good level of social capital and soft skills. These two constructs show a positive relationship where the increase in the level of social capital is seen to be able to improve the level of soft skills of students. Overall, the results show that the respondents have a high level of social capital and a high level of soft skills. All the elements that make up social capital and soft skills are seen to have a high mean. There were no significant differences for male and female students and for students with different educational backgrounds in the level of social capital but there were significant differences in the level of soft skills. There is a significant positive relationship for social capital with soft skills where high level of social capital is seen to have a better impact on students' soft skills. The results of this study provide exposure to the importance of social capital to students. The interactions that take place on a daily basis are seen to have a strong impact on students if they can be carried out well. Good social capital can not only help students in receiving social support at the university or college but also play a role in helping students achieve a common goal or goal. In addition, the trust and sense of togetherness that exists with each other gives a sense of value and a high sense of belonging to students which are seen as a good asset in producing better quality students.

REFERENCES

Bourdieu, P. 1986. The Forms of Capital. In Richardson, J. G., ed., *Handbook of Theory and Research for the Sociology of Education*, New York: Greenwood.

Bullen, P. & Onyx, J. 1998. *Measuring social capital in five communities in NSW. Overview of A Study.* (Online). http://www.mapl.com.au/Az.htm

Coleman, J. S. 1988. Social Capital in the Creation of Human Capital', *American Journal of Sociology*, 94, Supplement: 95–120.

Esa, A., Md Yunos, J., & Kaprawi, N. 2007. Persepsi pensyarah terhadap penerapan kemahiran kepimpinan menerusi kokurikulum di politeknik. *Jurnal Persatuan Pendidikan Teknik dan Vokasional*, 1(2): 50–61.

Fukuyama, F. 2000. *Social capital and civil society.* International Monetary Fund Working Paper WP/00/74. 1–19. Retrieved from https://papers.ssrn.com/sol3/papers.cfm?abstract_id=879582

Grootaert, C., Narayan, D., Jones, V. N., & Woolcock, M. (2004). *Measuring social capital: An integrated questionnaire.* The World Bank.

Hasbullah, J. 2006. *Sosial Capital (Menuju Keunggulan Budaya Manusia Indonesia).* Jakarta: MR United Press.

Hashim, H. R., & Ali, M. N. A. W. D. M. 2005. *Pembangunan sahsiah mahasiswa bersepadu: Konsep dan pelaksanaannya di Kolej Universiti Kejuruteraan.* In Paper in di Seminar Kebangsaan Kursus Sokongan Kejuruteraan (pp. 17–18).

Huzili, H., Azman, Z., & Shukri, M. 2008. Memperkasakan Mahasiswa Kejuruteraan Menerusi Penerapan Kemahiran Insaniah (Soft Skill). 583–596. Retrieved from http://103.86.130.60/handle/123456789/5836

Inayah. 2012. Peranan Modal Sosial dalam Pembangunan. *Jurnal Pengembangan Humaniora*, 12(1):43–49.

Kee, C. P., Ahmad, F., & Ibrahim, F. 2011. Hubungkait antara Kemahiran Insaniah Prasiswazah dengan Dimensi Hubungan Organisasi-publik. *Malaysian Journal On Student Advancement (Jurnal Personalia Pelajar)*, 14: 23–36

Kementerian Pengajian Tinggi. 2006. Modul Pembangunan Kemahiran Insaniah (Soft Skills) Untuk Institusi Pengajian Tinggi. Serdang, Malaysia: Universiti Putra Malaysia.

Madar, A. R., Abd Aziz, M. A., Abdul Razzaq, A. R., Mustafa, M. Z., & Buntat, Y. 2008. Kemahiran employability bagi memenuhi keperluan industri. *Prosiding SKIKS*, 8, 385–92.

Malek, M. A. N. A. 2009. Kesediaan Pelajar Pendidikan Teknikal dan Kejuruteraan dalam Menceburi Bidang Keusahawan. (Tesis, Universiti Teknologi Malaysia).

Prusak, L., & Cohen, D. (2001). How to invest in social capital. *Harvard business review*, 79(6):86–97.

Som, M. M. 2001. *Kepentingan mata pelajaran kokurikulum di kalangan pelajar institusi pengajian tinggi: tinjauan ke atas pelajar tahun akhir Ijazah Sarjana Muda Kejuruteraan Elektrik di KUiTTHO* (Doctoral dissertation, Kolej Universiti Teknologi Tun Hussein Onn).

Supriono, A., Flassy, D. J. & Rais, S. 2009. Modal Sosial: Definisi, Dimensi dan Tipologi. https://docplayer.info/70783746 - Oleh-agus-supriono-2-dance-j-flassy-3-sasli-rais-4-abstrak-pendahuluan.html

Walker, V. N. 2002. *Office of Co-curicular Life.* Sweet Briar: Sweet Briar College.

Wibowo, A. 2007. Menumbuhkembangkan Modal Sosial Dalam Pengembangan Partisipasi Masyarakat. *M'power,* 5(5).

Educational Innovation in Society 5.0 Era: Challenges and
Opportunities – Purnomo & Herwin (Eds)
© 2021 the authors, ISBN 978-1-032-05392-9

Perception and attitude of students to character education towards society 5.0 era

K. Fajriatin & A. Gafur
Graduate School, Yogyakarta State University, Yogyakarta, Indonesia

ABSTRACT: This research aims to investigate the students' perceptions and attitudes on the implementation of character education and the readiness of students towards the coming society 5.0 era in the Study Program of Pancasila and Civic Education, Mulawarman State University. This research is a descriptive research using quantitative approach. 114 students (50% of the population students of class 2017-2019) were taken as research samples using a proportional random sampling technique. The findings show the students perception the implementation of character education is quite good. This is indicated by 76% of the respondents who have "quite good" perception. Meanwhile, the students attitudes on the implementation of character education is quite good. This is indicated by 88% of the respondents who have "quite good" attitudes. And readiness from students toward the society 5.0 era is good. This is indicated by 51% respondents who have good readiness.

1 INTRODUCTION

In early January 2019, a new idea emerged from Japan, namely the idea of society 5.0 presented at the 2019 World Economic Forum in Davos, Switzerland. This idea emerged in response to industrial revolution 4.0 as a sign of the significance of technological development, but the role of the community is very much a consideration for the occurrence of this industrial revolution 4.0 (Puspita et al. 2020: 123)

The Government of Japan's society 5.0 initiative seeks to apply new technology to create a cyber-physical society. This will be done using super smart technologies that are more human-centered (Harayama 2017: 8) than our current information society. The goals of the society 5.0 initiative are very ambitious. As Bryndin (2018: 12) notes, the goal of society 5.0 is none other than the creation of equal opportunities for all and also to provide an environment for the realization of the potential of each individual; to that end, society 5.0 will use new technologies to remove physical, administrative and social barriers to one's self-realization. Likewise, Keidanren (2016: 10), the Japan Business Federation, imagines that in society 5.0, every individual, including parents and women, can live a safe and guaranteed comfortable and healthy life and each individual can realize the lifestyle he wants.

Technology in society 5.0 is expected to provide not only the minimum services required for individual survival but to make life more meaningful and enjoyable; In society 5.0, human-technology interactions will be utilized to provide a community-centered world that is sustainable, vibrant, and livable (Medina-Borja,

2017: 235). Thanks to the very high-tech cyber-physical nature, society 5.0 will be able to incorporate into the social structure and the types of dynamics of creatures previously not found in world society (Gladden, 2019: 4).

Educational institutions that are categorized as superior in Indonesia have not implemented this system. Starting from the education system, how to interact with educators and the educated, as well as the cultivation of modern thinking paradigms must be considered. In this case, the industrial system 4.0 towards society 5.0 needs preparation both from individuals, academics, and things that affect society.

In social life, we often find rampant juvenile delinquency, bullying, promiscuity, brawls and drug use that continue to grow and now attack the millennial generation. The character education crisis which is marked by signs of the times that brought destruction is still compounded by the mental block that has hit some Indonesian people. According to Zubaedi (2011: 69) the causes include: (1) bad self-image, (2) bad experience, (3) bad environment, (4) bad reference, and bad education.

Education at all levels during the new order era emphasized more cognitive aspects and less developed affective aspects such as emotional intelligence, besides that education also ignored the inculcation of values in students (Suyanto & Hisyam, 2000: 6). Such an education system causes the moral quality of Indonesian society to be low. Seeing these conditions, efforts are made to include the character values expected by students.

Higher education is an educational unit which is obliged to take part in shaping the character of

DOI 10.1201/9781003206019-5

the nation. Character development is very important for universities and their stakeholders to become a foothold in the implementation of character education in higher education (Dhiu & Bate, 2017: 173)

In connection with the course of character education, it is necessary to examine the perceptions and attitudes of students in the Pancasila Education Study Program and University Citizenship of Mulawarman. Perception is essentially a cognitive process experienced by everyone in understanding information about their environment, either through sight, hearing, appreciation, feeling, and smell (Thoha, 2010: 141-142). Meanwhile, according to Carole & Tarvis (2007: 193), perception is a set of mental actions that regulate sensory impulses into a meaningful pattern. According to Walgito (1989: 54-56) there are three conditions for the occurrence of perception, namely: 1) the existence of the object being perceived, 2) the presence of sensory organs or receptors, and 3) attention.

Meanwhile, the attitude described by Azwar (2010: 3) is defined as a reaction or response that arises from an individual towards an object which then leads to individual behavior towards that object in certain ways. Attitude is a tendency to react to a thing, person or object by liking, not liking or being indifferent (Sabri, 2010: 83). From that attitude, it is then used as material for reflection for teachers/lecturers to assess how the educational process has been going on.

Based on the description above, the researcher is interested in knowing what the perceptions and attitudes of students are like in the Pancasila and Citizenship Education Study Program towards the implementation of character education in the Pancasila and Citizenship Study Program at Mulawarman University.

2 METHODS

Method used in this study is a quantitative method because the data collected is in the form of numbers and processed using statistical analysis with the help of the SPSS 20 for Windows program. This research was conducted at the Pancasila and Citizenship Education Study Program, Mulawarman University from June to July 2020.

3 FINDINGS AND DISCUSSION

3.1 *Student perceptions of the implementation of character education*

The tendency of student perceptions towards the implementation of character education can be seen in the following Figure.

Figure 1 shows that students perceptions of the implementation of character education are dominated by students who have perceptions in the good category, which is 23%. Meanwhile, 76% were included

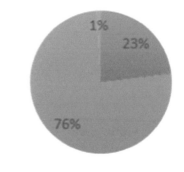

Figure 1. Student perceptions of the implementations of character education.

in the fairly good category and those included in the bad category had the smallest percentage at 1%.

This shows that the highest number of students perceptions of the implementation of character education is quite good, or it can be said that the students perceptions of the implementation of character education are quite good, because it is dominated by students who have quite good perceptions reaching 76%.

If you look at the trends above, it can be said that students' perceptions of the implementation of character education are classified as good. Students show obedience to rules and regulations in the study program environment, such as obedience to dress procedures or student appearance ethics, obedience to lecture activities, activeness in participating in various competitions and student creativity programs. Based on this, the Pancasila and Citizenship Education Study Program, Mulawarman University is expected to produce quality graduates not only intellectually but morally as shown by the positive character possessed by graduates of the Pancasila and Citizenship Education Study Program, Mulawarman University.

3.2 *Student attitudes towards the implementation of character education*

The tendency of student attitudes towards the implementation of character education can be seen in the Figure 2.

Figure above shows that the attitudes of students towards the implementation of character education are dominated by students who have attitudes in the fairly good category, which is 88%. Meanwhile, 11% were included in the good category and those included in the bad category had the smallest percentage at 1%. This shows that the attitudes of students towards the implementation of character education which are classified as good enough, or it can be said that the attitudes of students towards the implementation of character education are quite good, because it is dominated by students who have quite good attitudes reaching 88%.

The implementation of character education in the Pancasila and Citizenship Education Study Program

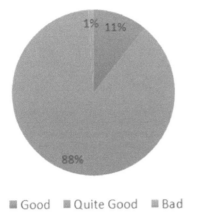

Figure 2. Student attitudes towards the implementation of character education.

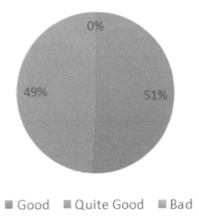

Figure 3. Student readiness in facing the era of society 5.

at Mulawarman University is quite good. Graduates of the Pancasila and Citizenship Education Study Program at Mulawarman University, graduates who are candidates for Pancasila and Citizenship Education teachers with good characters, are expected to be role models for the students they teach in the future, so being an example can bring changes to the character of the Indonesian nation.

The attitudes of students towards the implementation of character education that is not optimal can be caused by character education itself and is a process that is not easy and requires a relatively long time. Students must carry out a series of difficult processes in order to realize the goals of character education in the Pancasila and Citizenship Education Study Program, Mulawarman University such as adjusting the way of dress, communication ethics, ethics of attending lectures and so on. The attitude of students towards the implementation of character education must always be improved because basically the implementation of character education is beneficial for the students themselves and also for the Pancasila and Citizenship Education Study Program, namely the success of character education.

3.3 Student readiness in facing the era of society 5.0

This research not only seeks to determine the perceptions and attitudes of students in the Pancasila and Citizenship Education Study Program, the Faculty of Teacher Training and Education, Mulawarman University on the implementation of character education but also to determine the readiness of students in facing the era of society 5.0. In this study, the results of the descriptive analysis of readiness variables will be presented. Identification of the tendency for the high and low scores of students' perceptions towards the implementation of character education is determined based on ideal criteria. The tendency of student perceptions of student readiness can be seen in the following Figure:

Figure above shows that students' perceptions of the implementation of character education are dominated by students who have perceptions in the good category, namely 51%. Meanwhile, those included in the fairly good category were 49% and those included in the bad category had the smallest percentage at 0%. This shows that the highest number of students' perceptions of the implementation of character education is quite good, or it can be said that the students' perceptions of the implementation of character education are quite good, because it is dominated by students who have quite good perceptions reaching 51%.

If you look at the trends above, it can be said that students' perceptions of the implementation of character education are classified as good. This shows that the implementation of education has gone well. With the implementation of character education in the Pancasila and Citizenship Education Study Program, Mulawarman University which is well-perceived by students, it should be followed by increasing student character qualities by showing obedience to rules and regulations in the Prodi environment, such as obedience to dress procedures or student appearance ethics, obedience to lecture activities, activeness in participating in various competitions and student creativity programs. Based on this, the Pancasila and Citizenship Education Study Program, Mulawarman University is expected to produce quality graduates not only intellectually but morally as shown by the positive character possessed by graduates.

4 CONCLUSION

Based on the results of the research and discussion that has been described, the following conclusions can be drawn. Students' perceptions of the implementation of character education in the Pancasila and Citizenship Education Study Program, Mulawarman University are in the "quite good" category. This is shown by the results of the study which stated that 76% of the respondents had a fairly good perception. The attitude of students towards the implementation of character education is in the quite good category. This is shown in the results of the study which stated that 88% of the respondents had a fairly good attitude. And the readiness of students to implement character education towards the society 5.0 era is included in the good category. This is shown by the results of the research which states that 51% of the respondents have good readiness.

REFERENCES

Azwar, S. 2010. *Sikap Manusia: teori dan pengukuran*. Yogyakarta: Pustaka Pelajar.

Bryndin, E. 2018. System synergetic formation of society 5.0 for development of vital spaces on basis of ecological economic and social programs. *Annals of Ecology and Environmental Science* 2: 12–19.

Carole, W., & Tavris, C. 2007. *Psikologi Jilid 2*. Jakarta: Erlangga.

Dhiu, K. D., & Bate, N. 2017. The importance of character education in higher education: practical theoretical studies. *2 nd Annual Proceeding*, November 2017. 172–176.

Gladden, M. E. 2019. Who will be the members of society 5.0? towards an anthropology of technologically posthumanized future societies. *Social Sciences*, 8(5): 148. https://doi.org/10.3390/socsci8050148

Government of Japan. 2016. *The 5th Science and Technology Basic Plan. Provisional translation*. January 22. Available online: https://www8.cao.go.jp/cstp/english/basic/ 5thbasicplan.pdf (accessed on 25 June 2020).

Harayama, Y. 2017. Society 5.0: Aiming for a New Human-Centered Society. Japan's Science and Technology Policies for Addressing Global Social Challenges. *Interviewed by Mayumi Fukuyama*. Hitachi Review 66: 8–13.

Keidanren (Japan Business Federation). 2016. *Toward Realization of the New Economy and Society: Reform of the Economy and Society by the Deepening of 'Society 5.0'*. April 19. Available online: http://www.keidanren.or.jp/en/ policy/2016/029_outline.pdf (accessed on 25 june 2020).

Medina-Borja, A. 2017. *Smart human-centered service systems of the future. in future services & societal systems in Society 5.0. Edited by Kazuo Iwano, Yasunori Kimura, Yosuke Takashima, Satoru Bannai, and Naohumi Yamada*. Tokyo: Center for Research and Development Strategy, Japan Science and Technology Agency.

Puspita, Y., et al. 2020. Goodbye to the industrial revolution 4.0, welcome to the industrial revolution 5.0. *Proceedings of the National Education Seminar for the Postgraduate Program at the PGRI University of Palembang*. Palembang: 10 January 2020. 122–130. Retrieved from https://jurnal.univpgri-palembang.ac.id/index.php/ Prosidingpps/article/view/3794

Thoha, M. 2010. Perilaku Organisasi : Konsep Dasar dan Aplikasinya. Jakarta: Rajawali Press.

Sabri, M. A. 2010. Psikologi Kependidikan Berdasarkan Kurikulum Nasional. Jakarta: Pedoman Ilmu Raya.

Slameto. 2010. Belajar dan Faktor-faktor yang mempengaruhinya. Jakarta: Rineka Cipta

Suyanto, H. D. 2000. *Refleksi dan Reformasi Pendidikan di Indonesia Memasuki Millenium III*. Yogyakarta: Adi Cita.

Walgito, B. 1989. *Pengantar Psikologi Umum*. Surabaya: Bina Ilmu.

Widoyoko, E.P. 2004. *Penilaian hasil pembelajaran di sekolah*. Yogyakarta: Pustaka Pelajar.

Zubaedi. 2011. *Desain* pendidikan karakter*: konsepsi dan aplikasinya dalam lembaga* pendidikan. Jakarta: Prenada Media Group.

Educational Innovation in Society 5.0 Era: Challenges and Opportunities – Purnomo & Herwin (Eds)
© 2021 the authors, ISBN 978-1-032-05392-9

Implementation of civic virtue in character education in the era of Society 5.0

Y. Mahendra, Nasiwan & S.H. Rahmia
Graduate School of Yogyakarta State University, Yogyakarta, Special Region of Yogyakarta, Indonesia

ABSTRACT: The rise of youth social deviance indicates the weak idealism of the nation's character. The effort to create excellent and smart citizens is the civic education mission. Citizenship competence in psychosocial character education is reflected in knowledge, skills, determination, attitudes, and commitments that can be summarized as civic virtue. This study aims to determine factors supporting civic virtue implementation in character education in schools according to literature. This study uses a library research method by examining the previous research results on civic virtue in character education in schools. The results and discussion show that the implementation of civic virtue in schools can run effectively with the support of professional teachers' skills, students' attitudes and commitments, school policies, school culture, and values of local wisdom philosophy based on Pancasila. In conclusion, character education in schools according to the 2013 curriculum can be strengthened by civic virtue along with the development in the era of Society 5.0.

1 INTRODUCTION

Globalization has led to massive changes that make problems in the field of education increasingly complex in Indonesia. It has been a common knowledge that in the course of the era of Industry 4.0 and Society 5.0, students will be spoiled by increasingly sophisticated technology. As time goes by, it results in changes in students' behaviour patterns and perspectives in living their social lives.

Today, there are many phenomena of moral decline that occur as a result of technological advances. The loss of the citizenship character can be a question of fundamentalist attitudes and egoism that cause social deviance in the form of criminal acts committed by teenagers, such as drug and alcohol, promiscuity, and brawl. It has resulted in the character of citizenship crisis, which must be resolved structurally by educational institutions and stakeholders who are more creative, innovative, and proactive and can design the learning process in character education development, primarily civic character (Mulyono 2017). Based on this issue, character education needs to be designed in a holistic and contextual perspective to build critical-dialogical thinking in forming good character and intelligent citizens at the family, school, and state levels (Setiawan 2014).

One of the efforts that can be made to prepare ideal citizens is through education. The formation of good citizens is one of the main missions in Indonesia's national education through a character education program. According to Thomas Lickona (1991), in Education for Character: How Our School Can Teach Respect and Responsibility, character contains three interconnected parts, namely moral knowledge, moral feelings, and moral behaviour. Character is referred to as a person's identity formed in the life process in behaviour, thought patterns, and attitudes. Each individual can own these results because they are built by the socialization and education processes during his lifetime. Aspects of civic virtue must support the implementation of character building in students. It needs to be understood that the civic virtue conceptual-paradigmatically is a slice of or a prepared mix of civic confidence, skills, and commitment, which is the culmination of the whole psychological-pedagogical process or the civilization and empowerment of civic education (Udin 2016). Based on this understanding, civic virtue is the willingness of citizens to put aside their personal interests for the sake of public interest. Civic virtue consists of two elements, namely civic disposition and civic community. Civic dispositions are citizens' habits and attitudes in supporting the development of social functions to support political participation and a healthy political system for the public interest in a democratic system. Meanwhile, civic commitment is a commitment to an unconscious willingness, upholding democratic values and principles (Syarifa 2019).

Education in Indonesia is required to anticipate the rapid development of technology in the era of the industrial revolution 4.0. Curriculum design and educational methods must also adapt to the business climate, educational services, and industrial business, which continues to grow and is increasingly competitive. The changes occurring in the era of the industrial revolution 4.0 were also influenced by the emergence of society 5.0 era, which will significantly affect the human character, competition in the working world so that the skills required to change rapidly

DOI 10.1201/9781003206019-6

(Suryadi, 2020). Based on this explanation, the challenge we face is how to prepare and map the workforce of education graduates who are truly ready to compete in the working world following the expertise in their respective fields.

In response to the statement above, the education process should focus on artificial intelligence through connectivity in all things. However, it must also focus on the human component as the driving force for education, which has now entered the era of society 5.0, where this era offers a society that is centered on balance. The era of society 5.0 made the internet not only as a piece of information, but an essential supporting element in living life, so that in this era, technology becomes part of the human being, and the technological developments can minimize the existence of social disparities and economic problems in the future (Ely & Rizqi 2020).

2 METHOD

The method used in this study is a literature review with library research, which attempts to describe how the implementation of civic virtue in character education, especially, in the era of Society 5.0. In this literature review, the author uses various written sources such as articles, journals, and documents relevant to the issues examined in this study.

3 RESULTS AND DISCUSSION

3.1 *Building a good and smart citizen character*

The loss of citizenship character can lead to fundamentalist and selfish attitudes that cause students' social deviations. This character crisis must be overcome culturally by the Indonesian people. In this regard, the Indonesian nation recognizes the importance of national character building to maintain and form a good citizen's character. To form the character of a good citizen, of course, cannot be separated from education's vital role. Thus, the term good citizen is considered lacking, and it is necessary to add the word 'smart'. This word is closely related to education's role in the formation of a good citizen's intellectuality.

There are various ideas related to the concept of a good character. According to Thomas Lickona (1991), a virtuecan be divided into two categories: self-virtue and virtue of others. The explanation of self-righteousness (self-virtue) has to do with self-control and patience. Meanwhile, the explanation virtue of others relates to the willingness to share and feel happiness. In conclusion, the concept of virtue is not only in attitude, but also in knowledge and behaviour. Therefore, these characters include three things, namely, knowing what is good, feeling good, and doing good things.

The concept of character as a virtue is also explained by Aristotle (in Sunarso 2009), who states that being a good citizen is marked by civic virtue, including four components: simplicity, justice, courage, and wisdom. Therefore, it can be concluded that character is a habit because it can be taught through habituation. Thinking about the concept of a good and smart citizen, then what types of teachers will make students become excellent and smart citizens? In general, excellent and smart citizens for teachers are not much different from the criteria we have discussed earlier. However, there is a differentiating side, namely, teacher professionalism. Teacher professionalism can be characterized based on their attitude, behaviour, commitment, ability, knowledge, responsibility, having a work ethic and upholding the professional code of ethics of teachers as well as the character of good and smart teachers that are more inherent in their personality as an educator (Winarno 2012). As an educator, the teacher's personality is contained in Law Number 14 of 2005 concerning Teachers and Lecturers.

Based on the description mentioned earlier, how are the efforts to build the character of good and smart citizens in facing the Society 5.0 era? Education that discusses citizenship in facing the era of globalization and the era of Society 5.0 should develop civic competences with civic knowledge, civic skills, civic character and civic commitment. Therefore, it is believed to foster the character of good and smart citizens who are ready to compete in the era of Society 5.0.

3.2 *The urgency of Society 5.0*

In general, the concept of society 5.0 is the concept of a new life order in society. Japanese government has adopted the concept of society 5.0 as an anticipation of global trends resulting from industrial revolution 4.0. Japanese government defines society 5.0 as a human-centered society that balances economic progress and problem solving using a system that integrates the virtual and physical world (Hendarsyah 2019). According to Fukuyama (2018), the purpose of society 5.0 is to create a society where every human can enjoy the life to the fullest by utilizing big data to help human work that is currently developing significantly connected via the internet of things, and it is analyzed by using artificial intelligence for the welfare of society. The era of society makes society face technology that enables access to be integrated in a virtual space that feels like a physical space. Through society 5.0, artificial intelligence that pays attention to the human side will transform millions of data collected via internet on all aspects of life expected to form a new life order in society (Sabri 2019).

The way of Japanese in describing society 5.0 is as follows: (1) society 5.0 offers a human-centered society, (2) society 5.0 balances economic progress and the resolution of other social problems that can be connected through cyberspace and real world systems, (3) in society 5.0, it is no longer the capital that becomes the benchmark, but the data is the benchmark that can connect and move everything, (4) helping to fill the gap between rich and poor, (5) equitable distribution of health and education services throughout the region (Suryadi 2020).

In the field of education, a supportive educational climate is needed in facing the era of society 5.0. It requires a learning process that forces students to think critically, analytically, creatively and constructively so that the learning material can provide concrete solutions to every problem in everyday life by using knowledge and skills as an outcome of learning at school.

3.3 The value relevance of civic virtue in the era of Society 5.0

A nation that has a strong character will usually grow and develop to bring prosperity to its citizens. Based on this context, if we draw on the problems in Indonesian education, character building is the initial foundation for making individuals have good personalities and qualities, so that it is expected to make a civilized and advanced nation. Basically, education is an important part of the humanizing concept. It is expected that education can produce civilized students. Policies for character education, civic education (citizenship education), and religious education can be a place for inculcating students' virtue values. Civic education also has a fundamental role in character education including character development in the form of intelligence, attitudes, knowledge, commitment, skills and civic skills, all of which can crystallize back into civic virtue (Fauzi & Roza 2019). Holistically, character education, civic education, and religion aim to make every citizen have a sense of nationality and love for the country in the context of Pancasila values and morals, values and norms of the 1945 Constitution of the Republic of Indonesia (Yoga 2019).

Through this civic education, citizenship values can be developed to develop intelligence, knowledge, attitudes, skills, commitment, skills, participation and civic responsibility for every citizen (Juliati & Firman 2017). The urgency of civic virtue is very relevant and adaptive to changes or developments in industrial revolution 4.0 and society 5.0. The concept of civic virtue is apparently in line with Presidential Regulation Number 87 of 2017 concerning Strengthening Character Education (Penguatan Pendidikan Karakter) Article 3. It states that "Strengthening character education by applying Pancasila values in character education, especially including religious values, honesty, tolerance, discipline, work hard, creative independent, democratic, curiosity, spirit of nationality, love for the country, communicative, caring for the socio-environment, and responsible".

3.4 The challenges of character education in the era of Society 5.0

Currently, education in Indonesia has entered the era of the industrial revolution 4.0. The current trend for Indonesian education learning models is online learning and blended learning that uses the internet

to link educators and students (Ely & Rizqi 2020). The role of educators in the era of the industrial revolution 4.0 is that they must emphasize character, moral and exemplary in education during the process of transferring knowledge in schools. This statement is also supported by the idea according to Risdianto (2019) which explains that knowledge transfer can be replaced by technology, but the application of soft skills and hard skills cannot be replaced with any sophisticated tools and technology. In addition, educators' role in the era of industrial revolution and society 5.0 must have strong core competencies, including educational competence, competence for digital, competence in research, competence in globalization, and competence in future straties (Ely & Rizqi 2020). Along with the development of science and technology, learning process in industrial revolution 4.0 also experienced changes.

The learning process used to be one-way face-to-face between the teacher and students can now be done with online classes through social media or other online media that support the online learning process. In the era of the industrial revolution 4.0, students are required to think critically in analyzing a case and providing problem solving to the case, which must be relevant to the reflections of their daily lives. Thus, because of the existing of society 5.0, it is expected that technology in the field of education will not change the role of educators in teaching moral education and role models for students who have civic virtue to become good and smart citizens.

The development of the education field in Indonesia has entered a very important period, not only to provide optimal and qualified education services but also to pay attention to the sustainability of education itself in facing the increasingly fierce global competition. Currently, the challenges in Indonesia's education field are increasingly complex, requiring very serious preparation and thinking since education needs to get more attention considering that it is one of the important things in the order of people's lives concerning their readiness to face very fast social changes. The challenges of the education field in facing the Industrial Revolution era 4.0 and Society 5.0 need to have a very high degree of readiness. This readiness is intended for teachers who must have access and mastery of technology, pay attention to the level of media literacy among teachers, equal access to internet networks and school infrastructure in each region. Furthermore, the government's role in equitable development and educational facilities in the territory of Indonesia must be paid more attention, and it must become a priority so that the process of implementing internet-based learning and technology can be evenly distributed throughout Indonesia.

3.5 The efforts to develop character education in the era of society 5.0

Education is very important in building the good or bad of the human character following the normative standards. The Indonesian nation recognizes the importance of building national character to advance Indonesian education that produces human resources who will be able to compete in the era of Industrial revolution 4.0 and Society 5.0. Creating value-based characters during the learning process is very important for exploring and developing the character values of students. Related to the education field facing the era of Society 5.0, it is not enough for students to only understand the material or only be given a theory. A learning environment based on trust, caring, and mutual respect will naturally increase motivation, creativity, affection, and cognitive development (Lickona 2013). In this regard, the efforts that must be made in developing character education in facing the era of Society 5.0 are (1) the need to prepare students who can think critically, constructively and innovatively by applying HOTS (Higher, Order, Thinking, Skills) in the learning process in order to trigger students' ability to solve problems in a complex, critical and creative manner (Ramadhan & Dinna 2019).

Furthermore (2) a futuristic learning orientation renewal is needed by introducing learning that is not only about mastery of particular material, but it is also needed to connect the material to the utilization for the progress of society in the era of Society 5.0 (Indriyani 2019). Third (3) the need for selecting an appropriate learning model that adapts to the conditions and situations of the classroom and school culture or the applicable curriculum. It aims to provide space for students to find concepts of knowledge and creativity in expressing and telling their opinions. It also prepare more innovative learning systems, such as adjusting learning curricula and improving students' abilsociety'ses in datan Technology (IT), Operational Technology (OT), Internet of Things (IoT), and Big Data Analytic. Furthermore, it is also necessary to integrate physical, digital and human objects to produce competitive and skilled graduates, especially in data literacy, technological literacy and human literacy (Ely & Rizqi 2020). Fourth (4) in addition to choosing a learning model, it is also necessary to develop teacher competence in the cognitive, affective, psychomotor, attitude, skill and adaptive domains so that teachers can adapt in preparing for Society 5.0 era and uphold professionalism as teachers. Fifth (5) providing facilities and infrastructure as well as futuristic learning resources according to the needs in the form of IT-based smart building in classrooms, libraries, laboratories and other rooms supported by IoT and AI facilities that support learning resources and student learning media. In addition, rejuvenating and building education infrastructure to support the quality of education, research and innovation. Sixth (6) Reconstruction of institutional policies that are adaptive and responsive to the Industrial Revolution 4.0 and Society 5.0 in developing the required transdisciplinary and study programs. Creating a Cyber Learning program, such as a distance learning system, as a solution for residents in remote areas to access quality higher education. Seventh (7) Preparing human resources, especially cognitive, affective and psychomotor competences of teachers and other stakeholders who are responsive, adaptive, and reliable to face the Industrial Revolution 4.0 and Society 5.0 (Indriyani 2019).

4 CONCLUSION

The rise of social deviations committed by the younger generation shows the weak idealism of a nation's character. Efforts to create good and smart citizens are the mission of civic/citizenship education and character education. Civic competence in psychosocial character education is reflected in knowledge, skills, persistence, attitudes, and commitment, all of which can be summarized as civic virtue. The results of the study show that the implementation of the virtue of citizenship in schools can run effectively. It is supported by the ability of professional teachers, student attitudes and commitment, HOTS (Higher, Order, Thinking, Skills) implementation in the learning process, school policies, classroom/school culture, the right learning model election, futuristic educational facilities, and infrastructure, as well as philosophical values of local wisdom which are based on Pancasila. Character education in schools based on the 2013 curriculum can be strengthened by the civic virtue along with developments in the era of Society 5.0.

REFERENCES

Ely, N. F. & Rizqi, N.A. 2020. Kesiapan pendidikan Indonesia menghadapi Era Society 5.0. *Edcomtech: Jurnal Kajian Teknologi Pendidikan*. 5 (1): 61–66.

Fauzi. R. & Roza, P. 2019. Implementasi nilai kebajikan warga negara (civic virtues) di Institut Teknologi Bandung. *Journal of Moral and Civic Education*. 3 (2): 92–106.

Fukuyama, M. 2018. Society 5.0: Aiming for a new human-centered society. *Japan Economy Foundation Journal Japan Spotlight*.

Hendarsyah, D. 2019. E-commerce di era industri 4.0 dan Society 5.0. *Iqtishaduna: Jurnal Ilmiah Ekonomi Kita*. 8 (2): 171–184.

Hidayati, N. & Harmanto. 2017. Partisipasi masyarakat dalam membangun civic virtue di Desa Balun Kecamatan Turi Kabupaten Lamongan. *Kajian moral dan kewarganegaraan*. 5 (1): 106–120.

Indriyani, S. 2019. Memajukan inovasi pembelajaran di Era Society 5.0. Retrieved from: https://smol.id/2019/12/17/memajukan-inovasi-pembelajaran -di-era-society-5-0/ on September 8, 2020.

Juliati & M. Firman. 2017. Membangun "civic virtues" melalui nyanyian sebagai media pembelajaran untuk memotivasi proses belajar mengajar Kewarganegaraan (PKn). *Jurnal Kependidikan. XVII* (1): 17–36.

Lickona, T. 1991. *Pembentukan kepribadian anak, pesan moral, intelektual, emosional dan sosial sebagai wujud integritas membangun jatidiri*. Jakarta: Bumi Aksara.

Lickona, T. 2013. Pendidikan karakter: *Panduan lengkap mendidik siswa menjadi pintar dan baik*. Bandung: Nusa Media.

Mulyono, B. 2017. Reorientasi *civic disposition* dalam kurikulum pendidikan kewarganegaraan sebagai upaya membentuk warga negara yang ideal. *Jurnal Civics,* 14 (2): 218–224.

Ramadhan, P. S. & Dinna R.A. 2019. Peran pendidikan berbasis higher order thinking skills (HOTS) pada tingkat sekolah menengah pertama di Era Society 5.0 sebagai penentu kemajuan Bangsa Indonesia. *Equilibrium.* 7 (2): 137–140.

Risdianto, E. 2019. *Kepemimipinan dalam upaya dunia pendidikan di Indonesia di era revolusi industri 4.0.* Bengkulu: Universitas Bengkulu.

Sabri. I. 2019. Peran pendidikan seni di Era Society 5.0 untuk Revolusi Industri 4.0. *Seminar Nasional Pascasarjana.* Semarang: UNNES.

Samsuri. 2004. Civic virtue dalam pendidikan moral dan kewarganegaraan di Indonesia Era Orde Baru. *Jurnal Civics,* 1 (2): 225–239.

Setiawan, D. 2014. Pendidikan kewarganegaraan berbasis karakter melalui penerapan pendekatan pembelajaran aktif, kreatif, efektif dan menyenangkan. *Jurnal Pendidikan Ilmu-Ilmu Sosial.* 6 (2): 61–72.

Sunarso. 2009. *Warga negara dan pendidikan kewarganegaraan (kajian, konsep dan sejarahnya).* Yogyakarta: UNY.

Suryadi. 2020. Pembelajaran era distruptif menuju masyarakat 5.0. *Prosiding Seminar Nasional Pendidikan Program Pascasarjana Universitas PGRI Palembang.*

Syarifa, S. 2019. Konsep civic virtue dan pendidikan kewarganegaraan di Indonesia.

Udin S.W. 2016. Posisi akademik Pendidikan Kewarganegaraan (PKn) dan muatan/mata pelajaran Pendidikan Pancasila dan Kewarganegaraan (PPKn) dalam konteks sistem pendidikan nasional. *Jurnal Moral Kemasyarakatan.* 1 (1): 15–36.

Winarno. 2012. Karakter warga negara yang baik dan cerdas. *PKN Progresif.* 7 (1): 55–61.

Yoga, R.H. 2019. Relevansi pancasila era industri 4.0 dan Society 5.0 di pendidikan tinggi vokasi. *Journal of Digital Education, Communication, and Arts.* 2 (1): 11–20.

Educational Innovation in Society 5.0 Era: Challenges and
Opportunities – Purnomo & Herwin (Eds)
© 2021 the authors, ISBN 978-1-032-05392-9

The importance of using e-learning in teaching social science during and after the Covid-19 pandemic

E. Suprayitno & Aman
Graduate School of Yogyakarta State University, Yogyakarta, Special Region of Yogyakarta, Indonesia

J. Budiman
Tanjungpura University, Pontianak, West Kalimantan, Indonesia

ABSTRACT: The use of the information technology-based learning model (e-learning) in social science subjects answers problems during and after Covid-19. This research used a literary study technique to understand the importance of using e-learning in social science subjects during and after the Covid-19 pandemic. From the study using the literature review method, the following results were found: 1) the use of e-learning products has various models that can be empowered as guidance and instructions in training and teaching legally and meaningfully, such as web courses, web-centric courses, and web-enhanced courses, 2) as a supporting element in the social science subjects, e-learning is considered capable of providing real benefits, and 3) e-learning is able to support and enrich the substance in social science subjects to improve educational quality.

1 INTRODUCTION

As we all know, the main problem of the Covid-19 pandemic disaster in the education sector is the shift in the learning process from an offline system to an online system. This fact at the same time provides new opportunities and challenges for all educators to be able to take advantage of the sophistication of digital technology, communication facilities, and appropriate networks to learn such new skills such as: accessing information, managing learning media, integrating learning media, and evaluating student values effectively and efficiently. In line with that statement, Solihudin (2018) states that according to The Decree of Indonesian Ministry of Education and Culture (Permendikbud) Number 22 of 2016 concerning the standard of primary and secondary educational processes, one of the contents of the standard process is the use of information and communication technology to increase the efficiency and effectiveness of learning. Furthermore, Munir (2009) states that technology and information-based learning have informed the learning system from traditional forms to media formats, including computer media, with the internet's help, with which e-learning was created. This form of media for learning results in more fun, less boredom, and students more likely to be full of enthusiasm and motivation. Therefore, e-learning should be applied as a learning model, and it must exist as an answer to various problems in teaching and learning social science during and after the Covid-19 pandemic. Based on the explanation mentioned earlier, it is interesting for researchers to conduct further studies regarding the importance of using e-learning in social science subjects during and after the Covid-19 pandemic.

2 METHOD

This study uses the literature review method to increase understanding of the importance of using e-learning in social science subjects during and after the Covid-19 pandemic.

3 FINDINGS AND DISCUSSION

3.1 *E-learning as a learning model during and after the Covid-19 pandemic*

E-learning is conceptualized as an educational activity that empowers media in the form of computers. Therefore, all learning activities that use computers can be considered e-learning (Effendy & Zhuang, 2005). There are two essential perspectives related to e-learning, according to Munir (2009), namely: 1. electronic based learning is learning that empowers information and communication technology, especially devices in the form of electronics, which means that it is not only the internet, but also all electronic devices such as films, videos, cassettes, OHP, slides, electronics, LCD projectors, and tapes as long as they use a device, 2. internet-based learning is learning that uses internet facilities, which are online, as the main instrument. This means that e-learning must use the internet online, namely a computer facility connected to the internet. This is in line with Cisco's opinion (2001) that explains the philosophy of e-learning as follows: 1. e-learning is the delivery of information, communication, education, online training, 2. e-learning provides a set of tools that can enrich conventional learning values (conventional learning models, a study of textbooks, CD-ROMs, and computer-based training) so that they can answer the challenges of the globalization and development, 3. e-learning does not mean replacing conventional learning models in the classroom but strengthens the learning model through content enrichment and educational technology development, and 4. The capacity of students varies greatly depending on the form of content and the way it is delivered. The better the harmony between content and delivery tools to the learning styles, the better students' capacity, which in turn will give better results.

E-Learning has the following characteristics: 1. empowering the presence of electronic technology services; where educators and students, students and fellow students or teachers and fellow educators can interact quite easily without being limited by protocol matters, 2. empowering the advantages of computers in digital media or computer networks, 3. empowering independent teaching materials (self-learning materials), which are then stored on a computer to be accessed by educators and students whenever and wherever needed, 4. empowering learning schedules, curriculums, results of learning progress, and educational administration matters to be seen every time on the computer (Rusman, 2010). Bates (2008) reveals five principles for integrating technology into the social science learning model: 1. extending learning beyond what can be done without technology, 2. introducing technology in context, 3. providing opportunities for students to learn the relationships between science, technology, and society, 4. developing knowledge, skills, and participation as good citizens in a democratic society, and 5. contributing to research and evaluation of social and technological studies.

The use of e-learning products when applied to social science subjects during and after the Covid-19 pandemic has various models that can be used as guidance and instructions in training and teaching legally and meaningfully. Haughey's statement strengthens the development of e-learning as a learning model. According to Haughey in Saud (2019) there are three possibilities in developing an internet-based learning system: web courses, web-centric courses, and web-enhanced courses.

A web course is internet empowerment for educational purposes where students and educators are completely in two separate places, and face-to-face interaction is not required. All teaching materials, discussions, consultations, assignments, exercises, exams, and other learning activities are fully conveyed via the internet. In other words, this model uses a remote system.

A web-centric course is internet empowerment that combines distance learning and face-to-face (conventional) learning. Some of the material is conveyed via the internet, and some through face-to-face instruction. In this model, the instructor/the teacher can provide instructions for students to study the subject matter via the web that she/he has already prepared. Students are also given directions to find other sources from relevant websites. In face-to-face learning, students and teachers have more discussions about the material findings that have been studied via the internet.

The web-enhanced course is the internet empowerment to support improving the quality of learning conducted in the classroom. The internet's function is to provide enrichment and communication between students and teachers, fellow students, group members, students, and other sources. Therefore, the role of the teacher in this case is required to master the techniques of searching for information on the internet, guide students to search for and find sites that are relevant to learning materials, present material through attractive websites that they are also interested in it, and serve guidance and communication via the internet, among the other skills required.

3.2 *E-learning as a supporting element for teaching and learning social science during and after the Covid-19 pandemic*

An information technology-based learning model (e-learning) has a role as a supporting element in social science subject during and after the Covid-19 pandemic because it is considered capable of providing tangible benefits. This statement is in line with the opinion of Elangoan (1999), Soekartawi (2002), Mulvihil (1997) & Utarini (1997) in Suyanto (2005) that explain several benefits as follows: 1. providing e-moderating facilities where educators and students can communicate easily via internet facilities regularly or whenever the communication activity is carried out without being limited by distance, place and time, 2. educators and students can use structured and scheduled teaching materials or learning instructions via the internet to assess each other to what extent the teaching materials are studied, 3. both educators and students can study or review teaching materials anytime and anywhere when they need since the teaching materials are stored on the computer. If students need additional information about the material they are learning, they can access the internet more easily. Both educators and students can conduct discussions via internet that can be followed by many participants, so it may add broader knowledge and insight. The important point is that the role of learners, which is usually passive, becomes active. Furthermore, it is also explained that e-learning facilitates interaction between students and material/subject matter. Learners can share information or opinions on various matters relating to lessons or learners' self-development needs (Website Kudos, 2002). Besides, teachers can place learning materials and assignments that students must do in certain places on the web for students to access.

According to their needs, educators/teachers can also provide opportunities for students to access certain study materials and exam questions that can only be accessed by students once and within a certain period. In more detail, the benefits of e-learning can be seen from two angles, namely from the point of view of students and educators: 1. from the point of view of students, regarding the e-learning, it is possible to develop high learning flexibility. This means that students can access learning materials at any time and repeatedly. Students can also communicate with educators at any time. Under this condition, students can further strengthen their mastery of learning material. If infrastructure facilities are not only available in urban areas but have reached sub-districts and rural areas, e-learning activities will provide benefits (Brown, 2000) to students, including: a. studying in small schools in poor areas to follow certain subjects that cannot be provided by the school; b. allow home schoolers to study materials that parents cannot teach, such as foreign languages and computer skills; c. those feeling afraid of school, students treated at the hospital or home, those who have dropped out of school but are interested in continuing their education, those who are required to by the school, or students in various regions or even abroad can all use e-learning, d. those not accommodated in conventional schools can be allowed to get an education. 2. From the teachers' point of view regarding e-learning, the benefits obtained by teachers (Soekartawi, 2002) including: a. it is easier to update learning materials which are their responsibility following the demands of scientific developments that occur, b. self-development can be done by researching to increase their insights because they have relatively more free time, and c. controlling the learning activities of students. Moreover, teachers can also find out when students learn, what topics are studied, how long a topic is studied, and how many times a particular topic is reviewed, d. checking whether students have worked on practice questions after studying a certain topic, and e. checking students' answers and sharing their results with students. This is supported by Wulf's (1996) statement, which says that the benefits consist of four things, namely: 1. increasing the learning interaction process between students and educators/teachers (enhance interactivity). If it is designed carefully, electronic learning can increase the level of learning interaction, both between students and teachers, among students, and between students and learning materials (enhance interactivity). This is different from the conventional learning where not all students are able, dare or have the opportunity to ask questions or express their opinions in the discussion. Why is that? Because in conventional learning, the opportunities available or provided by educators to discuss or ask questions are very limited. Moreover, this limited opportunity often tends to be dominated by some students who are responsive and courageous. This situation will not occur in electronic learning. Students who are shy or hesitant or lack of courage have ample opportunities to ask questions or convey statements/opinions without feeling watched or under pressure from their classmates. 2. This allows learning interaction from anywhere and anytime (time and place flexibility). Given the learning resources that have been packaged electronically and available for access to students via the internet, students can interact with these learning resources anytime and anywhere. Likewise, the tasks of learning activities can be submitted to the teacher once they are done. No need to wait for an appointment to meet the teacher. Students are not strictly bound by the time and place of learning activities such as in conventional education. 3. Reaching out to students in a broad range (potential to reach a global audience). Because of the flexibility of time and place, the number of students who can be reached through electronic learning activities is increasing or expanding. Place and time are no longer obstacles. Anyone can study anywhere and anytime. Interaction with learning resources is done via the internet, and the learning opportunities are really wide open for anyone who needs them. 4. E-learning makes it easy to improve and store learning materials (easy updating of content as well as archiving capabilities). The facilities available in internet technology and a variety of software that continue to develop also help facilitating the development of electronic learning materials. Likewise, the improvement or updating of learning materials in accordance with the demands of the development of scientific material can be done periodically and easily.

On the other hand, as an improvement in the method of presenting learning material, it can also be done either based on feedback from students or on the results of the assessment of the teacher as the person in charge or guide of the learning material itself. Knowledge and skills for the development of electronic learning materials need to be mastered in advance by the teachers/educators who will develop electronic learning materials as well as regarding the management of their own learning activities. There must be a commitment from educators who will monitor the progress of learning activities and also regularly motivate their students. Based on this explanation, it can be said that the use of e-learning in the social science subject can provide answers to various problems in the education sector during and after the Covid-19 pandemic in terms of several useful aspects.

First, changing the learning paradigm in social science subjects that have tended to be conventional into media is necessary. Education is then not only focused on teachers, media, textbooks and printed media, but also on the use of information technology. E-learning can be a richer and more up-to-date source of social science learning. As we know, social science learning material includes social events and cases that occur in various regions. By using e-learning, students no longer study the cases or past events that usually exist in textbooks, and they are also no longer too busy to be given the task of finding newspapers. However, students are able to easily access all the information needed quickly and up-to-date in enriching the material they study. E-learning has the ability to facilitate discussion and collaboration activities consisting of a group of people, to organize face-to-face communication (teleconferences), to enable internet users to communicate audio-visually so that real-time verbal and non-verbal communication is possible.

3.3 The urgency of e-learning in social science subject during and after the Covid-19 pandemic

An information technology-based learning model or e-learning has an urgent value as a provider of facilities and infrastructure that can support or enrich the substance in social science subjects in order to improve the quality of education. This statement is in accordance with the purpose of technology in learning that is to solve problems and to facilitate student learning activities (Warsita, 2008). Furthermore, the urgent value can be seen in the comparison between the implementation of traditional learning and e-learning, which in traditional learning tends to assume that educators are knowledgeable people and are assigned to transmit knowledge to their students, while in e-learning, the main focus tends to be on students. Students will be more independent at certain times and are responsible for their learning, especially in social science subject. The learning atmosphere under the e-learning will force students to play a more active role in learning. Students design and look for materials by using their own efforts and initiatives (Rusman, 2010).

Reflecting on the empirical reality so far, both before the Covid-19 pandemic to the existence of Covid-19 pandemic, regarding the social science learning process at the school level, social science teachers seem less than optimal in empowering learning resources. The social science learning process tends to be teacher-centered, textbook centered, and one-media-focused. Therefore, it cannot be blamed if many students perceive the social science learning process as boring, monotonous, unpleasant, as having too much memorizing, too little variety, and various other complaints. However, the presence of an information technology-based learning model (e-learning) can provide a new atmosphere and alternative solutions in social science learning during and after the Covid-19 pandemic. From this explanation, it is increasingly clear and proven that the use of e-learning in teaching and learning social science subject is a breakthrough innovation in terms of method, source and learning media that contains values of renewal and difference or even change in the world of education, and it has specific goals.

4 CONCLUSION

The use of information technology-based learning models (e-learning) in social science subjects is an answer to problems during and after Covid-19. In this case, it is very important to use the e-learning in social science because there are various models that can be used as guidance and instructions in training and teaching legally and meaningfully, such as: web courses, web centric courses, and web enhanced courses. Furthermore, it is considered to be able to provide real benefits and have an urgent value as a provider of facilities and infrastructure that can support or enrich the substance in social science subject in order to improve the quality of education.

REFERENCES

Bates, A. 2008. Learning to Design WebQuests: An exploration in preservice social studies education. *Journal of Social Studies Research*, 32(1): 10–12.

Brown, M. D. 2000. *Education world: Technology in the classroom: Virtual high schools, part 1. The voices of experience*. Retrieved from www.educationworld.com/a_tech/tech0 52.shtml.

Cisco. 2001. E-learning: *Combines communication, education, Information, and Training*. Retrieved from http://www.cisco.com.

Effendi, E., & Zhuang, H. 2005. *E-learning: Konsep dan aplikasi*. Yogyakarta: CV. Andi Offset.

Jh, T. S. 2018. Pengembangan e-modul berbasis web untuk meningkatkan pencapaian kompetensi pengetahuan fisika pada materi listrik statis dan dinamis SMA. *Jurnal Wahana Pendidikan Fisika*, 3(2): 51–61.

Munir. 2009. *Pembelajaran jarak jauh berbasis teknologi informasi dan komunikasi*. Bandung: Alfabeta.

Rusman. 2010. *Model-model pembelajaran: Mengembangkan profesionalisme guru*. Jakarta: PT Raja Grafindo Persada.

Sa'ud, U. S. 2010. *Inovasi pendidikan*. Bandung: Alfabeta.

Soekartawi. 2002. *E-learning: Konsep dan aplikasinya. Seminar e-learning*. Jakarta: Balitbang Depdiknas.

Suyanto, A. H. 2005. *Mengenal e-learning*. asep_hs@yahoo.com. Retrieved from http://www.asep-hs.web.ugm.ac.id.

Tritularsih, Y., & Sutopo, W. 2017. Peran keilmuan teknik industri dalam perkembangan rantai pasokan menuju era industri 4.0. *Seminar dan Konferensi Nasional IDEC 2017* Surakarta, 8-9 Mei 2017.

Warsita, B. 2008. *Teknologi pembelajaran: Landasan dan aplikasinya*. Jakarta: PT. Rineka Cipta.

Website kudos on "what is learning". Retrieved from: http://www.kudosidd.com/learning_solutions/definition.

Wulf, K. 1996. Training via the Internet: where are we? *Training & Development*, 50(5), 50–56.

Educational Innovation in Society 5.0 Era: Challenges and
Opportunities – Purnomo & Herwin (Eds)
© 2021 the authors, ISBN 978-1-032-05392-9

The development of instrument analysis for elementary school children's social interaction patterns in the era of revolution 4.0

Firmansyah, A. Senen, Mujinem, Hidayati & S.P. Kawuryan
Universitas Negeri Yogyakarta, Sleman, Yogyakarta, Indonesia

ABSTRACT: This research aims to develop an instrument of social interaction pattern analysis of primary school children in the era of revolution 4.0. This study is part of a series of research development assistance programs of elementary school childrens' social interaction. The CFA test results for constructing childrens' social interaction obtained nine manifest variables, which have a count value greater than 1.96, and a factor coefficient value greater than 0.5. The test results of the students' social interaction construct meets the goodness of fit statistics' requirements.

1 INTRODUCTION

Technological developments are accompanied by scientific developments, indirectly changing the behavioral patterns of people's lives. The current level of mastery of technology has become an indicator of a nation's progress (Ngafifi, 2014). The development of information technology began in the 18th century with the first mechanical looms in 1784. In the year 1840, the construction of railroads began and with them the steam engine's invention (Schwab, 2016). The influence of technology has brought significant changes to human life. The changes that occurred at that time were known as the industrial revolution. Industrial developers are also trying to compete to create technological tools that they can use in various industrial fields.

Today, the world has entered the era of industrial revolution 4.0, characterized by increased connectivity, interaction, digital systems, artificial intelligence, and virtual systems. Khan and Turowski said that four factors drive the industrial revolution 4.0 and are characterized by increased digitization of manufacturing: 1) big data, computing power, and connectivity, 2) the emergence of the analysis capabilities of business intelligence, 3) the occurrence of new forms of interaction between humans and machines, and 4) improvement of the digital transfer instructions to the physical world (Khan & Turowski, 2016). Hermann also added that the principle applied in the industrial era, 4.0, is interconnected (connection) between machines, devices, sensors, and humans. This is so they can be connected and communicate through the Internet of Things (IoT) or the Internet of People (IoP) (Hermann et al., 2016).

Technological advances in the era of the very rapid industrial revolution 4.0 recently don't just open up interactions that occur widely through the internet world. They have also disrupted various areas of human life. The acceleration in online media use has changed people's behaviors when communicating, sharing ideas, interacting with the aim to affect others, and showing the behaviour transformation. People tend to want something that is detailed through the instant, rapid accessibility of information obtained through social media and mobile devices that support digital social networks (Sri Subawa & Wayan Widhiasthini, 2018). These things have triggered stimuli in the self to change thinking, lifestyles, self-existences, self-pride, social culture, and eliminate ethics in a world without boundaries and society.

People's dependence on instantaneous access has generated much social exclusion. Social exclusion is a condition where people are at a stage of dependence on information technology to ignore social interaction directly. Shifting social interaction is currently felt in the people who live in urban areas and has penetrated all aspects of society, both from children and adults. According to Shifa's research, today, children aged 5-12 years had become the primary users to take advantage of advances in information technology (Ameliola & Nugraha, 2015).

The utilization of existing technology, especially for children experiencing developmental stages of learning, should not be avoided. However, today's generation of young people are required to master various technologies to compete globally with the times. For example, in the world of education, the advancement of technology can indirectly increase children's knowledge, build relationships, make more friends without being limited by distance and time, and make it easier for children to find and find out the latest information widely.

What should be a concern for parents and educators today is to ensure that existing technological developments are not misused. This will indirectly harm children, especially in their social values. According to Wibhowo, the negative impact of technological development for children, especially in their social roles, is that children can experience a dependence on technology (Wibhowo, 2011). Children will indirectly prefer to do things individually rather than in groups. That makes children less sociable. Children today are more likely to be shy if they have to communicate directly in front of the crowd and prefer relationships through cyberspace.

Another effect is that the child feels complacent with the knowledge gained to assume that what they read or get in a virtual world is complete and final. It leads to the generations' tendency to become a generation of people that is easily satisfied and shallow-thinking. Coupled with the presence of various social media sites, gaming devices, and gadgets, people often forget the benefits of social interaction directly (Damayanti; dkk, 2020). If this is allowed to continue, then the worst effects arising from the lack of social interaction that occurs directly in the conflict on the people nearby can cause stress in children and adolescents.

As a response to developments in the increasingly rapid industrial revolution era 4.0, various countries have started to take strategic steps in minimizing the negative impacts they may cause. As a developed country in the field of technology, Japan has initiated the concept of society 5.0. The concept of society 5.0 focuses on the human role in transforming big data into new local wisdom that expects to improve the ability of humans to create an opportunity for humanity and the creation of a more meaningful life (Fukuyama, 2018). If the industrial revolution 4.0 allows us to access and share information through the internet, then the 5.0 era society is an era in which all existing technology is a part of man himself. Unlike the industrial revolution, 4.0 emphasizes more on business opportunities. Technology in society 5.0 expects to create new values that will eliminate social, age, gender, and language inequality, and provide products and services specifically designed for the diverse needs of individuals (Faulinda, 2020). These aspects require human beings who have a prominent character as the central controller of integrating the real world and the virtual world.

The community often forgets about technological developments in the era of the industrial revolution 4.0, and the era of society 5.0 is the strengthening of character for each individual. Currently, direct interaction changes when children and adolescents are preoccupied with their gadgets to surf in cyberspace. Research conducted by Aang Ghunaifi mentions that phenomena in children and adolescents in the 21st century often prefer to spend time with their gadgets and ignore the people around them (Ghunaifi, 2015). Tolerance towards others' feelings is reduced and indirectly influences individuals to withdraw from the surrounding environment. If this is allowed, the current generation will be more likely to experience gadget addiction.

Addiction to the need for the internet is very influential on the psychology of each individual. Such conditions can lead to loneliness, reduced self-control, and social comfort inside often encountered these days (Cardak, 2013). When the virtual world is more easily accessible by the public, then the addiction patterns associated with gadgets will continue to rise. The social problems that arise include; a lack of tolerance, withdrawal from the surrounding environment, a lack of self-control, and difficulty performing daily activities. Some studies have shown that some of today's elementary school students find the use of technology to be more fun than interacting with their peers (Novitasari, 2016). They are not free from the influence of various game applications contained in the gadgets designed for children. Most of the parents' gadgets to their children are based on their children's wishes and aim to introduce an earlier technology and make sure their children do not get bored quickly.

The values of local cultural wisdom, humanism, and socialism is increasingly eroded due to globalization's swift influences, namely an era that "affirms" technology. That is because technology promises ease, progress, speed, popularity, and even productivity. The modern era identifies with the era of the digital society. Every human activity will be driven through a series of digital technology innovations. This technology operates by pressing a few digits (numbers) that arranges with different sequences. Relationships created among individuals are digital exchange relationships; each man only performs a series of transactions or interactions via digital symbols. This started from trading operations, communications, interactions, and other things. This has all been driven digitally. Of course, this fact is very worrying for life to come. For example, in social relations, social media platforms like Facebook, WhatsApp, Instagram, and Twitter are considered a representation of social interaction like in the real world.

2 METHOD

This instrument's development refers to Gable's method, consisting of fifteen stages, grouped into three parts: pre-development, concept development, and testing (Gable, 1986). Development of the concept of social interaction of primary school children studied through focus group discussion (FGD) and validation of experts consisting of sociologists, measurement experts, education, language, and psychology. The trial phase tests the suitability of the concept of children's social interaction with field data. The limited trial conducts on five elementary school students in Wates District, Kulon Progo Regency, D.I. Yogyakarta, to determine the instrument's legibility, a field trial was conducted at SDN Pengasih with 200 students as respondents.

The analysis in this study used the content validity index (S-CVI) scale formula (Polit, D. F., & Beck, 2006). The quality of the instrument items for each dimension of primary school student social interaction was analyzed using the item response theory approach to the partial credit model or IRT PCM (Firmansyah et al., 2020). Confirmatory factor analysis (CFA) was employed to test the construction of social interaction analysis instruments for elementary school students (Kirsch, I. S., & Guthrie, 1980). The model built is feasible if it meets at least three criteria. Reliability of the instrument by using reliability is obtained (Mardapi, 2012).

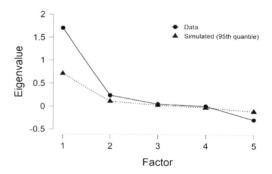

Figure 1. The form factor value.

3 RESULT AND DISCUSSION

Results obtained model social interactions of elementary school students formed on three dimensions: the social interaction of students at the family level, the students' social interaction with peers, and social interaction with the community. The construct of social interactions of elementary school students have a content validity of 0.98. Student social interaction at the family level consists of 3 factors: verbal communication, social contact, and imitation, with 14 questions. The dimensions of social interaction with peers consist of 3 factors: cooperation, openness, and frequency have content validities of 0.95. The total number of questions is 25. The dimension of social interaction with the community consists of 2 factors: positive feeling and empathy for the environment, with a content validity value of 0.93. The total number of questions on the dimensions of social interaction with the community consists of 11 items.

Field trials to prove the suitability of concepts and empirical data. The results of the item analysis carry out through the IRT PCM approach. The analysis results show that children's social interaction instrument's infit value with the family environment moves from 0.85 to 1.25, still within the range of the infit mean square (MNSQ) for items, between 0.77 to 1.3. Infit mean square obtained a mean of 1.03, a standard deviation of 0.09. Outfit mean square obtained a mean of 1.03 and a standard deviation of 0.07. Infit t had a mean of 0.09 and a standard deviation of 1.08. The estimated result of the infit mean square test is obtained by 1.05 with a standard deviation of 0.05.

Infit t had a mean of -0.05 and a standard deviation of 1.0. Outfit mean square was -0.08 and standard deviation of 1.04. The reliability coefficient is 0.98. The value of information on children's social interactions with the family environment ranges from -2.05 to $+1.4$.

The analysis showed that the infit value of the social interaction instrument between children and peers moved from 0.82 to 1.23, still within the range of the infit mean square (MNSQ) for items, namely 0.77 to 1.3. Infit mean square obtained a mean of 1.03, a standard deviation of 0.08. Outfit mean square obtained a mean of 1.03 and a standard deviation of 0.05. Infit t had a mean of 0.08 and a standard deviation of 1.04. The estimation result of the infit mean square obtained 1.02 with a standard deviation of 0.05. Infit t had a mean of -0.05 and a standard deviation of 1.0. Outfit mean square -0.05 and standard deviation of 1.02. The reliability coefficient is 0.90. The value of information on children's social interactions with peers is between -1.85 and $+1.2$.

The analysis results show that childrens' social interaction instrument's infit value with the community moves from 0.86 to 1.28, still within the range of the infit mean square (MNSQ) for items, namely 0.77 to 1.3. Infit mean square obtained a mean of 1.00, a standard deviation of 0.08. Outfit mean square obtained a mean of 0.97 and a standard deviation of 0.05. Infit t had a mean of 0.08 and a standard deviation of 0.98. The estimation result of the infit mean square obtained 1.02 with a standard deviation of 0.05. Infit t had a mean of -0.05 and a standard deviation of 1.0. Outfit mean square -0.05 and standard deviation of 1.02. The reliability coefficient is 0.90. The value of information on children's social interactions with the community is between -1.95 and $+1.4$. The three dimensions of primary school children's social interaction match the IRT PCM.

Table 1. CFA analysis results.

No	GOF Criteria	Cut off value	Result	Compatibility
1.	df	–	28	–
2.	Chi-Square	≤ 112	26.795	Good
3.	p-value	≤ 0.05	0.000	Good
4.	RMSEA	≤ 0.08	0.000	Good
5.	RMR	≤ 0.05	0.033	Good
6.	NFI	≥ 0.90	0.904	Good
7.	NNFI	≥ 0.90	0.952	Good
8.	IFI	≥ 0.90	0.910	Good
9.	GFI	≥ 0.90	0.910	Good
10.	AGFI	≥ 0.90	0.970	Good

The CFA test results for constructing children's social interaction obtained nine manifest variables, which have a count value greater than 1.96, and a factor coefficient value greater than 0.5. Analysis of the CFA model performed resulted in a Chi-Square value of 26,795, df 28, a p-value of 0,000, and an RMSEA value of 0,000. The test results of the student's social interaction construct meet the goodness of fit statistics' requirements. The relationship between the manifest variable and the student's social interaction analysis instrument's latent variable is present in Figure 2.

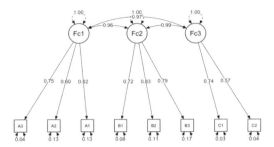

Figure 2. The results of the CFA test.

The loading factor for the latent variable of social interaction in the formed family environment is 0.96, with a p-value < 0.001. Simultaneously, the loading factor for the latent variable of social interaction with peers formed was 0.97, with a p-value < 0.001. Simultaneously, the latent variable factor of social interaction with the community environment formed is 0.99 with a p-value of < 0.001. Thus, the social interaction construct of elementary school children, which consists of social interaction in the family, between peers, and within the community environment, is supported by empirical data. The results of measuring the social interaction of elementary school students in the Wates sub-district, Kulon Progo district, Yogyakarta, are presented in Table 2.

Table 2. Frequency distribution

No	Interval	f
1	17–20	36
2	21–24	64
3	25–28	44
4	29–32	56
	Total	200

4 CONCLUSION

The analysis results and discussion of instruments' development for analyzing elementary school students' social interaction patterns can be concluded as follows. The instrument for analyzing students' social interaction patterns is classified as useful because it has a content validity index more significant than the predetermined minimum limit. Data that meets the goodness of fit statistics is supported and has a high-reliability coefficient.

Students' social interaction patterns have a good frequency distribution value, with the low category and the few high categories in Wates District, Kulon Progo Regency, D.I. Yogyakarta, while students with moderate category have the most number. The research does not cover the limit of children's social interaction patterns; therefore, further research needs to do on the range of students' social interaction cut scores.

REFERENCES

Ameliola, S., & Nugraha, H. D. 2015. Perkembangan media informasi dan teknologi terhadap anak dalam era globalisasi. *Prosiding the 5th International Conference on Indonesia Studies: "Ethnicity and Globalization"*, 2: 362–371.

Cardak, M., 2013. Psychological well-being and internet addiction. *Turkish Online Journal of Educational Technology*, 12(3): 134–141. https://files.eric.ed.gov/fulltext/EJ1016863.pdf

Damayanti, et al. 2020. Media sosial, identitas, transformasi, dan tantangannya (Damayanti Media (ed.); 2nd ed.). *Prodi Ilmu Komunikasi Universitas Muhammadiyah Malang*.

Nastiti, F. E., & Abdu, A. R. N. 2020. Kajian: Kesiapan Pendidikan Indonesia Menghadapi Era Society 5.0. *Edcomtech*, 5(1): 61–66.

Firmansyah, Herda, R. K., Damayanto, A., & Sidik, F. 2020. Confirmatory factor analysis to know the influencing factors of elementary school students' self-concept in Jetis Sub District, Bantul Regency. *JISAE: Journal of Indonesian Student Assessment and Evaluation*, 6(2): 196–202. https://doi.org/10.21009/jisae.062.010.

Fukuyama, M. 2018. Society 5.0: Aiming for a new human-centered society. *Japan Spotlight, 27(Society 5.0)*: 47–50. http://www8.cao.go.jp/cstp/%0A http://search.ebscohost.com/login.aspx?direct=true&db=bth&AN=108487927&site=ehost-live

Gable, R. K., 1986. *Instrument development in affective domain*. Kluwer-Nijhoff Publishing.

Ghunaifi, A. 2015. Merestorasi interaksi sosial pada era teknologi. *Prosiding Seminar Nasional Profesionalisme Tenaga Profesi*, 531–540.

Hermann, M., Pentek, T., & Otto, B. 2016. Design principles for industrie 4.0 scenarios. *Proceedings of the Annual Hawaii International Conference on System Sciences*, 2016-March, 3928–3937. https://doi.org/10.1109/HICSS.2016.488.

Khan, A., & Turowski, K., 2016. A perspective on industry 4.0: From challenges to opportunities in production systems. TBD 2016 - *Proceedings of the International Conference on Internet of Things and Big Data, IoTBD*, 441–448. https://doi.org/10.5220/0005929704410448

Kirsch, I. S., & Guthrie, J. T., 1980. Construct validity of functional reading tests. *Journal of Educational Measurement*, 1(7): 281–293.

Mardapi, D. 2012. *Pengukuran penilaian dan evaluasi pendidikan*. Nuha Litera.

Ngafifi, M. 2014. Kemajuan teknologi dan pola hidup manusia dalam perspektif sosial budaya. *Jurnal Pembangunan Pendidikan: Fondasi dan Aplikasi*, 2(1): 33–47. https://doi.org/10.21831/jppfa.v2i1.2616

Novitasari, W. N. K. 2016. Dampak penggunaan gadget terhadap interaksi sosial anak usia 5-6 Tahun. *Jurnal Online Program Studi S-1 Pendidikan Guru Pendidikan Anak Usia Dini*, 5(3): 182–186.

Polit, D. F., & Beck, C. T. 2006. The content validity index: are you sure you know what's being reported? Critique and recommendations. *Research in nursing & health*, 29(5):489–497.

Schwab, K. 2016. *The fourth industrial revolution. world economic forum*. www.weforum.org

Sri Subawa, N., & Wayan Widhiasthini, N. 2018. Transformasi perilaku konsumen era revolusi industri 4.0. *Jurnal Universitas Tarumanegara*, 131–139.

Wibhowo, C. R. S. 2011. *Stimulasi kecerdasan anak menggunakan teknologi informatika*. PT. Alex Media Komputindo.

Educational Innovation in Society 5.0 Era: Challenges and
Opportunities – Purnomo & Herwin (Eds)
© 2021 the authors, ISBN 978-1-032-05392-9

The effectiveness of batik learning activities on the improvement of character values in grade VII students of SMP PGRI 8 Denpasar

I.K. Mahendra, & N.M.M. Minarsih
Universitas Negeri Yogyakarta, Yogyakarta, Indonesia

ABSTRACT: This study aims to determine the effectiveness of batik learning activities on the improvement of character values in grade VII students of SMP PGRI 8 Denpasar. The research approach is quantitative with a quasi-experimental design, data collection using observation, questionnaires and documentation. Data analysis used a descriptive average, and the research subjects were 21 students. The results showed that the average value of patience in the pre-test was 79.36% and in the post-test 87.3%, there was an increase of 7.94%. The average value of pre-test hard work was 79, 38% and post-test 88.8%, an increase of 9.44%. The average score of responsibility in the pre-test was 76.19% and the post-test was 85.71%, an increase of 9.52%. Based on the overall data, the pre-test average was 78.31% and the post-test 87.27%, with an increase of 8.96%, so that the batik learning activities was effective in increasing the character values of patience, hard work and responsibility.

1 INTRODUCTION

Character values are very important for every student at every level of education. This is because character education has the essence of attitude and moral education. Positive character values need to be possessed by every student, including junior high school students. This is because junior high school students range from 12 to 15 years. Freud described that children at that age are included in the genital stage, where the child begins to enter puberty, which is marked by the maturation of the reproductive organs and sexual hormones (Kozier, 2011). In this phase students need support in the process of achieving independence and the ability to make decisions. Independence and decision making are important things, but in practice they require mastery of positive character values as a basis for making positive decisions.

One of the positive character values that support independence and decision-making is the value of patience, responsibility and hard work. These values help students in making positive decisions. Patience is the ability to restrain emotions, thoughts, words, and behaviors that are carried out actively with good intentions, and the act of obeying the rules (Hafiz, 2015: 1). Mastery of the character value of patience is needed by junior high school students. This is because students will meet and socialize with many other students, accept learning assignments so that the ability to hold back emotions when facing problems with good intentions will form good behavior in students.

Another positive character besides patience, what students need is a character of responsibility. Responsibility is the ability to understand positive and negative things, the ability to make effective decisions and the best choices within the limits of social norms and the ability to take risks (Parlina, 2016: 11), the values of the character of responsibility are important things to be controlled by junior high school students. This is because students experience a transition from elementary school age to junior high school.

Another positive character that is no less important is the ability to work hard. Hard work is a term that describes efforts that are continuously made in completing work to completion (Sukamti, 2014: 1). Mastery of the character of hard work abilities will help students to become strong students so that they are able to complete their tasks and responsibilities while studying at school. The three positive character values patience, responsibility and hard work are needed by students at the beginning of this puberty phase, this is because in this phase students try to find their identity and try to escape from parental control. This is very dangerous if students are not equipped positive character values.

The three characters or positive values above can be implanted explicitly in learning process at school. One of the activities in art learning is batik. Apart from being a cultural heritage of Indonesia, batik activities also increase students' creativity in addition to instilling positive character values. These were the reasons for the selection of batik activities in the process of developing positive character values in

DOI 10.1201/9781003206019-9

45

grade VII junior high school students at SMP PGRI 8 Denpasar.

2 LITERATURE REVIEW

2.1 Batik as a cultural heritage

Culture is the main source of character education and the many character values in it should be used as material in learning. Many cultures that exist in this nation become a sense of pride in character education because most of the existing cultures have become UNESCO heritage, where the values are inherited, such as batik. According to Kuswadji (1981: 2) Etymologically, batik means the ending. "Tik" in the word "batik" comes from the word drip. Thus, it can be said that "mbatik" is writing or drawing all complicated (small pieces) or it can be concluded that batik is literally carved night on cloth using canting. Batik has now become an extraordinary part of culture, especially the values that are derived from good local wisdom, so it is only natural that UNESCO designated batik as a world heritage on October 2, 2009 as an intangible heritage, meaning that it has inherited values, the values contained therein.

Culture, especially batik, has values which are usually called cultural values. Cultural values are abstract, invisible and intangible. But cultural values become a reference for students or community groups related to individual behavior. In order for the reference to be clear, community groups create norms, both written and unwritten, for example legal norms, norms of courtesy, norms of decency, and so on (Dyah Kumalasari, 2018: 62). However, in the application of culture to education there are several notes to establish the cultural insights of batik from our educational development, including: 1) culture is from and for humans, 2) with culture, humans build society and the environment, 3) human culture is used to build education, 4) education through culture occurs contextually, 5) education through culture occurs through processes, 6) the developing people through culture must involve the body, mind, and heart, 7) developing humans through culture, the cultural values must be integrated with him into his inner nuances, into his attitudes and behavior and become the basis for his way of thinking, 8) development through culture means convergent sustainability (Dick Hartoko, 1985: 37).

The inclusion of batik as a learning material has contributed to the goals of education in Indonesia as stated in UU No 20 Tahun 2003 Pasal 3 concerning the National Education System, where the goal of national education is to develop the potential of students to become human beings who believe and fear God, have noble character, are healthy, knowledgeable, competent, creative, independent, and become citizens who are democratic and have responsibility.

This explains that character education is the goal of national education so that value transfer-based education is very important, so students need to know what character education is. There are four types of character that have been known and implemented in the educational process, namely: 1) character education based on religious values, which is the truth of God's revelation (moral conservation); 2) character education based on cultural values, including in the form of character, Pancasila, literary appreciation, exemplary historical figures and national leaders; 3) character education based on the environment (environmental conservation); 4) character education based on self-potential, namely personal attitudes, the result of an awareness process of empowering self-potential which is directed to improve the quality of education (humanist conservation) (Yahya Khan, 2010: 2). Character education in point two, namely character education based on cultural values, is the discussion this time in the learning process. This is because the values of character, Pancasila, literary appreciation and exemplary figures can be learned more easily because they can be used as visual examples by students. For example, the ethical values that can be imitated from the ethical values in the environment, such as the value of working together. This confirms that character education can come from cultural values. Batik is a culture that contributes to character education, both from the process of making or the work achieved by students, because in the process of creating students are led to be more patient, responsible and hard working to achieve maximum batik results. Therefore, the values raised in this study are these three domains.

2.2 Character education

A paradigm shift in the national education system occurred around 2009 and character education became the focus of national education. The government, as mandated in the master character design, makes character education one of the priority programs for national development. This spirit is implicitly emphasized in the 2005-2015 RPJPN, character education is placed as the foundation for realizing the vision of national development, namely "creating a moral, ethical, cultured and civilized society based on the Pancasila philosophy. The process of making batik makes children have to be patient in every batik process, responsible for their obligations to complete it and work hard for the time given to complete the batik" (Mutiah, 2010: 45).

As for the explanation of the three domains to be discussed, namely:

2.2.1 Patience

Patience has several meanings, including: 1) Patience in maintaining the feeling of being careful, and not being in a hurry, because in a long batik process, the character expected by students is to be patient and enjoy everything that happens. 2) patience means persistence, determination, and steadfastness. 3) Organized/structured/systematic following of the batik process. For example, in the initial process, making a pattern with a pencil on cloth, if the participants are impatient, the students will immediately carve the night on the cloth without making a pattern first. 4) Being happy and understanding in a state of sadness, defeat, or suffering and not rebelling. If an accident occurs in the batik process, such as the unwanted drop of paint (wax) on the cloth, the expected character is that the students are patient and try to make it again. (Said, Abu-Nimer, & Sharify Funk, 2006).

2.2.2 *Hard work*

According to Kesuma (2012: 19) the characteristics of hard work are usually said to be the behavior of a person characterized by several tendencies. These tendencies include: 1) feeling worried if the work has not been completed. The long process of making batik requires more hard work, because every stage in the batik process has its own difficulties. 2) Checking on what to do or what is the responsibility of a position. Of course, in the learning process to make batik, students are required to check the results of their work before being assessed by the teacher, so if an error occurs, students are expected to work harder in revising their mistakes. 3) Time management. In the learning process, of course, there is a semester program that causes batik time to only have a few weeks. By managing the time they have, it is hoped that students will be able to complete batik tasks on time. 4) Able to organize existing resources to complete duties and responsibilities.

2.2.3 *Responsible*

According to Wibowo et al (2015: 171) Responsibility is the attitude and behavior of a person to carry out his duties and obligations to himself, society, the environment, the state, and God. The indicators of the character of responsibility used in this study are as follows: 1) Students carry out their duties wholeheartedly, meaning that every batik process that children go through is carried out seriously. 2) Students learn with high enthusiasm, meaning that every batik learning carried out by students is followed by a sense of enthusiasm for batik learning. 3) Students strive to achieve achievement, meaning that children try to achieve maximum results in the hope of getting the maximum value of batik. 4) Students are able to control themselves. 5) Students are accountable for the choices made. Accountable is that the chosen motive must be completed properly, because many students in the middle of making batik want to replace the batik motifs that they have previously chosen. 6) Students have discipline. Discipline in following the batik process, because in the process of making batik using tools that can endanger oneself and others. 7) Students do the task well, whether what is meant to be done is not carelessly done. The long process of making batik does not last forever and is not intended to bore the students, so following the process properly is a character education that should be received by

the students. 8) Students carry out their duties in an orderly way, meaning that they are students submitting assignments on time, and 9) Students make improvements if something goes wrong. The error is meant when the batik process makes an error or the night of the batik falls outside the motif, causing damage to the work.

3 METHOD

The research approach is quantitative with a quasi-experimental design. Quantitative research according to Creswell (2014: 32) is "quantitative research is an approach for testing objective theories by examining the relationships among variables. These variables, in turn, can be measured, typically on instruments, so that numbered data can be analyzed using statistical procedures." In short, quantitative research is an approach to testing the relationship between variables assessed using numbers so that it can be analyzed using statistical procedures, while quasi-experimental research according to Sugiyono (2010: 72) states that experimental research is a research method used to find the effect of certain treatments on those others under controlled conditions. Based on the above opinion, it is known that the statistical procedure used is an assessment using the numerical approach to find the effect of batik learning treatment on the character values of patience, hard work and responsibility.

Collecting data using observation, questionnaires and documentation. The data analysis from observation and documentation used descriptive analysis. Meanwhile, the questionnaire data analysis used descriptive data analysis mean. The descriptive mean is a group explanation technique that is based on the average value of the group (Sugiyono, 2016: 49). This analysis is used to make it easy for researchers to interpret the results of research conducted on the three aspects of the assessment of character values carried out on students.

The student subjects in this study were 21 people. A purposive sampling technique was used, the subjects were selected according to criteria, namely students who understand the learning art and culture of batik. The research was conducted four times in one month and was carried out in January 2020, namely on January 2, 2020, January 9, 2020, January 16, 2020 and January 23, 2020. Implementation at the first meeting was carried out with pretest and treatment 1, the second meeting was held with treatment 2, the third meeting was conducted with treatment 3 and the fourth meeting was carried out with treatment 4 and the posttest.

4 RESULT AND DISCUSSION

4.1 *Patience*

The histogram of the pre-test and post-test patience in the batik process can be seen as follows:

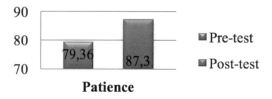

Figure 1. Histogram of patience in batik learning.

In Figure 1, it appears that the results of the students' patience in the pre-test class VII 1 is 79.36% which is classified as good and the results in the post-test class VII namely 87.3% which are also classified as good and there is an increase in the patience of the VII grade students 7.94%.

In addition to numerical data showing improvement, qualitative data was also obtained based on observations during before participating in batik learning and after participating in batik learning. Increasing patience through observation, it appears that after going through the batik process that requires patience, students become careful or not rushed in participating in learning cultural arts. Because in cultural arts, learning is more about practical lessons. Students also feel patient when problems occur in their work, either it needs to be revised or remade. Another thing that is also seen in the observation is that, now, students begin to be able to follow every process of making art, which, before participating in learning to make batik, students do not follow the directions systematically in creating art. Besides that, the students' patience seems to be getting better in case of an accident in their work, such as a color falling onto the paper when painting with the aquarel technique.

4.2 *Hard work*

The histogram for the pre-test and post-test hard work class VII 1 can be seen as following:

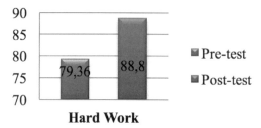

Figure 2. Histogram of hard work in batik learning.

In Figure 2, it appears that the results of the hard work of class VII students on the pre-test were 79.36% which was classified as good and the results on the post-test were 88.8% which was also classified as good and there was an increase in the hard work of students in grade VII 1 which was 9 , 44%

In addition to numerical data showing improvement, qualitative data were also obtained based on observations during before participating in batik learning and after participating in batik learning. Students feel worried if the tasks in learning cultural arts in particular are not completed, then students always consult the process that has been passed to the lesson teacher. This is done by students so that the results are made to be the best possible and on time

4.3 *Responsibilities*

The histogram of the pre-test and post-test responsibilities in making batik in class VII 1 can seen as follows:

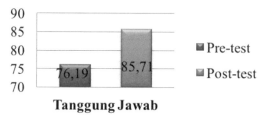

Figure 3. Histogram of responsibilities in batik learning.

In Figure 3, it appears that the result of student responsibility in the pre-test class VII 1 is 76.19% which is classified as good and the results on the post-test is 85.71% which is also classified as good and there is an increase in the responsibility of students in class VII 1. 9.52%.

In addition to numerical data showing improvement, qualitative data was also obtained based on observations during before participating in batik learning and after participating in batik learning. The increase in responsibility is an increase, this is shown from wholeheartedly participating in cultural arts lessons, where previously children were bored with learning cultural arts, now they can follow learning wholeheartedly and even enthusiastically. From this idea, it appears that students are trying to achieve things in which their seriousness can be demonstrated through effort and the seeking of maximum value. The responsibility arising from students is also responsible for the choices that are taken, such as the objects that are completed and do not displace them in the middle of the process as before the introduction of batik learning. From the sense of responsibility for the choice also arises a sense of order in completing tasks so that they are on time in collecting the assignments that have been given.

Based on the research results above, it is known that the three positive values, namely patience, hard work, and responsibility have increased after learning to make batik. This can be seen based on the histogram of the three aspects of the character's value when viewed from the average increase in the pretest and posttest scores. The histogram of the pretest and posttest values of the three character values above are as follows.

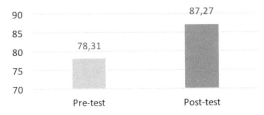

Figure 4. Histogram of effect of batik learning on character values.

Based on the overall data of the research results regarding the effect of batik learning on the character values of patience, hard work and responsibility in class VII students of SMP PGRI 8 Denpasar, it is known that the pre-test average was 78.31% and the post-test 87.27% with an increase of 8.96%, so based on the above data it is known that batik learning activities are effective in increasing the values of the characters of patience, hard work and responsibility.

ACKNOWLEDGEMENT

This research is fully supported by SMP PGRI 8 Denpasar and Universitas Negeri Yogyakarta.

REFERENCES

Creswell, J. W. 2014. *Research design: Qualitative, quantitative and mixed methods approaches*, 4 Edition. London: Sage

Hafiz, S. E. 2015. *Psikologi kesabaran. Buletin KPIN*: Vol. 1., No. 2, November 2015. Diakses dari https://buletin.k-pin.org/index.php/arsip-artikel/22-psikologi-kesabaran

Kesuma, D. 2012. *Pendidikan karakter kajian teori dan praktik di sekolah*. Bandung: PT Remaja Rosdakarya

Khan, Y. 2010. Pendidikan Karakter Berbasis Potensi Diri. Yogyakarta. Pelangi Publishing.

Kozier, Erb, Berman & Snyder. (2011). *Buku ajar fundamental keperawatan: Konsep, proses dan praktik* (7 ed., Vol. I). Jakarta: EGC

Kuswadji. 1981. *Mengenal seni batik di Yogyakarta*. Proyek Pengembangan Permuseuman Yogyakarta.

Mutiah, D. 2010. *Psikologi bermain anak usia dini*. Jakarta. Kencana

Parlina. (2016) Hubungan Antara Self Regulated Learning Dengan Tanggung Jawab Santri Tingkat Slta Di Pondok Pesantren Modern Zam - Zam Muhammadiyah Kecamatan Cilongok Kabupaten Banyumas. *thesis*, Universitas Muhammadiyah Purwokerto.

Said, A. A., Abu-Nimer, M., & SharifyFunk, M. (2006). *Contemporary islam: Dynamic, not static*. New York: Routledge

Sugiyono. 2010. *Metode penelitian pendidikan pendekatan kuantitatif, kualitatif, dan R&D*. Bandung: Alfabeta.

Sugiyono. 2016. *Statistika untuk penelitian*. Bandung: Alfabeta.

Sukamti, N. 2014 *Aspek kerja keras dan solidaritas sosial wanita tani pada kelompok wanita tani mekar sari di Desa Jurang Jero Kecamatan Karangmalang Kabupaten Sragen*. Diakses dari http://eprints.ums.ac.id/30012/2/04._BAB_I.pdf

Wibowo, et al. 2015. *Pendidikan karakter berbasis kearifan lokal di sekolah (Konsep, strategi, dan implementasi)*. Yogyakarta. Pustaka Pelajar.

Educational Innovation in Society 5.0 Era: Challenges and
Opportunities – Purnomo & Herwin (Eds)
© 2021 the authors, ISBN 978-1-032-05392-9

The need fulfillment of assistive technology for students with intellectual disabilities in Indonesia

Ishartiwi, E. Purwandari, R.R. Handoyo & A. Damayanto
Yogyakarta State University, Yogyakarta, Indonesia

ABSTRACT: Assistive technology is a broad umbrella of terminology related to all tools that is helpful, adaptive, and rehabilitative, which is used by individuals with special needs, including individuals with intellectual disabilities. This study aimed to know the need fulfillment of student with intellectual disabilities in terms of the (1) availability of assistive technology; (2) types of assistive technology available; and (3) use of available assistive technology. The research subjects were 415 teachers in special schools and inclusive schools throughout Indonesia. Data were collected through online questionnaire. The data in this study were analyzed through descriptive analysis technique. The results showed that: (1) assistive technology was available at all education levels, although in a limited number; (2) the types of assistive technology available are mostly low technology; (3) not all available assistive technology was used in learning, and it was mostly implemented at the primary school level.

1 INTRODUCTION

Assistive technology is a broad umbrella of terminology related to all tools that is helpful, adaptive, and rehabilitative, which is used by individuals with special needs, including individuals with intellectual disabilities. WHO (2016) mentioned that the elderly, disabilities, people with non-communicable diseases, people with mental conditions including dementia and autism and people who suffer from gradual decline in function are the people who need assistive technology the most. Assistive technology as tool, software, or system can be used to improve, maintain, or repair the functional ability of individuals with needs (Assistive Technology Industry Association, ATIA 2019). Today, World Health Organization has launched the Global Cooperation on Assistive Technology (GATE) program to increase access to high quality and affordable assistive technology for people with various disabilities, illnesses and age-related conditions.

As the first step, GATE has developed a list of product support, the list of priority aid products based on the greatest need at the population level. A certain group of people who get benefit from assistive technology are individuals with intellectual disability. Intellectual disability is disability that is characterized by significant limitation in intellectual function (IQ) and adaptive behaviour, which includes many social and practical daily life skills and this occurs before the age of 18 (AAIDD 2020). Intellectual disability becomes one of the groups that should utilize assistive technology since it can improve daily life and

communication skills of individuals with intellectual disabilities.

Research and practice related to the use of assistive technology for persons with intellectual disabilities offer considerable opportunities for the advancement of population health and the realization of human rights. Through the use of assistive technology, individuals with intellectual disabilities who have obstacles in verbal communication, they can take advantage of assistive technology to express their basic wants and needs and social interactions (James 2014). In addition, the use of assistive technology in the form of electronic portable (PEAT) is proven to be able to help individuals with intellectual disabilities manage their time in completing vocational tasks in the workplace. There are two major advantages of using the PEAT. First, video sequences help individuals with intellectual disabilities to remember the work sequence so they do not need to be reminded by the adults around them. Secondly, ethically, it is more ethical because individuals with intellectual disabilities will be independent and do not need to be accompanied continuously (Collins & Collet-Klingenberg 2018).

In contrast to the benefits derived from the development and use of assistive technology for individuals with intellectual disabilities, it is currently unknown how many persons with intellectual disabilities globally have access to appropriate assistive technology products and what factors affect their access. The fulfillment of need of assistive technology for individuals with intellectual disabilities in Indonesia needs to be immediately mapped so that they can immediately take

50

DOI 10.1201/9781003206019-10

advantage of the extraordinary speed of assistive technology developments in the world. There are various access dimensions to service from assistive technology that need to be considered, namely awareness, availability, affordability, adaptability, acceptability, quality, utilization, relevance and effectiveness (Levesque et al. 2013). This study focused on the fulfillment of need of assistive technology for individuals with intellectual disabilities in terms of the availability of assistive technology, the types of available assistive technology, as well as the use of available assistive technology.

2 RESEARCH METHOD

2.1 Research subject

The population in this research was special school teachers in Indonesia who educate students with intellectual disabilities. The research samples were determined purposively, they were 415 teachers who educated students with intellectual disabilities at elementary, junior and senior high school.

2.2 Instrument

The research instrument was developed based on the result of study held by Levesque et al. (2013) related to access dimension to service from assistive technology that should be considered. The instrument developed focuses on three aspects, namely availability, utilization and quality.

2.3 Data collection

The data gained through distributing online questionnaires using google form. Questionnaires were distributed simultaneously with online seminar related to assistive technology organized by the Laboratory of the Department of Special Education, Faculty of Education, Yogyakarta State University.

2.4 Data analysis

Data analysis was performed using quantitative descriptive. Quantitative data analysis was carried out through frequency distribution and descriptive statistics with SPSS to view and analyse the questionnaire data that had been filled in by the teacher.

3 FINDINGS AND DISSCUSSION

3.1 Findings

The results of the questionnaire related to the availability of assistive technology at each level of education in Indonesia can be seen in Figure 1 below:

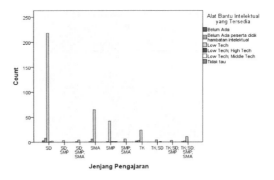

Figure 1. The availability of assistive technology for individuals with intellectual disabilities available at schools.

Based on Figure 1, it can be seen that the availability of assistive technology for all levels of education is limited to low technology. There are levels of education that are united because the education system in "SLB" still combines several students into one class with different education levels. This merger was carried out because of the limitations of teachers and classrooms, and a consideration that students' abilities are almost the same. Furthermore, the use of assistive technology for individuals with intellectual disabilities in Indonesia that has been available is presented in Figure 2.

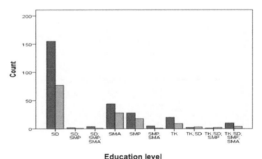

The utilization of assistive technology:
■ : No
■ : Yes

Figure 2. The utilization of assistive technology for individuals with intellectual disabilities in Indonesia.

Based on Figure 2, it can be concluded that the utilization of assistive technology for individuals with intellectual disabilities is still not optimal at almost all education levels. The types of assistive technology available in schools are still limited to the low technology category. In more detail, this type of low technology can be seen in Table 1.

Table 1 shows that the assistive technology available to students with intellectual disabilities is still limited as learning resource.

Table 1. The type of assistive technology in the low technology category is available at all levels of special schools in Indonesia.

	Frequency	Percent	Valid Percent	Cumulative Percent
Block Letters, puzzle, Picture	1	.2	.2.	2
Picture Series (giving instruction to do something which are followed by direct practice)	1	.2	.2	.5
Instruction in the form of pictures, for example the instructions about how to wash hands	381	91.8	91.8	92.3
Word Cards about brushing teeth	1	.2	.2	92.5
	1	.2	.2	92.8
Puzzle (professions, animals, fruits, etc.)	1	.2	.2	93
Schools only treat students with hearing and other impairments.	2	.5	.5	93.5
None	2	.5	.5	94
None	25	6.0	6.0	100
Total	415	100.0	100.0	

Source: The results of data analysis through SPSS

3.2 Discussion

The results of survey through questionnaires of 415 special school teachers in Indonesia indicate that the availability of assistive technology for students with intellectual disabilities is still limited to the low technology category. The types of low technology available are also limited for learning. The results of this study are relevant to previous research where it was stated that the provision and use of assistive technology in low- and middle-income countries is still limited (Matter et al. 2017). However, we can examine a number of factors that may be contributing to the limited supply and use of assistive technology in low- and middle-income countries. The first factor is the lack of awareness and knowledge about assistive technology for individuals with intellectual disabilities. Individuals with intellectual disabilities are generally still considered as undervalued and stigmatized group, where cultural perceptions can play an important role (Hatton & Emerson 2015). The society's view about individuals with intellectual disabilities who are still considered unable to be independent, troublesome, and will depend on others, this condition, will be the obstacle in reducing the access support toward assistive technology for them.

Educational orientation for students with intellectual disabilities that still refers to pure academic needs to be reviewed again so that the education is more directed towards functional academic. Education for students with intellectual disabilities who are still academically oriented will also contribute to the availability of assistive technology for them. Pure academic tends to encourage the provision of assistive technology that is oriented towards learning resources that has not yet led to a barrier. It would be more interesting if the educational orientation for students with intellectual disabilities is more directed at functional academic aspects. The motivation to meet the need of assistive technology will certainly be higher and will not only lead to learning resources but also as a barrier. The fulfillment of these two functions of assistive technology can be done by socializing the importance of assistive technology for individuals with intellectual disabilities. The aim is to increase knowledge and awareness of the importance of assistive technology for students with intellectual disabilities. The implementation is of course inseparable from the collaboration among schools and families or caregivers so that the use of assistive technology can be carried out in a sustainable manner (Codling & Macdonald 2011).

The second factor is finance. It is undeniable, the provision of assistive technology in the high technology category cannot be separated from costs that are comparable to the technology developed, especially the development of high technology category. The high technology in Indonesia itself is still minimal so that they still have to import high technology products from outside or make and or modify them at costs that are sometimes considered quite high by schools. This problem certainly requires a solution in the form of a government policy on the use of assistive technology. The special policy on assistive technology has an important role in overcoming the limitations of assistive technology and can be used as a basis for making assistive technology accessible to all those who need it (Sund 2016).

Specific policies related to assistive technology not only make it possible to provide assistive technology products, but also financial support to develop assistive technology according to individual needs. Of course, in practice, this financial service should be prepared related to the system how the service is applied (Andrich et al. 2013). The implementation of assistive technology policy programs has been carried out in developed countries such as Norway, which states that people who need assistive technology are entitled to receive financial support under a national insurance scheme (Sund 2016). Policies in high-income countries certainly can be a motivation for low- and middle-income countries to develop policies appropriate to their needs and abilities (Borg et al. 2011). Despite the recognition of the importance of providing and accessing assistive technology in low- and middle-income countries, there is still a lack of scientific evidence on specific ways to achieve this (Rohwerder 2018).

4 CONCLUSION

Assistive technology facilities already exist at all levels of education even though the number is limited and is still low technology. However, the available assistive technology facilities are not all optimized by teachers and the types of assistive technology are still limited. Assistive technology facilities already exist at all levels of education even though the number is limited and is still low technology. However, the available assistive technology facilities are not all optimized by teachers and the types of assistive technology are still limited. The results of this study still need to be studied further to determine the factors that the availability and use of assistive technology is not optimal in schools.

REFERENCES

AAIDD. 2020. *Definition of intellectual disability*. Diakses pada 8 September 2020 melalui https://www.aaidd.org/intellectual-disability/definition

Andrich R., Mathiassen N.-E., Hoogerwerf E.-J. & Gelderblom G. J. 2013. Service delivery systems for assistive technology in Europe: an AAATE/EASTIN position paper. *Technology and Disability* 25: 127–46

ATiA. 2019. Available from:https://www.atia.org/home/at-resources/what-is-at/

Borg J., Lindstrom A. & Larsson S. 2011 *Assistive technology in developing countries: a review from the perspective of the convention on the rights of persons with disabilities. Prosthetics and Orthotics International 35: 20–29.WHO.* Priority Assistive Products List. GATE Initiat [Internet]. 2016: 1–16. Available from: http://apps.who.int/iris/bitstream/10665/207694/1/WHO_EMP_PHI_2016.01_eng.pdf?ua=1

Collins, J. C., & Collet-Klingenberg Lana. 2018. Portable elektronic assistive technology to improve vocational task completion in young adults with an intellectual disability: A review of the literature. *Journal of intellectual disability (JOID)*. Vol. 22(3), 213–232.

Codling M. & Macdonald N. 2011. Sustainability of health promotion for people with learning disabilities. *Nursing standard (Royal College of Nursing (Great Britain) : 1987)* 25: 42–7.

Hatton C. & Emerson E. 2015. *International Review of Research in Developmental Disabilities: Health Disparities and Intellectual Disabilities.* Academic Press Elsevier

James S. E. 2014. Review of assistive technology: interventions for individuals with severe/profound and multiple disabilities. *Journal of Child and Family Studies*, 23: 169–71.

Levesque J.-F., Harris M. F. & Russell G. 2013. Patient-centred access to health care: conceptualising access at the interface of health systems and populations. *International Journal for Equity in Health*, 12: 1–9.

Matter R., Harniss M., Oderud T., Borg J. & Eide A. H. 2017 Assistive technology in resource-limited environments: a scoping review. *Disability and Rehabilitation: Assistive Technology* 12: 105–14.

Owuor J., Larkan F. & Maclachlan M. 2017. Leaving no-one behind: using assistive technology to enhance community living for people with intellectual disability. *Disability and Rehabilitation. Assistive Technology* 12: 426–8.

Rohwerder, B. 2018. *Assistive technologies in developing countries. K4D Helpdesk Report.* Brighton, UK: Institute of Development Studies.

Sund, T. 2016. *Assistive technology in Norway* [Online]. Available: https://www.nav.no/en/Home/About+NAV/Publications/ [Accessed November 2017].

Educational Innovation in Society 5.0 Era: Challenges and Opportunities – Purnomo & Herwin (Eds)
© 2021 the authors, ISBN 978-1-032-05392-9

Learning mathematics online during the COVID-19 pandemic: Is it without problems?

Desmaiyanti & Sugiman
Yogyakarta State University, Yogyakarta, Indonesia

ABSTRACT: This study aims to identify the problems faced by mathematics teachers in online learning during the COVID-19 pandemic. This study is descriptive with a survey method. The participants of this study were 185 mathematics teachers faced with online learning during the pandemic. The results were analyzed based on the problems found in the city and village. The problems found were 1) technical problems, 2) teaching preparation and methods, 3) assessment methods, and 4) student control. The problems were caused by 1) the limitation of infrastructure and environmental conditions, and 2) the lack of experience and insight of the teachers related to online learning. Thus, it can be seen that in the implementation of online learning, teachers have to face many problems. However, it is possible to overcome this so that learning can take place.

1 INTRODUCTION

Since 2011, the era of the industrial revolution 4.0 has been introduced by Germany to the world. This era is marked by new terms such as the Internet of Things (IoT), big data, artificial intelligence, cloud computing, blocks, etc. Moreover, one of the pillars in the 4.0 industrial revolution is the widespread use of the internet as a means of making it easier to connect machines, devices, sensors, and humans as well as being a source of information to create new product functions and features (Bassi 2017).

Not satisfied with the 4.0 industrial revolution, Japan has recently started to introduce the society 5.0 era which is marked by digitizing all aspects of human life. The key of this era is a combination of cyberspace and the real world to produce quality data then create new values ??and solutions to solve various challenges (Fukuyama 2018). Thus, these two eras changed the course of the world in all areas of life.

One of the fields changing is education. The changes that occurred in both eras led to the same thing. The focus of the industrial revolution 4.0 is the development and educational skills to make learning more adaptable, smart, portable, global, and virtual (Shahroom & Hussin 2018). Meanwhile, the mission in the era of society 5.0 emphasizes the use of the virtual world in all aspects of life including education. So, these two eras have one thing in common, namely the use of the internet in learning. This fact makes online learning a familiar breakthrough.

Unfortunately, this situation does not necessarily mean that the life of the Indonesian people can quickly keep up with the world's developments. Limited means of communication, transportation, and public knowledge make online-based activities rarely usable or even completely impossible to use. This limitation is especially evident in the border and remote areas. This limitation occurs in all fields, including education. This is why online learning is often avoided and is not a choice in improving the quality of education used by teachers in Indonesia.

However, the conditions in the world that are being attacked by a deadly virus known as COVID-19 have made things change drastically. Like it or not, ready or not, all countries in the world must avoid face-to-face activities and offline gatherings. In February of 2020, China was the first to close schools, with other countries shortly following. By the end of March, all 46 countries had closed some or all of their schools. And by the end of June, the duration of school closures ranged from 7 to 19 weeks across OECD and partner countries (Schleicher 2020). In order to ensure continuity of learning, circumstances require teachers to switch to the online delivery of lessons. There are no more options; it is necessity (Dhawa 2020). As a result, online activities carried out in all spheres of life must become an undeniable new normal.

The educational field is the area that most affected by the conditions of the world with COVID-19 problems. The new normality must be implemented in learning and teaching activities. In past times, students studied at the school offline or face-to-face with the teacher. But now this can no longer be done. Everyone must keep a distance to avoid spreading the virus. So, if the teacher wants the students to still achieve the learning goals, online learning must be done.

54

DOI 10.1201/9781003206019-11

Online learning is learning using infrastructure or hardware such as computers that are interconnected and useful for sending data, in the form of text, messages, graphics, or sound (Riyana 2013). Online learning was first implemented in 1981 by the Western Behavior Sciences Institute (Harashim in Sun & Che, 2016). Online learning is also known as e-learning.

The students can be taught with full online learning if they have several requirements. First, ICT literacy means that students must master the basics of ICT as a tool for learning. Second is independence, which means that students can learn independently without the need for direct guidance from the teacher. Third is creativity and critical thinking, which means students have the creativity to take advantage of all available tools such as browsing, chatting, group discussions, video conferencing, online quizzes, online drills, and others so that learning runs optimally (Riyana 2013).

However, since the Large-Scale Social Restrictions were implemented in all regions of Indonesia, learning in schools has been carried out online with all limitations though the students are not versed in the above requirements. For example, Indonesian students and teachers still have limited knowledge, are unfamiliar, and have little experience related to using ICT. Moreover, many students weren't ready to study independently without an explanation from the teacher. Then, Indonesian students still have low levels of creative and critical thinking.

Online learning has been carried out in all subjects including mathematics. All people know that mathematics is an unpopular subject because it is difficult to understand. And now, learning mathematics becomes more and more difficult because of the necessity of doing it online.

As such, it is natural that many problems occur, especially in mathematics. The problems will be different according to the level of education such as elementary school, junior high school, senior high school, vocational high school, and college. Moreover, the kind of problems occurring when learning online in villages will be different from those in cities. This study aims to show the problems faced by students and teachers in villages and cities.

2 METHODS

2.1 Research design

The methodology used in this research is descriptive research with a survey method. The population of this study was all mathematics teachers in Indonesia who taught online during the COVID-19 pandemic. The sample in this study was mathematics teachers who carried out online learning who taught on the islands of Sumatra, Java, Borneo, and Sulawesi. Sampling was carried out by accidental (convenience sampling) to obtain 185 teachers as the sample.

2.2 Participants

The subjects of this study were 185 teachers consisting of 92 teachers who teach in villages and 93 teachers in cities.

Table 1. Demographic background of participants

Demographic Percentages	background	Number of	Participants
Gender	– Male	47	25,4%
	– Female	138	74,6%
School*	– Elementary	20	10,8%
	– Junior High	53	28,6%
	– Senior High	60	32,4%
	– Vocational	37	20%
	– College	15	8,1%
Teaching Time	– 0–5 years	90	48,6%
	– 6–10 years	47	25,4%
	–> 10 years	48	25,9%

*School where teaching

2.3 Data collection and data source

The data was collected by distributing online questionnaires through the G-Form. These questionnaires contained several questions related to problems in teaching and learning mathematics online. Then, the data was analyzed based on the problems occurring when online learning in the cities and the villages. The questionaires was open for three days.

All data provided by participants was kept confidential and was used only for this study. Matters related to participant responses will not affect the future of the respondents.

3 RESULTS

3.1 Online media data used by mathematics teachers in the village

Table 2. Online media data used by mathematics teachers in the village

Media	Percentages
1. WA Group	77%
2. Zoom	7%
3. Google Classroom	10%
4. Google meet	7%
5. G-Drive	9%
6. Edmodo	3%
7. G-Form	4%
8. Others*	7%
9. Nothing	3%

*Kahoot, Qiuizziz, Supermath, Youtube, sms, phone

3.2 Online media data used by mathematics teachers in the city

Table 3. Online media data used by mathematics teachers in the city

Media	Percentages
1. WA Grup	57%
2. Zoom	32%
3. Google Classroom	20%
4. Google meet	13%
5. G-Drive	13%
6. Edmodo	2%
7. Be-smart	2%
8. Quipper	2%
9. Saba Banten, School Application, Video, LMS Canvas	5%
10. Others*	14%
11. Nothing	0%

*Edlink, Telegram, IG Live, Email, Schoology, Messenger

3.3 Data on problems faced by mathematics teachers in villages and cities

Table 4. Data on problems faced by mathematics teachers in villages and cities

Main Problems	Villages	Cities
1. Technical problems	77%	48%
2. Preparation for learning	24%	23%
3. The teacher has difficulty making interactive media (video, animation, ppt) with small data sizes	54%	44%
4. Difficulty assessing cognitive aspects	67%	49%
5. Difficulty doing affective aspect assessments	79%	78%
6. Difficulty conducting a psychomotor aspect assessment	84%	72%
Student Control	28%	35%
Students are not enthusiastic	57%	41%
Students experience anxiety	65%	61%
7. Student Control	28%	35%
8. Students are not enthusiastic	57%	41%
9. Students experience anxiety	65%	61%
10. Students have decreased mathematics performance	83%	72%

4 FINDINGS

According to the research results, as much as 77% of teachers in villages and 57% of teachers in the cities are using WA. Just 7% of teachers from the villages try to use synchronous media such as Zoom and Google Meet. Unfortunately, they must face ineffective learning due to timeouts to wait if the network suddenly becomes lost or unstable. Some teachers in other villages have also tried to use LMS, Google Forms, Edmodo, or Kahoot to facilitate assessment and assignment but they also have to be patient if the set time has to be delayed due to network reasons or student limitations in owning a data plan or cellphone.

On the other hand, as many as 32% of teachers in the city have used video conferencing media such as Zoom, Google Meet, and others. Unlike in a village that faces hardships, this makes learning more interactive. The use of video and media to control student activities is also widely used, such as Google classroom, Quipper, Edmodo, and so on. This option is quite good considering the availability of infrastructure in cities is better than in villages.

Apart from technical problems, the problems faced by teachers both in the village and in the city are closely related to the media they choose and the teachers' insights into online media. For this reason, from the data obtained, the reduction was carried out so that there were four main problems faced by teachers in online mathematics learning as follows.

4.1 Technical problems

From the data, technical problems are faced more by teachers in villages than in cities. These technical problems include a) a poor network, b) limited data access, c) the availability of communication tools, e) cost barriers. This is normal because the geographical and economic conditions of rural communities are full of limitations.

For reasons of very poor networking, as many as 3% of village teachers who should be doing online learning don't do it. As a solution, some teachers choose to visit students' homes. Another teacher chooses to divide the children into shifts and then come to school for a few minutes to be given assignments to collect at the end of the week. Elsewhere, teachers use radio as a medium to reach students who do not have cellphones or quotas.

For data issues, based on the provisions in the Regulation of the Minister of Education and Culture (Permendikbud Number 19 of 2020 Regular article 9A paragraph 1 part a) regarding changes to the Technical Instructions for School Operational Assistance, it is said that during the stipulation period for the COVID-19 Public Health Emergency status set Central Government, regular boss funds can be used for credit needs, data packages, and/or paid online education services for teachers and students. Also, if students do not have tools of communication, the PKH social service (the family hope program) has distributed funds to poor people. These funds can be used to purchase gadgets or laptops for children who attend school through online learning.

4.2 Teaching preparation and methods

The problems related to teaching preparation and methods faced by teachers both in villages and cities with almost the same percentage and include: a) teachers have difficulty designing lessons, b) teachers have trouble explaining the lesson via online media, c) teachers have trouble managing learning time, d) teachers have not mastered using online media, e) teachers have difficulty using mathematical symbols when using online media, e) the teacher has difficulty providing feedback when online learning;

In fact, if the teachers want the learning of mathematics during the COVID-19 pandemic to run well, teachers must be able to prepare various materials properly so that students do not feel bored with online learning, which sometimes can feel like the teacher only gives assignments to students (Wiryanto 2020). This problem is related to the weak ability of teachers to carry out online learning management. Besides, teachers still lack experience in online learning. As a result, teachers prefer to give assignments without first giving an explanation.

4.3 Assessment

For problems related to assessment, the data shows that the teachers in both rural and urban areas have the same opinion. They have difficulty making assessments. Teachers have difficulty finding how to be able to assess student progress while there are many opportunities to cheat. Teachers cannot be sure that assignments or exams that students do at home are their work. The task may be carried out by other people like their parents.

The data shows that teachers in both rural and urban areas experience difficulties in designing assessments for students in terms of cognitive, affective, and psychomotor aspects. Teachers are not used to assessing students' progress from a distance. For cognitive problems, the teacher only assigns mathematics tasks to be completed by students. Meanwhile, for the affective, the teachers use the portfolio as an assessment. For psychomotor skill problems, some teachers in the city ask students to make videos or create certain projects. Unfortunately, most of the teachers in both rural and urban areas prefer not to do affective and psychomotor assessments.

4.4 Student control

Data shows that the problem with student control is mostly faced by teachers in cities. These problems include: a) difficulties in making students understand explanations when presented in online media, b) difficulty in making students actively respond, c) difficulty controlling whether students follow the material well or not (discipline), d) difficulty controlling whether the task is done by the student or not (honesty), e) students become less enthusiastic about learning (laziness and lack of interest), f) students are not accustomed to learning independently, g) students are often absent (discipline), h) student activities are less controlled during online learning (discipline).

The next problem is related to the enthusiasm of students in learning. From the data, it can be seen that students in cities are a little more enthusiastic than students in villages. In addition to the spirit of psychological problems faced by students in learning online, there are anxiety problems. From the data, it can be seen that more than 50% of teachers in both rural and urban areas agree that students in rural and urban areas experience anxiety.

The other problem in online learning associated with student control is student achievement. From the data, it can be seen that teachers in the village agree that students have decreased their achievement when learning online. The teachers in the city also agree with this and many other teachers agree. However, the teachers from the city are not as many teachers in the villages.

5 DISCUSSION

The technical problems are the big problems that mathematics teachers in Indonesia must face while implementing online learning. The reality showed that the existing infrastructure in Indonesia does not always support online learning (Basuki 2007). There are only 56%, with only 150 million of its 268 million population in Indonesia having access to the internet (Jalli 2020). But we can overcome this situation by using the available infrastructure. In Sub-Saharan Africa, they used radio or television when a mobile phone or the internet is not available (Burns 2020). On other hand, if it is not possible to use the internet online, we can use it offline by saving the files on a flash disk and then giving it to students (Castelo 2020). In areas with limited connectivity, we can use more traditional distance learning modalities, often a mix of educational television and radio programming, and distribution of printed materials (De Giusti 2020, UNESCO 2020). Therefore, the teacher's creativity is the key to solving technical problems (Febrianto et al. 2020).

The problem of teaching and learning preparation is very closely related to how teachers understand how to prepare lesson materials properly. Online learning can run well if the teacher pays attention to the following: 1) setting learning objectives, 2) preparing learning materials including techniques and media so that the learning material is interactive and runs effectively, 3) designing the operational learning process, 4) improving computer multimedia-related skills, 5) understanding the operation and maintenance of the hardware used, 6) adapting online learning to the curriculum, 7) adding insight related to computer learning techniques, and 8) always being responsive to technological developments (up to date) (Hanum 2013). On the other hand, the collaboration between teachers is one way to increase teacher capacity in the use of technology in learning for preparing the learning processes. Ignorance and lack of insight should no longer be reasons that cause problems in learning (UNESCO 2020).

Online learning should be prepared by teachers maximally. Therefore, online learning goes in the right way according to expectations and learning goals. Teachers should not only burden students with many tasks. If necessary, the teacher comes with ideas in the door to the door of the students. The teacher is not only positioned as a transfer of knowledge, but still prioritizing *ing ngarso sung tuladha, ing madya mangun karso, tut wuri handayani* (in front, an educator should be able to set an example among students, educators should be able to give ideas, and in the back, an educator should be able to give encouragement) (Syah 2020).

Meanwhile, there are several notes that teachers must pay attention to when creating the design and manufacture of lesson materials in online learning: 1) the availability of hardware and software to be used, 2) availability of brainware to manage hardware and software, 3) the consideration to implement an interactive multimedia learning module that makes students actively learn, 4) materials are easy to manage and update, 6) accessible teaching materials, 7) teaching materials are designed interactively and provide menus, panels, and assistance facilities to clarify the mechanisms, the form or type, and the composition of the materials (Hanum 2013).

Furthermore, the problems about teaching and learning preparation are also closely related to teachers' knowledge of the use of ICT. Several studies show that ICT competencies of Indonesian teachers are unevenly spread across regions and largely inadequate (Widodo & Riandi 2013, Mailizar & Fan 2020). This fact should be an important note for teachers to continue to improve their competence so that they are not hindered in online learning due to the inability to operate media that require ICT.

For the problem about assessment, there are some alternative assessment tools can be used in online learning, namely authentic assessment, performance, and portfolio (Gaytan 2014). There are some aspects

Table 5. Alternative assessment indicators in online learning

Assessment	Indicators
Authentic	– The relationship between performance and the real world
	– Positions between students
	– Students' understanding of the assessment system
	– Student motivation in performance
Performance	– Product relationship with learning
	– Communication with parents regarding performance
	– possible performance complexity
	– Possible performance-related troubleshooting
	– The link between performance and active student responses
	– Possible performance with challenging, responsive tasks
	– Possible time required
Portfolio	– Possibility of storing tasks for development, processes, and final products
	– The importance of checking all tasks
	– The possibility of viewing the work as a collection of products (not just looking at the final product)
	– The possibility of join evaluation of students

that should be considered when preparing alternative assessment tools.

Assessing online learning is not easy because no technology can truly ensure the authenticity of student work. Therefore, online assessments must balance the various online components, assessment criteria, and tools needed to assess student learning effectively and comprehensively while meeting the challenges of accountability, reform, and student learning needs.

If we talk about student control, in online learning, the quality of content material is the teacher's key to being a success in controlling students, increasing student enthusiasm for learning, and automatically reducing students' anxiety levels. Besides, one of the standards in online teaching is the ability of the teacher to create assignments, projects, and assessments that are aligned with students' different visual, auditory, and hands-on ways of learning (Pape & Wicks 2011). Therefore, to create effective content, the teachers should pay attention to the following: 1) teacher-centered instructional content is procedural, declarative and well-defined and clear, 2) learner-centered content means that the content presents results (outcomes) that focus on developing creativity and maximizing independence, 3) providing work examples to make it easier to understand and practice, 4) providing additional content such as games (Hanum 2013).

6 CONCLUSION

The problems that arise when learning mathematics is carried out online consist of four main problems. There are 1) technical problems, 2) teaching preparation and methods, 3) assessment methods, and 4) student control. All of these problems were caused by 1) the limitation of infrastructure and environmental conditions, 2) the lack of experience and insight of the teachers related to online learning.

Therefore, it is right that online learning is the best way to deal with a pandemic period. However, many problems must be resolved in advance so that the learning system can run effectively. The situation of COVID-19 pandemic is so difficult and dramatic that we cannot afford to be pessimistic.

REFERENCES

Bassi, L. 2017. Industry 4.0: Hope, hype or revolution? *RTSI 2017 – IEEE 3rd International Forum on Research and Technologies for Society and Industry, Conference Proceedings.* https://doi.org/10.1109/RTSI.2017.8065927.

Basuki, S. 2007. IT and education, the case study of E-Learning in Indonesia . *Korea-ASEAN Academic Conference on Information Revolution and Cultural Integration in East Asia, Ho Chi Minh city, Vietnam, January 25–26, 2007.,* 1–13. Retrieved from http://eprints.rclis.org/9048/1/E-learning_-_Vietnam_-_2007.pdf.

Burns, M. 2020. School, interrupted: 4 options for distance education to continue teaching during COVID-19. Retrieved from dari https://www.globalpartnership.org/

blog/school-interrupted-4-options-distance-education-continue-teaching-during-covid-19.

Castelo, M. 2020. Continuing remote learning for students without internet. Retrieved from https://edtechmagazine.com/k12/article/2020/04/continuing-remote-learning-students-without-internet.

De Giusti, A. 2020. Policy brief: education during COVID-19 and beyond. *Revista Iberoamericana de Tecnología en Educación y Educación en Tecnología*, 26(e12). https://doi.org/10.24215/18509959.26.e12.

Dhawan, S. 2020. Online learning: a panacea in the time of COVID-19 crisis. *Journal of Educational Technology Systems*, 49(1): 5–22. https://doi.org/10.1177/0047239520934018

Febrianto, P. T., Mas'udah, S., & Megasari, L. A. 2020. Implementation of online learning during the covid-19 pandemic on Madura Island, Indonesia. *International Journal of Learning, Teaching and Educational Research*, 19(8): 233–254. https://doi.org/10.26803/ijlter.19.8.13.

Fukuyama, M. 2018. Society 5.0: aiming for a new human-centered society. *Japan SPOTLIGHT*, 27(Society 5.0): 47–50. Diambil dari http://www8.cao.go.jp/cstp/%0Ahttp://search.ebscohost.com/login.aspx?direct=true&db=bth&AN=108487927&site=ehost-live.

Gaytan, J. 2014. Effective assessment techniques for online instruction. *Information Technology, Learning, and Performance Journal*, 23(1): 25–33. Diambil dari http://search.proquest.com/docview/219816513/abstract/77B81AD0A634A49PQ/1%5Cn http://media.proquest.com/media/pq/classic/doc/1058985151/fmt/pi/rep/NONE?cit%3Aauth=Gaytan%2C+Jorge&cit%3Atitle=EFFECTIVE+ASSESSMENT+vTECHNIQUES+FOR+ONLINE+INSTRUCTION&cit%3Apub=.

Hanum, N. S. 2013. Keefetifan e-learning sebagai media pembelajaran (studi evaluasi model pembelajaran e-learning SMK Telkom Sandhy Putra Purwokerto). *Jurnal Pendidikan Vokasi*, 3(1): 90–102. https://doi.org/10.21831/jpv.v3i1.1584.

Jalli, N. 2020. Lack of internet access in southeast asia poses challenges for students to study online amid COVID-19 pandemic. Retrieved from https://phys.org/news/2020-03-lack-internet-access-southeast-asia.html.

Pape, L., & Wicks, M. 2011. *National standards for quality online programs. International Association for K-12 Online Learning* (2 ed.). Vienna: iNACOL. Retrieved from http://www.eric.ed.gov/ERICWebPortal/recordDetail?accno=ED509638.

Riyana, C. 2013. Produksi bahan pembelajaran berbasis online. *Modul Pembelajaran Universitas Terbuka Tangerang Selatan*, 1–43.

Schleicher, A. 2020. The impact of COVID-19 on education: Insights from education at a glance 2020. *OECD Journal: Economic Studies*, 1–31. Retrieved from https://www.oecd.org/education/the-impact-of-covid-19-on-education-insights-education-at-a-glance-2020.pdf.

Shahroom, A. A., & Hussin, N. 2018. Industrial revolution 4.0 and education. *International Journal of Academic Research in Business and Social Sciences*, 8(9): 314–319. https://doi.org/10.6007/ijarbss/v8-i9/4593.

Sun, A., & Chen, X. 2016. Online education and its effective practice: A research review. *Journal of Information Technology Education: Research*, 15: 157–190. https://doi.org/10.28945/3502.

Syah, R. H. 2020. Dampak Covid-19 pada Pendidikan di Indonesia: Sekolah, Keterampilan, dan Proses Pembelajaran. *SALAM: Jurnal Sosial dan Budaya Syar-i*, 7(5). https://doi.org/10.15408/sjsbs.v7i5.15314.

UNESCO. 2020. Education in a post-COVID world?: Nine ideas for public action International Commission on the Futures of Education, 26.

Widodo, A., & Riandi. 2013. Dual-mode teacher professional development: challenges and re-visioning future TPD in Indonesia. *Teacher Development*, 17(3): 380–392. https://doi.org/10.1080/13664530.2013.813757

Wiryanto. 2020. Proses pembelajaran matematika di sekolah dasar di tengah pandemi COVID-19. *Jurnal Review Pendidikan Dasar, kajian Pendidikan dan hasil Pendidikan, Universitas Negeri Surabaya*, 6(2)

Educational Innovation in Society 5.0 Era: Challenges and Opportunities – Purnomo & Herwin (Eds)
© 2021 the authors, ISBN 978-1-032-05392-9

Assessing the discriminant validity of the curiosity scale using confirmatory factor analysis

H. Sujati

Universitas Negeri Yogyakarta, Yogyakarta, Indonesia

ABSTRACT: Curiosity is one of the important characters that elementary school students need to have in Indonesia. Unfortunately, until now there is no standard instrument to measure it. Therefore, it is necessary to develop instruments to produce a curiosity scale that meets the psychometric criteria. One step that should not be ignored by the developer in the process of developing the scale is to assess the discriminant validity of the factors. Based on the results of theoretical studies, the variable of curiosity has four factors, namely: being interested in new things, questioning something, the courage to try, and being enthusiastic about solving problems. The results of the confirmatory factor analysis show that the AVE square value of the four factors is greater than the correlation value between the factors, so it can be concluded that the four factors meet the discriminant validity criteria.

1 INTRODUCTION

In Indonesia, curiosity is one of the characters that must be instilled in every educational institution, from preschool education to higher education. Determination of curiosity as mandatory teaching material is very reasonable. Kidd & Hayden (2015) states that curiosity is a basic and natural component of life. Curiosity is to encourage students to learn (Lindholm 2018). This statement illustrates the importance of curiosity in the context of the teaching-learning process. This is in line with research findings of Kang et al. (2009) that curiosity can improve learning. This is consistent with the theory that the main function of curiosity is to facilitate learning. Likewise, students who have high curiosity have higher performance compared to those who have low curiosity (Arnone, Grabowski, & Rynd 1994). To find out the learning achievement needs to be assessed, which is a set of procedures designed to obtain information about the growth, development and achievements of students to be compared with a standard (Shermis & Di Vesta 2011). Information about student achievement can be quantitative or qualitative. One instrument that can be used to obtain quantitative information is scale. Unfortunately, up to now there is no standardized measurement scale of curiosity. For this reason, it is necessary to develop a scale.

Developing a scale is the first step in any quantitative psychological research. A standard scale must meet various criteria, one of which is construct validity. According to Hair et al. (2019), construct validity can be divided into three, namely convergent, discriminant and nomological validity. In this study, discriminant validity was reported specifically. The question is "Does the scale developed meet the discriminant validity criteria?"

1.1 Curiosity

Etymologically, in the Spanish dictionary called curioso which means to do something with full perseverance (Manguel 2015). It was said by Engel (2015) that curiosity became a basic component in learning. High curiosity causes curiosity to know deeply about something. In addition, curiosity makes it more open to anything new that is not yet known. The statement is in accordance with Dewey's opinion (Jeraj & Mariè 2013) which states that curiosity is "the mother of all sciences". Curiosity is the tendency to seek new information and experiences. The same thing was stated by Benoit, Coolbear & Crawford (2008). They stated that curiosity was an internal motivation to gain knowledge.

According to Mustari (2014), curiosity is an emotion associated with natural prying behavior such as exploration and investigation, in which there is a sense of awe and wonder about something. Curiosity has encouraged scientists to conduct various studies so that various scientific innovations are found. A similar understanding was expressed by Reio, Petrosko, Wisell & Thongsukmag (2006) who interpreted curiosity as the desire to obtain new information and knowledge that encourages individuals to explore their environment. Prochniak (2017) stated that curiosity is an intrinsic human need to find new information that motivates individuals to explore.

Gulten (2011) argues that someone shows signs of curiosity since childhood as a process of learning new things so that curiosity will continue to emerge throughout human life. Curiosity is accepted as a driver of the learning process and has a positive effect on learning. Curiosity is a system of positive emotional and motivational orientation towards appreciation, search, and self-regulation of new ideas and challenging information and experiences. This is very important in the field of education which emphasizes students to learn better.

Gulten (2011) and Jones (2013) argue that curiosity is a driver of cognitive, social, spiritual, and physical development. curiosity has a number of benefits, including strengthening social relationships, helping prevent Alzheimer's disease, and helping learning. People who have curiosity are often considered to be good listeners and able to speak. Furthermore, Kashdan & Roberts (2006) said that someone who has curiosity tends to bring fun and novelty in social communication. Therefore, stimulating curiosity must be made as a center of education and learning.

Engel (2015) states that curiosity is an important component in learning. It can even be said that curiosity is the centerpiece of the learning candle. Therefore, curiosity is not just interested in something. Curiosity must be expressed in the form of a real act in an effort to answer a question about an uncertainty. Engel (2011) defines curiosity as a desire to obtain information or new experiences that trigger a reaction, can be in the form of exploration, using information, or a resolution of experience. It was said by Zuss (2012) that people who have curiosity when they receive a stimulus that is not in accordance with the cognitive map will be motivated to search for information with the aim of creating conformity with what was previously known.

Jones (2015) states that curiosity encourages students to work hard with full attention and continue to actively learn. In addition, Engel (2011) argues that curiosity is the child's need to try and make the world more logical. Curiosity as an impulse to explain unexpected things. Engel (2011) asserts that curiosity is the need to solve mysterious problems.

According to Al Farani (2013), someone who has high curiosity has the following signs: (1) likes to question everything related to the concept discussed, (2) uses all five senses to recognize or observe an object or object, (3) happy exploring books, maps, pictures, objects, and so on to look for new ideas, and (4) conducting experiments on an object or object. Jones (2013) holds that curiosity is symptomatic in the form of: (1) positive reactions to new, strange, and mysterious elements through exploration and manipulation activities, (2) indicating the need or desire to find out more about themselves or their environment, (3) recognize the environment quickly to look for new experiences, and (4) continue to explore and investigate to find out more about the environment. According to Pisula (2009), indicators of curiosity are: (1) paying attention to the search for new things, (2) not afraid to make mistakes when conducting experiments in everyday life, and (3) showing creativity in thinking and behavior.

Different opinions expressed by Harlen (2001). He states that the curiosity of elementary school-aged children is marked by four indicators, namely (1) enthusiastic in solving problems, (2) focusing on the observed objects, (3) enthusiastic about the science process, and (4) asking every step of the activity. Enthusiasm in solving problems is seen when students work on worksheets, and in working on problems given by the teacher. The focus on the observed object can be seen when students conduct experiments. Enthusiastic about the science process can be seen when students concentrate when experimenting, students pay attention to each work procedure well and do not play around during experimental activities. Asking each step of the activity is seen when students ask questions about things related to activities carried out by students. The results of a factor analysis study conducted by Reio et al. (2006) show that curiosity has three dimensions, namely searching for experience, seeking information and searching for physical sensations.

Based on the above study, four indicators of curiosity character can be summarized, namely being interested in new things, questioning something, daring to try and enthusiastic in solving a problem. These four indicators form the basis of developing a scale of curiosity.

1.2 *Discriminant validity*

Discriminant validity is one of three types of construct validity (Hair, Black, Babin, & Anderson 2019). Discriminant validity is often called divergent validity (DeVelis 2017). Discriminant validity assessment aims to prove that one construct is completely different from another (Voorhees et al. 2015). The assessment of discriminant validity is very important in research involving latent variables and using several items as a representation of the construct (Ab Hamid, Sami, & Mohmad Sidek 2017).

Discriminant validity occurs if there is a low correlation between constructs (Gefen 2005). In other words, it can be stated that discriminant validity describes the ability of each construct to differentiate itself from other constructs in a model (Hair et al. 2019). According to Fornell & Larcker (1981), discriminant validity can be determined by comparing the cross loading value and the Square Root of Average Variance Extracted (AVE) value. Ghozali (2014) states, if the root value of AVE is greater than the correlation value between constructs, it shows that the construct has discriminant validity.

2 METHOD

2.1 *Sample size*

Schumacker & Lomax (2010) argue that obtaining the right precision in confirmatory factor analysis requires 250 to 500 respondents, while Hoelter (1983) proposes 300 respondents. Based on these two suggestions, the researcher used 300 respondents who were drawn randomly from the various considerations above. In this study, 300 respondents were used randomly from 675 grade 5 elementary school students.

2.2 Data analysis

This research was analyzed using confirmatory factor analysis method. As stated by Thompson (2004) that confirmatory factor analysis (CFA) is one method of factor analysis. In general, factor analysis is defined as a statistical technique that can be used to reduce the many observed variables to a small number of latent variables by examining the covariance between the observed variables (Worthington & Whittaker 2014). Brown (2015) stated that through confirmatory factor analysis the validity of the scale construct can be determined, one of which is discriminant validity.

3 RESULT

It has been described above that discriminant validity refers to a construct's ability to differentiate itself from other constructs. Discriminant validity can be determined by comparing the correlation between constructs with the AVE square root of each construct. If the square root value is greater than the correlation value between constructs, it can be stated that the construct meets the criteria for discriminant validity (Hair et al. 2019). The results of the discriminant validity analysis are presented in Table 1.

Table 1. The results of the analysis of discriminant validity.

Construct	A	B	C	D
Interested in new things (A)	**0.710**			
Question something (B)	0.576	**0.813**		
The courage to try (C)	0.425	0.334	**0.783**	
Enthusiastic about solving problems (D)	0.563	0.524	0.336	**0.820**

Based on Table 1, it is known that all values of the square root of AVE are greater than the correlation values between constructs, so it can be stated that the four constructs have discriminant validity.

4 DISCUSSION

The purpose of this study was to assess the discriminant validity of the curiosity scale. Based on the literature review, it has been determined that the scale of curiosity consists of four constructs, namely being interested in new things, questioning something,

daring to try and enthusiastic to solve problems. After being analyzed, the four constructs were declared to have discriminant validity. These ideas are evidenced by the value of the square root AVE which is greater than the correlation value between constructs. These findings prove the theoretical assumption that in developing a scale overlapping between constructs should not occur (DeVellis 2017). Statistically, the differences between these constructs are evidenced by the low correlation between constructs (Hair et al. 2019). This correlation also proves that the constructs are mutually independent (Voorhees et al. 2015).

The low correlation between constructs as found in this study is also evidence of the fulfillment of an assumption that items in one construct should be highly correlated, while other items outside the construct have low correlation (Zaiþ & Bertea 2011). This finding is also evidence that the four constructs are unique, only measuring the constructs in question and not measuring other constructs (Hair et al. 2010). This means that the constructs are able to explain more variance in the observed latent variables.

5 CONCLUSION

Based on the results of the analysis and discussion above, it can be concluded that the four constructs that construct the curiosity scale have met the criteria for discriminant validity. This is as evidence that the items in the construct actually measure the constructs that should be measured and do not measure other constructs. Thus, the uniqueness of each construct as theorized is empirically proven.

REFERENCES

Ab Hamid, M. R., Sami, W., & Mohmad Sidek, M. H. 2017. Discriminant Validity Assessment: Use of Fornell & Larcker criterion versus HTMT Criterion. *Journal of Physics: Conference Series, 890:* 012163. doi:10.1088/1742-6596/890/1/012163.

Al Farani. 2013. Cita-Citaku: Buku Guru. Jakarta: Kementerian Pendidikan dan Kebudayaan.

Arnone, M. P., Grabowski, B. L., & Rynd, C. P. 1994. Curiosity as a personality variable influencing learning in a learner controlled lesson with and without advisement. Educational Technology Research and Development, 42(1): 5–20. doi:10.1007/bf02298167.

Benoit, D., Coolbear, J. & Crawford, A. 2008. *Encyclopedia of Infant and Early Childhood Development.* Toronto: Elsevier Inc.

Brown, T. A. 2015. *Confirmatory factor analysis for applied research,* 2nd Edition. New York: The Guildford Press.

DeVellis, R. F. 2017. *Scale development: Theory and applications,* 4th Edition. Thousand Oaks, CA: Sage.

Engel, S. 2015. *The hungry mind: The origins of curiosity in childhood.* London: Harvard University Press.

Fornell, C., & Larcker, D. F. 1981. Evaluating structural equation models with unobservable variables and measurement error. *Journal of Marketing Research*, 18: 39–50. DOI: 10.2307/3151312.

Gefen, D. 2005. A practical guide to factorial validity using PLS-Graph: Tutorial and annotated example. *Communications of the Association for Information Systems*, 16. doi:10.17705/1cais.01605.

Ghozali, I. 2013. *Aplikasi analisis multivariat dengan Program IBM SPSS 21 Update PLS Regresi*. Semarang: Badan Penerbit Universitas Diponegoro.

Gulten, D. C. et al. 2011. Investigating the relationship between curiosity level and computer self-efficacy beliefs of elementary teachers candidates. *The Turkish Online Journal of Educational Technology, 10*(4): 248–154.

Hair, J. F., Black, W. C., Babin, B. J., & Anderson, R. E. 2010. *Multivariate data analysis,* 7th Edition. Englewood Cliffs: Prentice Hall.

Hair, J.P., Black, J.P., Babin, J.P., & Anderson, R.E. 2019. Multivariate Data Analysis, Eighth Edition. Harlow: Cengage Learning.

Harlen, W. 2001. *Teaching, learning and assessing science*. London: Sage Publications Inc.

Hoelter, D.R. 1983. The analysis of covariance structures: Goodness-of-fit indices, sociological. *Methods and Research*, 11: 325–344. doi.org/10.1177/0049124183011003003

Jeraj, M., & Mariè, M. 2013. Relation between entrepreneurial curiosity and entrepreneurial self-efficacy: A multi-country empirical validation. *Organizacija*, 46(6). doi:10.2478/orga-2013-0027

Jones, J. B. 2013. *The creative imperative: School librarian and teachers cultivating curiosity together*. California: Libraries Unlimited.

Kang, M. J., Hsu, M., Krajbich, I. M., Loewenstein, G., McClure, S. M., Wang, J. T., & Camerer, C. F. 2009. The wick in the candle of learning. *Psychological Science*, 20(8): 963–973. doi:10.1111/j.1467-9280.2009.02402.x.

Kashdan, T. B., & Roberts, J. E. 2006. Affective outcomes in superficial and intimate interactions: Roles of social anxiety and curiosity. *Journal of Research in Personality*, 40(2): 140–167. doi:10.1016/j.jrp.2004.10.005.

Kidd, C., & Hayden, B. Y. 2015. The psychology and neuroscience of curiosity. *Neuron*, 88(3): 449–460. doi:10.1016/j.neuron.2015.09.010.

Lindholm, M. 2018. Promoting curiosity? *Science & Education*, 27(9–10): 987–1002. doi:10.1007/s11191-018-0015-7.

Manguel, A. 2015. *Curiosity*. London: Yale University Press.

Mustari, M. 2014. *Nilai karakter refleksi untuk pendidikan*. Jakarta: PT Raja Grafindo Persada.

Pisula, W. 2009. *Curiosity and information seeking in animal and human behavior*. USA: Brown Walker Press.

Prochniak, P. 2017. Development and testing of the elements of the nature curiosity scale. *Scientific Journal Publishers Limited*, 45(48): 125–1254. doi.org/10.2224/sbp.6130.

Reio Jr., T., Petrosko, J., Wiswell, A., & Thongsukmag, J. 2006. The measurement and conceptualization of curiosity. *The Journal of Genetic Psychology*, 167(2): 117–135. doi:10.3200/gntp.167.2.117–135.

Schumacker, R. E. & Lomax, R. G. 2010. *A beginner's guide to structural equation modeling*, 3rd Edition. New York: Taylor and Francis Group.

Shermis, M. D. & Di Vesta, F. J. 2011. *Classroom assessment in action*. New York: Rowman & Littlefield Publishers, Ink.

Thompson, B. (2004). *Exploratory and confirmatory: Understanding concepts and applications factor analysis*, 1st Edition. Washington: American Psychological Association.

Voorhees, C. M., Brady, M. K., Calantone, R., & Ramirez, E. 2015. Discriminant validity testing in marketing: an analysis, causes for concern, and proposed remedies. *Journal of the Academy of Marketing Science*, 44(1): 119–134. doi:10.1007/s11747-015-0455-4.

Worthington, R. L. & Whittaker, T. A. 2014. Scale development research: A content analysis and recommendations for best practices. *The Counseling Psychologist*, 34(6): 806–838. DOI: 10.1177/0011000006288127.

Zaiþ, A. & Bertea, P. E. 2011. Methods for testing discriminant validity. *Management & Marketing*, IX(2): 217–224.

Zuss, M. 2012. *The practice of theoretical curiosity*. New York: Springer.

Educational Innovation in Society 5.0 Era: Challenges and Opportunities – Purnomo & Herwin (Eds)
© 2021 the authors, ISBN 978-1-032-05392-9

Collaboration practices between educators in inclusive education before and during Covid-19

W. Hardiani & Hermanto
Graduate school, State University of Yogyakarta, Yogyakarta, Indonesia

ABSTRACT: Inclusive education is about collaboration. This paper reviews the collaboration practices in inclusive education before and during Coronavirus Disease (Covid-19). Education is expected to face any situation with collaboration practices between school professionals in inclusive education because that provides significant benefits for peers and all school professionals. During the Coronavirus outbreak, the government decided students should learn from home so that collaboration needs to be adjusted. Collaboration in inclusive education is working together on activities between the regular teacher and special education teacher where the contribution status by each of them is equal in order to discuss shared goals and problem solving to improve student performance. Collaboration practices expands the opportunity for all school professionals based on their field backgrounds and expertise which improves the successful potential of inclusive education before and during the pandemic.

1 INTRODUCTION

The Coronavirus (Covid-19) is a global challenge that causes significant problems in almost all sectors include education. Coronavirus affected nearly 1.6 billion learners in more than 160 countries (United Nations, 2020). In response to the status of the Coronavirus, Indonesia's government decided to implement a distance learning system. This forms a new habit of society so that quite a number of educational institutions have adjusted their learning and temporarily stopped face-to-face classes. Avgerinou & Moros (2020) state that this pandemic has set the new rhythm and changed the way education will look forever.

Education before the coronavirus was daunting due to finance challenges and the coronavirus made it worst. This this required that teachers collaborate and face these problems confidently so that education remained effective for students.

Without access to education, students are more vulnerable to losing their lives and futures. For the most vulnerable groups of children such as girls, children in refugee camps and immigrants, children with special needs and children who have experienced trauma, education is the savior of their lives (educationcannotwait.org 2020). Education is not only a protection, but also a hope for a brighter future for them.

Inclusive education is expected to be able to provide alternative solutions because no one can guarantee when this virus will end. Society needs a solution that still provides access to education for peers in this risky condition, to catch up and return them to their previous level of competence (educationcannotwait.org 2020).

Successfull inclusive environments have essential practices include assessing peers, performance-based assesment, use of visual tools and also collaboration (Budiarti & Sugito, 2018). The ability to collaborate is part of the basic skills that are suitable in diverse teams such as in inclusive education because it unites the visions of the participants involved in achieving common goals (Friend & Cook, 2003). To meet all children's learning needs in an inclusive setting, school professionals use a collaborative approach.

However, several studies have shown challenges in collaborative practices in inclusive schools before and during Covid-19. Prior to Covid-19, a general problem in collaborative practice was stated by Brown & Muschaweck (2004): that collaborative teamwork was often overlooked. Educators in inclusive schools feel dissatisfied with the school's efforts to create inclusive schools, their role is inadequate or there is no collaborative process. In addition, collaboration teams in inclusive schools are not formed naturally (mandatory), not based on the purpose of meeting student learning needs so that collaboration is limited to a formality and does not provide significant results. Another challenge that quite affects collaboration is the lack of self exploration such as (attitude, feelings and knowledge) which prevents professionals from defining their respective roles (Du Plessis, 2006).

During Coronavirus, collaboration in inclusive education is made more impossible due to the governments obligation to do everything from home and implement physical distancing. Coronavirus created new challenges in education, especially in the way teachers collaborate.

2 LITERATURE REVIEW

2.1 *Collaboration in inclusive education*

The successful implementation of inclusive education cannot be separated from its supporting components. Black-Hawkins (2014) in his paper states that there are 4 elements that characterize the inclusion: access, collaboration, achievement and diversity. Florian (2017) adds that the socio-psychological approach where there is multidimensional interaction between teachers, students, parents and representatives from various fields in the school is an important aspect of inclusive schools and is the "heart" of this education system.

Collaboration is a way for teachers to work together where the position of the teacher is equal and they are bound by the same goals and share problem solving. Cook & Friend (2010) states that collaboration is a kind of method used by professionals to achieve their common goals. The term collaboration is often used to describe shared activity. However, working together with other people in the same room does not guarantee collaboration occurs (Friend & Bursuck, 2012).

Conoley & Conoley (2010) stated that true collaboration is when all members work together and give their equal contribution value, there is a clear purpose in its implementation, they solve student problems together and they feel valued in that team. Collaboration team members set clearer goals, value other people's opinions of others and develop the dynamics needed for collaboration to occur (Swenson, 2000).

Collaboration is how professionals in inclusive schools interact with each other and work cooperatively to complete a task or series of tasks under certain conditions (Friend et al., 2010). Collegiality refers to a mutual respect and support between colleagues with a common goal. Collaboration provides significant benefits when each collaborator contributes, shares goals, achievements, works voluntarily, contributes equally and shares the same responsibility (Friend et al., 2010) and is based on the efforts of all members to achieve the same goals (Adams et al., 2016).

Collaboration practices in inclusive schools depend on the equality of its members (Suc et al., 2017), and the willingness to coordinate based on their competencies and field and to learn more than when they work individually or separately (Eccleston, 2010).

Bouillet (2013) states that the essence of inclusive education is a shared vision that results in change, transformation and improvement towards a better direction and becomes a guideline that benefits all students. Without the readiness of professionals to learn from each other, without recognizing each other's goals and roles, without joint planning and problem-solving activities, without taking joint responsibility for decisions taken from the results of discussions, collaboration in inclusive education is impossible (Mitchell, 2007).

2.2 *Collaboration practices in inclusive education before Covid-19*

It was quite difficult to find collaborative practices in inclusive schools between special educators and regular teachers before the Coronavirus in Indonesia. However, Budiarti and Sugito (2018) note that there are few schools in Yogyakarta practice collaboration and some of them implement it less.

Friend et al. (2010) notes that because the traditional education culture is still held by educators, they end up working separatedly so limit their chance to open opportunities and cooperate effectively which ultimately made several new problems occur. According to conversations with the researcher's friend (which is a regular teacher), they are even more confident working alone because feel free to teach without any feeling of assessment.

In some instances, other teachers are reluctant to give suggestions to each other for fear of disrupting their partner's teaching habits. In other cases, the classroom teachers emphasized that adjustments were not possible in their classroom, even though the special educator could be a teaching assistant for them. When the professionals in the class don't get along they become uncomfortable with each other to discuss problems directly and try to find solutions on their own (Friend & Bursuck, 2012).

Chomza (2017) suggests that teachers are expected to cooperate with shadowers to carry out special programs for children who are assisted so that the implementation of special programs can be more effective. Another critical condition is both the regular teacher and special educator came from different pedagogies and this make collaboration in inclusive class more challenging, as they look at education in different way. Different backgrounds need preparation program models such as Blanton & Pugach (2007) proposed, which includes discrete, integrated and merged. In a discrete model as an example, the education program is separated between general and special education. Teachers are provided only one special education course and then faculty collaborate to merge special and general pedagogy (Blanton & Pugach, 2007).

Effective collaboration needs to be built between regular teachers and special education teachers through communication (Friend & Cook, 2003). In one research that was conduct by Fullerton et al (2011) about Teacher Candidate Self Assesment (TCSC) on collaboration, it is found that the candidate teachers assessed collaboration and clear communication in instructional planning as very important. They also added the role of parents and communities who can be part of the inclusive school team.

However, based on resercher's who have been special educators in inclusive education, regular teachers and special educators never get opportunities to get along with each other or even accept the same program before they cooperate in the same class so that this makes communication harder for them both. Inclusive class was built by two different fields of teachers who struggle to work on behalf of students. In order to implement collaboration practices, readiness to learn from each other and an understanding the aims of education and teacher's roles is neccesary, as is sharing, planning and problem solving. This is the task of inclusive schools in Indonesia.

Emotional connection is also important to build, without being afraid to reveal each other's weaknesses and strengths, and to help each other without waiting someone else's to do it because through close collaboration, information will be disseminated, knowledge is shared and new skills will be built (Milteniene, 2012).

2.3 Collaboration practices in inclusive education during Covid-19

Coronavirus demands the ability of students to adapt to a new learning system with an unknown period of time. The learning system which is then shifted from traditional to online requires the availability of certain educational support tools such as the internet (Basilaia & Kvavadze, 2020) and awareness of the cultural, gender, socio-economic and geographical context is needed so that it does not worsen equality (Lancet Planet Health, 2020).

Huang et al. (2020) stated that this new education system is a big challenge for teachers who teach children with special needs. Especially for countries that have limited technology-such as Indonesia, they have not been able to maximize the implementation of online learning for all students (Sintema J, 2020).

Certain conditions such as differences in support from parents, the capacity of schools to serve online learning, differences in student's resilience in motivation and their ability to learn independently can exacerbate the situation (OECD, 2020). Comunityengaged and community-led are a key component in every strategy to addresses today and future challenges (UNESCO, 2020). This revealed that education during the3 Coronavirus needs the capacity of educator to collaboratively with resourcefulness.

If teacher carries out all duties alone, facing dozens of children who come from various abilities without collaborating, without meeting face to face and losing control of students directly, they would be overwhelmed and lonely. When teachers do their own assignments, it can trigger stress because of the teaching workload given to them. In addition, they have families at home who also needs attention.

This pandemic presents teachers with a brand new challenge including with facilities and the availability of infrastructure. This means teachers have a responsibility for ensuring that student learning continues as they themselves adapt to online or distance learning (consulting.learningforward.org, 2020). Teachers are the spearhead in providing learning during the Coronavirus and must be able to survive with all conditions (Rasmitadila et al., 2020).

Issues such as internet connection which doesn't reach rural areas, expensive internet packages (Rasmitadila et al., 2020), and the students who are not fully prepared for the new habit-learning from home are todays challenges and can make everything worse for the educational world in Indonesia.

The strategies that can be used is to focus on completing the current academic year and the second to provide alternative strategies for further learning if the Covid-19 vaccine has not been developed (OECD, 2020).

The key priority for education today is the welfare of students and the professional staff working in it. Maintaining effective social relationships between students, teachers and parents can enhance the acquisition of learning objectives and create strategies for maintaining that well-being. With an open curriculum that supports social, emotional well-being and is integrated with today's life so that it truly meets the specific needs of students during a pandemic (washingtonpost.com, July 2020).

Schools make efforts to support education during the pandemic by collaborating and communicating. Teachers can provide information about what parents can do and how to support their children during study from home. If possible, security arrangements are made, divisions of work between departments, mechanisms for teachers to continue to collectively connect and support each other, updates on the latest short and simple educational technology that can be found anywhere (John, 2020).

Teachers also can build partnerships between schools and higher education institutions or involve the government to facilitate a forum for developing adequate professionalism. Teachers' anxiety during a pandemic is how they can stay connected with students during a pandemic which can exacerbate equality problems (nytimes.com, April 2020).

Increasing the professionalism of teachers and providing access to online resources and platforms so they can collaborate and overcome the challenges during the Coronavirus and respond to the situation with relevant reactions is critical. In Indonesia, internet-based courses are offered in two ways. These are synchronously and asynchronously. Synchronous learning includes conferences and online chatting. Meanwhile asynchronous learning is indirect, using an independent learrning approach. Some subject matter is displayed on LMS on Moodle, blogs, onlne discussions, videos, articles and other platforms (Tarrman, 2020).

The school system needs to provide professional learning support for teachers, trainers, and leaders to ensure they are not working individually, but collaboratively. One of the websites that facilitates the needs of teachers during Covid-19 is Learning Forward. They provide virtual sessions mentored by experts and trainers, communities for teachers to share resources and examples of effective virtual learning practices, tools and strategies that can support online team learning and collaboration (consulting.learningforward.org, 2020).

During the Coronavirus, technology is a formidable thing that connects distant people through collaboration to implement sustainable education for all. IT Work Remotely & Teaching and Learning Services (mcgill.ca, 2020) states that one of the principles in technology that can be used in difficult situations like this is using simple technology.

The key to distance learning is the existence of multiple learning modes and using collaboration between professionals (teachers). Schools can support each other by developing sustainability plans and sharing information about these strategies.

Parents' involment is the most crucial thing when we talk about distance learning since they are the primary people supporting children with special needs at home. Schools needs parent involvement and must collaborate effectively with them. One research conduct by (Tindle et al., 2016) stated that some parents expressed concern at home during virtual school which was beyond their capacity and too demanding on their schedule. However, on the positive side parents presume that virtual school is easier and more flexible since they can do it in a variety of ways.

Coronavirus has taught parents how complex teachers' work is at school since they are the students' facilitators. In order to implement collaboration, parents think virtual school should have a communication system to stay in touch with teachers so that they can shared students progress (Tindle et al., 2016).

Online learning for students with disabilities has a few contribution factors including lack of experience in using technology, economic conditions, internet access and parental overload through housework (Daruka & Nagavci, 2020). However, there are several advantages when implementing virtual school during Coronavirus including creating new routines, improving their health, becoming more involved with family member activities and receiving more attention from them. These activities can increase parents commitment and support their child acquire new skills (Daruka & Nagavci, 2020) and also, make collaboration easier.

3 CONCLUSION

Collaborative practice in inclusive education is a complex system that requires the role of various available sources to make it successful. An educator who works alone will feel alone and directionless.

In difficult times such as the current Covid-19, we must ensure that education for children with special needs is guaranteed. The main priority for education during a pandemic is the welfare of teachers and students. This prosperity can be achieved by collaborating. Educators in inclusive school help each other to plan for the continuity of education during a pandemic by sharing information about effective distance education practices and adjusted to the diverse abilities of students such as the principles of inclusive education.

Certain support tools and platforms that provide space for professionals to share information and problem solving strategies in remote learning,

collaboration will be built more closely. Conventional media such as telephones allow parents to be able to stay connected with teacher so they still can communicate and share about students' problem during study from home.

In a difficult situation like this, where social interaction is limited and students are forced to be ready to accept conditions without prior preparation, collaborating is the key to its implementation. Educators in inclusive schools, in this case special educators and regular teachers must think about effective strategies so that inclusive education can continue to be relevant in all conditions. Covid-19 also teaches us that the best education system is not the most sophisticated one, but the education system that is most adaptable and can survive in all conditions. This education system, although easy, has a significant impact and is not affected by conditions.

REFERENCES

Adams, D., Harris, A., & Jones, M. S. 2016. Teacher-parent collaboration for an inclusive classroom: Success for every child. *MOJES: Malaysian Online Journal of Educational Sciences*, 4(3): 58–72.

Arcadis. 2020. *What COVID-19 can teach us about risk and uncertainty?* https://www.arcadis.com/en/global/arcadis-blog/joris-winters/what-covid-19-can-teach-us-about-risk-and-uncertainty/ (Accessed on 9 Augustus 2020)

Avgerinou, M. D., & Moros, S. E. 2020. The 5-phase process as a balancing act during times of disruption: transitioning to virtual teaching at an international JK-5 school. *Teaching, technology, an teacher education during the COVID-19 pandemic: Stories from the field. Waynesfield, NC, USA: Association for the Advancement of Computing in Education (AACE)*, Jun (15): 583–594.

Basilaia, G., & Kvavadze, D. 2020. Transition to online education in schools during a SARS-CoV-2 coronavirus (COVID-19) pandemic in Georgia. *Pedagogical Research*, 5(4): 1–9.

Black-Hawkins, K. 2014. Researching inclusive classroom practices: The framework for participation. *The SAGE handbook of special education*, 389–403.

Blanton, L. P., & Pugach, M.C. 2007. Collaborative programs in general and special teacher education. *Council of Chief State School Officers, Washington, DC.*

Bouillet, D. 2013. Some aspects of collaboration in inclusive education–teachers' experiences. *Center for Educational Policy Studies Journal*, 3(2): 93–117.

Budiarti, N. D., & Sugito, S. 2018. Implementation of Inclusive Education of Elementary Schools: a Case Study in Karangmojo Sub-District, Gunungkidul Regency. *Journal of Education and Learning*, 12(2): 214–223.

Chomza, N. 2017. Kolaborasi guru reguler dengan guru pendamping berkebutuhan khusus di sekolah inklusi kelas 1 SD Taman Muda Yogyakarta 1. Universitas Negeri Yogyakarta.

Conoley, J. C., & Conoley, C. W. 2010. *Why does collaboration work? Linking positive psychology and collaboration.* University of California Santa Barbara.

Daniel, S. J. 2020. Education and the COVID-19 pandemic. *Prospects*, 1–6.

Daruka, Z. H., & Nagavci, N. 2020. The impact of the COVID-19 pandemic on the education of children with disabilities. *StatCan COVID-19: Data to Insights for a Better Canada*, 45280001.

Consulting.learningforward.org. 2020. *Establish school-based communities of learners.* https://consulting.learning forward .org / consulting-services / professional- learning-communities/. (Accessed on 14 Agustus 2020).

Consulting.learningforward.org. 2020. *Ensure equity and excellence in teaching and learning…virtually!.* https://consulting. learningforward.org/virtual-learning/?_ga=2.242022663.403991543.1597270858-1218649352.1597270858 (Accessed on 14 Agustus 2020).

Du Plessis, S. 2006. Collaborative partnerships in an inclusive approach to the acquisition of english as language of learning and teaching.

Educationcannotwait. 2020. *Covid-19 and Education in Emergencies.* https://www.educationcannotwait.org/covid-19/ (accessed on 12 Agustus 2020).

Eccleston, St. T. 2010. Successful collaboration: Four essential traits of effective special education specialists. *The Journal of International Association of Special Education*, 11(1), 40–47.

Florian, L. 2017. The heart of inclusive education is collaboration. 12(2): 248–253.

Friend, M., & Bursuck, W. D. 2012. Including students with special needs: A practical guide for classroom teachers (sixth). *Library of Congress Cataloging in-Publication Data.* www.pearsonlightened.com

Friend, M., Cook, L., Hurley-Chamberlain, D., & Shamberger, C. 2010. Co-teaching: An illustration of the complexity of collaboration in special education. *Journal of Educational and Psychological Consultation*, 20(1): 9–27.

Friend, M., & Cook, L. 2003. *Collaboration skills for school professionals (4th ed).* Pearson Education.

Fullerton, A., Ruben, B. J., Mcbride, S., & Bert, S. 2011. Evaluation of a merged secondary and special education program. *Teacher Education Quarrterly*, *spring*, 45–60.

Huang, R., Tlili, A., Yang, J., & Chang, T. 2020. Handbook on facilitating flexible learning during educational disruption, *The Chinese Experience in Maintaining Undisrupted Learning in COVID-19 Outbreak. March.*

Lancet Planet Health. 2020. COVID-19 as a global challenge?: towards an inclusive and sustainable future. 5196(20), 19–21. https://doi.org/10.1016/S2542-5196(20)30168-6

Lawrence-Brown, D., & Muschaweck, K. S. 2004. Getting started with collaborative teamwork for inclusion. In *Journal of Catholic Education*, 8(2).

Milteniene, L. 2012. Teacher collaboration in the context of inclusive education. February 2017.

Mitchell, D. 2007. *What really works in special and inclusive education.* London and New York: Routledge

Reimers, F., Education, G., & Initiative, I. 2020. Supporting the continuation of teaching and learning during the COVID-19 Pandemic.

Sintema, E. J. 2020. Effect of COVID-19 on the performance of grade 12 students: Implications for STEM education. *Eurasia Journal of Mathematics, Science and Technology Education*, 16(7): 1851.

Suc, L., Bukovec, B., & Karpljuk, D. 2017. The role of inter-professional collaboration in developing inclusive education?: experiences of teachers and occupational therapists in slovenia. *International Journal of Inclusive Education*, 21(9): 938–955.

Swenson, N. C. 2000. Comparing traditional and collaborative settings for language intervention.

Tarman, B. 2020. Editorial: Reflecting in the shade of pandemic. *Research in Social Sciences and Technology*, 5(2): 1–4.

Tindle, K., Mellard, D., & East, B. 2016. *Online learning for student with disabilities: Recommendations for parent engagement.* Lawrrence, KS. Center on Online Learning and Students with Disabilities, Univerrsity of Kansas. *December*, 1–8.

UNESCO. 2020. Education in a post-COVID world?: Nine ideas for public action International Commission on the Futures of Education.

Educational Innovation in Society 5.0 Era: Challenges and Opportunities – Purnomo & Herwin (Eds)
© 2021 the authors, ISBN 978-1-032-05392-9

Investigating students' self-regulated learning and academic procrastination on primary school during distance learning

T. Nugraha, & S. Prabawanto
Universitas Pendidikan Indonesia, Bandung, Indonesia

ABSTRACT: A review of self-regulated learning (SRL) as a determinant of learning success and academic procrastination (AP) as a psychological barrier which is predicted to occur when autonomous learning needs to be done. The correlational survey research was conducted on 67 parents of primary school students which aimed to explore their perspectives on these attitudes that determine the success of distance learning. This study focused on the investigation of SRL & AP that appears in primary school mathematics learning. The study obtained the correlation between SRL and AP students with a statistical power of -0.634 with a significance level of α 0.05. This relationship indicates a significant negative influence between SRL and AP during distance learning. Empowerment of SRL during this pandemic can be used as the main capital in post pandemic learning and psychological learning barriers (AP) need to be minimized, so that learning objectives can be achieved efficiently.

1 INTRODUCTION

Education is faced with significant challenges in dealing with the COVID-19 pandemic. Almost all schools and colleges were closed during COVID-19. The closures occured because the school cannot carry out normal teaching and learning activities (Kong, 2020), so that all educators and school levels including primary education throughout the world are forced to adapt to challenging things (Berry & Kitchen, 2020). The challenge in question is to demand the acceleration of digital transformation in teaching and learning through the adoption of online learning (Rospigliosi, 2020).

Every education unit including primary school is demanded to be able to implement digital literacy learning. This is because the COVID-19 pandemic makes education an emergency issue which must rely on educational technology as a front-line emergency service (Williamson et al., 2020). It is even predicted that in some time after the pandemic, universities and schools might continue to implement online learning (Verma et al., 2020). Thus, there is a transformation of education from direct learning in the classroom to distance learning that takes place at home.

Distance learning as a student-centered method certainly involves independent learning that greatly contributes to the pandemic as a process of documenting experiences and insights stemming from radical change (Berry & Kitchen, 2020; Zhao et al., 2014). During online learning, students must set their own situations and conditions in accessing the learning management system which includes activating

learning, sustaining learning, and structuring learning (Montgomery et al., 2019). The character of learning independence is usually referred as Self-Regulated Learning (SRL), which the effectiveness of digital learning is highly dependent on the ability of students' SRL and the ICT capabilities of the SRL (Yen et al., 2018).

SRL is one of the things which has implications in cognitive and educational interventions where students are aware of their cognitive and knowledge processes and monitor or regulate their learning (Rutherford et al., 2018; Yen et al., 2018). In other words, SRL focuses on how students activate, change and maintain their learning practices selectively in a varied home and school context both in the realm of structure and content (Zimmerman & Martinez-Pons, 1988). Therefore, SRL is a learning attitude in which students are independent in setting learning goals, monitoring performance in achieving these goals, and also adjusting the needs of learning resources both knowledge and behavior to achieve these goals.

SRL operates through a number of processes including self-monitoring, standard setting, evaluative judgment, self-appraisal, and affective self reaction (Bandura, 1991; Zimmerman & Bandura, 1994; Zimmerman & Pons, 1986, 1988). The process is described as a cycle that involves feedback and adjustments (Zimmerman, 2000). Based on that, the aspect of SRL which plays a major role is the capability to mobilize, direct and maintain self-learning efforts (Zimmerman & Bandura, 1994), so students have the opportunity to make self-orientation to improve the quality of their learning (Zimmerman & Kitsantas, 2014). The SRL

DOI 10.1201/9781003206019-14

cycle is classified into three phases including planning, self-regulation and self-reflection (Schunk & Zimmerman, 2007; Zimmerman, 2000). The phase operates partly through internal self-standards and evaluative reactions to learning (Zimmerman & Bandura, 1994). Therefore, SRL involves three important components in learning, namely personal, behavioral, and environmental.

Distance learning by involving independent learning controls has opportunities and challenges to self-regulation deficits or academic task delays called academic procrastination (AP) (Ljubin-Golub et al., 2019; Romano et al., 2005; Wilkinson & Sherman, 1990). This is because learning done at home without supervision and direct assistance from the teacher will have the potential to lack concentration. In addition, learning without classroom management runs the risk of having distractions such as television, games and so on. Therefore, AP's attitude needs to be prevented because the procrastinating is the beginning of bad habits (Day et al., 2000). Various AP identification studies have been conducted on secondary and higher education students, but there is still no review of primary school students' AP, especially in the era of distance learning, even though bad habits like this need to be prevented as early as possible.

SRL has been identified as important for the achievement of academic achievement during online learning (Montgomery et al., 2019). Moreover, SRL is an important thing in obtaining mathematics competence in primary school (Mägi et al., 2016). However, this still leaves the question of correlation between SRL and AP, which certainly can determine academic achievement. Therefore, this study aims to identify correlation studies between self-regulated learning with academic procrastination in a sample of primary school students and investigate the contribution of distance learning to each of the indicators of learning attitudes. The aim is to improve the perspective of SRL endeavoring and the emphasis on students' AP in distance learning which is considered very important for future studies.

2 METHODS

This study is a non-experimental study, a correlational design, in which researchers have a relationship between SRL and AP students during distance learning due to the COVID-19 pandemic. Correlational research design predicts the degree of relationship between variables through statistical tests of relationships and explains the meaning of the relationships between these variables (Creswell, 2012). This study involved 67 respondents who were parents/guardians of primary school students in West Java, Indonesia. Explanation and approval are included at the beginning of filling out the questionnaire to minimize and secure a response rate high enough to provide credibility and reliability of data which are considered to be the main difficulties in web-based correlational survey research (Cohen et al., 2007; Creswell, 2012).

Data collection adapts research instruments developed in previous studies. The SRL students' instrument was adapted from a scale developed by Martinez-Lopez et al. (2017). Validity evidence was provided through a content validation index (CVI) and kappa coefficients. The purpose of this scale is to measure the level of students' SRL based on six SRL indicators, namely goal setting, environmental structuring, task strategies, time management, help seeking, and self-evaluation. As for measuring the student's AP used a questionnaire adapted from the Bashir & Gupta (2019) and Tuckman (1991) scale. The scale measures four AP indicators including time management, task aversiveness, sincerity, and personal initiative. The questionnaire was slightly modified to fit the context of distance learning implemented in Indonesian elementary schools. The data analysis test used is the Pearson Product Moment Product Parametric correlation test because the data is normally distributed and homogeneous.

3 RESULT AND DISCUSSION

3.1 Result

Our identification begins with the acquisition of variable indicator analysis findings from the results of a survey conducted. The study found that the percentage score of each indicator of self-regulated learning of primary school students during the implementation of distance learning is in the interval 61–70%. On the other hand, we found that the aspects of academic procrastination student learning is still quite high with a percentage of 48–56%. The findings are more clearly represented in Figure 1.

Figure 1. Percentage of SRL and AP indicators.

The single SRL and AP data that we obtained through the processing of instruments were arranged in a scatter plot. The resulting scatter plot can test the linearity and examination of the significant SRL and AP students' data distribution. We determine that SRL and AP data are linear on the basis that the data distribution forms a sloping line from left to right (negative linear). Scatter plot is also useful to see the bivariate normal distribution based on the oval lines (cigar shaped). Figure 2 shows that the majority of data scores are cigar shaped and only a few are outside the line which are assumed to have no significant impact on the overall significance of the correlation. The scatter plot review shown in Figure 2 with this almost linear trend opens our study to further analyze the correlation of the two variables which are seen as two important things in distance learning.

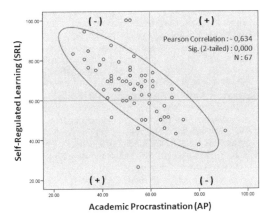

Figure 2. Scatter plot of students' SRL & AP.

Inferential statistical correlation test was performed to determine the degree of relationship and direction of the relationship as well as the significance between the two attitude variables studied namely SRL and AP. Determination of the type of correlation test is based on the results of the normality and homogeneity test of the data. We found that SRL and AP had fulfilled the data normality criteria performed by the Shapiro-Wilk test with p-values of 0.516 (SRL) and 0.464 (AP) that exceeded the significance level of α (0.05). We also found that the data met the assumption of homogeneity through the Levene test. As a result of the assumption of normality and homogeneity of the data, we decided to measure the strength and naturalness of the correlation between SRL and AP surveys through the Pearson Product Moment Parametric test. A Pearson test, obtained the degree of correlation of 0.634 with a negative relationship direction. The correlation coefficient obtained can be interpreted as a fairly high relationship. We have simplified this data through Table 1 below.

Table 1. Analysis correlation between students' SRL & AP survey.

Learning Aspect	Normality	Homogeneity	Correlation
Self-Regulated Learning	0.517	0.172	−0,634
Academic Procrastination	0.464		
Interpretation	Normal	Homogenous	High Enough

3.2 Discussion

Incidental learning which has not been predicted beforehand makes a concern how students can achieve learning goals in this emergency. Moreover, primary school students are students who are still in the concrete operational stage in which they still need the context of real learning practices, especially in developing mathematical concepts. Therefore, autonomous learning needs to identify how the level of self-regulated learning that occurs during a pandemic has learning potential. Understanding SRL's initial conditions is needed as a starting point in supporting student academic achievement, in line with a study conducted by Gunzenhauser & Saalbach (2020) which focuses mathematics learning in primary education because SRL and academic performance seem very strong in the case of mathematics. However, a better understanding of the aspects of SRL functions in students and their relationship to academic skills is important in order to consider other possible interventions or predictors (Mägi et al., 2016).

The study found that in the early stages of distance learning, the average achievement of SRL indicators for primary school students exceeded 50%. This indicates that primary education already has a high enough potential in learning independence. Nevertheless, this potential must be re-optimized, especially some indicators that are still not optimal, such as help seeking and self-evaluation. The challenge faced by students in distance learning is the amount of information with varying quality, so it is necessary to concentrate on goals and evaluate the achievement of learning objectives (Yen et al., 2018). This also indicates that distance learning still does not provide space for students in the development of social aspects (Grau & Whitebread, 2012), so the design or context provided by teachers and parents need to be considered because it can affect SRL (Zimmerman & Bandura, 1994). When teachers cannot design assignments, choice strategies, and time management, they will limit students' SRL (Boekaerts, 1997). The discussion and guidance room needs to be optimized for both teachers and parents, because group work and collaborative guidance have been recognized as important in developing metacognition and SRL (Grau & Whitebread, 2012).

The correlation coefficient that we found has a negative relationship with the degree of relationship in the category being quite high, this means that students having high learning independence tend to have low academic postponement attitudes and vice-versa. This study adds a wider AP space where in previous studies AP was known to be negatively correlated with academic achievement (Kim & Seo, 2015). The opportunity to grow AP seeds needs to be tackled because AP since the primary school age can have an impact on poor academic performance, late assignments, anxiety, difficulty following learning, depression and a decline in self-esteem in the future (Shih, 2017). Given that one of the causes of AP is the rejection of the task (Day et al., 2000), at the age of primary school positive parenting needs to be raised to minimize the development of AP (Batool, 2020). The limitation in this study is the lack of analysis of the causes of AP, so we need a deeper review of the predictors of the high AP score. It is hoped that technological advancements in the era of digital learning that accelerate the needs of SRL (Yen et al., 2018) will be increasingly empowered after the pandemic.

4 CONCLUSION

The success of distance learning can be caused by the attitude of learning independence in students. But in reality these attitudes still have challenges in the form of learning delays. Therefore, learning designers must be able to strive for the development and optimization of students' self-regulated learning during learning and reduce the potential for learning barriers in the form of academic procrastination. The potential of SRL students during distance learning which is done incidentally is already considered quite high, especially the attitude of SRL can be viewed as the main learning modality of lifelong education that needs to be developed from primary school. This limited correlational study needs to be developed or improved in the future through various methods such as the identification of other variables that affect the significance of correlation or the existence of development and research efforts on online learning design that can optimize SRL and minimize student AP.

ACKNOWLEDGEMENT

This research was supported by the Educational Fund Management Institute (LPDP) through the Indonesian Education Scholarship. We also thank the parents, teachers, and primary school principals who participated in this study.

REFERENCES

Bandura, A. 1991. Social cognitive theory of self-regulation. *Organizational Behavior and Human Decision Processes*, *50*(2): 248–287. https://doi.org/10.1016/0749-5978(91)90022-L

Bashir, L., & Gupta, S. 2019. Measuring academic procrastination: Scale development and validation. *Elementary Education Online*, *18*(2): 939–950. https://doi.org/10.17051/ilkonline.2019.562076

Batool, S. S. 2020. Academic achievement: Interplay of positive parenting, self-esteem, and academic procrastination. *Australian Journal of Psychology*, *72*(2): 174–187. https://doi.org/10.1111/ajpy.12280

Berry, A., & Kitchen, J. 2020. The Role of Self-study in Times of Radical Change. *Studying Teacher Education*, 1–4. https://doi.org/10.1080/17425964.2020.1777763

Boekaerts, M. 1997. Self-regulated learning: A new concept embraced by researchers, policy makers, educators, teachers, and students. *Learning and Instruction*, *7*(2): 161–186. https://doi.org/10.1016/S0959-4752(96)00015-1

Cohen, L., Manion, L., & Morrison, K. 2007. *Research methods in education* (Sixth). Routledge.

Creswell, J. W. 2012. *Educational Research: Planning, Conducting and Evaluating Quantitative and Qualitative Research* (4th ed.). Pearson.

Day, V., Mensink, D., & O'Sullivan, M. 2000. Patterns of academic procrastination. *Journal of College Reading and Learning*, *30*(2): 120–134. https://doi.org/10.1080/10790195.2000.10850090

Grau, V., & Whitebread, D. 2012. Self and social regulation of learning during collaborative activities in the classroom: The interplay of individual and group cognition. *Learning and Instruction*, *22*(6): 401–412. https://doi.org/10.1016/j.learninstruc.2012.03.003

Gunzenhauser, C., & Saalbach, H. 2020. Domain-specific self-regulation contributes to concurrent but not later mathematics performance in elementary students. *Learning and Individual Differences*, *78*(February): 101845. https://doi.org/10.1016/j.lindif.2020.101845

Kim, K. R., & Seo, E. H. 2015. The relationship between procrastination and academic performance: A meta-analysis. *Personality and Individual Differences*, *82*: 26–33. https://doi.org/10.1016/j.paid.2015.02.038

Kong, Q. 2020. Practical Exploration of Home Study Guidance for Students during the COVID-19 Pandemic: A Case Study of Hangzhou Liuxia Elementary School in Zhejiang Province, China. *Science Insights Education Frontiers*, *5*(2): 557–561. https://doi.org/10.15354/sief.20.rp026

Ljubin-Golub, T., Petričević, E., & Rovan, D. 2019. The role of personality in motivational regulation and academic procrastination. *Educational Psychology*, *39*(4): 550–568. https://doi.org/10.1080/01443410.2018.1537479

Mägi, K., Männamaa, M., & Kikas, E. 2016. Profiles of self-regulation in elementary grades: Relations to math and reading skills. *Learning and Individual Differences*, *51*: 37–48. https://doi.org/10.1016/j.lindif.2016.08.028

Martinez-Lopez, R., Yot, C., Tuovila, I., & Perera-Rodríguez, V. H. 2017. Online Self-Regulated Learning Questionnaire in a Russian MOOC. *Computers in Human Behavior*, *75*: 966–974. https://doi.org/10.1016/j.chb.2017.06.015

Montgomery, A. P., Mousavi, A., Carbonaro, M., Hayward, D. V., & Dunn, W. 2019. Using learning analytics to explore self-regulated learning in flipped blended learning music teacher education. *British Journal of Educational Technology*, *50*(1): 114–127. https://doi.org/10.1111/bjet.12590

Romano, J., Wallace, T. L., Helmick, I. J., Carey, L. M., & Adkins, L. 2005. Study procrastination, achievement, and academic motivation in web-based and blended distance learning. *Internet and Higher Education*, *8*(4 SPEC. ISS.): 299–305. https://doi.org/10.1016/j.iheduc.2005.09.003

Rospigliosi, P.' asher.' 2020. How the coronavirus pandemic may be the discontinuity which makes the difference in the digital transformation of teaching and learning. *Interactive Learning Environments*, *28*(4): 383–384. https://doi.org/10.1080/10494820.2020.1766753

Rutherford, T., Buschkuehl, M., Jaeggi, S. M., & Farkas, G. 2018. Links between achievement, executive functions, and self-regulated learning. *Applied Cognitive Psychology*, *32*(6): 763–774. https://doi.org/10.1002/acp.3462

Schunk, D. H., & Zimmerman, B. J. 2007. Influencing children's self-Efficacy and self-regulation of reading and writing through modeling. *Reading and Writing Quarterly*, *23*(1): 7–25. https://doi.org/10.1080/10573560600837578

Shih, S. S. 2017. Factors related to Taiwanese adolescents' academic procrastination, time management, and perfectionism. *Journal of Educational Research*, *110*(4): 415–424. https://doi.org/10.1080/00220671.2015.1108278

Tuckman, B. W. 1991. The development and concurrent validity of the procrastination scale. *Educational and Psychological Measurement*, *51*(2): 473–480. https://doi.org/10.1177/0013164491512022

Verma, G., Campbell, T., Melville, W., & Park, B. Y. 2020. Science Teacher Education in the Times of the COVID-19 Pandemic. *Journal of Science Teacher Education*, *00*(00): 1–8. https://doi.org/10.1080/1046560X.2020.1771514

Wilkinson, T. W., & Sherman, T. M. 1990. Perceptions and Actions of Distance Educators on Academic Procrastination. *American Journal of Distance Education*, *4*(3): 47–56. https://doi.org/10.1080/08923649009526716

Williamson, B., Eynon, R., & Potter, J. 2020. Pandemic politics, pedagogies and practices: digital technologies and distance education during the coronavirus emergency. *Learning, Media and Technology*, *45*(2): 107–114. https://doi.org/10.1080/17439884.2020.1761641

Yen, M. H., Chen, S., Wang, C. Y., Chen, H. L., Hsu, Y. S., & Liu, T. C. 2018. A framework for self-regulated digital learning (SRDL). *Journal of Computer Assisted Learning*, *34*(5): 580–589. https://doi.org/10.1111/jcal.12264

Zhao, H., Chen, L., & Panda, S. 2014. Self-regulated learning ability of Chinese distance learners. *British Journal of Educational Technology*, *45*(5): 941–958. https://doi.org/10.1111/bjet.12118

Zimmerman, B. J. 2000. Chapter 2: Attaining self-regulation A social cognitive perspective. *Handbook of Self-Regulation*, 13–39.

Zimmerman, B. J., & Bandura, A. 1994. Impact of Self-Regulatory Influences on Writing Course Attainment. *American Educational Research Journal*, *31*(4): 845–862. https://doi.org/10.3102/00028312031004845

Zimmerman, B. J., & Kitsantas, A. 2014. Comparing students' self-discipline and self-regulation measures and their prediction of academic achievement. *Contemporary Educational Psychology*, *39*(2): 145–155. https://doi.org/10.1016/j.cedpsych.2014.03.004

Zimmerman, B. J., & Martinez-Pons, M. 1988. Construct validation of a strategy model of student self-regulated learning. *Journal of Educational Psychology*, *80*(3): 284–290. https://doi.org/10.1037//0022-0663.80.3.284

Zimmerman, B. J., & Pons, M. M. 1986. Development of a Structured Interview for Assessing Student Use of Self-Regulated Learning Strategies. *American Educational Research Journal*, *23*(4): 614–628. https://doi.org/10.3102/00028312023004614

Educational Innovation in Society 5.0 Era: Challenges and
Opportunities – Purnomo & Herwin (Eds)
© 2021 the authors, ISBN 978-1-032-05392-9

Exploring e-learning platforms used by students in Indonesia during the Covid-19 pandemic

S. Rahayu & Supardi

Graduate School of Yogyakarta State University, Yogyakarta, Special Region of Yogyakarta, Indonesia

ABSTRACT: Covid-19 is a disease that has been a big problem for the development of a country, especially in the education field. One of the most worrisome things is a decline in the quality of human resources, so that some efforts were made to overcome it, one of which is learning through e-learning. E-learning provides effectiveness and efficiency in learning because it provides a new learning atmosphere that is not limited by time and place. This research aims to theoretically explore the implementation of e-learning in Indonesia during Covid-19 according to the literature review. Students from elementary school to higher education use many platforms, such as WhatsApp group, Zoom, Schoology, and Edmodo. However, the most widely used platform is Google Classroom. The challenge of implementing e-learning is on the internet connection as the main source and educators' role in providing materials.

1 INTRODUCTION

E-learning has been implemented in several schools in Indonesia, but the implementation is still in limited development. The limited development is because there are still many basic problems in educational institutions that have not been optimized, such as the lack of school buildings and educators. This condition can be found in several places in Indonesia, one of which is in Jember, specifically in SDN Tegalwaru 4 (Tegalwaru 4 Elementary School). There are only three classrooms for grades 1 to 6, and the students are only taught by two teachers (Mulyono, 2019). Meanwhile, the application of e-learning requires maximum preparation. According to Teddy & Swatman (2006), there are several kinds of readiness for e-learning, namely 1) student readiness, 2) teacher readiness, 3) infrastructure, 4) management support, 5) school culture, and 6) face-to-face learning tendencies.

The portrait of the education gap in Indonesia is obvious where private and public schools show the quality of education services and students in the schools reflecting their economic conditions. This condition will not be able to equalize the results of using e-learning. Based on research conducted by Shofiyah (2016), it was found that there is no positive effect of using e-learning on social science learning outcomes as indicated by a value (0.961 <2.01) with a significance value of 0.342.

The results portrayed are due to the application of e-learning that is still simple, the limited availability of social science materials, teacher-centered learning, and other influencing factors such as physical, psychological, family, etc. Another finding was

conveyed by Muhdi and Nurkolis (2020) in their research, which says that e-learning in PAUD (Preschool) has been effective, but e-learning policies are still less effective due to three obstacles, including pedagogy, technology, and economics.

The current development of e-learning becomes a new topic due to the Covid-19 outbreak. The application of social distancing requires students to study at home. This is confirmed by the circular letter of the Minister of Education and Culture (Mendikbud) Number 4 of 2020 that the learning process from home is carried out with the following provisions: a) learning from home through online learning, and it is implemented in order to provide meaningful learning experiences for students, b) learning from home can be focused on life skills education including regarding the Covid-19 pandemic c) learning activities and tasks from home can vary among students and d) evidence of learning activities from home is by giving qualitative feedback (Mendikbud, 2020).

Online learning apparently cannot be done in all schools due to educators' limitations in using technology. This condition requires educators to come to students individually or in groups since the condition needs to be solved by educators for the learning process that must not stop halfway. According to Putra (2020) in Kompas daily news, he states that learning had been made using Google Classroom and Whatsapp, but not all educators mastered the online system. It is a fact that cannot be denied that there are still many educators who need skills and insights about technology for the education quality improvement. According to Supangat et al. (2018), the first step that can be taken to improve the quality of education is to increase

74

DOI 10.1201/9781003206019-15

educators' skills in using information technology for student learning.

Barriers to education during the Covid-19 pandemic in Indonesia include economic factors. Most people in the village prioritize their daily needs rather than providing educational facilities such as smartphones, wifi, internet, etc. This factor must be overcome by educators so that learning can continue. According to the Kompas daily news written by Taufiqurrahman (2020), a teacher at SDN Batu Putih (Batu Putih Elementary School), Sumenep Regency, comes to his student's house one by one after several weeks of unsuccessful online learning because many students do not have smartphones. The factors of educators' ability and students' readiness actually influence each other in online learning continuity.

Educators will continue to face various kinds of problems and obstacles in learning for their students' future. However, they certainly need support from the government, the community, and the parents to mutually create good cooperation and lead to successful learning.

2 METHOD

The method used in this study is a literature review to find out what e-learning platforms are used by educators during the current Covid-19 pandemic. In this literature review, the authors use various written sources such as articles, journals, and documents relevant to the issues studied in this study.

The data obtained are then collected, compiled, reviewed, analyzed, and concluded to obtain recommendations regarding literature studies.

3 FINDINGS AND DISCUSSION

3.1 *Online learning (e-learning)*

The existence of the learning transformation makes conventional learning decline in function. This condition is due to the existence of e-learning as a new learning model. E-learning influences the process of change or transformation of conventional education into digital form, both in terms of content and systems (Agustina, 2013). Silahudin (2015) suggests that e-learning is a learning approach mediated by a computer connected to the internet, where students seek to obtain learning materials that can meet their needs. In line with Silahudin, Priansa (2017: 147) adds that e-learning is a type of teaching and learning activity that allows the delivery of learning materials to students by using internet media. Thus, it can be concluded that e-learning is a learning tool whose learning system utilizes internet media to obtain learning materials.

The response of students about online learning can be said to be effective. This situation is due to the high motivation of educators supporting the learning and adequate learning facilities. Mustakim (2020), in his study, asserts that online learning by using online media was rated and considered to be very effective (23.3%), effective (46.7%), normal (20%), ineffective (10%), and no one considered it as very ineffective. The same issue was conveyed by Saifuddin (2017) that shows 98.8% of students know e-learning, 86.3% supported the implementation of e-learning, and 77% said that they are satisfied with the implementation of e-learning, then there was also found that students' perceptions on e-learning are positive. They find it is useful, increases motivation, makes the materials easier to understand and improves their readiness for the lesson. The participants' perceptions or assumptions can prove the description of the learning effectiveness as evidence of their learning experiences.

The principle of e-learning is reflected in learning, which is characterized by prioritizing the role of the internet and technology. According to Kusuma (2011), e-learning is learning that's implementation is supported by technology services such as the telephone, audio, videotapes, and computers. In Indonesia, the internet has become a basic necessity to access various social media activities, buying and selling, and education. An example is shown in a study conducted by Mujib (2012), a Yogyakarta portrait as a student city. He found that most respondents who get unsatisfactory learning outcomes rarely use the internet as a learning medium, and vice versa. If the use of internet itself is wise, it will actually have a positive impact.

The experience of learning online will be felt differently by students. The difference is due to a new habitual or way that must be lived out of the ordinary, how students are trained to learn independently, patiently, thoroughly, and persistently in understanding the material on their own. Ratnawati, et al. (2019) stated that the benefits of e-learning for teachers are as a supplement, complement, and to complete the conventional learning. Moreover, e-learning trains the teachers to provide a healthy learning environment and train students' independence, and e-learning itself can be a source of learning information.

The impact of e-learning in the teaching and learning process is an increase in student motivation. Based on the findings in Suwastika's study (2018), the application of e-learning that is long and stable at STIKOM Bali is one factor that improves student motivation. High motivation felt by students due to the easy accessibility. E-learning is a learning medium that does not recognize distance and time, so that the level of learning flexibility will be high. According to Susilawati & Irfandi (2019), the application of e-learning at the Web-based SMK Negeri 1 Bandar Lampung (Bandar Lampung 1 Vocation School) can facilitate student access in getting materials without having to be constrained by distance and time. The perceived convenience will give an effect that has a positive influence on building motivation.

The use of e-learning is not only limited to the school environment, but it also has been widely applied in universities. It is assumed that higher education is a place of education that already requires students to be more independent in learning, so they are expected to learn without the need for educators. At present, many universities are implementing e-learning because students are the agent of change who have the greatest influence. It is also supported by the large tuition fees so that it is comparable to the facilities obtained. Based on the research conducted by Firman & Rahman (2020), it was found that: university students have the basic facilities needed to take part in online learning, online learning has flexibility in its implementation and can encourage independence and motivation to learn, distance learning will be able to be more effective in implementing social distancing.

3.2 *E-learning platforms used during the Covid-19 pandemic*

Educators do various kinds of ways to continue the learning process during the pandemic. It is proven by educators' efforts to prepare online learning or what is commonly known as e-learning. Furthermore, during pandemic conditions, maintaining distance is very important to keep the coronavirus transmission away. E-learning has the goal to make a greatly effective distance learning. Suharyanto & Mailangkay (2016) states that the more intensively e-learning is used, the more quality of student learning will also increase, indirectly improving learning achievement. The purpose of implementing e-learning is to expand access to public education in improving the quality of learning.

Learning in a pandemic situation like this does require the teacher's role in guiding the learning process and parents' role at home for the fluency of their children in understanding the material and completing the assignments. Both teachers and parents have significant roles in succeeding in the learning process. According to Dewi (2020), the impact of Covid-19 on the implementation of online learning in elementary schools can be carried out well if there is a collaboration among teachers, students, and parents in learning at home.

The use of e-learning implemented in various levels of education, including Elementary School, Junior High School, Senior High School, and Higher Education, proves educators' concern is to maintain the quality of human resources. Many online learning applications can be used by educators today. Pratiwi (2020) states that online learning at Satya Wacana Christian University has been effective by using Zoom, Google Classroom, Schoology, and Edmodo applications, but there are obstacles found regarding the internet connection, which is less supportive. Astini (2020) adds that it is more effective to use the Google Classroom Zoom application at the elementary school level, and it is different for the lower classes that the most effective application for them is WhatsApp group.

Based on the literature review, there are several e-learning platforms used by educators. One of the applications used by most educators is Google Classroom. Thus, the first discussion will be related to Google Classroom. Google Classroom is an online learning application that is often used by educators during the pandemic. It is easy to apply, and it also has provided various tabs regarding assignments and assessments. The simplicity of this application makes it easy for educators and students to understand the learning objectives. According to Astini (2020), online learning at STKIP Agama Hindu uses the WhatsApp group application and Google Classroom because it is effortless to use. Tabiin (2020) asserts that Google Classroom is an application for online learning that makes it easier for teachers to create, share, and group each assignment without using paper. Besides, Hakim (2016) emphasizes that Google Classroom is an internet-based service provided by Google as an e-learning system. Besides, the service is designed to help educators to create and distribute paperless assignments to students.

In almost all level of schools, Google Classroom is chosen and applied as their learning platform. Many studies show that it has positive effects, and it is suitable for the current conditions. According to Tinungki & Nurwahyu (2020), in terms of planning the lesson, Google Classroom is considered quite effective with a percentage of 79%, and in terms of implementation, it has been implemented according to the plan of 86.7%. Then, the accessibility of lecturers and students to google classroom is considered constant (77.2%). Fulfillment of process and content standards also meets National Education Standards (76.3%). The documented aspects of learning are also considered constant (78,2 %). Thus, it can be concluded that the application of google classroom is very effective and good.

Basically, the e-learning platforms' features have similarities, and they are fairly easy to use. The only difference lies in an understanding of the user and how to implement it. According to Abidin et al. (2020), many online platforms and media can be accessed by educators and students. Several platforms that have proven their effectiveness in learning are Google Classroom and Edmodo. In line with Abidin, Fauzan (2020) conveys that Google Classroom is currently very effective and efficient in space and time to provide teaching materials that students can use to be studied continuously. The effectiveness felt by educators or students in utilizing e-learning during Covid-19 illustrates that there is a success in the learning process.

This second e-learning application is also often used by educators in delivering material, namely Zoom. Zoom makes it easy for the learning process to run, which is usually applied through face-to-face video broadcasts, but the problem is that the internet signal must be good or reliable so that the communication can be effective, not intermittent, which will result in less conveying the intent of the audience. According to Fitriyani et al. (2020), the Zoom application is a new class in the learning process, and it can be a solution for effective learning in school. Meanwhile, according to Utami (2020), synchronous online learning with the Kahoot-assisted Zoom Meeting can be used as an alternative learning and an effort to increase mathematics learning achievement.

The Zoom application can be a problem if it cannot be properly connected to the internet. This condition will provide a meaningless learning experience, and there will also be boring for students due to obscurity. According to Syswianti et al. (2020), there is a perception that students are not good at learning midwifery care courses with the zoom application during Covid-19 because of the following reasons: 1) The Zoom application requires a strong internet connection; 2) There is noise due to microphone quality problems; 3) The students lack focus because of bad signals, and there is no supervision from the lecturers. The existence of these problems has an impact on the ineffectiveness of learning achievement.

The third e-learning application that is often used by educators is Edmodo. According to Zwang (2020), Edmodo is an educational site based on social networking consisting of various educational contents. In Edmodo, there are links, videos, pictures, documents, and a presence list that may facilitate the learning process. Besides, Umaroh (2012) states that Edmodo makes the class more dynamic, and the learning does not depend on time so that activity and flexibility between educators and students can run smoothly.

The research results show that Edmodo has a big impact on the progress of the development of students. Latif, et al. (2013) suggest that Edmodo in learning has a potential effect on student learning outcomes. Setyono (2015) adds that the difference in learning outcomes between them and those not applied Edmodo is very drastic. Trisniawati, et al. (2020) convey another fact that the use of Edmodo in this pandemic period shows good results, which is shown by learning outcomes in 2D, 2E, and 2I classes that are in the good category, so that the use of Edmodo in learning can be applied properly, and it can also be modified with the other platforms. Edmodo, which has a similar look as Facebook, allows educators and students to collaborate, connect, and share content (Nu'man, 2013).

The fourth application used in e-learning is Schoology. According to Azizah et al. (2017), there is a significant effect of using Schoology on learning outcomes as evidenced by an M-gain of 0.619 (moderate category). The results of other studies were also presented a similar thing as Utami et al. (2017) found that the use of Schoology has a significant effect with the difference in the average N-gain of the experimental class of 0.76 and the control class of 0.66. Students' understanding of concepts in learning in the category of understanding the concept of 67.57%, in the category of guessing of 0%, in the category of not understanding the concept of 30%, and the category of misconception 2.43%. According to Murni (2016), there is a significant and positive influence between Schoology and student learning outcomes.

Essentially, all e-learning platforms have similar functions and benefits. Dewantara (2018) supports that there is no difference in analytical skills gained by students who use Edmodo and Schoology in learning so that both platforms can be used to train and improve students' analytical skills. Besides that, there are advantages and disadvantages of Schoology conveyed by Haryanto (2018). The advantages include: 1) it is cheap and easy to get, 2) it is efficient, 3) it is fun, 4) it is long-distance, 5) it has complete features, 6) it is paperless, 7) it has a discussion forum, 8) it is able to send files, images, etc., 9) it is able to submit anywhere, 10) it has easy task control and 11) it easily prints the results of the job. However, the drawbacks are: 1) it depends on the internet, 2) waste of internet quota, 3) the user must have an Android phone, 4) there is a risk of plagiarism, 5) it is easy to copy and paste the work, 6) students can collaborate with people in class or out of the class.

The fifth e-learning application is a WhatsApp group. This application is easy to use, and it is simpler compared with other aforesaid platforms. It also does not consume a huge quota. Even though by only using the chat quota, we can receive messages in the WhatsApp group. This is what makes the positive point of using the group WhatsApp application. According to Anhudasar (2020), it is the preferred application in online lectures out of 56 university students, consisting of 91.8% of students who chose the Whatsapp group application, 4 students (6.5%) chose Zoom, and 1 student (1.6%) chose email. In addition, the students' opinions regarding the understanding obtained showed that 1 student answered "understand very well," 23 students answered "understand," 34 students answered "sometimes understand," and 4 students answered "do not understand." This study illustrates that most students can understand the materials provided by educators through the WhatsApp web application to students. However, the drawback is that you cannot make video calls for more than 6 people, unlike in Zoom, which can effectively deliver live learning with larger groups.

WhatsApp group is more suitable for primary school students where smartphones can be more controlled and accessible to parents. Thus, when there is an assignment, the parents will immediately accept it and then convey it to their children. According to Hutami (2020), there must be good communication with parents to take advantage of information technology in online learning.

The effectiveness of using WhatsApp Group can be seen from the learning development process. Yensy (2020) asserts that the lectures in statistics courses in Mathematics Department of FMIPA FKIP, Bengkulu University that use WhatsApp group have higher learning outcomes than before use any application. There are weaknesses, namely signals that are difficult to reach by students who live in a particular area, and the chat history makes memory full. The existence of weaknesses will definitely become a learning barrier, for that it is better to use a supported internet card even in local area and to delete files that are not important so that the cellphone does not work slowly.

4 CONCLUSION

The use of e-learning platforms during the pandemic is indeed very supportive to make learning effective and successful. By applying e-learning, controlling the learning process can be done in anywhere and anytime. Either using Google Classroom, Zoom, Schoology, Edmodo or Whatshapp Group platforms, all is beneficial and very helpful for smooth learning process. However, the big problem is that the internet connection must be reliable. Thus, if there is an e-learning platform that does not positively impact, there is a lack of compatibility in using the internet network or even in using the platform. It is necessary to have a high degree of readiness in using e-learning platforms as it requires paying attention to the readiness of students, parents, teachers, and the environment.

REFERENCES

Abidin, Z., Rumansyah & Arizona, K. 2020. Pembelajaran online berbasis proyek salah satu solusi kegiatan belajar mengajar di tengah Pandemi Covid-19. *Jurnal Ilmiah Profesi Pendidikan*. 5(1): 64–70.

Agustina, M. 2013. Pemanfaatan E-Learning sebagai media pembelajaran. *In Seminar Nasional Aplikasi Teknologi Informasi (SNATI)*. 1(1).

Anhusadar, L. O. 2020. Persepsi mahasiswa PIAUD terhadap kuliah online di masa Pandemi Covid-19. *Journal of Islamic Early Childhood Education*. 3(1): 44–58.

Astini, N.K.S. 2020. Pemanfaatan teknologi informasi dalam pembelajaran tingkat sekolah dasar pada masa pandemi Covid-19. *Jurnal Lampuhyang*. 11(2): 13–25.

Astini, N.K.S. 2020. Tantangan dan peluang pemanfaatan teknologi informasi dalam pembelajaran online masa Covid-19. *Cetta: Jurnal Ilmu Pendidikan*. 3(2): 241–255.

Azizah, A.R., Suyatna, A., & Wahyudi. I. 2017. Pengaruh penggunaan e-learning dengan schoology terhadap hasil belajar siswa. *Jurnal Pembelajaran Fisika*. 5(2): 127–138.

Dewantara, D. 2018. Perbedaan kemampuan analisis mahasiswa antara pembelajaran berbantuan schoology dan edmodo pada mata kuliah Fisika Biologi. *Jurnal Pengkajian Ilmu dan Pembelajaran MIPA IKIP*. 6(1): 1–8.

Dewi, W.A.F. 2020. Dampak Covid-19 terhadap implementasi pembelajaran daring di sekolah dasar. *Jurnal Ilmu Pendidikan. Edukatif: Jurnal Ilmu Pendidikan*. 2(1): 55–61.

Fauzan, F.A. 2020. Utilizing google classroom as an interactive learning medium in middle of impact corona virus diseas 19 for teachers. *Jurnal Borneo Akcaya*, 6(1), 93–102.

Firman & Rahman, S.R 2020. Pembelajaran online di tengah pandemi Covid-19. *Indonesian Journal of Educational Science (IJES)*. 2(2): 81–89.

Fitriyani, D., Fauzi, I,. & Sari, M.Z. 2020. Motivasi belajar mahasiswa pada pembelajaran daring selama pandemik Covid-19. *Jurnal Kependidikan*. 6(2): 165–175.

Hakim, A.B. 2016. Efektifitas penggunaan e-learning moodle, google classroom dan edmodo. *I-STATEMENT*. 2 (1): 1–6.

Haryanto, S. 2018. Kelebihan dan kekurangan e-learning berbasis schoology (Studi PTK dalam pembejaran mata kuliah academic listening). *Seminar Nasional GEOTIK 2018, Universitas Muhammadiyah Surakarta*.

Hutami, M.S., & Nugraheni, A.S. 2020. Metode pembelajaran melalui whatsapp group sebagai antisipasi penyebaran Covid-19 pada AUD di TK ABA Kleco Kotagede. PAUDIA: *Jurnal Penelitian dalam Bidang Pendidikan Anak Usia Di ni*. 9 (1): 126–130.

Kemendikbud. 2020. Surat Edaran Nomor 1 Tahun 2020 tentang Pencegahan Corona Viris Disease (Covid-19). Retrieved from http://lldikti3.kemdikbud.go.id/v6/2020/04/21/surat-edaran-direktur-jenderal-pendidikan-tinggi-kementerian-pendidikan-dan-kebudayaan-nomor-1-tahun-2020 - tentang - pencegahan - penyebaran - corona - virus-disiase-covid-19-di-perguruan-tinggi-kementerian/.

Kusuma, A. 2011. E-learning dalam pembelajaran. *Jurnal Ilmu Tarbiyah dan Keguruan*. 14(1): 35–51.

Latif, Y., Darmawijoyo., Putri, R. I. I. 2013. Pengembangan bahan ajar berbantuan camtasia pada pokok bahasan lingkaran melalui edmodo untuk siswa MTs. *Jurnal Matematika Kreatif-Inovatif*. 4(2): 105–114.

Mujib, M. 2012. Pengaruh penggunaan internet terhadap hasil belajar siswa sekolah menengah atas di Kota Yogyakarta. Unpublished Thesis. *Universitas Islam Negeri Sunan Kalijaga Yogyakarta*.

Mulyono, P. 2019. Ironis potret sebuah SD di Jember dengan 3 ruang kelas dan 2 guru. Retrieved from https://news.detik.com/berita-jawa-timur/d4388248/ironis-potret-sebuah-sd-di-jember-dengan 3-ruang-kelas-dan-2guru?_ga=2.212772369.486050691.1596856887-839351739.1491434637.

Murni, C.K. 2016. Pengaruh e-learning berbasis schoology terhadap peningkatan hasil belajar siswa dalam materi pernagkat keras jaringan Kelas X Tkj 2 Pada SMK Negeri 3 Buduran, Sidoarjo. *Jurnal Information Technology and Education*. 1(1): 86–90.

Mustakim. 2020. Efektivitas pembelajaran daring menggunakan media online selama pandemi Covid-19 pada mata pelajaran matematika. *Journal of Islamic Education*. 2(1): 1–12.

Nu'man, A.Z. 2014. Efektifitas penerapan e-learning model edmodo dalam pembelajaran pendidikan agama islam terhadap hasil belajar siswa (Studi kasus: SMK Muhammadiyah 1 Sukoharjo). *Duta.com: Jurnal Ilmiah Teknologi Informasi dan Komunikasi*. 7(1).

Nurkolis, M. 2020. Keefektifan kebijakan e-learning berbasis sosial media pada PAUD di masa pandemi Covid-19. *Jurnal Obsesi: Jurnal Pendidikan Anak Usia Dini*. 5(1): 212–228.

Pratiwi, E.W. 2020. Dampak Covid-19 terhadap kegiatan pembelajaran online sebuah perguruan tinggi kristen di Indonesia. *Jurnal Perspektif Ilmu Pendiidkan*. 34(1): 1–8.

Priansa. D.J. 2017. *Pengembangan strategi danmodel pembelajaran*. Bandung: Pustaka Setia.

Putra, H.P. 2020. Kisah perjuangan guru jemput bola tugas sekolah di kala pandemi. Retrieved from https://foto.kompas.com/photo/read/2020/6/18/15924898473ba/1/kisah-perjuangan-guru-jemput-bola-tugas-sekolah-di-kala-pandemi.

Ratnawati, N.K.M., Utama, I.D.G.B., Dewantara, I.P.M. 2019. Pemanfaatan e-learning pada mata pelajaran Bahasa Indonesia. *Jurnal Pendidikan Bahasa dan Sastra Indonesia Un diksha*. 9(1): 46–56.

Saifuddin, M.F. 2017. E-learning dalam persepsi mahasiswa. *Jurnal Varidika: Kajian Penelitian Pendidikan*. 29(2): 102–109.

Setyono, E.Y. 2015. Pengaruh penggunaan media jejaring sosial edmodo terhadap hasil belajar mahasiswa pada topik pembuatan Kurva-S menggunakan microsoft excell. *Jurnal Sosial dan Humaniora*. 5(1): 42–49.

Shofiyah, S. 2016. Pengaruh penggunaan android dan e-learning terhadap hasil belajar mata pelajaran IPS siswa Kelas VIII SMPN 3 Kepanjen Malang. Unpublished Thesis. *Universitas Islam Negeri Maulana Malik Ibrahim Malang.*

Silahuddin. 2015. Penerapan e-learning dalam inovasi pendidikan. *Jurnal Ilmuah Circuit.* 1(1): 48–59.

Suharyanto & Mailangkay, A. 2016. Penerapan e-learning sebagai alat bantu mengajar dalam dunia pendidikan. *Jurnal Ilmiah Widya.* 3(4): 17–21.

Supangat, Amna, A.R., & Sulistyawati, D.H. 2018. Analisis pemahaman guru tentang teknologi informasi (Studi kasus guru di SD dan SMP Sekolah Shafta Surabaya). *Seminar Nasional Call For Paper & Pengabdian Masyarakat.* 1(01): 458–468.

Susilawati, B & Irfandi, A. 2019. Penerapan e-learning pada SMKN 1 Bandar Lampung berbasis web. *Jurnal Cendikia.* XVII: 218–226.

Suwastika. I.W.K. 2018. Pengaruh e-learning sebagai salah satu media pembelajaran berbasis teknologi informasi terhadap motivasi belajar mahasiswa. *Jurnal Sistem dan Informatika.* 13 (1): 1–5.

Syswianti. D., Suryani, N., & Wahyuni, T. 2020. Evaluasi pembelajaran daring dengan menggunakan aplikasi zoom di masa pandemi Covid-19 pada mata kuliah pengantar asuhan kebidanan. *Jurnal Medika Cendikia.* 7(1): 40–50.

Tabiin, Q.A. 2020. Google classroom sebagai alternatif e-learning pembelajaran akidah ahlak di masa pandemi Covid-19 di Madrasah Aliyah Hidayatullah Pringsurat. *Academia.edu.* Retrieved from http://www.academia.edu/download/63464681/1903018014_Qoerul_Ahmad_Tabiin_mini_research20200529-92220-77ii95.pdf.

Teddy, S. & Swatman, P. M. C. 2006. E-learning readiness of Hong Kong Teachers. *The Journal of Education Research University of South Australia.*

Taufiqurrahman. 2020. Kisah viral guru Avan, datangi satu per satu rumah murid untuk mengajar di tengah pandemi Corona. Retrieved from: https://regional.kompas.com/read/2020/04/18/14595211/kisal-viral-guru-avan-datangi-satu-per-satu-rumah-murid-untuk-mengajar-di?page=all

Tingungki, G.A & Nurwahyu, B. 2020. The implementation of google classroom as the e-learning platform for teaching non-parametric statistics during COVID-19 pandemic in Indonesia. *International Journal of Advanced Science and Technology.* 29(4): 5793–5803.

Trisniawati,. Rhosyida, N., & Muanifah, M.T. 2020. Eksplorasi hasil belajar mahasiswa melalui penggunaan kuis dan tugas pada e-learning edmodo di era pandemi Covid-19. *Science Tech: Jurnal Ilmu Pengetahuan dan Teknologi.* 6(2): 39–47.

Umaroh, S. (2012). Penerapan project based learning menggunakan microblogging edmodo untuk meningkatkan prestasi belajar siswa. Unpublished Thesis. *Universitas Pendidikan Indonesia Bandung.*

Utami, N.P.G.K. 2020. *Pengaruh pembelajaran daring sinkronus dengan zoom meeting berbantuan kahoot terhadap prestasi belajar Matematika siswa Kelas X SMA Negeri 1 Negara.* Unpublished Thesis. *Universitas Pendidikan Ganesha.*

Utami, R.R., Rosidin, U & Wahyudi, I. 2017. Pengaruh Penggunaan e-learning dengan schoology materi gravitasi newton terhadap hasil belajar siswa. *Jurnal Pembelajaran Fisika.* 5(2): 81–91.

Yensy. N.A. 2020. Efektifitas pembelajaran statistika matematika melalui media whatsapp ditinjau dari hasil belajar mahasiswa (Masa pandemik Covid 19). *Jurnal Pendidikan Matematika Raflesia.* 5(2): 65–74.

Zwang, J. 2010. Edmodo: A free, secure social networking site for schools. Retrieved from: https://www.eschoolnews.com/2010/12/15/edmodo-a-free-secure-social-networking-site-for schools/#:~:text=Edmodo%20doesn't%20require%20any, education%2Dbased%20social%20networking%20site.

*Educational Innovation in Society 5.0 Era: Challenges and
Opportunities – Purnomo & Herwin (Eds)
© 2021 the authors, ISBN 978-1-032-05392-9*

Pre-service teacher education reform in Indonesia: Traditional and contemporary paradigms

A. Mustadi & P. Surya
Yogyakarta State University, Yogyakarta, Indonesia

Mei-Ying Chen
National Chiayi University, Chiayi City, Taiwan

ABSTRACT: This study aims to explore the experiences and views on pre-service teacher education in Indonesia. It is a case study. The sample is five full-time teacher educators established the non-probability purposive sampling technique. The data were collected through semi-structured interviews and document analysis. The data were reviewed, interpreted, and organized into categories or themes. The findings show that one of the solutions to the deterioration of Indonesian students' achievement is by raising the teacher professionalism. The traditional professionalism lies on reality and theory-based body of knowledge, skills, and culture. This is integrated in the Bachelor of Education (B.Ed.) graduates. In contrast, the contemporary professionalism implies that individuals can be a professional when they acquire a set of skills through competency-based-training. This is realized in the new graduate programme of Pre-service Teacher Education (PsTE). The PsTE programme emphasizes the professional skills through the subject-specific workshops at the university and teaching practice in school.

1 INTRODUCTION

The Indonesian government attempts to improve teacher professionalism through teacher education. This is a response to the deterioration of Indonesian students' achievement both in national and international assessments, such as PISA and TIMSS (Mullis, Martin, Foy, & Arora 2012; OECD 2014; OECD 2016; Pritchett 2015). Beside educational equity and standards-driven accountability, improving teacher quality is a globally *legitimized* yet locally adaptable policy response to international education assessments (Wiseman 2013). However, these assessments are *criticized* and misused as an instrument of intervention into the governance of national education of each participating country (Meyer & Benavot 2013). Within this governance, a teacher is the most important factor in the educational process (OECD 2005). However, this does not view teachers as thinking, judging, and acting professionals (Priestley, Biesta, & Robinson 2015). Instead, it *emphasizes* teaching as technical accomplishment and measurement (Menter 2010).

The Indonesian government centralized pre-service teacher education policy to deal with this challenge (OECD/Asian Development Bank 2015). Teacher education in Indonesia needs fundamental reform. It is important to construct teacher professionalism orientated to the recent economic development

(OECD 2005; OECD/Asian Development Bank 2015) and the traditional independent thinking as well as civic participation (Meyer & Benavot 2013). The national education reform includes the redesigning of the pre-service teacher education curriculum. This curriculum development is an implementation of the national policy change increasing teacher qualification from bachelor degree to graduate professional education (Teacher Act No. 14, 2005). This one-year professional education programme, called Pre-service Teacher Education (PsTE) programme, is intended to upgrade teacher education to the graduate level. This programme has never existed before in Indonesia; and it just started in 2014. The PsTE programme is provided for both Bachelor of Education (B.Ed) holders and other bachelor degree holders. The curriculum of PsTE programme seems to be redundant, overlapping, inefficient, and ineffective to improve teachers' quality (Indriyani 2015; Sutoyo 2014). This is based on the possibility that the four-year-pre-service teacher education of the B.Ed. programme will be ineffective due to the implementation of the PsTE programme (Margi 2013; Suara Merdeka 24th June 2014; Subkhan 2011). Meanwhile, Ningrum (2012) suggests that the curriculum structure and content of the B.Ed. programmes do not have relevance and continuity with the PsTE curriculum. In addition, the teaching practices at the B.Ed. programme and that of the PsTE are overlapping.

In Indonesia, both teacher professionalism and teaching quality are associated with the efforts to improve student performance (Utanto & Gunawan 2017). Then, rigorous teaching license requirements can contribute to good teachers having ideal professionalism and practicing a good teaching. The professional teaching certification in Indonesia refers to the PsTE programme. It is argued that the PsTE's student teachers have both good grounding in developing their pedagogical competence (Anita & Rahman 2013; Maryati Prasetyo, Wilujeng, & Sumintono, 2019) and better ability in terms of subject-specific pedagogy (Anwar 2012). This indicates that the PsTE programme can be a potential medium to develop teacher professionalism and it may subsequently elevate Indonesia's poor achievement in both TIMSS and PISA (Mullis et al. 2012; OECD 2014).

Since teachers are the main resources for every nation to ensure this global competitiveness (OECD 2005), both pre-service teacher education programme and teacher professional development programme have become more crucial for every national education system. Furthermore, both teacher education and teaching profession have become a common major area of government policy intervention within international environment (Furlong 2013). Meanwhile, a proposal to revitalize teaching profession status in Indonesia was made by the Indonesia Teacher Union (ITU). The proposal combined increasing teacher's qualification, quality, and welfare. This was to raise the low status of teaching profession in Indonesia at that time (Chang 2014). The Indonesian government issued the national standard of teaching stipulated on the Teacher Act (2005) and its implemention regulations. The political influences both from the ITU and the parliament significantly shifted the policy preference or paradigm to increase teacher welfare rather than to develop education quality through professional teachers (Raihani & Sumintono 2010; Wibowo 2011). This blurs the urgency of redesigning the pre-service teacher education curriculum both at the B.Ed. programme and the PsTE programme to certify the pre-service teachers. In addition, the abolishment of the Teaching Licensing Programme IV implies distrust from the government in the quality of pre-service teacher education graduates from teacher education institutions at higher education level (Margi 2013; Raihani & Sumintono 2010).

The Teacher Act (2005) has raised crucial issues concerning teacher management and development. This current teacher management and development indicates the effect of the new managerialism on public services. The new managerialism institutionalizes market principles in the public education and it focuses on the outputs measured in terms of performance indicators. However, it often disregards the inputs or resources (Lynch 2012). However, the performance indicators only measure what can be counted and tend to neglect the intangible aspects during the process to achieve that output. Lynch (2012) argues that this relentless output-led monitoring also undermines the care and nurturing dimensions of teaching and learning. This is because of their immeasurability within the confined time frames of performance indicators. Therefore, imperative strategic planning, performance-led measurement, audit culture at school, and the competence-based teacher are emphasized in the new managerialism (Connell 2009). Furthermore, Hoyle (2001) argues that professional teachers as individuals are more important than the teaching profession itself. In addition, individuals can be a professional when they acquire a set of skills through competency-based training. Actually, the current movement of new professionalism can be viewed as a chance to professionalize the teaching profession. The Indonesian government has made a set of policies on teacher competencies, incorporation national teacher standards, and role of various ministry units and agencies to support teacher competencies, teacher certification, and special and professional allowances for teachers (Chan 2014).

With reference to the background and the literature review, this research aims to explore the experiences and views of teacher educators with regard to the national policy change to teacher education provision after the promulgation of Teacher Act No. 14/2005. It explores how teacher educators at an Indonesia university design the teacher education curriculum. The activity of designing a curriculum is a response to reality, which, in this case, is the national policy change regarding teacher education structure and teacher professionalism principles in Indonesia.

2 RESEARCH METHOD

2.1 Research design

This research is qualitative in nature with a case study design to explore deeply the case of the pre-service teacher education reform. It took place in a teacher education institution in Indonesia. The setting was chosen with the consideration of its concerns on education.

2.2 Participants

A non-probability purposive sample of five participants is derived from a group of full-time teacher educators of the teacher education institution. They have experiences regarding the process of teacher education curriculum design for either the B.Ed. programme or the PsTE programme, or both of these programmes. Therefore, the choice of research participants is based on whether they have been involved in the designing of the ITE curriculum. The details of the research participants is shown in Table 1.

Table 1. Participant's education and work experiences.

No	Participants	Involvement in designing ITE curriculum
1	Mr Epsilon (60 years old)	• Head of the Centre for the Development of Educational and Non-Education Profession which is organizing the PPG at Sunny University. • Formulating the PPG policy at the national level.
2	Mrs Gamma (60)	• Designing the "genuine" PPG curriculum in the national level. • Designing the KKNI-based curriculum 2014 for the B.Ed. programme at the national level.
3	Mrs Delta (34)	• The PPG Coordinator at the Science Education Department. • Designing the PPG Curriculum in the national level. • Designing the KKNI-based curriculum 2014 for the B.Ed. in the faculty level.
4	Mrs Beta (40)	• Coordinator of the B.Ed. in History Study Programme • Designing the KKNI-based curriculum 2014 for the B.Ed. in the national level.
5	Mr Alpha (29)	• Designing the KKNI-based curriculum 2014 for the B.Ed. in the departmental level.

2.3 Data collection

The research data were collected through semi-structured interviews and document analysis. The data are verbal in nature, consisting of transcripts of interviews and documents regarding the natural context in which the teacher education curriculum policy change occurs.

2.4 Data analysis

This study employed qualitative thematic analysis by adopting the Framework approach from Ritchie & Spencer (1994). The data analysis was applied in the following order: familiarization, identifying a thematic framework, indexing, charting, and mapping and interpretation. An index of central themes and subthemes, which was represented in a matrix with its display of cases and themes, was developed.

3 FINDINGS AND DISSCUSSION

3.1 Findings

Teacher educators' thought about the current development of PsTE system in Indonesia and interview data on the implementation of teacher education reforms at the university level are presented in Tables 2 and 3.

Table 2. Teacher educators' thought about the current development of pre-service teacher education system in Indonesia.

Codes, Key Features, and Interpretation

Mr Epsilon

• He believes that at the PPG, the non-education graduates should do a matriculation programme at least for one year, because they have deficit in the pedagogic competence. However, the non-education and the B.Ed. are not differentiated on the current PPG. He perceives that this kind of PPG is not fair.
• He believes that the PPG should be selective and prioritizing the B.Ed. graduates.
• Since the competence is standardized into four aspects: pedagogic, professional, personal, and social competence, to achieve wholeness of the four competencies, he believes that ideally the teacher education is a boarding scheme. In the campus is mainly to strengthen the pedagogic and professional competences. Meanwhile in the dormitory is to strengthen the personal and social competences.

Researcher's interpretation: the PPG likely adopts what the established profession applying; however, this is not consistent. The established profession in Indonesia, such as physician, lawyer, pharmacist, only can be enrolled by the bachelor of degree graduates on the same discipline, but the PPG can be enrolled by any bachelor degrees. Although the argument to minimize the non-education enrolment has been proposed, the decision refers the teaching profession as an open profession. The boarding-school scheme that is applied in the PPG is closer to the managerial-controlled professionalism from the Government. I addition, the appreciation for the Post-SM3T PPG graduates who are given the civil servant status is bonding the graduates to be loyal to the Government.

The information from the participants about teacher educators' thought about development of pre-service teacher education system in Indonesia run by PPG programme was inconsistent. The holders of bachelor degrees outside teacher's profession could enter this programme, but they were suggested that they should take a one-year matriculation programme. The holders of bachelor degrees in non-education programme did not lack content knowledge but they did not have educational science to teach student. Other opinions from participants were appreciation about boarding scheme that was applied for Post-SM3T PPG. The boarding scheme also played a role in establishing professional teacher characters. It helped the government to know the loyalty from the SM3T PPG graduates.

Table 3. The implementation of teacher education reforms at university level.

Codes, Key Features, and Interpretation

Mrs Gamma

- Her beliefs in English Language Initial Teacher Education are that the ITE of English Language Teacher does not makes the student teachers become good in English. Instead, the ITE accepts those having good basic of English, then they refine and polish to be more sophisticated. Then, it is added with both the English linguistics (the English as scientific discipline, not as a skill) and the English pedagogy (a science of teaching English).

- Researcher's interpretation: The ITE should be viewed not only as preparing the teachers with the content and the teaching skills, but also the academic foundations and attitudes. This takes time. The PPG should be provided for the high-quality academic student teachers, who have strong drive to be professional teachers.

Mrs Delta

- The B.Ed. programme emphasizes the instructional theories and subject knowledge content. This applies peer teaching on microteaching and supervised or accompanied teaching. This does not produce teacher candidates, in terms of the graduates is obtaining the B.Ed. title, not as licensed teacher. The B.Ed. programme should be rearranged so that the educational courses are offered in the final semesters. The initial semesters give more portions to the subject knowledge content. However, it is not ignoring the development of teacher professionalism through the habituation since the beginning of the programme.

- Researcher's interpretation: The B.Ed. programme provides the fundamental theories and concepts of education and subject knowledge content. The beliefs about ITE in the B.Ed. that placed the educational courses at the final semesters can be called as the "combined B.Ed.", in which the student teachers learn the content first for several semesters, and then it is added with the educational courses. In the past, the non-education programme continued the study with the Akta IV. Her background in Biology Education and Science Education likely affects this view.

Mrs. Beta

- The PPL (teaching practice) at the B.Ed. should be abolished at all and moved to the PPG, since the graduates are not yet eligible to teach.

- More portions for the subject knowledge content in the B.Ed.

Researcher's interpretation: this ITE attempt to copy the scheme of professional education of the established profession, such as physician. More theoretical content at the bachelor degree, then it is added with the professional education. However, since the PPG also recruits the non-education graduates, it is not consistent. It has been modified, from exclusive to selective. However, the matriculation as the equalization for those from the non-education programme is not implemented. This is not fair enough.

The information from the three participants about the implementation of teacher education reforms in university level is different from one another. The first participant suggested that ITE should drive student teachers to be professional teachers. PPG was the place to establish it. The second participant focused that educational courses were given in the final semester after the student had learned about content. And the third participant suggested that teaching practice had better move to PPG courses and B.Ed. programme should focused on teaching about content.

3.2 Discussion

3.2.1 The development of pre-service teacher education system in Indonesia

The need for quality enhancement for pre-service teacher education in Indonesia was more focused on quantitative expansion in the New Order era in 1966–1998 (Raihani & Sumintono 2010). At that time, the quantity of pre-service teacher education graduates was emphasized to anticipate the rise of student population across the country. Before 2005, secondary school teachers had to obtain a four-and-half-year-undergraduate degree (the B.Ed.; or B.A.) with the Teaching Licensing Programme IV to be a licensed teacher. In contrast, the new requirement to undergo the PsTE is made for all school level teachers in Indonesia. Currently, the Teacher Act (2005) mandates that teachers must have the necessary talent, interest, zeal, idealism, commitment, faith, devoutness, and noble character. For teachers to be able to develop these key attributes, it is perceived that they must undergo appropriate formal pre-service teacher education before they are certified as qualified teachers. Therefore, universities are expected to employ an entry level assessment as a significant overhaul of pre-service teacher education provision (Heryadi 2007). This assessment only recruit's student teachers who really have basic aptitude and interest in teaching.

Among other issues, pre-service teacher education in Indonesia has been bedeviled by an outdated curriculum. "The curriculum in pre-service teacher education is too traditional and not aligned with new, more modern expectations for what teachers should know and do" (USAID 2009: viii). In addition, pre-service teacher education in Indonesia has not shown significant improvement in the education process, which influences the quality of the graduates. This is forcing the graduates to start learning the real encounters once they are in the workplace (Raihani & Sumintono 2010). Furthermore, in Indonesia, not many teacher educators seem to demonstrate enough interaction and recognition of their students' level of understanding (Maison 2013).

The Indonesian government has initiated the new PsTE; it may be viewed as an effort to develop teaching as graduate level-profession. This is contributing to the teacher professional enhancement (OECD 2005). The Indonesian universities are running the PPG programme for the B.Ed. degree holders. The new policy of teaching qualifications in Indonesia is a move away from the traditional pre-service teacher education practice through the four-year-B.Ed programme. The right to teach of the B.Ed. graduates has been reduced since the implementation of the Teacher Act (2005). The B.Ed. degree holders no longer enjoy the automatic status of licensure. Although the B.Ed. programme is still offered, it is not different from the other non-education majors: in the sense, all graduates from both programmes can automatically become teachers. Furthermore, all graduates wanting to become teachers must obtain the teaching profession certificate through the PsTE programme. This teaching certification through the PsTE is organized and delivered by selected and accredited teacher education institutions at HEI, known as the teacher education institutions. However, most universities still run the B.Ed. programme, even though it is uncertain that B.Ed degree holders would proceed to undertake the PsTE programme, in which it would ensure that they are licensed as qualified teachers upon their graduation (Indriyani 2015). This is because the PsTE programme is run on a limited scale. It would be interesting to find out the current nature of the B.Ed. programme in the aftermath of Teacher Act (2005) in relation to the PsTE programme.

The central government provides a full scholarship for the selected B.Ed. graduates to enroll on the PsTE programme, which is in a boarding school format. This is due to both the educational cost of the programme and the subsequent professional allowance costs (Chang 2014). Nevertheless, this boarding-school-format PsTE programme can build student teachers' social-personal characters more comprehensively since it provides broad learning experience than merely class activities (Suresman 2015). What is equally interesting about the new requirement for licensing teacher is that the B.Ed. graduates are not automatically accommodated to the PsTE programme. Therefore, there are intended characteristics that are needed by both the B.Ed. and the PsTE graduates.

Indonesian National Qualification Framework (INQF). The competencies of the B.Ed. graduates should be equal to Learning Outcomes Level 6 of the INQF standard, while the competencies of the PsTE graduates should be equal to Learning Outcomes Level 7 (Table 4).

Table 4. The Level 6 and Level 7 generic learning outcomes.

The Level 6 Learning Outcomes for a Bachelor Degree	The Level 7 Learning Outcomes for a Graduate Professional Education
• Able to apply his or her field and optimize his or her own field of knowledge, technology, and/or arts on problem solving; and able to adapt to the faced situation.	• Able to plan and organize resources under his or her responsibility, and evaluate his or her work comprehensively with optimizing science, technology, and/or arts to take the steps of strategic organizational development.
• Able to master theoretical concepts on a particular field generally; able to master theoretical concepts on a specific field deeply; and able to formulate procedural-problem solving	• Able to solve the problems of science, technology, and/or art under his or her field through mono-disciplinary approach.
• Able to make right decisions based on data and information analysis, and able to give guidance on choosing alternative solutions independently and collectively.	• Able to conduct research and make a strategic decision of all aspects which are under his or her field authority accountably and responsibly.
• Responsible for his or her own tasks; and be given responsibility of organizational performance-outcomes.	

3.2.2 The implementation of the teacher education reforms at the university level

The pre-service teacher education reforms at the university level are manifested in the curriculum development. Basically, the first step of the curriculum design is the formulation of the purpose of the pre-service teacher education programme (Fish & Coles 2005; Tyler, 1971; Utomo Suminar, & Hamidah, 2019). While formulating the purpose, teacher educator embeds teacher professionalism principles stipulated in the Teacher Act (2005) and the national standard of teacher competence (2007) as the ideal curriculum of the teacher education in Indonesia. In addition, at the implementation level, the teacher education curriculum should be designed based on the

3.2.3 The pre-service teacher education at the undergraduate level of B.Ed. programme

Since the B.Ed. programme is organized by the autonomous department at the university level, its curriculum development much relies on the professional role of teacher educators. The embedded teacher professionalism in the pre-service teacher education is also to sharpen the mind and intellectual processes. Such a personal competency is about "self-conducting in accord with the norms of religion and the Indonesian national and social law and culture" (the National Standard of Teacher Competence, 2007). Furthermore, the B.Ed. graduate profile and learning outcomes should refer to the generic learning outcome standard of the INQF. The INQF-based curriculum of the B.Ed. programme must have at least 144 credits. The department embeds the teacher professionalism principles stipulated in the Teacher Act (2005) and the National Standard of Teacher Competence (2007) by designing courses. The pedagogic competence is mainly embedded into the general basic educational courses (*Mata Kuliah Dasar Kependidikan*) and the teaching and learning skills courses (*Mata Kuliah Keterampilan Proses Pembelajaran*). The professional competence, in the sense of mastering the subject-matter knowledge, is mainly embedded by teacher educators in the core subject courses and the educational development courses at the departmental level. Furthermore, these groups of courses for the B.Ed programme encompass the four teacher education aspects as the process of education rather than training for becoming a teacher (Ovens 2000). This implies that teacher professionalism embedded in the B.Ed. programme tends to follow the classical requirements of such a true profession which must have theory-based body of knowledge, skills, and culture which are mastered through a lengthy period of specific education and training (Etzioni 1969).

Now, after the introduction of the PsTE programme, the graduates of B.Ed programme are likely to be a half way to becoming teachers. This indicates that the curriculum is an ideological, social, and aspirational document that reflects local circumstances and needs (Grant 2014). Furthermore, the intended teacher education curriculum at the university level is interpreted from the ideal teacher education curriculum (Goodlab & Associates, 1979, quoted in Westbury 2008). This ideal curriculum refers to the National Standard of Teacher Competence (2007). There is a divergent purpose of the teacher education since the B.Ed. programme is not the specific professional education that produces teachers anymore.

With reference to the "what works" (Wiles & Bondi 1989: 48), the B.Ed. curriculum shows the transferability of the taught knowledge, skills, and attitudes. This means that the B.Ed. programme continues the traditional role as a pre-service teacher education graduating novice teachers regardless their status. It is argued that the standardization of the pre-service education throughout the B.Ed. and the PsTE programmes has impacted upon the course design and structure, and particularly the emphasis on the development of subject knowledge at the B.Ed. curriculum (Burgess 2000). The adjustment of the B.Ed. curriculum purpose prepares the B.Ed. graduates to be able to compete in the job market. To prepare the role of their graduates in the society, the department embeds teacher professionalism including the social and personal competencies (the National Standard of Teacher Competence, 2007) into the B.Ed. curriculum. These consist of attitude, knowledge acquisition, specific, and general skills. These are the common B.Ed. learning outcomes, especially the attitude and general skills aspects, as shown in Table 5.

Table 5. The common B.Ed. learning outcomes.

No.	B.Ed. learning outcomes
1.	Devouting to the God and ability to demonstrate religious attitude.
2.	Upholding the humanistic values while conducting task based on religion, morals, and ethics.
3.	Contributing to developing life quality in the society, nation, state, and civilization based on the Pancasila.
4.	Playing a role as citizens who are proud and patriotic, nationalistic, and responsible to the state and nation.
5.	Respecting the diversity of culture, perspectives, beliefs, and religions, and other original thought/findings.
6.	Working together and having social sensitivity and care for the society and environment
7.	Obeying the law and discipline in life within the society and State.
8.	Internalizing the academic values, norms, and ethics.
9.	Demonstrating a responsible attitude in the field of work independently.
10.	Internalizing the spirit of independence, entrepreneurship, and resilience.
11.	Recognizing and care for local wisdom, and ability to adapt to the diversity of social-culture both on the national and global scale.

The expected attitudes in the B.Ed. learning outcomes represent the social and personal competencies (the National Standard of Teacher Competence, 2007). The pre-service teacher education programme is a medium to build the Pancasila ethos of the student teachers. This represents the perennial and idealist philosophies (Wiles & Bondi 1989) that education programmes including the B.Ed programme should teach the eternal truths and wisdom of the ages. The department embeds the teacher professionalism principles, especially on the pedagogic and professional competencies (the National Standard of Teacher Competence, 2007). There are common specific abilities developed in the B.Ed programme graduates, including: (1) the ability to plan, implement, and evaluate the teaching and learning activity suitable for the students' characteristics and the learning materials' characteristics; and (2) the ability to conduct the student-centered teaching and learning process by using the various instructional media and ICT to make the teaching and learning effective, creative, and contextual.

Basically, in addition to educating students to become future teachers, another purpose of the B.Ed. programme contradicts its traditional vision as the genuine pre-service teacher education. The traditional B.Ed programme should specifically prepare student teachers to deal with teaching as complex and demanding intellectual work involving specialized knowledge and skills (Zeichner 2014). The B.Ed. programme in Indonesia employs "two visions of the B.Ed. and two purposes of the B.Ed. curriculum role". Firstly, the B.Ed. programme prepares the students to become teachers, though they have to undertake the PPG programme, and secondly, it also anticipates the students to be able to get an alternative occupation.

3.2.4 *The pre-service teacher education at the graduate level teacher profession education programme*

In addition to the B.Ed. programme, the PsTE programme is a professional educational programme specifically preparing student teachers to be professional teachers. There are some significant barriers to teacher education reform in order to support future teachers' competencies (Pujaningsih & Ambarwati 2020). Thus, the curriculum design should answer the question of "what makes for good practice in the teaching profession" (Fish & Coles 2005). According to the Ministry of Education and Culture (2014), the Post-SM3T PsTE programme is designed to produce graduates who have integrated competencies, which are strong character and leadership. The PsTE programme is directed to providing students with the real experience in order to become professional teachers. The teacher professionalism principles stipulated in the National Standard of Teacher Competence (2007) are embedded into the two curricular activities of the PsTE programme. They are both the academic curriculum in the campus and the teaching practice at school, and the boarding-life education curriculum in the campus dormitory. The PsTE curriculum enriches the academic pedagogy or subject knowledge and the professionalism taught in the bachelor degree. The pedagogic, personal, social, and professional competencies stipulated in the Teacher Act (2005) and the National Standard of Teacher Competence (2007) are embedded in the PsTE curriculum throughout the standard of the graduate's competencies, sub-competencies, and indicators. Thus, the PsTE student teachers should achieve the intended learning outcomes.

The teaching profession which has specific values especially on the personal and social competencies derived from the national philosophy of Pancasila, which is according to religious values. Those represent the perennial philosophy in education (Wiles & Bondi 1989). Furthermore, the teacher professionalism principles are embedded in details in the PsTE learning outcomes. The government has set the generic learning outcomes of the PsTE programme, which are synchronized from the graduate's sub-competencies (Ministry of Research, Technology, and Higher Education, 2017). These learning outcomes cover all the teacher professionalism principles stipulated in both the National Standard of Teacher Competences (2007) and the INQF Standard of Learning Outcomes Level 7 (The National Qualification Framework, 2012). These standards of the PsTE learning outcomes answer the next questions from Fish and Coles (2005: 51), namely: "What kind of person should the curriculum seek to create/cultivate?" "What should be the aims of the initial teacher education curriculum?" and "What

capabilities, characteristics, knowledge, and capacities should be developed by the teacher education curriculum?" Furthermore, the learning materials of the PsTE programme are derived from these learning outcomes.

The PsTE curriculum seems to imitate what the established profession applies on its professional education. According to the Ministry of Education and Culture (2014), since the student teachers of the Post-SM3T PsTE are B.Ed degree holders, the curriculum structure of the PsTE programme consists of workshops on organizing or developing the teaching and learning materials for their subject (subject-specific pedagogy). These activities are made similar to the authentic teacher's professional tasks at school. Thus, this confirms the principles that the developed activities, tasks, and experiences must be authentic, real world, relevant, constructive, sequential, and interlinked (Meyers & Nutty 2009). The PsTE programme is professional, practice-oriented in which the curriculum content emphasizes the training of teaching skills and techniques at advanced level and instruction about facts and information (Ovens 2000). Thus, the PsTE programme activities at campus employ the role of the initial teacher training, rather than the initial teacher education.

Teacher professionalism principles, especially the pedagogic and professional competencies stipulated in the Teacher Act (2005) and the National Standard of Teacher Competence (2007) are embedded through the subject-specific pedagogy workshops in the PPG programme (Table 6). This is the workshop on developing teaching and learning materials and it is specifically adjusted to each subject. Furthermore, the activity of the workshop is also synchronized with certain topics, subjects, and grades in the school curriculum. The schools in Indonesia now employ Curriculum 2013. Thus, the workshop in the PsTE programme refers to this school curriculum. In addition, the embedding teacher professionalism that encompasses pedagogic, professional, personal, and social competences continues in the teaching practice at school. It is in line with the opinions of Suryandari, Fatimah, Sajidan, Rahardjo, & Prasetyo (2018) that the modern society is very dependent on the development of science and technology, including social problems.

Although the teacher professionalism in Indonesia is nationally prescribed, there is a broad or extended teacher professionalism applied in the designed teaching practice curriculum. This professional teaching practice does not solely focus on the teaching and learning activity in class promoted by the effective teacher model (Menter 2010), or individualistic and restricted professionalism merely in the classroom context (Bair 2014). The teaching practice also includes professional commitment, showing passion and open-mindedness, leadership and organization of the learning environment, personal competence related to self-knowledge and personal characteristics, and social competence (Malm 2009). In addition, the classroom action research aims to

Table 6. The PPG graduate's competence and indicators.

Competence	Sub-competence	Indicators
Pedagogic	1. Planning the teaching and learning	1. Formulating pupil's competence indicators and learning outcomes based on the standard of graduate competence. 2. Organizing teaching and learning materials, process, sources, media, assessment, and evaluation. 3. Designing teaching plan according to the syllabus and implementing the principles of Techno-Pedagogical Content Knowledge (TPACK).
	2. Conducting the teaching and learning	Making the teaching and learning atmosphere and process which is educative and make the pupil bright based on pedagogic principles to facilitate pupils' potential and character.
	3. Assessing and evaluating the teaching and learning	1. Conducting authentic-holistic learning assessment that measures attitude, knowledge, and skills. This is assessment of learning. 2. Conducting assessment as a learning process. 3. Conducting assessment's results to develop the teaching and learning quality. This is assessment for learning.
Personal	Behaving according to religion norms, law norms, social norms, ethics, and cultural values.	1. Practicing the teaching of his/her own religion, as a human that believes in God, devotion, and having a noble character. 2. Having the nationalist ethos and patriotism based on the Pancasila, the Constitution 1945 of the Republic of Indonesia, commitment to the State of Republic of Indonesia and the spirit of Bhinneka Tunggal Ika (Unity in Diversity). 3. Demonstrating obedience to the law by implementing the norms according to the law and regulation in education and the teaching profession. 4. Performing a role model who is honest, has a noble character, diligent, responsible, and proud of being a teacher. 5. Willing to conduct self-development independently and continuously.
Social	Having ability to communicate, to interact, and to adapt with pupils, colleagues, parents/guardian, and society effectively and efficiently.	1. Communicating and interacting with pupils, colleagues, educational personnel, parents, and society through oral and written forms which are polite, effective, and productive. 2. Participating in the national development as a good citizen. 3. Having commitment to adapt and to use information and communication technology on conducting the professional tasks.
Professional	1. Mastering the learning materials in widely and deeply	1. Analysing the intended pupil's competence (learning outcomes) as the base of selecting materials. 2. Implementing and evaluating learning materials, structure, concept, and scientific paradigm that support the development of science, technology, and arts
	2. Mastering and finding concept, approach, technique, and method of the relevant science, technology, or arts.	1. Mastering concept, approach, technique, or method of the relevant science, technology, or arts. 2. Finding new concepts, approaches, techniques, or methods of the relevant science, technology, or arts.

fix the teaching practice and to develop the teaching and learning quality in school (Wiyarsi & Purtadi 2017). In addition to the core of the professional teaching training at the university campus and school, teacher professionalism principles are also embedded into the boarding-life education in the campus dormitory. The teacher professionalism built in this boarding-scheme is especially the personal and social competencies (National Standard of Teacher Competence, 2007; Teacher Act, 2005). The boarding-life education is a comprehensive-holistic education programme (Ministry of Education and Culture 2014). This boarding-life education is attached to the teacher education programme based on the stipulation of the National Education System Act (2003), the Teacher Act (2005) and their implementation regulations.

The personal and social competencies as part of teacher professionalism (National Standard of Teacher Competence, 2007; Teacher Act, 2005) have been embedded through the structured-life education in the university dormitory. Furthermore, this indicates that the extended teacher professionalism is not focused on student achievement. This extended teacher professionalism contradicts the relentless output-led approach that undermines the care and nurturing dimensions of teaching and learning (Lynch 2012).

4 CONCLUSIONS

In the pre-service teacher education curriculum, teachers must have four basic competencies: pedagogic, personal, social, and professional competencies. Teacher educators have embedded and synchronized such competencies in the B.Ed. programme. The purposes of B.Ed programme now cover both the education of students to become future teachers or take up other occupations. The curriculum design specifically addresses what makes for good teaching practice. The teacher professionalism principles are embedded in selected and organized learning experiences, which encompass the four basics competencies. These learning experiences are actualized into the theoretical and practical courses which emphasize the combination of attitude, knowledge, and skills. Groups of B.Ed. courses essentially focus on the process of education rather than training for becoming a teacher. Meanwhile, selected and organized learning experiences in profession and practice-oriented teacher education, such as the PsTE programme, tend to emphasize the training of teaching skills and techniques at the advanced level, and instruction on facts and information. Thus, the PsTE activities on campus employ the role of initial teacher training, rather than the initial teacher education. However, another aspect of professional education is to foster commitment and conformity to the norms and values of the teaching profession. This aspect is cultivated through the boarding-life education activities and teaching practice in school.

REFERENCES

Anita, N., & Rahman, A. 2013. Penilaian peserta PPG SM3T prodi PPKn UNESA terhadap pelaksanaan programme pendidikan profesi guru (PPG) Tahun 2013. *Kajian Moral dan Kewarganegaraan, 3*(1): 409–423.

Anwar, Y. 2012. Kemampuan subject-specific pedagogy calon guru biologi peserta programme pendidikan profesional guru (PPG) yang berlatar belakang basic sains pra dan post workshop. *Jurnal Pendidikan IPA Indonesia, 1*(2): 157–162. doi: https://doi.org/10.15294/jpii.v1i2.2133.

Bair, M. A. 2014. Teacher professionalism: What educator can learn from social workers. *Mid-Western Educational Researcher, 26*(2): 28–57.

Burgess, H. 2000. What future for initial teacher education? New curriculum and new directions. *The Curriculum Journal, 11*(3): 405–417. doi: https://doi.org/10.1080/09585170050200594.

Chang, M. C. 2014. *Teacher reform in Indonesia: the role of politics and evidence in policy making*. Washington, DC: The World Bank.

Connell, R. 2009. Good teachers on dangerous ground: towards a new view of teacher quality and professionalism. *Critical Studies in Education, 50*(3): 213–219. doi: https://doi.org/10.1080/17508480902998421.

Etzioni, A. 1969. *The semi-professions and their organization*. New York: Free Press.

Fish, D., & Coles, C. 2005. *Medical education: Developing a curriculum for practice*. Open University Press: Berkshire.

Furlong, J. 2013. Globalization, neoliberalism, and the reform of teacher education in England. *The Education Forum, 77*: 28–50. doi: https://doi.org/10.1080/00131725.2013.739017.

Grant, J. 2014. *Principles of curriculum design*. in T. Swanwick, ed, Understanding medical education: Evidence, theory, and practice. 2nd edn. Somerset, NJ: John Wiley & Sons, Ltd., pp. 31–46.

Heryadi, D. 2007. Entry level assessment sebagai salah satu programme dalam upaya meningkatkan kualitas lulusan lptk: kajian literatur tentang konsep dan pengaplikasiannya. *Jurnal Pendidikan dan Kebudayaan, 13*(65): 330–313. doi: http://dx.doi.org/10.24832%2Fjpnk.v13i65.339.

Hoyle, E. 2001. *Teaching as a profession*. in N. J. Smelser & P.B. Baltes, eds, International encyclopedia of social behavioral sciences. Elsevier, Ltd., pp. 15472–14746.

Indriyani. 2015. Persepsi mahasiswa kependidikan fakultas ekonomi Universitas Negeri Yogyakarta terhadap pendidikan profesi guru (PPG). *Pelita – Jurnal Penelitian Mahasiswa UNY, X*(1): 1–10.

Kementerian Pendidikan dan Kebudayaan. 2014. *Panduan programme pendidikan profesi guru prajabatan pasca programme SM-3T (*2nd ed.). Jakarta, Indonesia: Kementerian Pendidikan dan Kebudayaan.

Kementerian Pendidikan dan Kebudayaan. 2014. *Pedoman pendidikan berasrama bagi peserta programme pendidikan profesi guru prajabatan pasca programme SM-3T.* Jakarta, Indonesia: Kementerian Pendidikan dan Kebudayaan.

Kementerian Pendidikan dan Kebudayaan. 2014. *Programme sarjana mendidik di daerah terdepan, terluar, dan tertinggal (SM-3T)*. Jakarta, Indonesia: Kementerian Pendidikan dan Kebudayaan.

Kementrian Riset, Teknologi dan Pendidikan Tinggi. 2017. *Pedoman penyelenggaraan pendidikan profesi guru (the implementing guidance of teacher profession education).*

Jakarta, Indonesia: Kementrian Riset, Teknologi dan Pendidikan Tinggi.

Lynch, K. 2012. On the market: Neoliberalism and new managerialism in irish education. *Social Justice, 12*(5): 88–102. doi: https://doi.org/10.1080/17508487.2014.949811.

Maison. 2013. *Curriculum evaluation in higher education: A case study of a physics pre-service teachers' curriculum in Indonesia*. PhD Dissertation. Western Australia: Curtin University.

Malm, B. 2009. Towards a new professionalism: Enhancing personal and professional development in teacher education. *Journal of Education for Teaching: International Research and Pedagogy, 35*(1): 77–91. doi: https://doi.org/10.1080/02607470802587160.

Margi, I. K. 2013. Programme pendidikan profesi guru prajabatan dalam perspektif darwinisme sosial. *Jurnal Pendidikan dan Pengajaran, 46*(1): 87–95. doi: http://dx.doi.org / 10.23887 / jppundiksha.v46i1.1695.

Maryati, M., Prasetyo, Z. K., Wilujeng, I., & Sumintono, B. 2019. Measuring teacher's pedagogical content knowledge using many-facet rasch model. *Cakrawala Pendidikan, 38*(3): 452–464. doi: https://doi.org/10.21831/cp.v38i3.26598.

Menter, I. 2010. *Literature review on teacher education in the 21st century*. Edinburgh: Scottish Government Social Research.

Meyer, H., & Benavot, A. 2013. *PISA and the globalization of educaiton governance: some puzzles and problems*. In: H. Meyer and A. Benavot, eds, pisa, power, and policy: The emergence of global educational governance. Oxford: Symposium Books Ltd, pp. 9–26.

Mullis, I. V. S., Martin, M. O., Foy, P., & Arora, A. 2012. *TIMSS 2011 international results in mathematics*. Boston: TIMSS & PIRLS International Study Center and International Association for the Evaluation of Educational Achievement (IEA).

Ningrum, E. 2012. Membangun sinergi pendidikan akademik (S1) dan pendidikan profesi guru (PPG). *Jurnal Pendidikan Geografi, 12*(2): 49–55. doi: https://doi.org/10.17509/gea.v12i2.1783.

OECD. 2005. *Teacher matter: Attracting, developing, and retaining effective teachers*. Paris: OECD.

OECD. 2014. *PISA 2012 results in focus: What 15-year-olds know and what they can do with what they know*. OECD: Programmeme for International Student Assessment.

OECD. 2016. *PISA 2015 result in focus*. Paris: OECD.

OECD/Asian Development Bank. 2015. *Education in Indonesia: Rising to the challenge*. Paris: OECD Publishing.

Ovens, P. 2000. Becoming scientific and becoming professional: Towards moderating rationalism in the initial teacher education curriculum. *The Curriculum Journal, 11*(2): 177–197. doi: https://doi.org/10.1080/0958517005004519.

Peraturan Menteri Pendidikan Nasional Tahun 2007 No. 16. Standar kualifikasi akademik dan kompetensi guru. [Standard of teacher academic qualification and competence].

Peraturan Pemerintah RI Tahun 2012 No. 8. Kerangka kualifikasi nasional Indonesia. [Indonesian national qualification framework].

Priestley, M., Biesta, G., & Robinson, S. 2015. *Teacher agency: An ecological approach*. London: Bloomsbury Academic.

Pritchett, L. 2015. The Majority of The World's Children Are in School. So Why Aren't They Learning? [Homepage of The Guardian], [Online]. Available: http://www.theguardian.com/global-development / 2015/sep / 29/ majority-of-world-children-in -school-so-why-arent-they-learning? CMP=share_btn_tw [October 10th, 2015].

Pujaningsih, & Ambarwati, U. 2020. Self-efficacy changes in collaborative course for inclusive education preservice teachers. *Cakrawala Pendidikan, 39*(1): 79–88. doi:10.21831/cp.v39i1.26669.

Raihani & Sumintono, B. 2010. *Teacher education in Indonesia: Development and challenges*. In: K.G. Karrasand C.C. Woluter, eds, International handbook of teacher education worldwide. 1st edn. Athens: Athens-Atrapos, pp. 181–197.

Ritchie, J., & Spencer, L. 1994. *Qualitative data analysis for applied policy research* by Jane Ritchie and Liz Spencer in A. Bryman and R. G. Burgess, eds. "Analyzing ualitative data". London: Routledge, pp.173–194.

(PDF) Framework Analysis: A Qualitative Methodology for Applied Policy Research. Available from: https://www.researchgate.net /publication /267678963 _ Framework_ Analysis_A_Qualitative_Methodology _ for _ Applied_ Policy_Research [accessed Feb 19 2020].

Undang-Undang R. I. Tahun 2003 No. 20. Sistem pendidikan nasional [National Educational System].

Undang-Undang R. I. Tahun 2005 No. 14. Guru dan dosen. [Teacher and Lecturer].

Suara Merdeka. 2014. 24 June 2014. *Kerancuan pendidikan profesi guru*. Suara Merdeka.

Subkhan, E. 2011. *Kritik atas pendidikan profesi guru dan solusinya*.

Suresman, E. 2015. Studi deskritptif analisis model pembinaan sikap religius di LPTK penyelenggaraan pendidikan profesi guru pasca SM3T di Indonesia. *Edutech, 1*(3): 437–459. doi: https://doi.org/10.17509/edutech.v14i3.1389.

Suryandari, K. C., Fatimah, S., Sajidan, S., Rahardjo, S. B., & Prasetyo, Z. K. 2018. Project-based science learning and pre-service teachers' science literacy skill and creative thinking. *Cakrawala Pendidikan, XXXVII*(3): 345–355. doi: https://doi.org/10.21831/cp.v38i3.17229.

Sutoyo. 2014. Meningkatkan kualitas guru melalui pendidikan profesi guru. *Widya Wacana, 9*(1): 18–25. doi: http://dx.doi.org/10.33061/ww.v9i1.947.

USAID. 2009. Teacher education and professional development in Indonesia: A gap analysis. USAID/ Indonesia: GEM II - Aguirre Division of JBS International, Inc.

Utanto, Y., & Gunawan, D. 2017. Kurikulum pendidikan guru yang memberdayakan: Pembelajaran dari programme keteladanan, Heribertus. Paper presented at the Seminar Nasional Kupas Tuntas Kurikulum 2013, Himpunan Pengembang Kurikulum Indonesia (HIPKIN) Jawa Tengah. file:///C:/Users/ASUS/AppData/Local/Packages/ Microsoft.MicrosoftEdge_8wekyb3d8bbwe / TempState/ Downloads/8-8-1-SM%20(3).pdf

Utomo, B. U., Suminar, D. S., & Hamidah. 2019. Capturing teaching motivation of teacher in the disadvantaged areas. *Cakrawala Pendidikan, 38*(3): 398–410. doi: http://dx.doi.org/10.21831/cp.v38i3.26411.

Westbury, I. 2008. Making curricula: Why do states make curricula, and how? In: F.M. Connely, ed, The SAGE Handbook of Curriculum and Instruction (1st ed). California: Sage Publications, Inc., pp. 45–65.

Wibowo, U. B. 2011. Intensitas peran aktor kebijakan dan kekuatan politis dalam perumusan kebijakan sertifikasi pendidik. Jurnal Penelitian Ilmu Pendidikan, 4(1): 1–16. doi: http:// dx.doi.org / 10.21831 / jpipfip. v6i1.4736.

Wiles, J., & Bondi, J. 1989. *Curriculum development: A guide to practice (3rd ed.)*. Columbus, Ohio: Merril Publishing Company.

Wiseman, A. W. 2013. Policy responses to PISA in comparative perspective. In: H. Meyer and A. Benavot, eds, PISA, Power, and policy: the emergence of global educational governance. Oxford: Symposium Books Ltd, pp. 303–322.

Wiyarsi, A., & Purtadi, S. 2017. Chemistry teacher's ability to design classroom action research in hybrid learning programme. *Cakrawala Pendidikan, XXXVI*(2): 192–200. doi: http:// dx.doi.org / 10.21831 / cp.v36i2. 11586.

Zeichner, K. 2014. The struggle for the soul of teaching and teacher education in the USA. *Journal of Education for Teaching: International Research and Pedagogy, 40*(5): 551–568.

Educational Innovation in Society 5.0 Era: Challenges and Opportunities – Purnomo & Herwin (Eds)
© 2021 the authors, ISBN 978-1-032-05392-9

Envisaging Montessori visions on a K-2 learning environment in a digital form

M.H. Ismail
Universitas Pendidikan Indonesia, Bandung, Indonesia

ABSTRACT: This research is conducted to analyze the possibility of Montessori visions in digital learning environment management for K-2 Children. Its aim is not just to picturize the framework of Montessori visions for future development of learning with standardized services, but also to analyze the possibility of overcoming the recent problem of distance learning in Early Childhood Education due to the COVID-19 pandemic. The Integrated Literature Review is applied as a method to analyze and synthesize various articles gathered using an opportunistic sampling technique. Based on the study, there are several Montessori conceptions that are seemingly hard to integrate with recent digital worlds. However, there are some possibilities to alter the traditional learning environment with the use of AI to support the learning process. In summary, the advanced ICT, like immersive media, might not be appropriate for early children while the AI can't single-handedly become the reliable solution.

1 INTRODUCTION

The outbreak of COVID-19 has been taking a great toll on our daily lives. Most of every aspect of human activity these days has halted due to the social distancing or lockdown policy to limit the extent of the pandemic spread. Referring to the latest data from the Indonesian Health Ministry in August, 14th 2020 the number of COVID-19 Prevalence reached 135,123 cases with 6,021 deaths and 89,618 healed. They're also huge possibilities that prevalence can rise as long as the people weren't given the right cure to the disease.

What was stated about the recent event has been some kind of delicate matter especially within the scope of early childhood education. For the early children, the design of a learning environment can give them a great boost to their learning process. The pandemic that has been forcing most of the students to learn from their homes has caused a major problem in how the learning environment setting can support the children in 2020. On the other hand, these problems have opened other possibilities in the learning system that are perceived to be the most effective ways to support the learning from home program. The study in this area is considered to have good importance regarding the Indonesian educational policy no. 20 in 2003 (Kemendikbud, 2015; Kusnadi & Ismail, 2020) about the aims of education in Indonesia that encourage various explorations on the effort of facilitating students' learning process. Those stated above is also similar to Dewey's and Chomsky's visions of education that was cited by Wattimena (2012; Kusnadi & Ismail, 2020) in that it is important to design some good educational processes that will encourage various student development based on their characteristics, styles, and interests.

Several of the Montessori visions on early childhood learning as it is written in Sujiono (2009: p.108) that seem to be halted by the pandemic are the freedom for the children to learn within their own time, ways, and materials that are chosen by the children after being adjusted to their interests and competencies. The recent regulation urging most of the schools to replace their direct meeting in the class with distance learning system due to the effort of halting the COVID-19 pandemic has given some additional problems because it will eliminate the importance of learning environmental settings in the early childhood education practice.

This article is going to describe the result of a literature review that was trying to find the relevant correlations between Montessori visions and distance learning system and analyzing its learning environment design in meeting the challenge of distance learning, and also to envisage the possibility of Montessori learning environmental design in a digital form.

2 METHOD

Most of the goal of this integrative literature review is to criticize and synthesize the proper application of Montessori conception in early childhood classroom management during the social crises these days due to the Coronavirus pandemic. Various data from the collected articles will be analyzed to criticize and

DOI 10.1201/9781003206019-17

synthesize to gain further perspectives and theoretical frameworks on the case.

Data collection techniques that were applied within this research are using data acquired through personal research findings during analyzing various articles based on research and other literature studies that were published in some journals by using the opportunistic sampling technique that was stated by Creswell (2012: p. 209), allowing the researcher to purposefully collect the sample after the research begins to help answer the research questions. The data evaluation phase is required to discriminate whether the collected data is appropriate for further analysis to construct the study. The measurement of validity and reliability are conducted through some methodological evaluation lists to avoid various threats to validity and to ensure reliability by making limitations of variations of the data sets. Various gathered data are analyzed by using thematic analysis which is stated by Ely (1991: p. 150) as the widely used approach to conducting the final analysis.

3 RESULTS AND DISCUSSION

3.1 *Results*

The search for the literature had been examining 16 articles including research articles and other literature reviews, non-research articles, and policy statements. Those articles were being evaluated regarding the quality and how it related to the study.

Table 1. Article level of evidence rate

No	Authors	Rate
1	The Wildflower (2016)	13
2	Drigas, A.S. & Gkeka, E.G. (2016)	15
3	Zosh, J.M., Hirsh-Pasek, K., Golinkoff, R.M. & Morris, J.P. (2016)	9
4	Mariyana, R. & Setiasih, O. (2018)	10
5	State of Victoria (2012)	22
6	Fukuyama (2018)	8
7	Sobel, K. (2019)	8
8	Brown, N., Te Riele, K., Shelley, B. & Woodroffe, J. (2020)	18
9	NAEYC and the Fred Rogers Center (2012)	8
10	Khanzode, K.C.A. & Sarode, R.D. (2016)	3
11	Chapagain, N.K. & Neupane, U. (2020)	9
12	Jim Prentzas (2013)	11
13	Williams, R., Park, H.W., Oh, L. & Breazeal, C. (2019)	11
14	Priyono, D. (2020)	11
15	Natalie Kocour (2019)	7
16	Powell, M. (2016)	15

It was then divided into certain categories related to the research question. Those articles are grouped as follows.

Table 2. Article group of patterns

No	Authors	1	2	3
1	The Wildflower (2016)	x	x	x
2	Drigas, A.S. & Gkeka, E.G. (2016)	v	v	x
3	Zosh, J.M., Hirsh-Pasek, K., Golinkoff, R.M., Morris, J.P. (2016)	0	v	v
4	Mariyana, R. & Setiasih, O. (2018)	0	0	x
5	State of Victoria (2012)	0	v	v
6	Fukuyama (2018)	0	0	v
7	Sobel, K. (2019)	0	v	x
8	Brown, N., Te Riele, K., Shelley, B. & Woodroffe, J. (2020)	x	x	v
9	NAEYC and the Fred Rogers Center (2012)	0	x	x
10	Khanzode, K.C.A. & Sarode, R.D. (2016)	0	0	v
11	Chapagain, N.K. & Neupane, U. (2020)	x	x	v
12	Jim Prentzas (2013)	0	v	v
13	Williams, R., Park, H.W., Oh, L., Breazeal, C. (2019)	0	v	v
14	Priyono, D. (2020)	0	0	v
15	Natalie Kocour (2019)	v	v	x
16	Powell, M. (2016)	v	v	v
	Supporting	3	8	10
	Opposing	3	4	6
	Not mentioning	10	4	0

Note.
1. Supporting Montessori Conceptions and Distance Learning integration.
2. Supporting Montessori Learning Environment and distance learning integration.
3. Altering projections of Montessori Visions on digital learning environment.

As it can be seen on the table, there is not much of a correlated conception between Montessori visions and the distance learning system as there are only 3 articles that seem to support the integration and another 3 articles are seem to oppose it by describing some real degradation values of children's learning. If those contradicting tendencies are compared with their value of the evidence, the opposing articles hold reasonable statements. This can also be interpreted that most of the Montessori conceptions are irrelevant to the current situations that require the applications of the distance learning system. Meanwhile, the chances of correlated articles that supporting Montessori Learning Environment and distance learning integration is higher with 8 articles that are supporting the concept, 4 articles that seemingly opposing it, and the others that are not mentioning the issue. Regarding the data, apart from the whole of Montessori conceptions, recent technologies such as AI and immersive media are taking most of the issues by reporting how that advanced ICT can alter the various issues in a distance learning system.

The 3rd group that reported on the various chances that can alter the future shape of the Montessori learning environment reached 10 articles which is the highest correlation among the 3. This also means that most of the articles agree with the need for a new shape up from the Montessori traditions especially in the case of a pandemic. However, most of the topics from those articles were in doubt about the use of recent ICT especially the immersive media that must hold themselves due to the government regulations on forbidden use of immersive media to the children under 13 years old. Meanwhile, the other topic about the use of AI was having no restriction on government policies. Concerning those findings, the most relevant advanced ICT that applicable to this situation is the AI.

3.2 *Discussion*

Many aspects of the Montessori conceptions are usually related to the strong attachment with its traditional value of learning activity with a concrete learning environment. The Wildflower (2020) emphasizes the important role of children, parents, and designed learning environment in supporting children development as it is also known as the Montessori trinity. The role of the teacher is to provide the children with a good learning environment so they can freely explore and observe to construct their own knowledge. However, those conceptions as stated earlier might have been downgraded as it has been replaced by a distance learning system. Mark Powell (2016: p. 156) by citing AMI (Association Montessori Internationale) and AMS (American Montessori Society) statement that was urging some precautions to give a digital world to the children that it was thought won't be able to replace the importance of concrete learning environment and the social environment it follows. Those statements somehow correlate with Brown, Riele, Shelley, Woodfoffe, (2020) that were reporting about some degradations in children's learning value when they were forced to learn from home and using a distance learning system.

Other than what was stated earlier, Powell also mentions the importance of children these days to attain the new literacy era of technology. With the right capacity of literacy, children can have many benefits in their interactions when they were in the learning process. Furthermore, when it comes to the learning activity, sometimes the important matter is how the children can focus themselves and have good engagement in their learning process. Some experts call it "the flow," that in Kusnadi & Ismail (2020) is stated as a definition of how well one can focus and endure the task. Rogaten & Moneta (2016; Kusnadi & Ismail, 2020) also cited a statement from Csikszentmihalyi (1991) that made a further explanation: the "flow" is not only task absorption but also the form of cognitive efficiency, and some deep intrinsic motivations that make a person feels integrated within their activity. This state of behavior can probably occur when the children actively engage in the activity. This correlates to what Zosh, Hirsh-Pasek, Golinkoff, Morris, (2016) stated, that regarding the use of ICT, the learning occurs when the children are mentally engaged, actively making physical participations, making meaning within their learning process, and have some chances to make social interactions and self-actualization, and having some guided explorations in their interactions with the ICT based instructional media. The state of "flow" can alter the missing learning environment in the children learning process. Considering what was stated previously, sometimes what matters most is not the availability of a physical learning environment but the active engagement of the children in their learning activity. On the other hand, it will be requiring some advanced integration of ICT to give the children the social environment to allow them to make some good interactions with each other.

Another literature review that had been conducted by Victoria state (2012) reporting several case studies that show some success in implementing distance learning. One of the reports was stated that in the learning process the children were separated and only connected online. To this point, the freedom of learning like that of Montessori conceptions can be altered by the blended learning system. However, all of those statements that support the new ICT was not explicitly describing the form of a learning environment that was designed in the learning system. The existence of the learning environment is a kind of important thing because, during the social distancing period, the concrete learning environment that usually available to the kids when they were in the kindergartens was lost and replaced by whatever that was installed in their home. Although various alternatives are usually done by the teacher in facilitating the learning environment to support children learning process, presumably still it won't be sufficient enough to replace the natural environment.

Meanwhile, another hope comes up from the recent progress of ICT that was named web 5.0 can facilitate much more various interactions. Khanzode & Sarode (2016) also mention it as the symbiotic web that facilitates personal server, data storage, smart communicator, and other progressions of AI and the immersive media, and also other interaction systems that perform communication between avatar in the web. Several implementations of AI in early childhood education actually had been done like what was reported by Williams, Park, Oh, Breazeal (2019). Those AI technologies were called the KIBO robot, Scratch Jr., which facilitated the children to learn to develop some animation projects, and the social robot SoRo. All those technologies were designed to meet the needs of children between 4 to 8 years old. Using those AIs, children are not only encouraged to develop their knowledge but also attain computational thinking, social skills, and creativity by building some projects.

Other than AI, the immersive media that usually known as the presence of Virtual Reality (VR) that gives an immersive visual experience, Augmented Reality (AR) which augmented the physical world, Mixed Reality (MR) that somewhat combines VR and AR that users can interact with the object, Cross Reality (CR) which combines AR, VR, and MR. Despite the promising development of immersive media technology that can alter physical reality, Sobel (2019) also mentions the age restriction of using this technology. In UAS, there is certain regulation that forbidds the use of immersive media for children under the age of 13 years old due to the considerations of possible negative impacts of this technology. While others still consider that the lens from the tools themselves can reduce eye health.

Other than the regulation issue that halts the use of immersive media, there are theoretical arguments that can probably be opposing the use of this technology. Several statements from Vygotsky (2004) brought good considerations about how the immersive media could probably halt children's thinking skill development. Imagination always comes up based on the knowledge of reality that had been experienced by certain people. The important parts of his theory are about dissociation, association, and the combination of those two in the imagination construction process. Dissociation is the process when someone captures small fragments of reality impression and subjectively stores it in their mind as knowledge. While association is the process when someone is capturing impression forms, many fragments of reality are attained from various sources. And the third is the combination of the two processes. The knowledge from this process, later on, being the main source of imagination construction that on the following process is the crystallization of the external image. If it is related to the use of immersive media, the children will receive massive stimulation of naturalistic images that will have no space for reality distortions. Technically it is supposed that children will have a comprehensive understanding of the stimulations even though it will take a naturalistic view. This process can enrich children's understanding and help them to cope with the poor and simple imagination as stated by Vygotsky. But later on, the concern is could the children build their imagination from super naturalistic impressions? Which on the following process, would the children still be regarded as creative with that kind of stored knowledge? Another concern also stated by Bailey (2017; Sobel, 2019) is that taking a sample on the sesame street character that was set up with VR. Within his study, he found that the children might be overwhelmed by the stimulation that was received from VR.

Priyono (2020) mentioning some challenges that were in store especially for some developing countries in Southeast Asia in preparing society 5.0. The lacking physical infrastructure like the early childhood center and other supporting infrastructure making the gap seemingly wider in the city and the rural places or even the villages. Besides those physical infrastructures, the varied teacher competencies in the city and other places are could also prohibit the utopia of society 5.0. The effort to embrace the emerging modern technologies will bring various changes to the role of educational practitioners. State of Victoria (2012: p. 35) mentions that the teachers should be able to design and implement the curriculum that is delivered online and having a good understanding of the distance learning systems and the various technologies

that can be attached to implement it. Besides the teachers, the students are also required to have some good ICT literacies so that it will help them to study beyond the classroom border. Furthermore, Chapagain & Neupane (2020) discussing the changing of teacher roles that should facilitate and flexibly support children's self-construction knowledge.

Taking a more neutral statement, Brown, N., Riele, K., Shelley, B., Woodfoffe, J., (2020) stated that ICT literacy is important both to the children and to the parents more than the device itself because the computer and the internet alone won't be enough to support children's learning. Besides the technological availability, the appropriate stage of ICT literacy can help much in that learning program. Sobel (2019) citing Kleeman said that in using the immersive media, the thing that should be noted is that the use of this media is primarily to stimulate and to maintain the ongoing meaningful experiences that supports and meets children's needs in their learning process. Moreover, Sobel (2019: p.18) also citing some recommendations from Curtis Wong on how to implement the immersive media in the following steps such as; (1) Engage the audience with the story within; (2) Help the audience in building their mental model through some interactive explorations and multisensory stimulation; and (3) Give validation to the mental model that has been constructed by using data references. However, those recommendations still cannot guarantee the risks on the early age children's safety either on their physical growth or on their mental skills development and the children learning freedom as it was in Montessori's conception.

So, much on the changing roles and the required basic knowledge to effectively implement the latest generation of ICT in the early childhood education setting. Still, there are some additional considerations about the availability of the technological equipment stored, the considerations regarding the risks and the benefits that being brought, and also other possibilities of various technological integrations. Moreover, the fact that there are still many unpredicted risks regarding the use of immersive media, NAEYC and the Fred Rogers Center (2012) reporting about the equality of access that still became the major problem even in the developed nations like the USA not to mention the challenge for the South East Asian Nations as was stated earlier.

Based on those reports, on the state of utopia, we can imagine that the form of Montessori learning environmental design in a digital form will consist of the use of several technologies that are integrated to equip the teacher's efforts to develop the children's learning outcomes. In their learning process, perhaps the children won't have to use some kind of binocular lens or HoloLens to experience the immersive media stimulations regarding its risks on children's safety. Perhaps all those stimulations of immersive media can be projected to some other media like a TV screen if there will be another way to maintain the interactions between the children other than virtual being. Perhaps the segmented use of AI attached to the whole learning process designed by the teacher that integrated with some VR visualizations or video conference if both teachers and students can lessen the communication lag that often comes out and halting the expressivity like that of the concert interactions.

This study had brought various perspectives in discussing the relation of Montessori conception and distance learning. Although both concepts encouraging the freedom to learn for the children, it might not meet the Montessori concept of a physical learning environment that allows the growth of the social environment during children's learning activities. It was an important insight both on the Montessori visions and the recent ICT capabilities.

4 CONCLUSION

Although the fact that blended learning is considered flexible, it won't be sufficient to facilitate the children's needs for the learning environment support during their study from home which is also considered as the future needs of flexible learning. The use of AI, Immersive media, or the combination of both might meet the Montessori trinity requirements but it is currently facing obstacles either on the government regulation, the children's equality of access, or mainly how it can halt the children's thinking skills and safety risks on their health. Much more studies are required to explore other possibilities of the use of advanced ICT, especially on the aspect of how it could be negatively affect the children's health and their thinking skill development.

REFERENCES

Brown, N., Te Riele, K., Shelley, B. & Woodroffe, J. 2020. *Learning at home during COVID-19: Effects on vulnerable young Australians. Independent Rapid Response Report. Hobart: University of Tasmania, Peter Underwood Centre for Educational Attainment*. Retrieved from; http://www.icponline.org: 5–45.

Chapagain, N., K., & Neupane, U., 2020 Contextual learning: rethinking education for Nepal in the wake of COVID-19 crisis. *Applied Science and Technology Annals* DOI: https://doi.org/10.3126/asta.v1i1.30281: 100

Creswell, J., W. 2012. *Educational research; Planning, conducting, and evaluating quantitative and qualitative research 4th ed.* Pearson Education, Inc. Boston.

Drigas, A., S., & Gkeka, E., G., 2016 Montessori method and ICTs. *International Journal of Recent Contributions from Engineering, Science & IT (iJES)*, 4(1): 25–30.

Ely, M., 1991 *Doing qualitative research; circles within circles*. Taylor & Francis e-library.

Fukuyama, M., 2018 Society 5.0: *Aiming for a new human-centered society*. Retrieved from http://www.jef.or.jp.

Kemenkes, 2020 Covid-19 Update. *Accesed on august 14th, 2020*, on http://www.Covid19.kemenkes.go.id.

Khanzode, K. C. A., & Sarode, R. D. 2016. Evolution of the World Wide Web: from web 1.0 to 6.0. *International journal of Digital Library services*, 6(2): 1–11.

Kocour, N. 2019 *How blended learning impacts student engagement in an early childhood classroom*. https://nwcommons.nwciowa.edu/education_masters

Kusnadi, U., & Ismail, M., H. 2020. Improving the quality of children's development in kindergarten by stimulating the creative cognitive skills using the pragmatic approach to creativity. *Proceeding of the international Conference on Early Childhood Education and Parenting (ECEP 2019)*, Atlantis press: 264–265.

Mariyana, R., & Setiasih, O. 2018. Desain lingkungan belajar untuk mengoptimalkan multiple intelligences anak usia dini. *Jurnal Pendidikan Usia Dini*, 12(1): 141–152.

NAEYC and the Fred Rogers Center. 2012. *Technology and interactive media as tools in early childhood programs serving children from birth through age 8*. Retrieved from http://www.naeyc.org: 1–10.

Powell, M. 2016. Montessori practices: options for a digial Age. The Wildflower Foundation. *The NAMTA Journal*, 41(2): 156.

Prentzas J. 2013. Artificial Intelligence Methods in Early Childhood Education. In Yang XS. (eds) Artificial Intelligence, Evolutionary Computing and Metaheuristics. Studies in Computational Intelligence, vol. 427. Springer, Berlin, Heidelberg. https://doi.org/10.1007/978-3-642-29694-9_8

Priyono, D. 2020. *The challenge of ECCE inclusive education for society 5.0 in Southeast Asia*. SEAMEO CECCEP. Retrieved from http://www.criced.tsukuba.ac.jp/math/seameo/2020/presentations/14Feb/21-Dwi_P-2020.pdf

Sobel, K. 2019. *Future of childhood salon on immersive media and child development at Arizona state university*. Retrieved from; http://www.joanganzcooneycenter.org: 6–19.

State of Victoria. 2012. Blended learning a synthesis of research findings in Victorian education 2006–2011. *Ultranet and Digital Learning Branch Department of Education and Early Childhood Development*. Retrieved from; http://www.education.vic.gov.au:14–20

Sujiono. 2009. *Konsep Dasar PAUD*. Jakarta: PT. Index.

The Wildflower. 2020. *Authentic Montessori and contemporary considerations*. Retrieved from; http://www.wildflowerschool.org: 4–8.

Vygotsky, L. S. 2004. Imagination and creativity in childhood. *Journal of Russian and East European Psychology*, 42(1): 1–38.

Williams, R., Park, H., W., Oh, L., & Breazeal, C. 2019. PopBots: Designing an artificial intelligence curriculum for early childhood education. *MIT Media Lab, Massachusetts*. Retrieved from; http://www.robotic.media.mit.edu

Zosh, J., M., Hirsh-Pasek, K., Golinkoff, R., M., & Morris, J. P. 2016. Learning in the digital age: Putting education back in educational apps for young children. *Encyclopedia on Early Childhood Development*, 1–5.

Educational Innovation in Society 5.0 Era: Challenges and
Opportunities – Purnomo & Herwin (Eds)
© 2021 the authors, ISBN 978-1-032-05392-9

Online learning feedback for elementary school during the Covid-19 pandemic

F.N. Ismiyasari, E. Rahmawati, W. Kurniawan, Sutama, C. Widyasari, Z. Abidin & Z. Arifin
Universitas Muhammadiyah Surakarta, Surakarta, Indonesia

ABSTRACT: In facing the 5.0 society era through learning during the Covid-19 pandemic, there is a need for improving and strengthening the learning process between teachers and students by giving feedback. The research objectives describe the teacher's feedback for long-distance learning during the Covid-19 pandemic. The research used a descriptive qualitative approach. The research design was the study case of SD Muhammadiyah 16 Surakarta in the 2019/2020 academic year. The subject included the teachers and high-class students. The data collection employed used the techniques of observation, interviews, and documentation. The data were validated by the data triangulation. The data analysis was qualitative. The results showed that the teachers' online learning feedback was positive and negative. The positive feedback consisted of giving praise, passion, and reward or prize; the negative one was verbally warning for building and motivating the students learning. Positive and negative feedback in online learning can be given through applications such as Zoom, Google Forms, and WhatsApp.

1 INTRODUCTION

The 5.0 society era is coined by the Japanese government with the concept of technology-based human resources (Wibawa & Agustina 2019). This concept based on big data is collected by the Internet of Things (IoT), which is transformed Artificial Intelligence (AI) into something that can improve the ability of people to achieve a meaningful life (Özdemir & Hekim 2018). One of them is the education sector. This is because technology is an inseparable part.

The technology in the education field can be used to improve the quality of learning following the demands of the education paradigm. In this regard, on October 31, 2019, the President of Republic Indonesia Joko Widodo in an internal meeting of the Indonesian cabinet instructed the Ministry of Education and Culture to provide the right solutions to education following the conditions of technologies to the world (Hidayat 2019).

However, the solutions take a problem of the covid-19 virus infecting almost all over the world. The covid-19 virus first appeared in Wuhan, China (Shi et al. 2020). Corona (covid-19) is a new virus that causes respiratory disease (Kemenkes RI 2020). This virus delivers an impact on the original implementation of learning takes place to distance learning conducted online. The distance learning is a learning process by using various media information and communication technology (Kemendikbud RI 2013).

The implementation of distance learning can be supported by using various platforms (Bensalem 2018), such as Google Classroom, WhatsApp, Schoology,

Edmodo, Quipper School, Room Master, Zenius, & Cisco Webex (Adit 2020). By employing these technologies, teachers and students can carry out activities such as learning in the classroom. This is in line with the concept of society 5.0 which allows technology to provide access in a virtual space that looks like a physical space (Nastiti & Abdu 2019).

In welcoming the era of society 5.0 through learning amid the covid-19 pandemic, it is very necessary to improve and strengthen the learning process between teachers and students, namely through providing feedback. Through the feedback, the student will gain certainty regarding the achievement and ability of the learning objectives. The focus of the feedback is not merely on academic students, but also at the same time on building the students' character and behavior and attitude.

During the covid-19 pandemic, the feedback is very essential to monitor and motivate students during the learning activities at home. This is relevant to the research by (Anggraini, Hudiono & Hamdani 2015) that students who get feedback from the teacher have a better average of learning outcomes than students who do not get feedback.

One school that has implemented feedback during the covid-19 pandemic is SD Muhammadiyah 16 Surakarta. Based on the results of observations and interviews at SD Muhammadiyah Surakarta 16, providing feedback by the teachers through learning online in the middle of the covid-19 pandemic is positive and negative. Providing feedback is usually provided through learning applications such as Zoom, Google forms, and WhatsApp. The research objective is to describe the feedback provided by teachers to elementary school students during distance learning amid the covid-19 pandemic.

2 LITERATURE REVIEW

2.1 *Online learning*

Online learning is the learning is done online with the implementation of technology, including of synchronous and asynchronous (Dwi et al. 2020). Synchronous means the learning process happens at the same time between teachers and students enabling their direct interaction through learning online. Meanwhile, asynchronous is a learning process that does not occur at the same place or at the same time.

During the covid-19 pandemic, e-learning is used as an alternative to implementing the learning process. The use of e-learning in learning allows students to carry out learning activities from their homes by paying attention to health protocols.

In practice, the learning process based on e-learning by using adequate infrastructures can develop effective and efficient learning quality (Rustiani et al. 2019). Effective learning quality can be conducted by giving feedback. Giving feedback is carried out by teachers and students either individually or in groups (Kemendikbud 2016).

2.2 *Feedback*

Feedback is an action that is given to students by teachers to assist them in understanding learning in responses to their learning outcomes discussed (Windarsih 2016). There are four levels of feedback, namely (1) right or wrong information; (2) response giving; (3) explanation; and (4) concept strengthening (Roper in Windarsih 2016). The feedback has several functions namely warning, strategy improvement, informational functions, communication and motivation (Windarsih 2016).

The implementation of feedback for every class is different. It is due to adapting to the circumstances of students and teachers. Moreover, giving feedback on online learning takes some problems such as limited assistance when parents accompany the child to learn, lack of communication with parents because of busywork, and limited internet quota. After the finished learning, giving assignments to the students should immediately be done so that errors on the students' worksheets can be directly corrected. It is relevant to the research by Sofyatiningrum, Ulumudin & Perwitasari (2019) that giving feedback would be more meaningful if students who answer incorrectly can be corrected immediately by the teacher.

2.3 *Covid-19 pandemic*

The coronavirus or also known as covid-19 was first reported by WHO China. This virus is endemic in the city of Wuhan, China, and has spread in several countries, included in Indonesia.This virus is easily spread and transmitted through droplets of saliva or mucus or sneezing and air, and less likely to be affected by variations in climate in the world (MacKenzie & Smith 2020). As a step towards anticipating the spread of the virus covid-19, the Indonesian government appealed to all communities to carry out physical distancing (keeping a physical distance) and social distancing (keeping social distance). This is supported by the policy of President Joko Widodo on April 9, 2020 Number 083 /Sipres/A6/ IV/ 2020 regarding the appeal to the people to work, worship at home in an attempt to cut off the spread of the virus covid-19. As a form of support to Government efforts to anticipate the spread of the virus covid-19 Kemendikbud through the Ministry of Education Number 3 of 2020 on the prevention covid-19 in the educational field and No. 36 962/MPK.A/HK/2020 on online learning and work from houses in the context of preventing the spread of the covid-19 virus, urged all levels of education to carry out distance learning by paying attention to health protocols.

3 METHOD

This research used qualitative research with a case study research design. The subjects of this research were teachers and students in grade 4, grade 5, and grade 6. The research location was at SD Muhammadiyah 16 Surakarta for the 2019/2020 academic year. The data collection techniques used observation, interview, and documentation. Meanwhile, the research instrument sheets are:

3.1 *Observation*

The observation sheet is used as the guidance for observation activities by teachers and students in the learning process. In other words, sheets observation in this study is the observation sheet activities of teachers and students during the process of giving feedback in learning.

3.2 *Interview guidelines*

Interview guidelines are used to explore information from informants orally. The informant in this research is the teachers and students in grade 4, grades 5 and 6 of SD Muhammadiyah 16 Surakarta. The informants were randomly selected according to the level of education and teachers' activity. Meanwhile, the informant students are selected based on their level of activity in online learning. The interview process took place face to face and through the application WhatsApp.

3.3 *Documentation*

Documentation of data is used as the supporting documents in the form of the attendant list, a list of student grades, student worksheets (online), document the results of interviews and photographs as supporting feedback of online learning amid a pandemic covid-19 in primary schools.

This research procedure begins with case finding which is then continued by making research instruments, selecting informants who will be involved in the research as well as making observations, interviews, and documentation. The data analysis has been done after the necessary data has been obtained. The data were presented and concluded.

The data analysis used a qualitative analysis technique with the interactive model of Miles and Huberman (Figure 1). The activities in data analysis are data reduction, data display, and conclusion drawing and verification.

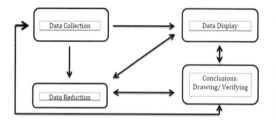

Figure 1. Data analysis techniques by Miles and Huberman.

Data were obtained from such as observations and interviews. After the data is reduced, they are presented and concluded.

4 RESULTS AND DISCUSSION

This research was conducted at SD Muhammadiyah 16 Surakarta. The research subjects were teachers and students in grade 4, grade 5, and grade 6, totaling 60 students and 3 teachers. Based on the interviews, it is found some feedback from the teachers about online learning during the pandemic covid-19. The following are some of the components discussed in this study.

4.1 Feedback on learning in class 4

According to interviews with teachers during online learning regarding the understanding of feedback, Mr. Ricky Darmawan as a grade 4 teacher revealed that "the feedback is a teacher activity to tell the students' learning problem so that students will be motivated to make an improvement and don't repeat it." The teachers provide feedback on each end of the subtheme. It is supported by the results of the interview with the students who showed the percentage of 65.4% of teachers always giving feedback after executing online learning (the end of the sub-themes). Meanwhile, the results of the interview with 20 students in grades 4, it is concluded that feedback is given by the teachers in the form of scoring, giving praise (spirit or motivation), motivation, reinforcement, reward, and punishment. In addition to this form of feedback, the teacher also provides feedback in the form of remedial and enrichment. Remedial activities are aimed at students who have not yet achieved mastery learning. Meanwhile, enrichment is aimed at students who have achieved complete learning. Remedial and enrichment activities are carried out at the end of the sub-theme. The following percentages are the remedial and enrichment activities of Grade 4.

In Table 1, it is reported that 26.9% said teachers always give the remedial and 38.5% showed that the teachers sometimes provide the remedial. The percentage stating that the teacher always gives the remedial and enrichment is lower than the students who declare that they sometimes provide remedial and enrichment.

Table 1. Percentage of class 4 remedial and enrichment activities.

Statement	Percentage			
	Always	Often	Sometimes	Never
The teacher provides remedial if the value is below the minimum completeness criteria	26.9%	11.5%	38.5%	23.1%
Teachers provide enrichment when a grade below the minimum achievement criteria (KKM)	30.8%	11.5%	42.3%	15.4%

4.2 Feedback on learning in class 5

According to the interviews with teachers during online learning regarding the understanding of the feedback, Apriyanto as a 5th-grade teacher said that "feedback is an activity provided by the teacher to see the weaknesses and strengths of students when taking learning materials so that learning objectives can be achieved to the maximum." The teachers provide feedback on each learning end. It is supported by the results of the interview with the students who show the percentage of 85.7% of the teachers always give feedback after implementing online learning (learning end). Meanwhile, the results of interviews with 20 students in grades 5, it can be concluded that the feedback is given by the teachers in the form of scoring, giving praise (spirit or motivation), correct answers, the strengthening of the answers, and the provision of reward. When grade 4 used feedback of punishment, in grade 5 the results of the interview with the students showed a percentage of 27% teachers give punishment to the students, but only the provision of warning or motivation. In 5th grade, the remedial feedback forms and enrichment are also given to the students. The remedial and enrichment activities were held at the end of the lesson. Here are the percentages of remedial and enrichment activities of 5th grade.

In Table 2, it can be seen that 42.9% said teachers always give the remedial and 38.1% showed the teacher sometimes provide the remedial. Meanwhile, 23.8% of students stated that teachers always give enrichment and 33.3% of students stated the teacher only occasionally provides enrichment.

Table 2. Percentage of class 5 remedial and enrichment activities.

Statement	Percentage			
	Always	Often	Sometimes	Never
The teacher provides remedial if the value is below the minimum completeness criteria	42.9%	4.8%	38.1%	14.3%
Teachers provide enrichment when a grade below the minimum achievement criteria (KKM)	23.8%	28.6%	33.3%	14.3%

The percentage stating the teachers who always provide the remedial is higher than students who declare that the teachers sometimes give the remedial, while the percentage who stated the teacher always gives a lower enrichment than students who expressed enrichment to students.

4.3 Feedback learning in class 6

According to interviews with teachers during online learning regarding the understanding of feedback, Ms. Mandy Ujiowati as a class 6 teacher revealed that "the feedback is an act of the teacher to determine the level of students' success by providing feedback on the results of learning that have been done." The teachers provide feedback at each end of the lesson or every day. It is supported by the results of the interview with the students who showed the percentage of 73.9% of the teachers always give feedback after implementing online learning (learning end). Meanwhile, the results of interviews with 20 students in grades 6, it can be concluded that the feedback is given in the form of scoring, giving praise (spirit or motivation), correct answer, granting rewards, and providing reinforcement of answers. Meanwhile, in grade 6 the feedbacks are in the form of punishment shows a percentage of 0%, which means that the teacher never gives punishment to students, but only gives verbal warnings to students. In grade 6, the remedial and enrichment were also given to students. The remedial and enrichment activities were conducted at the end of the lesson. Here are the remedial and enrichment percentage of Grade 6 students.

In Table 3, it can be seen that the activities of the remedial or when students get grades below KKM percentage. The students stated that the teachers who always give more remedial are lower than students who stated that teachers sometimes give the remedial. Meanwhile, the percentage of the students states that the teachers who always give a higher enrichment than students who stated that the teachers sometimes provide the enrichment. This is evident in the results

of the percentage of Table 3. It is shown that 30.4% said that the teachers provide the remedial and 39.1% showed that teachers sometime provide the remedial. Meanwhile, 34.8% of students stated that the teachers always provide the enrichment and 30.3% of students stated that the teachers sometimes provide the enrichment.

Table 3. Percentage of class 6 remedial and enrichment activities.

Statement	Percentage			
	Always	Often	Sometimes	Never
The teacher provides remedial if the value is below the minimum completeness criteria	30.4%	21.7%	39.1%	8.7%
Teachers provide the enrichment when a grade below the minimum achievement criteria (KKM)	34.8%	28.4%	30.3%	14.3%

The percentage stating that the teachers always provide the remedial is lower than students who declare that the teachers sometimes give the remedial, while the percentage reported that the teachers always give a higher enrichment than students who declared the enrichment.

Learning feedback is the focus of this study. Feedback is provided through online learning to upper-class students at SD Muhammadiyah 16 Surakarta. The result shows that there are six forms of learning feedback: (1) the provision of grade (score/number); (2) giving praise (enthusiasm/motivation); (3) giving rewards, (4) giving punishment; (5) correcting the correct answer; (6) providing answers. Here are the percentages of the feedback forms of grade 4, 5, and 6 (Table 4).

Table 4. Percentage of feedback from grade 4.

As Feedback	Percentage		
	Grade 4	Grade 5	Grade 6
Grading (score/number)	92.3%	81%	73.9%
Giving a compliment/spirit	65.4%	71.4%	56.5%
Giving compliment/ motivation, spirit	65.4%	71.4%	56.5%
Giving *rewards*	15.4%	9.5%	21.7%
Giving *punishment*	11.5%	27%	0%
Correction jaw a correct tire	26.9%	71.4%	30.4%
Granting answers	7.7%	19%	4.3%

Meanwhile, there are two additional feedbacks given to students in the form of remedial and enrichment. The remedial will be given if the student's score is below the minimum achievement criteria (KKM). Meanwhile, the enrichment will be conducted if the students' scores are below the minimum achievement criteria (KKM) (see Table 5).

Table 5. Percentage of the remedial and enrichment of Class 4, 5 and 6.

Description	Class student	Percentage			
		Always	Often	Sometimes	Never
Teacher	4	26.9%	11.5%	38.5%	23.1%
give	5	42.9%	4.8%	38.1%	14.3%
remedial	6	30.4%	21.7%	39.1%	8.7%
Teacher	4	30.8%	11.5%	42.3%	15.4%
give	5	23.8%	28.6%	33.3%	14.3%
enrichment	6	34.8%	28.4%	30.3%	14.3%

Based on Tables 4 and 5 above, it is known that the feedback is very essential in online learning during the Covid-19 pandemic. Giving feedback is carried out at the ends of the lesson and sub-theme. Of course, it needs to consider the situation and condition of students and teachers in carrying out learning (Seruni & Hikmah 2015). The percentage of learning feedback is the highest form of grade (92.3%), while the percentage of giving feedback is the lowest form of strengthening students' answers.

5 CONCLUSION

5.0 society era is centered on and technology-based human resources. In facing the era, learning amid the covid-19 pandemic is an indispensable effort to improve and strengthening the learning process. It can be conducted through the provision of feedback. Based on the results of the research, the feedback of online learning during the covid-19 pandemic at elementary school students of Muhammadiyah 16 Surakata shows that giving positive and negative feedback is very beneficial for teachers and students. Giving positive feedback shows a higher percentage than negative feedback. The positive feedback in the forms of giving a grade, giving compliments, encouragement and reward or prize, rectification correct answers, and providing the remedial and enrichment. Meanwhile, the negative feedback is in the form of punishment or verbally constructive warning that can be motivated the students to take a more active in learning.

Giving positive and negative feedback in online learning can be implemented through the applications of Zoom, Google forms, and WhatsApp. In practice, several factors are faced by the teachers to provide feedback to students such as limited assistance when parents accompany the child to learn, limited communication with parents because of busywork, and limited internet quota.

REFERENCES

Adit, A. 2020. *12 Aplikasi pembelajaran daring kerjasama kemendikbud, gratis!, Kompas.com*. Available at: https://edukasi.kompas.com/read/2020/03/22/123204571/12-aplikasi-pembelajaran-daring-kerjasama-kemendikbud-gratis?page=all.

Anggraini, W., Hudiono, B. & Hamdani. 2015. Pemberian umpan balik (Feedback) terhadap hasil belajar dan self-efficacy matematis siswa kelas VII SMP, *Jurnal Pendidikan Matematika FIKP Untan*, 4(9): 1–13.

Bensalem, E. 2018. The impact of WhatsApp on EFL students' vocabulary learning. *Arab World English Journal*, 9(1): 23–38.

Dwi, B. *et al.* 2020. Analisis keefektifan pembelajaran online di masa pandemi Covid-19, *Journal Pendidikan Guru Sekolah Dasar*, 2(1): 28–37.

Hidayat, T. 2019. *Trend teknologi revolusi industri 4.0*. Available at: https://unida.ac.id/teknologi/artikel/trend-teknologi-revolusi-industri-40.html.

Kemendikbud. 2013. Permendikbud RI Nomor 109 Tahun 2013: 1–8. Available at: https://lppmp.uns.ac.id/wp-content/uploads/2018/03/Permen-Nomor-109-tahun-2013-ttg-PJJ.pdf.

Kemendikbud. 2016. *Panduan gerakan literasi sekolah*. Jakarta: Kemendikbud.

Kemenkes. 2020. *Tentang novel corona virus (NCov), Kementerian Kesehatan RI*. Available at: https://www.kemkes.go.id/resources/download/info-terkini/COVID-19 (Accessed: 1 May 2020).

MacKenzie, J. S., & Smith, D. W. 2020. COVID-19: A novel zoonotic disease caused by a coronavirus from China: What we know and what we don't, *Microbiology Australia*, 41(1): 45–50.

Nastiti, F. E., & 'Abdu, A. R. N. 2019. Kesiapan pendidikan Indonesia menghadapi era society 5.0. *Jurnal Kajian Teknologi Pendidikan*, 5(1): 61–66.

Özdemir, V. & Hekim, N. 2018. Birth of industry 5.0: Making sense of big data with artificial intelligence, "the internet of things" and next-generation technology policy. *OMICS A Journal of Integrative Biology*, 22(1): 65–76.

Rustiani, S. *et al.* 2019. Measuring usable knowledge: Teacher's analyses of mathematics for teaching quality and student learning: 13–14.

Seruni, S. & Hikmah, N. 2015. Pemberian umpan balik dalam meningkatkan hasil belajar dan minat belajar mahasiswa, *Formatif: Jurnal Ilmiah Pendidikan MIPA*, 4(3): 227–236.

Shi, H. H., Jiang, N., Cao, Y., Alwalid, O., Gu, J., Fan, Y., & Zheng, C. 2020. Radiological findings from 81 patients with COVID-19 pneumonia in Wuhan, China: a descriptive study, Available at: https://www.thelancet.com/action/showPdf?pii=S1473-3099%2820%2930086-4.

Sofyatiningrum, E., Ulumudin, I. & Perwitasari, F. 2019. Kajian umpan balik guru terhadap hasil belajar siswa, *Indonesian Journal of Educational Assesment*, 2(2): 56.

Wibawa, R. P. & Agustina, D. R. 2019. Peran pendidikan berbasis higher order thinking skills (Hots) pada tingkat sekolah menengah pertama di era society 5.0 sebagai penentu kemajuan bangsa Indonesia, *Equilibrium*, 7(2): 137–141.

Windarsih, C. A. 2016. Aplikasi teori umpan balik (feedback) dalam pembelajaran motorik pada usia dini, *Tunas Siliwangi*, 2(1): 20–29.

Educational Innovation in Society 5.0 Era: Challenges and
Opportunities – Purnomo & Herwin (Eds)
© 2021 the authors, ISBN 978-1-032-05392-9

Multiliteracy education challenges: Will narrative texts used in textbooks open students' minds to critical literacy?

H.N. Fadhlia & W. Purbani
Yogyakarta State University, Yogyakarta, Indonesia

ABSTRACT: This study investigates critical literacy dimensions implemented in narrative texts and how they are presented. Narrative texts in junior high school and senior high school textbooks were taken as the subject of study since they are compulsory materials taught in the classroom but are regularly considered as a non-powerful material regardless of their multiple insights. A qualitative content analysis was conducted to explore critical literacy dimensions in narrative texts, whereas datasheets consisting of the development of critical literacy dimensions were utilized as the data collection method. The result revealed that the dimensions consisting of disrupting the commonplace, interrogating multiple viewpoints, focusing on sociopolitical issues, taking action, and promoting social justice were covered in narrative texts presented through learning unit and learning activity. Those covered skills indicated that narrative texts assist students to develop their critical literacy skills but teachers' multimodalities and strategies are still needed to give students guidance.

1 INTRODUCTION

Critical literacy has become a crucial discussion which has obtained more attention in recent decades (Gustine, 2018; Kim, 2012; Ko, 2010). This topic has been discussed in both English-speaking countries and non-English speaking countries (Ko, 2010; Gustine, 2018; Mbau, 2019) as many scholars believe that critical literacy stimulates students to investigate hidden messages, explore the status quo, and take action for more equal societies (Dozier et al., 2006). It is an act of understanding the message to raise consciousness. Moreover, critical literacy promotes multicultural perspectives in education. Yoon (2016) reveals that multicultural perspectives can be promoted through critical literacy where it is implemented by providing more insights in making common invisible and visible issues. This relates to the concept of multiliteracy education that involves cultural and linguistic diversities which school should encourage teachers to use knowledge of local literacies, both the language and its cultural meaning (Heath, 1983).

Even though critical literacy has been attempted for several years in the EFL context (Gustin, 2014), this concern still gets less attention in their classroom practices due to a lack of understanding on critical literacy urgency and knowledge of its implementation in non-western settings (Kim, 2012). Kim (2012) further asserts that textbooks can be a tool to involve critical literacy in Asian classrooms by examining a text's construction and students' social/political contexts. By following this setting, students are allowed to share and

develop their ideas to reconstruct the meaning of the text and raise their consciousness towards their everyday lives. This idea supports evidence from previous research that learners can be encouraged to be critically literate through several texts, visual images, and spoken texts (Freire & Maccedo , 2005).

In relation to this study, according to curriculum 2013, junior high school and senior high school students are presented to learn various types of texts in their textbooks, such as procedural, descriptive, recountant, narrative, and report (Permendikbud no 37, 2018). However, the analysis conducted by Archakis (2012) reveals that most of the time, narrative is understood as the literary narrative which is not taken as a serious concern to create a meaningful relation between text and the readers' everyday lives. The narratives tend not to be sufficiently familiarized with students' context. This issue is regrettable since they contain some sociocultural aspects which motivate learners to investigate and discuss several values related to their daily lives (Schiffrin, 1996). Considering the implementation of narrative texts in the curriculum, this present study utilizes narrative texts to be investigated referring to some critical literacy aspects. Narrative texts are taken to be the context of this study since they potentially invite teachers, students, and other practitioners to understand any culture which influences their behavior in understanding diversities (Holland & Kilpatrick, 1993). Thus, this study aims at exploring critical literacy dimensions in narrative texts presented in junior high school and senior high school textbooks published by the Indonesian government. It is conducted to see whether it

DOI 10.1201/9781003206019-19

consists of critical literacy dimensions which facilitate students to view differences within their sociocultural conditions.

2 LITERATURE REVIEW

2.1 *Critical literacy and critical literacy dimensions*

The meaning of critical literacy is derived from the work of Freire (1972) which relates to reading for justice. Freire insists that education can be used as a tool to raise students' awareness towards oppressed societies. It is in line with Janks' statement (2014) that education should connect to the world they are living in. Freire's statement strengthens educators to develop critical literacy in their classroom so that everyone can be critical towards their surrounding issues and understanding their roles as society in a global context by reading. Furthermore, it enables students to see how language carries some ideological issues on a text. Kim (2012) conclusively demonstrates that it is important to scrutinize how dominant discourse is transformed into a text. That is, critical literacy considers language learning a social practice where students do not only learn to improve their language skills development but also learn to understand themselves as social agents skilled in questioning the assumptions of society (Parker, 2013).

Lewis et al. (2002) proposes four critical literacy dimensions to put it into practice. These dimensions help language teachers to carry out critical literacy implementation in their teaching and learning environment (Mbau, 2019). Those dimensions consist of (a) disrupting the commonplace, which refers to viewing routines in a new point of view (Lewinson et al., 2002). It implies that language and any other sign systems are utilized to investigate perception in one context and comprehend experiences; it elicits a question: "how is the text trying to position me?" (Mbau, 2019). The second dimension, (b), is interrogating multiple viewpoints which relate to seeing experiences and texts through their own perspectives and others (Lewis et al., 2002). It facilitates students to have skills of reflecting on multiple perspectives, utilizing those voices to question texts, and finding out whose voice is silenced, then making it noticeable (Mbau, 2019). (c) is focusing on sociological issues where it uses language to be perceived to be never neutral or always influenced by systems and power connections (Lewison et al., 2002). It encourages students to be capable of understanding sociopolitical issues and unequal power structures in societies (Mbau, 2019). (d) is taking action and promoting social justice, which is a step ahead from those three dimensions. It enables learners to reflect and act to promote social justice (Lewinson et al., 2002). In this part, students will be involved in reflecting and acting based on what they have read. It allows students to do praxis—doing reflection and action towards the world they are living at (Freire, 1972).

2.2 *Narrative as a means of promoting multicultural education*

Multicultural Education is a movement to remove students' gaps in education to have equal opportunities regardless religions, genders, cultural backgrounds, and socioeconomic statuses (Ndura & Dogbevia, 2013; Yilmaz, 2016). It is a concept to promote justice within culturally diverse societies in our educational settings. This statement is confirmed by Ameny-Dixon (2013) stating that multicultural education is an approach based on democratic values which promotes cultural pluralism.

Moreover, narratives can be used as a means of promoting pluralism by understanding that literature will sharpen students' knowledge on the human condition. This fact is found by Connell et al. (2004) claiming that the narrative is a presentation of other social constructions consisting of values, shared meanings, and languages. It expands students' knowledge on valuing diverse cultures since narrative is a way of systemizing human's experiences and their environment where knowledge can be shared, exchanged, and transferred (Souto-Manning, 2014; Connell et al., 2004). It can be inferred that a narrative is a process to deepen our understanding regarding our sociocultural environment since Holland & Kilpatrick (1993) note that a narrative invites the readers to understand others' cultures and helps them to behave and build respectful relationships.

Including multicultural perspectives in our educational field facilitates students to be aware of their social environment. They will utilize a critical lens including global and multicultural perspectives in their learning process (Yoon, 2016). It is confirmed by Bobkina & Stefanova (2016) that the integration of literature in an academic curriculum is used as a means of raising students' critical awareness. Conversely, this is learned through narratives since it emphasizes how people in any culture develop values in their daily habits and stimulate insight to understand others' identities (Holland & Kilpatrick, 1993). This claim supports Connel (2004) asserting that narrative is a sociocultural construction exchanging knowledge. Discourse analysis and sociolinguistic aspects emerge Archakis (2013). That being said, narrative does not merely allow students to read a text, but it also expands their perspectives to see a certain context through different angles.

3 RESEARCH METHOD

This present study applied a qualitative content analysis adapted from Mayring (2014). Qualitative content analysis was selected due to the aim of the study to analyze critical literacy aspects in narrative texts published on government textbooks. This research design was suitable since qualitative content analysis characterized the material from one passage (Marying, 2014). It led the researchers to interpret the narrative text setting using some objectives utilized in analyzing critical literacy dimensions. It was conducted in order to figure out whether narrative texts applied on students' textbooks carried critical literacy dimensions to students. The procedure of this qualitative content analysis consisted of 7 steps. First, formulating the research question: in this present study the research proposed two research questions. They are "what critical literacy dimensions are included in textbook narrative texts?" and "how do textbook narrative texts open students' minds to critical literacy?" Second, categories were defined. The researcher used some literature related to critical literacy values and determined four dimensions of critical literacy as proposed by Lewinson et al. (2012). Third, formulating coding guidelines, this step leads the researcher to create analytical constructions to code the text based on some categories defined. In this step, the researcher developed two categories for each dimension in Lewinson et al. (2002). Next, coding: the researcher coded the texts based on the coding formula. Fifth, in the revising step, the researcher revised the coding guideline after the analysis was running at about 30%. Sixth, presenting the final work: the researcher revealed the result of the analysis in the analysis table. Last, analyzing the results: in this step the researcher reanalyzed the texts comprehensively to generate the conclusion.

English textbooks of Junior High School grade VIII and English textbooks of Senior High School grade X were collected as the source of the data collection. The researcher chose the textbooks purposively utilizing several reasons. The main reason of choosing these textbooks was because narrative text becomes one of the materials in this book. The textbooks chosen here were the revised version of English Book "When English Rings a Bell" Class VIII and "*Buku Bahasa Inggris*" Class X. They were selected as the instrument since they are textbooks for those who implement the 2013 curriculum. In this study, the researcher gathered several narrative texts and their activities taken from those compulsory books.

The data gathered was analyzed applying qualitative data analysis. The researcher was required to code the data after the categories were determined (Cohen, Manion, & Morrison, 2011). Four critical literacy dimensions proposed by Lewinson et al (2002) were utilized to code the data. The categories consisted of disrupting the commonplace, interrogating multiple viewpoints, focusing on sociopolitical issues, and taking action and promoting social justice. However, these categories were still developed into eight criteria based on critical literacy skills proposed by Lewinson (2002). The criteria used to code the data were (1) analyzing how the readers are positioned by the text, (2) developing the language of critique and hope, (3) including multiple perspectives and voices, (4) making difference visible, (5) challenging the unequal power relationship, (6) using literacy to involve in the politics of daily life, (7) engaging reflection and action upon the world in order to transform it, and (8) challenging how social action can redefine the existing discourse. Then, the data was classified based on the source of the data such as the learning objective, the learning input, and the learning activity. Those criteria were utilized to obtain comprehensive data.

4 FINDING AND DISCUSSION

The finding of this research supports Archakis's (2013) statement that narrative texts are not only limited to the aim of language teaching and literary value materials, but they motivate readers to criticize their own everyday lives. The following is the analysis of each of the skills categorized based on its critical literacy dimension. The first dimension, disrupting the commonplace, is a concept of seeing everyday context through new views (Lewinson et al., 2002). It is divided into two skills, namely analyzing how the readers are positioned by the text to support or disrupt the status quo and developing the language of critique and hope. First, the skill of analyzing how readers are positioned in narrative text of junior high school textbook is presented through its learning input. In the story of Miss Crow and Mr. Fox, students are able to see how Miss Crow's bad behavior in stealing cheese is immediately paid off by Mr. Fox's unfavorable attitude by taking over the cheese. This context positions the readers to consider the notion of "You reap what you sow." They can criticize the story and relate it to the notion existing in our society. Students might support or disrupt the notion regarding the story context as Archakis (2013) claims that this dimension projects values and identities to argue.

Moreover, this skill positioned the readers to see a status quo related to female stereotypes. It presents females' sensitivity and their tendency toward beauty. From the story, we can infer that the female actor drops the cheese when she opens her mouth to show that she is able to sing and she is the most beautiful bird in the world. This setting rouses readers' critical literacy skills since it includes female issues referring to her beauty. This issue can be meaningful to discuss in the classroom to help learners to acknowledge the status quo of females in the society. This type of learning input can be a source of doing reflection relating the learning setting and their daily context. Finally, they can determine whether they agree or disagree with the statement of status quo. This practice facilitated a critical literacy activity since critical literacy empowers learners to oppose the message created by the author (Lee, 2015). Meanwhile, the senior high school textbook presents this skill through its learning activity. It is interesting since they carry the readers to this skill by giving questions. This fact supports Bobkina & Stefanova (2016) claiming that critical literacy empowers reader-response criticism. The activity stimulates the readers to position themselves on a certain belief. The questions facilitate learners to reflect on how our society treats those of disadvantaged physical or psychological condition. The following activity is the example of the implementation of this skill in the learning activities of "Issumboshi" story.

POINTS TO PONDER

If you have a friend with disadvantaged physical or psychological conditions, would you not befriend with him/her? Would you laugh at him/her? What should you do? Why?

Figure 1. Learning activity in Issumboshi story.

Second, the critical literacy skill of developing language critique and hope can be seen on the narrative texts of junior high school and senior high school textbooks. It is taught through the learning activities. In the junior high school textbook, it is implied in the activity where learners are invited to investigate and develop the story in their surroundings. In this context, they are given chances to put their language of critique and hope by developing a story. Similarly, it is covered in the senior high school textbook by requesting learners to complete a story. This activity promotes critical literacy since they are given chances to criticize the story and enhance the narration. It can be seen in the following example:

Production

Activity 1

In groups of five, discuss a legend from your surrounding. Then, make an outline to help develop the story. Follow the steps of writing process for developing the story.

Figure 2. Learning activity in Junior High School textbook.

Task 3:
This is the beginning part of a story about Kanchil. The text is not written properly. Edit the 15 words in the text so that the story makes sense.

Kanchil, the small and clever mousedear, had many enemies in the forest. fortunately, he was quick-witted, so that evry time his life was threatened, he managed to escape.

One of his greatest enemies was Crocodile, who lived in the rivr that bordered the forest. many times Crocodile had tried to capture Kanchil. Crocodile was big, but he was not very clever. Kanchil was abel to trick him every time.

One day it was vry hot. There was no wind at all to refresh the thirsty plants and tres of the forest. it was in the middel of the dry season. For many weeks no rain had fallen, so the littel creeks where the small animals usd to drink had dried up. Kanchil was walking alone in the forest. he was very thirsty. He had walked a long way, looking for a brook where he could quench his thirt.

Figure 3. Learning activity in Senior High School textbook.

The second dimension, interrogating multiple views, means understanding or experiencing a context from our own and others' perspectives (Lewinson et al., 2002). It is categorized into two critical literacy skills namely multiple perspectives & voices and making differences visible. First, for the skill of including multiple perspectives and voices can be seen in the learning activities both in junior high school and senior high school textbooks. For instance, the learning activities presented in the junior high school textbook facilitates learners to arrange a legend based on a picture. In this activity, they are given chances to construct the story by considering many sources. It lets students engage in multicultural perspectives activities since they can find many versions of story then they scrutinize it to organize a comprehensive and meaningful one. This skill is also found in other activities both in junior and senior high school textbooks where learners have opportunities to discuss their work with their peers. It encourages learners to grasp multiple perspectives since in handling the discussion they will possibly have different voices with their partners. They are expected to enlarge their views to find the hidden messages, implied meanings, and opinions separated from facts (Bobkina & Stefanova, 2016).

Second, the critical literacy skill on making differences visible is embodied in the learning input of junior high school and senior high school narrative texts. They are given texts where they can see differences. For instance, in "The Lost Caterpillar" story, it is exposed that the baby caterpillar is different from the other worms. This difference does not mean to say that she is not beautiful. It shows that her dissimilarity is a sign that she is different and she is beautiful. The same thing occurred on the story of "Issumboshi," it tells us that he lives with disadvantaged physical condition. It shows that his differences do not stop him to hope and dream, and he still can get his blessing. In this context, learners require to familiar themselves with dissimilarities to make them recognize that we live in world where we embrace diversity. This situation also supports multicultural education since it implies a movement to respect differences in terms of gender, sociocultural status, cultural background, and others (Yilmaz, 2015).

The third dimension, focusing on sociopolitical issues, helps learners to put their attention on sociopolitical systems and power (Lewinson et al., 2002). It is developed into two skills namely challenging the unequal power relationship and using literacy to involve in the politics of daily life. However, these skills are limited to be implemented in the textbooks. It is only the skill of challenging unequal power relationships presented in narrative texts of the senior high school context. In senior high school textbooks, challenging the unequal power relationship is covered by examining how injustice issues occur in a family consisting of three daughters where the younger daughter always gets unequal opportunities. It gives a picture on how those two elder sisters treat their younger sister unequally. Students can see and examine the unequal power relationships in the story. This inequality issue is included as the topic of narrative text to redefine a justice relationship. This promotes reading for justice which implies the practice of sociocultural theorist concerning on inequality and injustice issues (Bobkina & Stefanova, 2016).

The fourth dimension, taking action and promoting social justice, is concerned on doing reflection and action towards oppression (Lewinson et al., 2002). It is developed into two critical literacy skills. They are engaging reflection and action upon the world in order to transform it and challenging how social action can redefine the existing discourse. First, engaging reflection and action upon the world, is implemented in the learning input of the "Goldilocks and the Three Bears" story. This occurs since it tells us that that our actions can hurt others. It is illustrated by the Goldilock's decision in trespassing into someone else's house. From this story, readers can have a reflection that our selfishness brings someone else sorrow. Goldilock is merely focused on her own satisfaction and concerned on what is "just right!" for her. This topic invites students to do reflection on what is actually "just right" and influence their action since their definitions will carry their own values in their daily life.

Meanwhile, in the senior high school context, this critical literacy skill can be seen through its learning activity. The activity invited readers to do reflection about the stories that exist in their surroundings. All the questions on the activity lead learners to a comprehensive reflection. They can start reflecting to what happens in the story chosen and generate the message. Learners are encouraged to do reflection and reveal a solution towards a problem (Bobkina & Stefanova, 2016).

Second, the critical literacy skill on challenging how social action can redefine the existing discourse can be received on the narrative text of senior high school textbook in its learning activities. This skill occurs on an action of writing a script fragmented from the "Strong Wind" story and performing it in the classroom. This type of activity can be generated as an action to promote social justice since the story represents injustice issues. Learners also add another insight in the story to encourage others to act equally in this world. The script will be presented in front of the class to remind and to persuade others to create justice in ours surroundings.

5 CONCLUSION

The aim of the study is to explore critical literacy skills in textbooks and how they are presented. Based on the analysis, seven critical literacy skills are covered in those narrative texts, they are (1) analyzing how the readers are positioned by text, (2) developing the language of critique and hope, (3) including multiple perspectives and voices, (4) making difference visible, (5) challenging the unequal power relationships, (6) engaging reflection and action upon the world in order to transform it, and (7) challenging how social action can redefine the existing discourse. It is only using literacy to involve in the politics of daily skill which is not covered on the texts. Relating to those skills, they represent critical literacy skills in narrative texts of junior high school and senior high school textbooks, they are introduced through learning units and learning activities. The skills imply this in the texts, in discussion, writing, and role play activities. Even though it is stated that critical literacy is covered in the narrative texts, teachers' multimodalities and strategies in teaching are still needed to lead their students to see the message behind the text. Therefore, since this research finding is still limited, future researchers are suggested to do further research including teachers' perception and practices in the classroom.

REFERENCES

Archakis, A. 2012. *The construction of identities in critical education*. Palgrave Macmillan.

Ameny-Dixon, G. M. 2004. Why multicultural education is more important in higher education now than ever: A global perspective. *International Journal of Scholarly Academic Intellectual Diversity*, 8(1): 1–9.

Bobkina, J., & Stefanova, S. 2016. Literature and critical literacy pedagogy in the EFL classroom: Towards a model of teaching critical thinking skills. *Studies in Second Language Learning and Teaching*, 6(4): 1–20. https://doi.org/10.14746/ssllt.2016.6.4.6

Connell, N. A. D., Klein, J. H., & Meyer, E. 2004. Narrative approaches to the transfer of organisational knowledge. *Knowledge Management Research & Practice*, 2(3): 184–193.

Fajardo, M. F. 2015. A review of critical literacy beliefs and practices of English language learners and teachers. *University of Sydney Papers in TESOL*, 10: 29–56.

Freire, P. 1972. *Pedagogy of the oppressed* (30th ed.). Herder & Herder.

Freire, P., & Macedo, D. 2005. *Literacy: Reading the word and the world*. Routledge.

Gustine, G. G. 2018. A Survey on Critical Literacy as a Pedagogical Approach to Teaching English in Indonesia. *Indonesian Journal of Applied Linguistics*, 7(3): 531–537.

Heath, S.B. 1983. *Ways with words: Language, life, and work in communities and classrooms*. Cambridge University Press.

Holland, T. P., & Kilpatrick, A. C. 1993. Using narrative techniques to enhance multicultural practice. *Journal of Social Work Education*, 29(3): 302–308.

Janks, H. 2012. The Importance of Critical Literacy. *English Teaching: Practice and Critique*, 11(1): 150–163.

Kim, S. J. 2012. Critical literacy in East Asian literacy classrooms. *Perspectives on Global Development and Technology*, 11(1): 131–144.

Ko, M.-Y. 2010. *Critical literacy development in a college-level English reading class in Taiwan* (Doctoral dissertation). Indiana University, Indiana.

Kuo, J. 2014. Critical literacy in the EFL classroom: Evolving multiple perspectives through learning tasks. *The Journal of Asia TEFL*, 11(4): 109–138.

Lee, Y. J. 2015. First steps toward critical literacy: Interactions with an English narrative text among three English as a foreign language reader in South Korea. *Journal of Early Childhood Literacy*, 17(1): 26–46.

Lewis, C. 2001. Literacy practices as social acts: Power, status and cultural norms in the classroom. *Journal of Literacy Research*, 35(3): 941–946.

Lewison, M., Flint, A. S., & Van Sluys, K. 2002. Taking on critical literacy: The journey of newcomers and novices. *Language Arts*, 79(5): 382–392.

Mbau, A. T., & Sugeng, B. 2019. Critical Literacy for ELT in Indonesia: What EFL Teachers should be aware of. *Journal of English Language Teaching and Linguistics*, 4(2): 143–156.

Marying, P. 2014. *Qualitative content analysis: Theoretical foundation, basic procedures and software solution*. GESIS-Leibniz Institute for the Social Sciences. Retrieved from http://nbn-resolving.de/urn:nbn:de:0168-ssoar-395173.

Morrell, E. 2008. *Critical literacy and urban youth: Pedagogies of access, dissent, and liberation*. Routledge.

Ndura, E., & Dogbevia, M. K. 2013. Re-envisioning multicultural education in diverse academic contexts. *Procedia-Social and Behavioral Sciences, 93*: 1015–1019.

Parker, J. K. 2013. Critical literacy and the ethical responsibilities of student media production.*Journal of Adolescent & Adult Literacy, 56*(8): 668–676. https://doi.org/10.1002/JAAL.194

Schiffrin, D. 1996. Narrative as self-portrait: Sociolinguistic constructions of identity. *Language in Society, 25*(2): 167–203.

Souto-Manning, M. 2014. Critical narrative analysis: The interplay of critical discourse and narrative analyses. *International Journal of Qualitative Studies in Education, 27*(2), 159–180.

Yılmaz, F. 2016. Multiculturalism and multicultural education: A case study of teacher candidates' perceptions. *Cogent Education, 3*(1): 1172394. https://doi.org/10.1080/2331186X.2016.1172394

Yoon, B. 2016. *Critical literacies: Global and multicultural perspectives*. Springer.

Educational Innovation in Society 5.0 Era: Challenges and Opportunities – Purnomo & Herwin (Eds)
© 2021 the authors, ISBN 978-1-032-05392-9

Adaptive learning 4C skills during and after Covid-19 in elementary school

A. Yatini & B. E. Mulyatiningsih
Universitas Negeri Yogyakarta, Yogyakarta, Indonesia

ABSTRACT: The enGauge 21st Century Skill Framework requires progressive education to upgrade 4C skills that consist of creativity, critical thinking, collaboration, and communication. The emphasis on learning skills based on the reason that educators and students still have many weaknesses in the field of learning and in the teaching process. UNESCO (2020) reported that during the Covid-19, 93.1% of the 1.3 billion students in the world carried out learning activities at home. 45 million of these are students from Indonesia (Statistics Indonesia, 2020). Although teaching and learning occur in virtual conditions, the teacher should construct 4C skills among the students. This research aims to investigate the obstacles and opportunities the students may have, then give recommendations about adaptive learning 4C skills for the student during and after a pandemic. The subject is a student at an elementary school in Yogyakarta. Data was collected through observation and interview.

1   INTRODUCTION

The framework of enGauge 21st Century Skill demands progressive education to prepare students who can compete in a globalized world. Various strategies and government efforts were carried out such as: the implementation of thematic learning, HOTS model learning, character education, and literacy education. However, programs that have been promoted will not succeed without being supported by 21st century skills. 21st-century skills are divided into three categories, namely: learning skills, literacy skills, and life skills. The learning and teaching process in schools is more directed so that students have good learning skills. Learning skills or what is known as 4C consists of creativity, critical thinking, collaboration, and communication.

The government concern to learning skill because both educators and students still have many weaknesses in the field of learning and teaching. The problems of educators in schools include not having pedagogical skills to carry out 4C skill-based learning (creativity, critical thinking, collaboration, and communication). Students at school also experience difficulties in understanding abstract problems, working in teams, being unable to determine credible information and may have limitations in communicating.

The information above is supported by the results of PISA in 2019 released by The Organization for Economic Co-operation and Development (OECD). The data show that the Indonesian students' reading ability achieved a score of 371. This score is below the average of 487. Mathematics received 379 points from an average of 487 points. Science got a score of 389. This figure is 100 points different from the average of 489 points.

In the PISA survey, in the ability to read, only 30 percent of the total Indonesian student respondents reached level two proficiency. Ability level to identify the main ideas in medium and long text. The ability to search for information based on explicit criteria that can reflect the purpose and form of the text.

Data on mastery of Mathematics also shows that only 28% of students can achieve level two proficiency. In this stage, students can interpret and recognize without instruction, and can present mathematically. However, only 1% of Indonesian students master mathematics at a high level (level five and above). The Covid-19 pandemic has influenced learning activities that initially took place in the classroom into a remote system. Adaptive learning is one of the life skills that students really need, especially to undergo learning activities during the Covid-19 pandemic. The adaptability required includes the ability of students to do school assignments independently of the teacher's supervision, the ability to study seriously with or without supervision, the ability to use technology for learning activities, and the ability to apply knowledge to survive.

The adaptive learning concept discussed in this study is about 4C skills. The researcher intends to find

DOI 10.1201/9781003206019-20

out how educators apply the creative process, critical thinking, collaboration and student communication to distance learning, especially in elementary level education. This research also provides recommendations about adaptive learning that is appropriate to students.

2 LITERATURE REVIEW

Adaptation is the ability possessed by living things to adapt to the environment (Shute, V. J. & Zapata-Rivera, 2008). Charles Darwin and Herbert Spencer argue that survival depends on the ability of individuals to adapt to their environment; those who have the ability to survive will succeed. The theory of evolution is in line with James and Dewey's claim that human success depends on intelligence; humans must be able to adapt to the environment to meet needs. With these habits, humans are seen as independent and creative creatures (Cronbach, 1975).

The ultimate goal of adaptive learning systems is to personalize student learning processes and accelerate outcomes. This goal can be obtained if students can identify what they do not know and then look for appropriate material content to study. Things that must be considered in adaptive learning: (1) knowledge, (2) dexterity with the mouse, (3) language, (4) learning difficulties, and (5) text information (Petersen et al. 2017). The benefits of adaptive learning are: being able to adapt to student learning needs, learning being able to take place individually and personally, learning being more interesting because it implements information technology assistance in interacting with students, making students not only passive receptors but partners in the educational process.

2.1 Century skills

The development of cognitive science shows that the expected results in learning will increase significantly when students are involved in the learning process through authentic real-world experiences. 21st Century enGauge skills are built on the results of continuous research and answer learning needs that clearly define what students need to thrive in the digital era. 21st-century enGauge skills include digital age literacy, inventive thinking, adaptability, and ability to manage complexity, effective communication, and high productivity (Ariyana et al. 2018). The 4Cs skills competency (creativity, critical thinking, collaboration, communication) are four skills that have been identified as 21st-century skills (P21) as very important and necessary skills for 21st-century education (Zubaidah, 2018). Creativity, thinking and innovation: students can produce, develop, and implement their ideas creatively, either independently or in groups. Critical thinking and problem solving: students can identify, analyze, interpret, and evaluate evidence, argumentation, claims, and data that is widely presented through in-depth study, and reflect on these in life. Communication: students can communicate the ideas effectively using oral, written, and technological media. Collaboration: students can work together in a group.

2.2 Creativity

Mel Rhodes (Fatmawiyati et al., 2011) states that creativity is a phenomenon, where a person (person)communicates a new concept (product) that is obtained as a result of a mental process (process) in generating ideas, which is an effort to fulfill a need (press) ecological pressure. Creativity contains four things: person, process, press, and product. Meanwhile, Stenberg, Kaufman, and Pretz (2002) define creativity as an ability to produce new products with high quality. According to Weisberg (2006), creative thinking is a way of thinking that produces something new (innovation). A creative person is someone who can innovate. Munandar (2009) revealed that there are person, press, process, and product strategies to develop individual creativity. This strategy can be expressed as follows: (1) Person, creativity is an expression and uniqueness of individuals with their environment. (2) Pusher (Press), creative talents will be achieved if there are encouragement and support (3) Process, creativity can be developed by providing opportunities for individuals to make creative inventions. (4) Product, a condition that allows a person to create meaningful products for himself and others.

2.3 Critical thinking

Type According to Robert H. Ennis (Zakiah, 2019), critical thinking is a reflective thought process that focuses on deciding what to believe or do. According to Redecker, critical thinking skills include the ability to access, analyze, synthesize information that can be learned, trained, and mastered (Nyakito et al., 2018). According to Ratna, someone is said to be able to think critically if that person can think logically, reflectively, systematically, and productively in making considerations and making decisions (Safira, 2019). Emily R. Lai (2011) mentions several characteristics possessed by the ability to think critically, namely: (1) analyzing arguments, claims, or evidence, (2) making conclusions using inductive or deductive reasons, (3) assessing or evaluating, (4) make decisions or solve problems. John Butterworth (2013) states that the main activities of critical thinking include three things, namely: (1) Analysis is identifying the main parts of a text and reconstructing it in a way that fully and accurately captures it. (2) Evaluation means judging how successful a text is: for example, how well an argument supports its conclusion; or how strong is some of the evidence for a claim that should be supported. (3) Further argument: the opportunity for students to give their own responses to the text in question, by presenting a reasoned case against the claims made.

2.4 Collaboration

John Myers (1991) explains the definition of collaboration which focuses on the cooperation process rather than the word cooperation which focuses on the product of the collaboration. Collaborative learning emphasizes the existence of work principles. Important principles that need to be considered in collaborative learning are as follows. 1) Each member works together to achieve common goals and interdependence, 2) individuals are responsible for the basis of their own learning and behavior, 3) cooperative skills are learned, practiced and feedback is given based on how best the skills training should be applied, and 4) the class or group is encouraged towards the implementation of a cohesive group work activity.

2.5 Communication

Communication according to Karman (2013) is a process, not something static. Communication requires a place, is dynamic, produces a change to achieve results, involves mutual interaction, and involves a group. Abdul Ghaffar (Lanani, 2013) suggests teachers/lecturers design interesting learning messages by paying attention to the following principles: (1) readiness and motivation; readiness includes mental and physical readiness. (2) Attention grabbing: human attention/concentration often changes and moves (is not focused). (3) Active student participation: educators must try to make students active in the learning process. (4) Repetition: students will receive and understand the material well if the material is repeated. (5) Feedback: feedback is important. The right feedback from educators becomes a trigger for enthusiasm for students. (6) Avoiding irrelevant material: educators must avoid material that is not relevant to the topic being discussed so that students do not feel confused.

3 RESEARCH METHODS

This paper is qualitative descriptive research. This study collected information about learning during Covid-19 through interviews with several teachers and students at the elementary school level in Yogyakarta. Besides, researchers also conducted documentation studies from related journals.

4 RESULT AND DISCUSSION

4.1 Adaptive learning for elementary school

Elementary school is a phase where students need a lot of space to explore and consolidate knowledge. To strengthen knowledge, children need intensive assistance from parents and teachers. However, the Covid-19 pandemic has indirectly forced students to solve their learning difficulties. Some of the adaptation factors faced by elementary school students include (1) adaptation to new learning spaces. Students are required to be able to study in their own homes with all existing conditions and limitations. This adaptation has a positive side because students have the opportunity to explore their own learning preferences. On the other hand, the parents often find it difficult to supervise their children. (2) Learning content adaptation. The learning material that students receive may be less than usual. Students will not receive a detailed explanation from the teacher about the material being studied. This adjustment requires synergistic cooperation from parents and teachers. Parents should be a liaison between students and teachers. Parents ask the teacher the material they didn't know then explain it to their children. (3) Adjustment in the field of technology. The essence of adaptive learning is the use of technology as a differentiator from traditional learning systems. Students who are keen to use their gadgets to play games have to turn this desire toward learning activities. Besides, parents are also expected to be good operators for their children to face technical problems during study.

(4) Adaptation of learning motivation. Motivation to learn is difficult for students because there are no binding rules when studying at home. Students would ask their parents to do all the assignments or they may not do it at all. This is like the confession of Eni (English teacher): "Many children do not collect daily and weekly assignments so I find it difficult to monitor their learning progress." This case was not only experienced by Eni, but almost all teachers found that there were children who did not want to collect assignments on time. Several reasons appeared, such as the lack of understanding of the assignment given, the internet data plan running out, and not checking incoming info. Bima, one of the students revealed the reason, "I don't collect the assignment because I didn't understand it. The tasks are hard to do."

(5) Adaptation of learning resources. Before the pandemic, students received a lot of information from the teacher. Students can ask the teacher directly what they want to know. At this stage of adaptation, students need to be trained in information searching skills with the internet. (6) Adaptation of learning activities. Learning activities are very different from what they encounter in the house. Students quickly feel bored because there are no friends to discuss or joke with. Student boredom while studying at home was expressed by Dzakka', a grade 1 student: "studying at home is boring, I want to meet friends. Hopefully, the coronavirus will disappear soon."

Things that need to be considered in adaptive learning in elementary schools are (1) Knowledge. Every child has a different level of understanding of the material. Teachers should be more sensitive to each student's condition by assigning assignments according to their abilities. In this aspect, the teacher also needs to group students into several parts to facilitate the distribution of tasks based on target. (2) Skills in using technological devices. The assignments given must also be tailored to the students' skills in technology. Students who may not know how to complete assignments via Google Form should be replaced with PDF or JPEG forms.

(3) Language. The use of language in delivering assignments should be kept simple and avoid ambiguous words so that students' learning motivation is not damaged after reading the instructions. (4) Learning difficulties. Parents need to be open with teachers about conditions that make learning difficult for their children. (5) Text information. The teacher needs to include text in the form of subtitles on the recording or video. Therefore, students can read a text through the screen to match what was heard and what was seen. This activity will indirectly stimulate the sensitivity of their sense of sight and hearing.

4.2 Creativity

Figure 1. Creativity factors

As shown in Figure 1, creativity in adaptive learning can be run by paying attention to the following factors:

4.2.1 Personal

As individuals, students have their own uniqueness. The assignment about creativity is not always about pictures but can be a text containing student poetry, song lyrics, or short stories. In this case, the teacher should provide a theme related to the assignment, while students are eligible to choose in what way they want to express it.

4.2.2 Motivation

During the Covid-19 pandemic, students experienced a decline in their enthusiasm for learning. Speeches can be made in fun activities such as projects to make mini-performances, then it would be assessed and witnessed by the teacher on a home visit. It is a good idea for the teacher to group students who live in one district for the project.

4.2.3 Process

The parents' creative process during learning at home should be supported by creating comfortable and pleasant conditions. This method can improve the students' mood to discover and create new things. Besides, parents also need to provide supporting facilities and infrastructure for students to actualize their creativity. (4) Product: meaningful creativity for students provides benefits. The teacher is better off giving a form of assignment whose benefits can be felt directly by students and their surroundings. It is better if the teacher asks students to install their posters on the roadside as a campaign movement to the community. The teacher can ask students to produce masks to share with neighbors in need.

4.3 Critical thinking

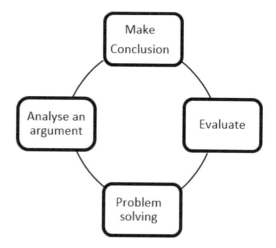

Figure 2. Process of critical thinking

As shown in Figure 2, critical thinking skills during a pandemic are improved by (1) analyzing arguments: students are given assignments in the form of descriptions by avoiding rote answers. The teacher needs to provide a minimum of words that students must write so that students have more responsibility for elaborating their analysis. For elementary school, critical thinking can be started by asking each student to make a question about why and why not something happens. (2) Making conclusions using inductive or deductive reasons. This method has been widely applied by teachers in making questions in the form of descriptions where students have to provide explanations and arguments.

However, if the critical thinking activity is only carried out through question sheets, then students can experience heavy pressure. Therefore, in a simple way, students need to watch a story from a video on YouTube. (3) Assessing or evaluateing. This activity is quite easy for students in elementary schools. If they get limited internet data, parents only need to accompany students to watch television that contains educational values. Through these television shows, parents ask students to rate the behavior of the character presented. Student assessments are followed by suggestions he gives for the character in question. This, of course, must be under the guidance and direction of the parents. (4) Make decisions or solve problems. During online learning, students faced various problems. This situation can be a medium for parents and teachers to introduce problem-solving concepts in a simple way. Parents and teachers can ask for better solutions for online learning problems.

4.4 *Collaboration*

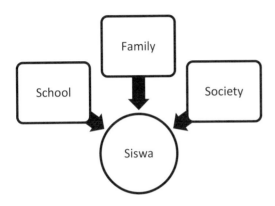

Figure 3. Three education center.

Collaboration in adaptive learning can be adapted from the concept of the three education centers introduced by K.H Dewantara (see Figure 3). Students collaborate with family (parents), school (teachers), and the environment (community) to carry out learning from home. The main education comes from the family so that during this pandemic the family has plenty of time to educate their children. Parents are expected to be able to become partners of teachers in explaining lesson materials to children. In addition, with lots of time at home, parents have the opportunity to transfer character values ??to their children. The better the collaboration between the child and the parents, the greater the chance for students to succeed. Likewise, collaboration among students, parents, and teachers can be actualized by monitoring and providing feedback on student learning progress. Collaboration with the community can be accomplished by collaborating with the community to propose learning services in a village environment where children in one area can work to learn together. Village-library would be a good choice for tutoring.

4.5 *Communication*

Communication is an important skill for children at primary age. Teachers can instill student communication skills during the Covid-19 through the following rules: (1) readiness and motivation: the teachers ought to know the readiness of students one by one. Based on observations, some teachers made a presence through daily journals, WhatsApp groups, and school websites. (2) Attention drawer: some teachers were observed using animated videos to explain their material. For example, Astri, an English teacher who explains English substance using the Tik-Tok application. "I chose the Tik-Tok because many students loved it and it was easy to use." (3) The student's participation can be monitored from the tasks submitted to the teacher. The teacher occasionally needs to make video calls to students so that communication goes in both ways. (4) Repetition. Repetition is carried out by giving the recording to students, then in another chance, the teacher asks a question related to the content. (5) Feedback. If the assignment is collected through the WhatsApp group, the teachers directly give the stickers as a token of appreciation. They also need to provide a review of the assignment that has been submitted. Besides, communication skills can be practiced verbally and non-verbally. Sending correspondence with their peers related to their experiences while studying at home is a kind of verbal method. Correspondence activity can be carried out via WhatsApp messages or by sending letters (offline). The teacher divides the students into pairs. The non-verbal way can be actualized by giving a picture and then the students are asked to tell the object through a recorder.

5 CONCLUSION

Adaptive learning 4C skills that applied by teachers are as follows: (1) creativity: by paying attention to the uniqueness of individuals, processes, driving factors, and products. (2) The critical thinking process while studying at home can be actualized by analyzing arguments and asking questions, making conclusions through video shows, providing ratings from television shows and solving problems independently while studying online. (3) The collaboration process can adopt the concept of a tri-center of education where students learn through their families, schools, and community. (4) Communication can be carried out with the help of technology such as video calls and recordings.

REFERENCES

Andrian, R., & Fauzi, A. 2019. E-Learning model to support industrial based adaptive learning for student vocational high school. *Jurnal Online Informatika*, 3(2): 86. https://doi.org/10.15575/join.v3i2.258

Cronbach, L. J. 1975. Beyond the two disciplines of scientific psychology. *American Psychologist*, 30(2): 116–127. https://doi.org/10.1037/h0076829

Dochy, F., & Dochy, F. 2018. The future of learning. *Creating Impact through Future Learning*, 1–11. https://doi.org/10.4324/9781351265768-1

Fatmawiyati, J., Psikologi, M., & Airlangga, U. 2011. *Jati Fatmawiyati | Magister Psikologi Universitas Airlangga Page 1*. 1–21.

Lanani, K. 2013. Belajar berkomunikasi dan komunikasi untuk belajar dalam pembelajaran Matematika. *Infinity Journal*, 2(1): 13. https://doi.org/10.22460/infinity.v2i1.21

Nyakito, C., College, N. T., & Box, P. O. 2018. *Baraton Interdisciplinary Research Journal)*, 8(Special Issue): 1–10.

Safira, N. 2019. *Berpikir kritis dalam keperawatan.* https://doi.org/10.31219/osf.io/7dakf

Shute, V. J. & Zapata-Rivera, D. 2008. *Handbook of research on educational communications and technology.* Springer-Verlag.

Verdú, E., Regueras, L. M., Verdú, M. J., De Castro, J. P., & Pérez, M. Á. 2008. An analysis of the research on adaptive Learning: The next generation of e-learning. *WSEAS Transactions on Information Science and Applications*, 5(6): 859–868.

Zakiah, L. & I. L. 2019. *Berpikir kritis dalam konteks pembelajaran.* Erzatama Karya Abadi.

Zubaidah, S. 2018. *Mengenal 4c: Learning and innovation skills untuk menghadapi era revolusi industri 4.0.* https://www.researchgate.net/publication/332469989

Educational Innovation in Society 5.0 Era: Challenges and
Opportunities – Purnomo & Herwin (Eds)
© 2021 the authors, ISBN 978-1-032-05392-9

Multiliteracy pedagogy challenges: EFL teachers' multicultural attitudes in the literacy classroom practices

U. Sholihah & W. Purbani
Yogyakarta State University, Yogyakarta, Indonesia

ABSTRACT: Students' cultural diversity and background commonly contribute as one of the issues in the teaching and learning process in the EFL context. Teaching language and teaching culture are two important things that cannot be separated in teaching EFL. Thus, teachers have a crucial role in promoting multicultural attitude in the literacy classroom practices. This study investigated EFL teachers' multicultural attitudes in their classroom practices; teachers' literacy classroom is one of the effective ways. There were five EFL teachers participated in the study. Teacher Multicultural Attitude Survey (TMAS) by Ponterotto (1998) was adapted in the study as an interview protocol to collect the data. The data were obtained from a structured interview and analyzed qualitatively using thematic analysis. In analyzing the data, teachers' multicultural attitude in the literacy classroom practices then was divided into three sub-themes, which are: (1) perception, (2) challenges, and (3) coping strategies. The findings discover the EFL teachers are aware of the multicultural reality of society showing a good multicultural attitude in literacy classroom practices

1 INTRODUCTION

The prominence of literacy and literacy practices as a tool of conceptual and cognitive development is well recognized (America 2014). In the twenty-first century, advanced literacy is a prerequisite to adult success (Murane, et al. 2012). Being critically literate is not only able to read but also to see, act, understand, challenge different perspectives and human practices and that promote an unequal power relation around the world (Yoon 2006). In this context, teachers have a crucial role in promoting the development of students' critical literacy in the classroom (Janks 2014) since Freire (1970, 1997) demonstrates that critical literacy involves critical consciousness and social action.

Literacy is a complex embodiment of reading, writing, and thinking in which through its meaning is created socio-cultural context. Norton (2010: 10), in his research conclusion, that literacy is beyond reading and writing. Further, it is about the text and the reader, student and teacher, classroom and community, in local, regional, and international areas. It can be said that learners' engagement in literacy practice will also show engagement in acts of identity.

On the other hand, Yoon (2016:5) states that critical literacies and global and multicultural education are related fields. Each of them has greatly contributed to our understanding of social justice by including the issues of racial and cultural minorities into curricula.

In Yoon's study (2016) students' teachers might have a good intention in teaching diversity; the lack of cultural understanding might have hindered him from revealing those cultures accurately and with less reproduction of cultural stereotypes and prejudice. As Yoon (2016: 5) said critical literacies and multicultural education are related fields, these need to be paid attention to in the classroom practices. It is stated that critical literacy and global and multicultural education have the intention to support social justice.

An extensive range of skills and knowledge to meet students' heterogeneous needs is required to accommodate the progressively diverse student population (Cole 2008; Gibson 2005). Berry, Smylie, & Fuller (2008); Lickona (2009) argues that it is intolerable for teachers to have little respect for their students because of their diverse cultural differences. Teachers have to be proficient in preparing and adapting to the school environment with all possible students' alterations, and further distinguish the effects that cultural differences can have on their students' education. Teachers' awareness of these effects is known to be more tolerable and open-minded towards students' different cultures in the classroom.

Related to the background above, it is essential to investigate EFL teachers' multicultural attitudes in literacy classroom practice. Accordingly, the research question of the study is how EFL teachers perceive multicultural attitudes in literacy classroom practices. The result of the study can contribute to the plans for preparing EFL teachers to be knowledgeable in multiculturalism. Furthermore, the qualitative phenomenological of this study can provide an understanding of EFL teachers on multicultural education and approach in teaching diverse student population.

DOI 10.1201/9781003206019-21

113

Finally, researchers, academicians, and scholars may also develop their understanding of the significance of multicultural education for schools to develop humanity and treat everyone equally.

2 LITERATURE REVIEW

2.1 *Literacy in the classroom practices*

Literacy is not about reading and writing (Guthrie 2010). Moreover, literacy is beyond reading and writing, is also about the text and the reader, student, and teacher, classroom and community, in local, regional, and international areas. That is why if learners engage in literacy practice, they also engage in acts of identity. It means that learners identify the sense of making meaning. At the final point is that Guthrie's statement, he (2004: 26) claims that at the present, English teachers live in threatening times. Besides, especially in the part of engaging students in literacy activities, teachers have to take their concentration to innovate the literacy activities based on the curriculum standards.

Kennedy (2002) as the conclusion of her study about classroom practices for literacy development of English language learner, she suggested some strategies for teachers in promoting literacy in the classroom. The strategies are, (a) learn as much as possible about students' culture and background to establish a direct connection, (b) place students into flexible learning groups and utilize tiered assignments, (c) create an instructional program that provides abundant and diverse opportunities for speaking, listening, reading, and writing while utilizing a variety of different teaching methods, (d) incorporate reading materials that stress cultural diversity and emphasize the positive aspects of the various culture, (e) use multiple measures for assessing and evaluating interest, attitudes, self-perception, and overall progress.

2.2 *Multicutural education*

Multicultural education is the alteration movement aiming to ensure that all students enjoy equal education and to create equal educational conditions for all students regardless of race, gender, culture, language, religion, social class (Kaya & Aydin 2013). It intends to create an educational environment for all students in equal situations that respect diversity (Basbay & Kagnici 2011). To create an egalitarian, correct, fair, inclusive pluralist, and transformational society, under the guidelines of multicultural education, students need to be facilitated to communicate actively and interactively with different cultures through social, civic, and political activities (Gay 2018).

Multicultural education allows students to understand the culture of their community, to remove cultural boundaries with other cultures, and to build a society that is common to all (Banks 1993). It is important in terms of generating a system where everyone will be satisfied, will be living in peace, and will be contributing to social peace. There is a need for qualified teachers who can understand the problems of these different groups, who can communicate with them correctly, and help them to raise their academic success so that teachers have important roles in this context (Wells 2008). Multicultural education aims to ensure equality of opportunity in education for all individuals and groups despite differences (Banks 1993).

2.3 *Multicultural attitude*

Teachers' cultural attitudes, values, hopes, and beliefs are brought to the classroom. (Banks 2004). Both knowledge and attitudes are closely linked and influence classroom practice (Banks 2004; den Brok, Bergen, & Brekelmans 2006). As stated by Vollmer (2000), teachers' perceptions and attitudes have a strong effect on the classroom's educational and social climate. More specifically, beliefs about inclusivity and heterogeneity in classrooms are one of the greatest predictors of teaching effectiveness (Stanovitch & Jordan 1998). Often, teachers are unaware of their ideological assumptions, which have been 'naturalized' to such an extent, that they are seen as part of the 'common sense'. Among teachers who work with immigrant students, these ideological assumptions include societal beliefs regarding newcomers' acculturation and the role of the school in this process.

3 METHOD

The method used in the research is qualitative research. Creswell (2009) states that qualitative research is a process of scientific research that is intended to understand human problems in a social context by creating a comprehensive and complex picture presented, reporting detailed views of sources of information, and carried out in natural settings without any intervention from researchers.

The sampling technique was utilized to select participants in the study. Sampling means the smaller number of cases, units, or sites chosen from a much larger population. Some samples are supposed to be representative of the wider population (Hammond & Wellington 2013, p. 174). In rigorous research, sampling is another important issue that researchers need to pay attention to (Cohen et al. 2000).

Purposive non-probability sampling is usually chosen by researchers as it helps in the understanding of the ideas and perceptions of the participants in a given context in a particular time (Patton 2002). Purposive sampling includes the choice of a sample that the researcher already knows something about where he/she chooses particular participants that are likely to provide the most valuable data (Denscombe 2007).

Five in-service teachers, who were currently enrolled in a teacher training program, participated in this study. They were secondary school English teachers. The sample being drawn was appropriate for the researcher due to the limited time to obtain the data.

Data were obtained during August 2020. Data were in the form of interview data. Online interviews were done through some platforms such as google meeting, WhatsApp chat, and WhatsApp voice note. The data gathering instrument of this research was an interview guideline adapted from the questionnaire Teacher Multicultural Attitude Survey (TMAS) by Ponterotto (1998),

This study used thematic analysis (Braun & Clark 2014). Thematic analysis is an exploratory approach where analysts code or mark their sections of a text according to their patterns contributing to some relevant themes (Schwandt 2007).

The first stage of analyzing the data was transcribing the interview recordings and it was then followed by coding stages. Initially, the researchers read the transcripts carefully to identify potential themes. The second level of analysis involved reviewing these initial codes. The researchers considered particularly how to retain the diversity of the initial codes while producing overall elements, or sub-themes. Next, the authors reviewed themes and sub-themes before defining and naming them. Finally, once themes were finalized, the final report was written.

4 FINDINGS AND DISCUSSION

Through the thematic analysis, the researcher found the answer to the research question. The researcher identified one emergent theme, which is the EFL teachers' multicultural attitude in the literacy classroom practices then it is divided into three sub-themes which are: (1) perception, (2) challenges, (2) coping strategies.

4.1 Perception

According to Sue and Sue (2003), teachers must be aware of his/her biases and assumptions about human behavior to become culturally effective and responsive. Moreover, they add that teachers should acquire knowledge of the particular group of students, and should be able to culturally strategies in working with students from diverse cultural backgrounds. From the previous statements, EFL teachers' multicultural attitude can be seen from the teachers' perception of multiculturalism in the classroom. Teachers' perception of multicultural awareness are as follows:

"They have different backgrounds such as cognitive, behavior, social and economy, and their place." (T_1)

"I can see clearly that they have different cognitive and personality. Some extroverted students wanted to have an online meeting although they didn't have a schedule on that

day. On the other hand, some students didn't want to activate their cameras during the online class. The majority of the students have good English speaking skills since they have learned English from kindergarten" (T_2)

To understand the students' diverse backgrounds in the classroom, teachers tried to acquire information from some sources such as the students' identity from colleagues, the students' cultural background from the internet, magazine, newspaper, and the students' cognitive background from the result of the IQ test in the school. Every teacher has their strategies to obtain information before they did teaching practice in the classroom. The information than was followed-up in developing appropriate materials for the multicultural classroom. Some of their words are as follow:

"They are different personalities, motivation in learning, needs, interest, ability, and their learning style. I used the result from the IQ test from the school to know all of the students' backgrounds. As the follow-up of the IQ test, I gave the students pre-assessment to know their ability in English." (T_5)

The participants did a lot of things to understand students' homogeneity before they taught in the classroom as a means to create a comfortable environment during the teaching and learning process. Gutierrez-Gomez (2002) states that teachers' multicultural skills can be manifested if he/she can create and maintain a safe and caring environment. The following is the way the teacher understand students' homogeneity:

"I used questionnaire also to know the students' motivation, needs and interest. The result of the questionnaire will be useful for me to give scaffolding to each student in the classroom." (T_5)

4.2 Challenges

4.2.1 Challenges related to students' diverse background

Powerful teaching and learning are obligated for teachers' and students' activity to create in the classroom since the static and standardize curriculum policy neglect the current presence of students' diversity. However, teachers and students distinguish to easily adapt and innovate curriculum and instruction in their classroom to fulfill the academic and other objectives they wish from literacy education (Skerret 2015).

Teachers faced come challenges teaching students with diverse backgrounds in the classroom. The students often refused to collaborate or to work together with a certain classmate. Challenges that appeared in the classroom are below:

"I experienced in a classroom, there is a student often rejected by some groups due to his/her level of cognitive relatively low." (T1)

"One of my students has descendent from Eastern, their friends often refused him to join in a group. I tried to discuss with all of the students that they have to work with everyone since they are equal as students and as human beings."

4.2.2 Challenges related to students' personality

Beliefs about inclusivity and heterogeneity in classrooms are one of the greatest predictors of teaching effectiveness (Stanovitch & Jordan 1998). To promote multicultural awareness, problems appeared in the classroom since there were students with diverse personalities. Some students were not confident to work in a group or shy. When problems appeared in the teaching and learning process, teachers tend to solve them by giving consultation to the students outside the teaching and learning process. Moreover, discussing with the homeroom teacher, and the psychology of the school became an alternative to deal with students' problems that teachers cannot solve by themselves.

Moreover, all of the teachers usually having discussions or share their problems with colleagues so that they can learn about the challenges that appeared in other classes. The participant' experience is stated in the excerpt below:

"I usually handle the cognitive problem of students in the class. If I found other problems of students that I can't handle, I collaborate with counseling teacher, or psychology at school." (T_2)

4.2.3 Coping strategies

Coping strategies to reveal the multicultural attitudes in the classroom involves many aspects, such as strategies to increase the students' awareness about multiculturalism, preparing appropriate materials and method for diverse students in the classroom, and assessing the students' performance. Banks (1991) says that educators use several approaches to integrate cultural content into the school and university curriculum. He adds that teachers can contribute an approach to integrate cultural content, concepts, and themes into the curriculum without changing its basic structure, purpose, and characteristics.

As Banks (1991) suggests, the participants selected theme, topic, and text that meet with the students' diverse background so that their multicultural awareness will also be increased. Selecting appropriate materials was believed as an essential thing to do since in the literacy classroom practice goal is to develop students' critical consciousness and social action as Freire (1970, 1997) said. One of the strategies in preparing appropriate materials in the classroom is as follow:

"I tried to look for information about cultural resources they brought. I read from newspapers, magazines, and the internet to understand the non-linguistic part. Therefore, I can use the information as resources to select appropriate materials, such as selecting a theme or topic. For me, English class is beyond linguistic." (T₃)

Teachers also raised their multicultural awareness in assessing students' performance. They would give various scaffolding and treatment to the students who needed it. Accommodating diverse students requires the teachers to pay more attention to the students' cognitive level. Varying methods, approaches, and activities in the classroom need to be done to accommodate students' learning styles. Grant and Gomez (2001) say that in multicultural teachers build on students' learning styles, adapt instruction to the existing skills of students in the class, and involve students in thinking through and analyzing the life situations of different people. Some of their words are as follows:

"I used to apply a teaching model that raised students' cultural awareness, such as cooperative and collaborative learning. They discussed topics related to cultural diversity also their project. Therefore, they understood their diversity yet had the same goal to finish the project. At the end of the class, they had a reflection of their learning process, the skills they learned, and their group work. It is also important to add a rating scale or questionnaire about intercultural awareness. For assessment, the different cultural background would influence their expectation toward their learning. Scaffolding is needed to bridge their diversity in terms of input knowledge." (T₃)

"I often applied tiered classroom. The students were grouped according to their cognitive level, and they had a different level of the task." (T₂)

Cooperative and collaborative learning were used in the classroom activities since it made the students work in a group to share, discuss, and collaborate with their friend that has a different background. Moreover, tiered instruction was also used to vary the level of assignment based on the cognitive level of students. Teachers showed their decision making and social action approach as states Bank (1994) that these strategies of allowing students to pursue projects and activities to take personal, social, and civic action extend in the curriculum.

Working in a group is a good moment in the classroom to raise students' cultural awareness. In pair work and group work, teachers gave the responsibility to the students to contribute, and help each other so that they can solve the problem in the group. All of the strategies done by the teachers have promoted multicultural education in the classroom. Multicultural education aims to create and develop equal educational opportunities for students from diverse racial and cultural groups. It also helps students to acquire the knowledge, skills, and attitudes needed to function in a democratic society (Navita 2014). Following are some of the words of the teachers:

"I usually apply a learning method that involved the students in group work. Through this method, I expected the students would learn

*their friends' characteristics to raise their multi-
cultural awareness, learn the materials together
with their friends, and work collaboratively.
Moreover, I wanted them to improve their
responsibility toward their task, and their interaction in the group helped them to understand
the materials well." (T_4)*

*"I often give chance to students to work in
group so that they can interact and collaborate
each other in group work." (T1)*

The role of a teacher as motivator, facilitator, and
evaluator worked well since he/she can monitor each
student and support them to do the best on their work.
By treating students equally, teachers also develop
students' motivation and self-confidence. Rewards
also were given to all students that made progress.
This activity is believed to increase a good atmosphere in the classroom since they can treat students
equally. The following are the statements from the
teachers:

*"Before the class begin, I used to motivate the
students, treated them equally, and objectively.
So that the atmosphere of the class before they
learned are comfortable." (T_4)*

*"At the end of the class, we do a reflection
on what materials and skills we have learned,
learned from teaching and learning process."
(T_2)*

The strategies promoted by Banks (1991, 1994)
about multicultural awareness in the classroom are
in line with strategies for teachers in promoting literacy by Kennedy (2002). She suggests teachers (a)
learn as much as possible about students' culture and
background to establish a direct connection, (b) place
students into flexible learning groups and utilize tiered
assignments, (c) create an instructional program that
provides abundant and diverse opportunities for speaking, listening, reading, and writing while utilizing a
variety of different teaching methods, (d) incorporate reading materials that stress cultural diversity and
emphasize the positive aspects of the various culture,
(e) use multiple measures for assessing and evaluating interest, attitudes, self-perception, and overall
progress.

Overall the findings revealed that the participants
acknowledge multiculturalism through their teaching in literacy classroom practice. Moreover, Navita
(2014) suggests that teachers' multicultural awareness
can be gauged from some aspects such as professional development, competence, and school policies. Related to the previous statement, all of the
participants have joined a teacher training program,
practiced doing reflective teaching, and had a discussion with colleagues to improve their performance.
In the area of professional development and competence, teacher multicultural awareness can be further
enriched if teachers with specialization or background
in multicultural education will be tapped to help in the

promotion of multicultural awareness programs in the
school (Navita 2014).

5 CONCLUSION

This qualitative study explored teachers' multicultural
attitude in literacy classroom practices. The findings
showed that an experienced teacher's classroom practices were more visibly related to having a good
multicultural attitude although the strategies of each
participant to deal with diverse students in the classroom were not the same. Furthermore, the findings
show that all of the participants consider having a good
multicultural attitude. Although this study of just five
professional English teachers as such may have limitations as to how the result can be generalized. The
purpose of exploring teachers' multicultural attitude
in the literacy classroom practices is not to look at the
best practices of multicultural teaching in the classroom, yet, the idea is to see what motivates a teacher
to do multiculturalism to achieve the goal of literacy
practice in the classroom. As Yoon (2016) stated that
global and multicultural education was in the related
field. Moreover, Freire (1970, 1971) stated that critical literacy involved critical consciousness and social
action. A good multicultural attitude of teachers promotes students' critical consciousness and further will
become social action.

REFERENCES

America, C. 2014. Integrating literacy practices in business
education: Pedagogic intentions for teacher training. *A
Journal for Language Learning* 30(3):16–25.
Banks, J. A. 1991. *Teaching strategies for ethnic studies.*
Boston: Allyn and Bacon.
Banks, J. A. 1993. *Diversity within unity: Essential princi-
ples for teaching and learning in a multicultural society.*
Washington D.C.: Office of Educational Research and
Development (ERIC Document Reproduction Service
No. ED 339 548)
Banks, J. A. 1994. *An introduction to multicultural education.*
Boston: Allyn and Bacon.
Banks, J. A. 2004. *Multicultural education: Characteristics
and goals.* In J. Banks & C. Banks (Eds.), Multicultural
education: Issues and perspectives: 3–30. San Fransisco:
Jossey-Bass
Basbay, A., & Kagnici, D. Y. 2013. Cokkulturlu yeterlik algi-
lari olcegi: Bir olcek gelistirme calismasi. *Egitim ve Bilim,*
36(161): 199–212.
Braun, V., & Clarke, V. 2006. Using thematic analysis in
psychology. *Qualitative Research in Psychology* 3(2):
77–101.
Braun, V., & Clarke, V. 2014. What can "thematic analysis" offer health and wellbeing researchers? *International
Journal of Qualitative Studies on Health and Well-Being*
9: 1–2.
Brok, P. D., T. Bergen, & M. Brekelmans. 2006. "Con-
vergence and Divergence between Teachers' and Students' Perceptions of Instructional Behaviour in Dutch
Secondary Education." In Contemporary Approaches to
Research on Learning Environments: World Views, edited

by D. L. Fisher and M. S. Khine, 125–160. Singapore: World Scientific.

Cohen, L., Manion, L., & Morrison, K. 2000. *Research methods in education. 5th edition.* London: Routledge Falmer.

Creswell, J. W. 2009. *Research design: Qualitative, quantitative, and mixed method approach third edition.* Sage Publications: USA.

Denscombe, M. 2007. *The good research guide for small-scale research projects.* Berkshire: OUP.

Freire, P. 1970. *Pedagogy of the oppressed.* New York: Continuum.

Freire, P. 1972. *Pedagogy of the Oppressed.* New York: Herder & Herder.

Freire, P. 1997. *A response.* In P. Freire, J. Fraser, D. Macedo, T. McKinnon, & W. Stokes (Eds.), *Mentoring the mentor: A critical dialogue with Paulo Freire* (pp. 303–329). New York: Peter Lang.

Gay, G. 2018. *A synthesis of scholarship in multicultural education.* Retrieved from www.Ncrel.org: http://www.ncrel.org/sdrs/areas/issues/educatrs/leadrshp/le=gay.htm

Guthrie, J. T. 2004. *Teaching for literacy engagement.* Retrieved from jlr.sagepub.com: Journal of Literacy Research

Gutierrez-Gomez, C. 2002. Multicultural teacher preparation: establishing Safe Environments for Discussion of Diversity Issues. *Multicultural Education* 10(1): 31–39.

Hammond, M., & Wellington, J. 2013. *Research methods: The key concepts.* London: Routledge.

Janks, H. 2014. *Doing critical literacy: Texts and activities for students and teachers.* New York: Routledge.

Kaya, I., & Aydin, H. 2013. *Turkiye'de anadilde egitim sorunu: Zorluklar, deneyimler ve ili dilli egitim modeli onerileri.* Istanbul: Ukam.

Kennedy, T. J. 2002. Classroom practices for literacy development of english language learners. *Northwest Journal of Teacher Education* 2 (1): 6.

Murane, R. J, et al. 2012. Literacy challenges for the twenty-first century: Introducing the issue. *The Future of Children* 22(2).

Navita, N. C. 2014. Teachers' multicultural awareness of the school environment: Basis for a proposal for multicultural awareness enhancement program for teachers. *Asian Pacific Journal of Education, Arts, and Sciences* 1(4).

Norton, B. 2010. Identity, literacy, and english-language teaching. *Canada: TESL Canada Journal* 28.

Patton, M. 2002. *Qualitative research and evaluation methods.* London: Sage Publications.

Ponterotto, J. G., Baluch, S., Greig, T., & Rivera, L. 1998. Development and initial score validation of the teacher multicultural attitude survey. *Educational and Psychological Measurement, 58*(6):1002–1016.

Schwandt, T. 2007. *The sage dictionary of qualitative inquiry.* London: Sage Publications.

Skerret, A. 2015. A framework for literacy education in multicultural, multilingual, and multiliterate classrooms. *Multicultural Education Review, 7*(1–2): 26–40.

Stanovitch, P. J., & A. Jordan. 1998. Canadian teachers' and principals' beliefs about inclusive education as predictors of effective teaching in heterogeneous classrooms. *Elementary School Journal, 98*: 221–238.

Vollmer, G. 2000. Praise and stigma: Teachers' constructions of the 'typical ESL student'. *Journal of Intercultural Studies, 21*(1): 53–66.

Wells, R. 2008. Global and multicultural opportunities, challenges, and suggestions for teacher education. *National Association for Multicultural Education, 10*(3): 142–149.

Yoon, B. 2016. *Critical literacies: Global and multicultural perspectives.* New York: Springer.

Educational Innovation in Society 5.0 Era: Challenges and
Opportunities – Purnomo & Herwin (Eds)
© 2021 the authors, ISBN 978-1-032-05392-9

Implementation of wooden craft vocational learning to improve life skills in students with disabilities

A. Sulistyo & Kasiyan
Yogyakarta State University, Yogyakarta, Indonesia

ABSTRACT: Mentally retarded students have low intellectual ability and adaptive behavioral obstacle so that the learning process is formulated into vocational self-development form to improve life skills. This study aims to describe the implementation of learning process, product results, and barriers in vocational wood crafts learning. This research is a qualitative research with a case study approach. The data were collected through structured interviews and participatory observation. Furthermore, the data were analyzed by qualitative descriptive. The results of this study are as follows. (1) The implementation of wooden craft vocational learning is carried out in 26 lesson hours with a proportion of 60% of the 42 lesson hours. Wood craft subject matters include bench work techniques, scrolling techniques, massive wood materials knowledge, artificial materials, painting, sanding, and finishing. (2) Product results of wood skills learning are Educational Teaching Aids, puzzles, wooden paintings, miniatures, souvenirs, calligraphy, cutting boards, and wooden batik motifs. (3) Barriers to learning vocational wood crafts include lack of air circulation in the wooden studio, barriers to mentally retarded students behavior, marketing constraints, and obstacles to working with the business world.

1 INTRODUCTION

Wooden craft skills are one of the vocational study to develop life skills, as well as for mentally retarded students. This ability is expected to be a business that can produce something in their life. Vocational skills learning is very appropriate for mental retardation in order to have provisions for independent living (Huang & Cuvo, 1997; Mumpuni Arti, 2007: 18). In addition, learning skills and life skills for mental retardation is needed by every individual in an effort to survive (Nur Wahyuni 2018: 137). The vocational learning process emphasizes training to master a competency through practice. Plus the additional internship program in the business world provides a more real learning experience outside the classroom. This allows them to acquire direct knowledge.

Vocational learning is more appropriate for mental retardation than learning that emphasizes theory. This is because they are mentally equivalent to 12 years of age, lack ability to think abstractly and logically, have less control over feelings, easy to forget, difficult to remember terms, lack of terms understanding, and bad concentration (Werry & Aman 2013:339). The cognitive limitations of mental retardation allow an impact on weak motor skills (Astati 2001: 5), while learning vocational skills is more practice that requires motor skills. Mentally retarded is a condition in which individuals who have intelligence that are significantly below the average and are accompanied by an inability to adapt to behavior that appears during development (Lynn, 2001; Berkson, 2013;

Wijaya 2013: 21). Meanwhile, according to the American Association on Mental Deficiency (AAMD) as quoted by Purtranto (2015: 208) defines mental retardation as a disorder that includes intellectual function below average, namely with IQ 84 and below based on tests before the age of 16 years. Retardation is also synonymous with IQ below the average of normal person. The classification used in Indonesia is the same as Government Regulation no. 72 years 1991, with mild mental retardation (able to educate) their IQ is 50–70, moderate mental retardation (able to train) their IQ is 30–50, and severe and very severe mental retardation (able to care for) their IQ is less than 30. Mild mental retardation can be educated and trained because mentally they can reach or equal to students aged 8–10 years. In their social life, they are able to socialize and adapt to the social environment and are able to do semi-skilled jobs (Kemendikbud, 2012: 7). Through supervision and a little attention, mild mental retardation can be placed in jobs that are simple, repetitive, and do not require much complex thinking.

The implementation of wooden craft vocational learning is divided in two parts, practical learning and theoretical learning. The emphasis for mentally retarded students is more on practical learning because mentally retarded students when faced with theoretical learning will become abstract without learning equipments (Sugiyartun 2009: 30). The learning equipments in question are in the form of modules, books, media, or tools that are real and easy to learn by students. The learning equipments contains guidelines, instructions, and material content that is made simple and easy to

DOI 10.1201/9781003206019-22

understand. The goal is for teachers to arrange learning programs to be adapted to the cognitive development of mentally retarded students. This is because cognitive abilities and everything related to thinking activities are different for each development, so a teacher must adjust the learning material according to the stage of student development (Abdurrahman 2003:255).

The principle of vocational learning in students with mild mental retardation has implications that a learning program should be structured from simple to more complex stages (Mumpuniarti 2007: 139). So the learning material for students with mild mental retardation is arranged in stages, especially in practice. Vocational learning of wood crafts emphasizes mastery of certain techniques for the manufacture of wood craft products.

The learning equipment for wood craft skill practice are divided into two, namely manual tools and machinary tools. Both are quite dangerous if used without teacher supervision. In addition to being sharp and using high electrical voltage, these tools also require special expertise for the process of using them. Therefore, the teacher's role in practical learning is very important, especially to demonstrate how to use tools correctly and safely. In addition, teachers are required to supervise students using wooden skill tools during practice so that there are no work accidents.

This is very interesting to research considering that mild mental retardation has limitations in complex thinking such as difficulties in abstract thinking, imagination, memory, concentration, and difficulty relating one thing to another (Fuller & Sabatino 1998; Mumpuniarti 2007:5). However, even though students with mild mental retardation have complex cognitive limitations, their independence to live in society is the main goal. Life skills for mentally retarded students through vocational learning are very important, because by mastering life skills they will not depend on their parents to fulfill their daily life (Putri & Murtiningsih, 2013: 157; Ning Suryani 2018: 104). In addition, skills and life skills make the life of mentally retarded students more productive every day (Mechling 2013: 260).

SLB Negeri 2 Yogyakarta is one of the education units that organizes vocational learning of wood crafting to improve mentally retarded students's life skills. Currently, there are not many studies that examine the wooden crafts vocational learning process for mentally retarded students. Therefore, this study aims to determine (1) the implementation of wooden craft vocational learning in mentally retarded students, (2) the product results of wooden craft vocational learning for mentally retarded students, and (3) identificating the barriers to wooden craft vocational learning in mentally retarded students.

2 RESEARCH METHOD

This research is a qualitative research with a case study approach. The main instrument of this research is the researcher himself as a human instrument (Brian J. Brown & Sally Baker 2007: 71). The informants in this study were one woodcraft teacher and 4 mild mental retardation students at the high school and high school levels. The data were collected using participatory observation techniques and structured interviews. Participatory observation is an observation that is carried out by directly observing an activity (Wina Sanjaya 2013: 77), in this case it is the wood crafts vocational learning process. Meanwhile, interview is used to collect data which is used to reveal information from the subject orally (Borg and Gall 2007: 246). Structured interviews were carried out on teachers and students, carried out in a structured manner so that the direction and objectives were in accordance with the guidelines. Then, for data analysis techniques using qualitative descriptive, in particular, using the model Matthew B. Miles, A. Michael Huberman, & Johnny Saldana (2013). It includes activities: data collection, display and discussion, and drawing conclusions.

3 RESULTS AND DISCUSSION

3.1 *Implementation of wooden craft vocational learning in students with disabilities*

Vocational learning of wood crafts is one of the areas of developing life skills that are taught to mentally retarded students. Learning activities are carried out with complete learning facilities ranging from manual to machinal tools. These tools are also supported by adequate high voltage electricity. In addition, the materials used to learn to make wood products are also complete, such as massive wood, artificial wood, and plywood. Meanwhile, complete finishing materials are also available such as mowilex, impra, thinner, and others. Overall, the facilities to support wood crafts vocational learning are very complete. Vocational learning of wood crafts begins with assessment activities to measure students abilities. Teachers must have diagnostic abilities, and assessment is a must in the implementation of inclusive learning in schools (Dolgova et al. 2017; Mumpuniarti 2018: 57). This assessment activity is carried out to filter out mental retardation abilities; classification, placement, and determination of mental retardation education programs; determine the direction and needs of mental retardation; individualized educational program development; and determining learning strategies (Moh. Amin 1995). The results of this assessment are used as a guide for teachers to develop optimal learning, especially in making wood craft product designs.

The development of wooden craft product designs for mentally retarded students is based on the results of student assessments. Assessment activities are carried out by observation method, interviews, and discussions with class teachers. The results of the assessment conducted by the teacher are compiled to see general abilities, special abilities, obstacles, and potential.

General skills include the ability to read, count, write, draw, and measure. Special abilities consist of cutting, gluing, assembling, sanding, and finishing abilities. Obstacle notes were made by the teacher regarding specific student limitations, such as deviant behavior, moving difficulties, and others. Meanwhile, the potential notes seen from the point of view of students' abilities that can be developed, for example, students have a deeper ability to sand but are weak in other abilities. Overall the results of the assessment have shown the disabilities, potentials and needs of students (Dewi Dian Puspa 2018). The results of this student ability assessment are then used as the base for developing a wood craft vocational learning plan and product design to be studied. The division of wooden craft vocational lesson hours refers to the Regulation of the Ministry of Education and Culture of the Republik of Indonesia Number 40 year of 2014 concerning the basic framework and curriculum structure of special high schools, the content of the 2013 curriculum for the SMPLB education unit consists of 40% academic aspects and 60%–70% contains vocational aspects. The allocation of wooden craft vocational learning hours consists of 26 learning hours from a total of 42 lesson hours. In its implementation, the learning process is divided into 2 stages, namely practical and theoretical learning.

Learning activities are based on the subject matter of learning that can be taught to students. The following is the main material for wood crafts based on the results of the student's ability assessment:

a. Bench working technique: making simple products using manual tools.
b. Scrolling technique: chop wood with a scroll tool with simple shapes such as puzzles, calligraphy, mosaic, and others.
c. Massive wood materials knowledge: studying the characteristics of teak, Dutch teak, sengon wood, mahogany, and rosewood
d. Artificial materials: study the characteristics of Medium-density fibreboard (MDF), plywood, and multiplex materials.
e. Painting: demonstrate painting with the brush and spray technique.
f. Sanding: practicing how to level and smooth wood surfaces with coarse sandpaper and fine sandpaper.
g. Finishing: finishing the product until it is ready to be sold.

Each learning material is developed by the teacher to increase student motivation and creativity. One of the product development carried out by the teacher is combining wood with resin. Resin is an exudate (sap) that is released by many plants, especially by conifers. This sap can freeze for some time and form a hard mass and make a transparent effect. The resin material was chosen because it is easy to apply and use. The colors on the resin are also not monotonous and can be combined with several colors. In addition, the resin is easy to combine, has a low shrinkage rate and a low failure rate. The main purpose of using this resin is to create a pleasant learning atmosphere and arouse student interest in learning (Wakiman 1998). A pleasant learning atmosphere increases the motivation of mentally retarded students in learning. In addition, this product development will increase students' creativity in learning. In the implementation of learning wood craft skills using a project based learning model with demonstration methods. In the implementation of wood craft skills learning using a project based learning model with demonstration methods. According to Kemis (2013: 96) the demonstration method for mental retardation is to show a process of how an object or tool works. Teachers are required to be active in demonstrating and supervising students when they practice learning using tools. Meanwhile, the project based learning approach is used to improve students' thinking skills to be more critical and creative. This is in line with research conducted by Fiyola Triana Eldiva (2018: 348) which shows that project based learning can improve the critical thinking skills of mentally retarded students. In practical learning, students are required to learn independently according to the teacher's instructions. The teacher shows how and the students follow it step by step. If found, students who have difficulty in one stage, the teacher will instruct other students to help. Meanwhile, if there is a failure and an accident, the teacher immediately makes incidental instructional actions, for example making a new design or making first aid actions in an accident. Instructional incidental actions are decisions made quickly and intelligently as a form of mental retardation learning services (Ishartiwi 2012: 1). After the product is finished, the product is ready to be assessed and displayed in an exhibition.

The evaluation of wooden craft vocational learning is intended to measure students' abilities based on the process and results. The evaluation carried out is based on the initial assessment of learning. The teacher sees the development of students' abilities before and after learning. This evaluation is also a base for further learning assessment. Moh. Amin (1998) states that assessment during and after mental retardation is given lessons, it is necessary to design advanced programs. In vocational learning, wood crafting are used to formulate new product designs that optimize mental retardation's life skills.

3.2 Results of wood crafts vocational learning for students with disabilities

The product of wooden craft vocational learning is planned by the teacher and produced by mentally retarded students. Products made by mentally retarded students's processing time and the final result basically cannot be predicted. In accordance with the results of research conducted by Nurian Anggraini (2016: 87), it is concluded that the work of mentally retarded students cannot be predicted when the completion time is carried out and the final results cannot be as perfect as expected according to their abilities. The average processing time for a single product takes 3 months.

The products of mental retardation students are divided into 3 types, namely Educational Teaching Aids, wall hangings, and functional products. Educational Teaching Aids are made and used by teachers in schools as learning media. The colors used tend to be colorful to attract students' attention. The final finishing of the product also uses a paint that is safe and does not fade easily so it does not harm students' health. The material used to make Educational Teaching Aids is mahogany. Wall decorations made in wood craft learning consist of several kinds, namely wooden paintings, souvenirs, calligraphy, and batik motifs. The techniques used to make the product are the scrolling technique and the bench working technique. Almost all products are made of teak wood with a spray finish to accentuate the original wood grain shape. Especially for wood painting products using Medium-density fibreboard (MDF) and plywood. The functional products produced by the students are cutting boards, pencil boxes and tissue boxes. The technique of making this product is by using bench work and scrolling techniques. The materials used for this product are teak, Dutch teak, and mahogany.

3.3 Identification of barriers to wood craft vocational learning in mentally retarded students

Wood crafts vocational learning is arranged and adapted to every obstacle and potential of mentally retarded students. This is intended so that students are able to work independently. In the learning process there are obstacles that interfere with students' work process. These obstacles come from the mentally retarded students themselves, barriers to wood craft learning facilities and infrastructure, barriers to product marketing, and barriers to cooperation with the business world. In the following, we will discuss the barriers to wood crafts vocational learning for mentally retarded students at SLB Negeri 2 Yogyakarta.

First, the biggest learning barrier in wooden craft vocational learning is from the mentally retarded students themselves. Students are often absent from school, motor problems at work, difficulty understanding instructions and directions from the teacher, and behavioral disorders during studies. Overall, this barrier is the characteristic of mentally retarded students themselves. This obstacle is basically a challenge to develop effective and attractive vocational learning. Tony Wagner (2008) quoted by Agus Sutarna (2020: 73) mentions seven skills so that vocational learning can survive in the new world order, namely (1) Critical Thinking and Problem Solving, (2) Collaboration Across Networks and Leading by Influence, (3) Agility and Adaptability, (4) Initiative and Entrepreneurialism, (5) Effective Oral and Written Communication, (6) Accessing and Analyzing Information, and (7) Curiosity and Imagination. Thus, the quality of teachers as planners and implementers of vocational learning must continue to be developed. Steps to increase teacher competence in wooden craft vocational learning can be done with a teacher competence increasing program with a focus on abilities, attitudes, and skills; and carry out training with level 3 expertise certificates in accordance with the areas of expertise being handled (Nur Wahyuni 2018: 144). With the teacher focusing on developing specifications for wood craft expertise, especially wood craft production techniques, vocational learning will be more meaningful for mentally retarded students. The main context in addition to providing skills for mentally retarded students to live in society is to form the attitude of mentally retarded students who are responsible, communicative, cooperative, and honest. It is more urgent to see the main problem in wood crafts vocational learning which is dominated by mental retardation behavior.

Second, the barriers to facilities and infrastructure in wooden craft vocational learning are related to supporting the learning process. The learning process is carried out in a wooden craft studio with tools and materials stored in it. Based on the observations, it shows that the wooden craft studio room lacks air circulation, messy arrangement of tools, and unorganized storage of materials and works. This will greatly affect the safety and health of students in learning wood crafts.

The learning process of wood crafts produces a lot of dust and sawdust waste. If there is not enough air circulation, it will cause health problems for students and teachers, especially in breathing. Research conducted by Anik Lestari (2010: 27) concluded that subjects who worked in the sanding section for a long time experienced restrictive lung disorders and mixed pulmonary function disorders. Given that the practice of making wood products generates a lot of wood dust, especially in wood sanding and cutting. Therefore, the most urgent development of facilities and infrastructure in learning wood crafts is to provide ventilation that is able to remove dust and dirt from the manufacture of wood products. The arrangement of wooden equipment also needs to be rearranged to provide moving access when studying in a woodcraft studio. Arrangement of tools must also pay attention to body posture when working with tools, because it will affect students' posture in the long term. The wrong work position due to the disproportionate position of the tool must be corrected immediately to make it easier for students to work. In addition, access to electricity cable arrangement is also an urgent and important consideration to immediately followed up. This is because it is very dangerous considering the various cognitive limitations in students who tend to pay less attention to the concept of cause and effect.

Third, difficulties in marketing to sell wood craft products that have been made by students. So far, the products that have been made are stored in school display windows. Marketing is carried out by participating in every exhibition held by the Yogyakarta Special Region government. If production is carried out continuously without any clear marketing action it will only result in stockpiling of products. Exhibition activities are one way to develop students' with mental retardation entrepreneurial spirit. The role of teachers in this exhibition is as edupreneurs to develop creative and innovative businesses in the education sector (Agrawal 2013: 5). It is hoped that when mentally retarded students graduate they will be able to help their parents develop a business or enter the world of work, in this case including developing programs, selling products, services or technology in the education sector.

Fourth, schools still have difficulty collaborating with the business or industry, especially in establishing student internships. Fourth, schools still have difficulty collaborating with the business or industry, especially in establishing student internships. During this time, the internship program for students who take self-development of wood crafts has undergone a welding skills internship. Internships in the business world are a means of updating competencies. What has been learned in school is then implemented in the industrial world as work experience. If there is a mismatch between learning in schools and the industrial world, it will cause the inability of students to study optimally in the business world. The business world still looks down the ability of mentally retarded students so that work programs are difficult to implement. The main reason is the inability of the business world to teach mentally retarded students. Lack of communication between schools and the business world creates lack of understanding of the obstacles and potentials of mentally retarded students. The internship program essentially provides work experience and career development for mentally retarded students. Some of the lessons learned from mentally retarded career development are being able to get used to a good work attitude; able to make career decisions, finding job, and job-finding skills; and able to integrate work values into the structure of personality values (Mumpuniarti 2007: 10).

Establishing a cooperative network relationship with the business world and industry must be continued as an effort to alleviate poverty and provide services to mentally retarded students. Schools must make efforts to strengthen vocational programs including strengthening networks with the business world in developing life skills (Luqman Hidayat 2018: 40). One way to establish school networks with the business world is a teacher apprenticeship program. According to Nur Wahyuni (2018: 143) teacher apprenticeship activities in the business world can strengthen vocational education cooperation in the form of internship activities. He further explained the need for communication to integrate the school curriculum with the business world. The business world can provide input and advice on the competencies taught in schools, the hope is that when students graduate they can be absorbed and employed by business.

4 CONCLUSION

Wooden craft vocational learning is held as a service to mentally retarded students so that they have life skills with skill competencies in the community. The achievement of wooden craft learning outcomes is in accordance with the obstacles and potentials of mentally retarded students, and this is in accordance with their capacity. The main obstacle in the learning process comes from within the mentally retarded child itself, this is a big challenge for teachers to design vocational learning that is creative, motivating, and independent of mentally retarded students. In addition, the development of a wooden craft vocational program is also focused on improving facilities and infrastructure to create safe and healthy learning, establishing relationships for product marketing as a means of developing entrepreneurship education, and collaborating with the business world to create internship programs, develop vocational curriculum, and distribution for mentally retarded students graduates. Thus, improving the life skills of mentally retarded students can be achieved optimally through the development of wooden craft vocational learning.

REFERENCES

Abdurrahman, M. 2003. *Pendidikan Bagi Anak Berkesulitan Belajar*. Jakarta: PT. Rineka Cipta.

Agrawal, R. 2013. How to identify and Select a Business Opportunity and Then Implement the Business Idea? A Case on Edupreneurship in India. *Journal of Business Studies Quarterly*, 4(4): 176–182.

Amin, M. 1995. *Ortopedagogik Anak Ttunagrahita*. Jakarta: Departemen Pendidikan dan Kebudayaan.

Anggraini, N. 2016. *Vocational Skills Development for People with Intellectual Disabilities by Institution BBRSBG Kartini Temanggung Central Java*. Proceedings of the 2nd International Multidisciplinary Conference, Nov 2016.

Astati. 2001. *Persiapan Pekerjaan Penyandang Tunagrahita*. Bandung: CV Pandawa.

Borg, W. R., & Gall, M. D. 2007. *Educational Research (Six edition)*. New York: Longman.

Brown, B. J. & Baker, S. 2007. *Philosophies of Research into Higher Education*. London, United Kingdom: A&C Black.

Dewi, D. P. 2018. *Asesmen Sebagai Upaya Tindaklanjut Kegiatan Indentifikasi Terhadap Anak Berkebutuhan Khusus*. Jurnal Wahana, 70(1): 17–24. https://doi.org/10. 36456/wahana.v70i1.1563

Eldiva, Fiyola Triana. 2018. *Project Based Learning in Improving Critical Thinking Skill of Children with Special Needs*. Prosiding disampaikan pada International Conference on Special and Inclusive Education (ICSIE 2018). https://doi.org/10.2991/icsie-18.2019.64

Hidayat, L. 2018. Urgensi Guru dan Kompetensi Edupreneur dalam Dukungan Pendidikan Vokasional di Sekolah Luar Biasa. *Widya Wacana, 13*(2): 40–45. http://dx.doi.org/10. 33061/ww.v13i2. 2259

Ishartiwi. 2012. Kompetensi Guru Pendidikan Khusus dan Model Pengembangannya dalam Upaya Peningkatan Kualitas Generasi Bangsa Penyandang Difabel. *Jurnal Pendidikan Khusus*, IX(1): 1–11.

Kementerian Pendidikan dan Kebudayaan. 2014. *Kurikulum 2013. Pedoman Pelaksanaan Bagi Peserta Didik Berkebutuhan Khusus*. Jakarta: Direktorat Pendidkan Khusus dan Layanan Khusus.

Kementerian Pendidikan dan Kebudayaan. 2012. *Integrasi Pendidikan Karakter dalam Pembelajaran bagi Peserta didik Tunagrahita*. Jakarta: Kemendikbud Dikrektorat PKLK

Kemis. 2013. *Pendidikan Anak Berkebutuhan Khusus Tunagrahita*. Jakarta Timur: PT. Luxima Metro Media.

Lestari, Anik. 2010. *Pengaruh Debu Kayu terhadap Gangguan Fungsi Paru Tenaga Kerja di CV. Gion & Rahayu, Kec. Kartasura, Kab. Sukoharjo, Jawa Tengah.* Skripsi: Fakultas Kedokteran: UNS. https://digilib.uns.ac.id/dokumen/detail/ 14379/

Mechling, et al. 2013. *The Use Of Mobile Technologies To Assist With Life Skills/Independence Of Students With Moderate/SevereIntellectual Disability And/Or Autism Spectrum Disorders: Considerations For The Future Of School Psychology. Psychology in the Schools*, 50(3), https://doi.org/10.1002/pits.21673

Miles, Matthew B. A., Michael Huberman, & Johnny Saldaña. 2013. *Qualitative Data Analysis.* London: Sage Publication.

Mumpuniarti. 2006. Manajemen Pembinaan Vokasional Bagi Tunagrahita di Sekolah Khusus Tunagrahita. *Jurnal Pendidikan Khusus*, 2(2): 1–17.

Mumpuniarti. 2007. *Pembelajaran Akademik Bagi Tunagrahita*. Yogyakarta: FIP UNY.

Mumpuniarti 2018. Kesiapan Guru Sekolah Reguler untuk Implementasi Pendidikan Inklusif. *Jurnal Pendidikan Khusus, 14*(2): 57–61.

Suryani, N. 2018. *Kekuatan Kognitif Siswa Tunagrahita Ringan terhadap Kegiatan Pembelajaran Keterampilan Budidaya Hortikultura*. Jurnal Ilmu Pendidikan. Vol. 2 (2). ISSN: 2549–9114. http://dx.doi.org/10.26858/pembelajar. v2i2.5760

Purtranto, B. 2015. *Tips Menangani Siswa yang Membutuhkan Perhatian Khusus*. Yogyakarta: Diva Press.

Sanjaya, W. 2013. *Penelitian Pendidikan: Jenis, Metode, dan Prosedur*. Jakarta: Prenadamedia Group.

Sugiyartun. 2009. *Pendidikan Tunagrahita*. Bandung: Alfabeta.

Sutarna, Agus, et al 2020. *Manajemen Pendidikan Vokasi.* JawaTengah: CV. Pena Persada.

Wahyuni, N. 2018. Peran Pendidikan Vokasi Bagi Anak Berkebutuhan Khusus dalam Menghadapi Tantangan Zaman. *Jurnal Keluarga, 4*(2): 137–147.

Wakiman. 1998. *Menumbuhkan Kesenangan Belajar Matematika Melalui Permainan.* Jurnal Kependidikan, Nomor 2. Page 209–222

Werry, J. S., & Aman, M. G. 2013. *Practitioner's Guide to Psychoactive Drugs for Children and Adolescents*. Berlin: Springer Science & Business Media.

Wijaya, A. 2013. *Teknik Mengajar Siswa Tunagrahita*. Yogyakarta: Imperium.

Educational Innovation in Society 5.0 Era: Challenges and Opportunities – Purnomo & Herwin (Eds)
© 2021 the authors, ISBN 978-1-032-05392-9

Support of parents and schools for online learning during the Covid-19 pandemic

I.W. Liasari & A. Syamsudin
Early Childhood Education Department, Yogyakarta State University, Indonesia

ABSTRACT: The addition of Coronavirus cases in Indonesia has increased very sharply. In addition to the economic impact that is making people suffer, there is an unprecedented destruction of the education system in Indonesia and even the whole world. After the outbreak of the COVID-19 pandemic in Indonesia in mid-March of 2020, the central and regional governments set policies in the world of education, which is to temporarily eliminate direct face-to-face learning and replace it with online learning. This study aims to determine the support of parents and schools for online learning during the COVID-19 pandemic. This research was conducted at the Kindergarten Budi Mulia Dua Sedayu Bantul Yogyakarta. The research uses the qualitative approach with in-depth interviews in examining data with the parents and teachers, and distributes questionnaires to parents. The results in this study revealed the following: creating a comfortable environment, facilitating the child in learning, maintaining and ensuring the child to implement a clean and healthy life, accompanying children in learning, activities and playing activities together, guiding and motivating children. A guide is needed for parents so that children can understand and enjoy learning activities while at home in a fun way. Providing teacher learning materials by creating video materials and giving children the tools and materials they need, schools also make face-to-face programs twice a week, as well as one-on-one programs (one teacher one child), but there are still some parents who disagree with the program. Parent collaboration with schools is needed to support the ongoing online learning activities during the COVID-19 pandemic period.

1 INTRODUCTION

Coronavirus (COVID-19) attacked Indonesia in early 2020, until August 2020 there were 137,468 people exposed to COVID-19. A total of 40,076 people did the treatment (https://covid19.go.id/). Right at the beginning of August 2020 those who recovered experienced an increase of 91,321 people, for those who died the total number was 6,071 people. In response to the Coronavirus disease 2019 (COVID-19) pandemic, 107 countries had implemented national school closures by March 18, 2020. The application of restriction on community activities to break the spread of the COVID-19 virus occured in various sectors including the education, economic, religious, social and community sectors. One of the sectors most influenced by lockdown is education. Various attempts are made by the government to break the chain of the spread of the virus, among which is by issuing the law government Number 21 year 2020. Large scale social restrictions resulted in the limitation of various activities including schools. Meanwhile, the activity of learning from home or BDR is officially issued through the circular letter of the Minister of Education and Culture number 36962/MPK. A/HK/2020 about learning online and working from home. This was done in order to prevent the spread of the Coronavirus. This policy forces teachers and students to keep working and learning from home from the Early Childhood Education or ECE level to higher education. This policy certainly affects not only the relationship of teachers and pupils during the learning from home (BDR), but also the importance of optimizing the role of parents in the implementation of learning from home (BDR).

Research on the effects of prior pandemics and disasters clearly indicates that there will be both immediate and long-term adverse consequences for many children, with particular risks faced during early childhood, when brain architecture is still rapidly developing and highly sensitive to environmental adversity (Shonkoff et al., 2012). Developmental psychology researches largely founded that learned experiences through environmental factors during early childhood engender the fundamentals for lifetime behavior and success as it is a crucial phase for cognitive, emotional and psychosocial skill development. During a severe pandemic like COVID-19, community-based mitigation programs, such as closing of schools, parks, and playgrounds will disrupt children's usual lifestyle and can potentially promote distress and confusion. While online classes and assignments have been the only effective way for continuing education in this situation, experts have already cautioned about being over-burdened. Specific psychological needs, healthy

DOI 10.1201/9781003206019-23

life-styles, proper hygiene advice, and good parenting guides can be addressed through the same online platform (Wang et al., 2020).

The notion that older adults and people with serious comorbidities are particularly vulnerable to worse outcomes from COVID-19 can create considerable fear amongst the elderly (Dubey et al., 2020). Other psychological impacts may include anxiety, irritability and excessive feeling of stress or anger. Those older adults with cognitive decline may become much more anxious, agitated, and socially withdrawn, thus their specific needs demand specific attention. Indoor physical exercise might be a potential therapy not only to maintain robust physical health, but also to counteract the psychological impact in this trying time (Wang et al., 2020). Today's self-centered, busier-than-ever human race could potentially appreciate home-confinement during COVID-19 as an opportunity to promote healthier parent and child relationships by correct parental strategies and strengthen family bonding by spending more quality time together with older parents/dependent members residing in the same household. For populations in general, the government should create real-time, online tracking maps regarding COVID-19 updates to alleviate stress, anxiety and confusion.

Discussing education, according to the Law of the Republic of Indonesia number 20 year 2003 on national education system, Education pathways in Indonesia consist of formal, non-formal, and informal education. Early Childhood Education (ECE) is included in non-formal education. Although not including formal education, but in regulation of the Minister of Education and Culture Republic of Indonesia No. 146 year 2014 about the 2013 curriculum for Early Childhood Education (ECE) that the ECE is the most fundamental education level that is considered the most fundamental cause of future child development will be determined by a wide range of stimulation, meaningful stimulation given from an early age, where early childhood is the most appropriate time for such stimulation to allow the child to develop optimally.

One of the education levels most influenced by lockdown application is the Early Childhood Education (ECE) level. The learning process of children especially the process of learning activities in schools is one of the ways to improve children's developmental aspects. However, after a COVID-19 pandemic that forces children to study at home so as to make structured children's activities during school to support the child's developmental achievement level. Children's activities that have been stuck in the school are finally back to the education process of parents with the implementation of the term home learning that creates meaningful educational processes. Application of learning to achieve the six development aspects are currently hindered due to COVID-19. The implementation of activity restrictions carried out by the government through a lockdown policy to suppress the spread of the COVID-19 virus pandemic forces learning activities for children to be carried out at home. The activities of children at home with the habit of going to school in the morning require a process of changing new learning activities by prioritizing the role of parents in implementing learning. The role of parents in accompanying children's success while studying at home is very central, in relation to this, the WHO (2020) released various guidelines for parents in accompanying their children during this pandemic which includes parenting tips to be more positive and constructive in accompanying children during activities at home.

Children need support in early stimulation and learning, health and nutrition, care and protection. Parents giving children the best start in life means ensuring them good health, proper nutrition and early learning. Children need to grow up in a nurturing and secure family that can ensure their development, protection, survival, and participation in family and social life. The mechanisms for building parental support according to Cohen & McKay (2008) are, namely: 1. real Support, although actually everyone can provide support in the form of money and attention, real support is most effective if it is well appreciated by the recipient. Providing real support which results in feelings of disorder and unacceptability will actually add to individual stress and stress in the life of the parents. Forms of tangible support include, among other things, attention and materials. 2. Support expectations: support groups can influence individual perceptions of threats. Expecting individuals in the same person to have been in the same situation for advice and assistance. Expectancy support can also help improve individual strategies by suggesting alternative strategies that build on previous experiences and getting people to focus on the more positive aspects of the situation.

Sarafino & Smith (2011) that parental support is very influential on the success of their children in learning. Therefore, parents who care about the progress of their children's education will always try to provide the best for their children, provide the facilities they want in order to achieve good children's achievements, and pay attention to their children's development. This is in line with the opinion of Hasbullah (2001) which states that parents are the first and foremost people who are responsible for the survival and education of their children.

The role of parents in assisting the success of the child during study at home is very central, as related by the WHO (2020) when they released various guides for parents to accompany their kids during this pandemic that includes parenting tips to be more positive and a constant in the accompanying child during the day at home (Lilawati, 2020). But this role became widespread as a companion to academic education. Prabhawani (Lilawati, 2020) The challenges in implementing online learning are faced by Early Childhood Education (ECE) institutions. Beside experiencing confusion in adapting to the online learning system, teachers are also required to continue their learning activities to school supervisor, parents, students, and general public. So, the teachers still do their jobs well even in the outbreak condition of COVID-19. This is a form of responsibility for the duties given to the teacher as educators and teachers.

Pandemic COVID-19 teaching and learning activities must keep running. Fulfillment of each nation's rights to education must remain fulfilled. Therefore, teachers as an education mobilizer must keep fulfilling their duties well. A study environment was at school with teachers and classmates, Now change only at home accompanied by parents. This will certainly require adaptation and impact the learners. The learning environment should be made as positive as possible with the school environment, so that student motivation can grow. So, there is a strong sense of community and engagement between children, parents, teachers, principals, and staff where those members feel strongly about caring for one another. These provisions reflect good practices in engaging families in helping to educate their children, because students do better when parents are actively involved in the education process, both at home and at school.

2 RESEARCH METHODS

This research uses a qualitative approach. In qualitative research it is common to use different data-gathering methods in the same investigation (Sadler & Doo, 2016). This allows the researcher to deepen the non-exhaustive feedback collected through asynchronous tools such as the questionnaire and integrate those data points with additional tools like individual interviews.

The subject of research is the parents of Budi Mulia Dua Sedayu Bantul Yogyakarta. The data generated in this study is qualitative descriptive. This qualitative description of research obtains information and results from parents and school support for online learning. The data collection techniques used in this study were with interviews and observations with the teachers in the kindergarten. Another data collection technique is using questionnaire. The data collection instruments used are human instruments or researchers themselves who function to establish research focus, choose data sources, informants, data collection and data analysis and find conclusions. The questionnaire used in this study in the form of a closed questionnaire which is presented in the form of a question.

The technique of triangulation of sources is triangulation techniques, methods, and the techniques of triangulation theory. The technique of triangulating data collection that is sourced from the results of observations and interviews. Data analysis when the empirical data obtained is considered qualitative data. The activities of the data analysis are carried out through three activity streams that occur simultaneously, namely data reduction, data presentation, and conclusion/verification. The analysis of data such as a process cycle and interactions at the time before, during, and the time of data collection to describe the flow of research conducted is described in the image below:

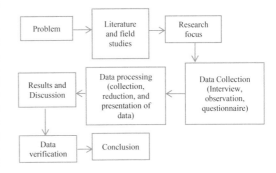

Figure 1. The design of the study.

3 RESULTS AND DISCUSSION

Based on the data that concluded, here are the parts of a Google form that has been distributed, the results of the research will be outlined below:

Figure 2. Image research survey online (the link in google form is https://forms.gle/jpeg6NLJRjctuM92A).

The results of the questionnaire of parental support during the process of online learning here are the results of the questionnaire that has been given a rating by parents of students who come to guide her child in doing home learning (study from home), see Table 1.

Table 1. The Results of the Questionnaire

No	Statement	The choice of the person	
		Yes	No
1	Whether the Father/Mother provide education or understanding about COVID-19 to the child?	56	0
2	Whether Learning Online is the Father/Mother closer to the Child?	55	1
3	The Pandemic COVID-19, Whether the Father/Mother to agree with Online Learning?	55	1
4	Whether the Father/Mother to agree with the Program of Online Learning conducted the school?	53	3
5	Whether the school authorities are very Cooperative in providing Learning Online?	49	6
6	Online Learning courses a given school is burdensome Father/Mother?	34	22
7	Whether Online Learning this further adds to the burden of busy days when the Father/Mother at home?	42	14
8	Whether the Father/Mother give encouragement to the child to be always optimistic and not easily give up in learning at home?	56	0
9	Whether the Father/Mother still gives the spirit/praise the child when he gets good results that can maintain it?	56	0
10	Whether your Father/Mother show the steps that must be done in the study?	55	1
11	Whether the Father/Mother to facilitate the child learning to use a computer/ Laptop? Whether the accompanying when using a Computer/Laptop?	24	32
12	Whether the Father/Mother to facilitate the child learning to use a Smartphone? Whether the accompanying when using a Smartphone?	46	10
13	Whether the Father/Mother take the time to teach the child to learn?	54	2

(Continued)

Table 1. (Continued)

No	Statement	The choice of the person	
		Yes	No
14	Whether the Father/Mother prepared a special room for children's learning that is comfortable?	42	14
15	Whether the Father/Mother to establish good communication with the child when at home? For example?	56	0
16	Whether the Father/Mother take the time to play with the Child?	56	0
17	Whether your Father/Mother do activities together when at home during the Pandemic COVID-19 this?	55	0

Responding to the rising cases of Coronavirus (COVID-19) in Indonesia, especially in Yogyakarta, a parent in TK Budi Mulia 2 Sedayu Bantul Yogyakarta agrees with the idea that education is conducted online. This is to break the chain of spread of the virus COVID-19. The vigilance of the parents against the virus COVID-19 requires that parents provide an understanding of whether it is COVID-19, signs of exposure to the virus, and what the symptoms are. The results of the questionnaire stated that the 56 parents agree and are already providing education about the virus COVID-19 for the sake of the safety of children and 55 parents agreed to do so if learning is done online. The learning system is usually implemented directly in school, but now must be implemented with the system remotely. In terms of the methods of learning, because of varying conditions and limitations, the learning situation is entirely unique when compared to anything that has come before.

In accordance with the results of the study that the 46 parents facilitate the child to online learning, one of the facilities is a smartphone that is given and should be with the parents whenever possible. Other facilities given to parents include are a desk study, a room that is conducive and comfortable. All parents agree that online learning in the pandemic COVID-19 is generating free time for parents to play with children so as to produce communication that better benefits the child. 55 parents also feel that online learning is making the parents more close to the child at home. In addition to parents, the school strongly supports the policy of the government. 53 of the 56 parents agree with the program provided by the school in the course of online learning. Such changes occur in Early Childhood Education (ECE).

Based on the data that has been shown in Table 1, it can be seen that older people feel learning at home is very effective during the pandemic but that does not mean learning in schools is not more effective compared to learning activities at home.

3.1 Create a comfortable environment

Create a comfortable environment, facilitate the child in learning, maintain and ensure the child is to apply clean and healthy lifestyle, to assist children in learning, doing activities and playing together, guide and motivate the child. Parents who are faced with the competing demands of limiting social interactions and remaining at home with their children may be particularly vulnerable during this time; research shows that continual close contact under stress is a risk factor for aggressive behaviors and violence (Brown et al., 2020). Furthermore, some families are experiencing other challenges, such as working from home while also caring for and educating their children. Given that school and childcare professionals are central to identifying concerns of abuse and neglect, children who may have once been identified as at risk in these settings may be more vulnerable to maltreatment as they spend most of their time at home.

3.2 Facilitating children in the learning

The fulfillment of the basic needs of the family is an attempt to meet care, upbringing and education needs. In this case, there is an obligation and family responsibility which is to meet the basic needs of the child, such as nurture, nourishment, education, and protection of the child (Undang-Undang Republik Indonesia Nomor 35 Tahun 2014 Tentang Perubahan Atas Undang-Undang Nomor 23 Tahun 2002). The basic needs of the family being met is a condition for the prosperity or harmony of the family (Puspitawati, 2013). The provision of a place of learning is a reasonable goal to make sure the child is the focus. The parents also explained that a cozy study room will motivate children to learn. A means of learning that is conducive and comfortable will have a part in the achievement of the learning results of children, which is the measure of the achievement of national education goals (Attha, 2005).

3.3 Maintain and ensure the child is to apply clean and healthy lifestyle

Every parent would want his child to be always in a healthy state, especially with the current conditions i.e. during the pandemic. The parents become increasingly worried about it. The parent is reminding the child to always maintain a healthy lifestyle and clean in order to avoid various diseases and to teach children to follow a protocol of health. In line with that expressed by the Kurniawati et al., (2020), which states that parents have a very important role in terms of educating children, one of them is to become and provide a good example for children, in addition, providing warning and advice on the child is also an important thing that parents should do to always live a clean life

with the children. Apart from that, the role of parents is to ensure that children eat nutritious food, sleep regularly, wash hands diligently, play actively, etc (Kurniati et al., 2020)

Kurniawati et al. (2020) stated that parents provide affection their children not only in the form of education and trust, but also to always control the progress of the child. As we know, children can quickly learn by imitation, so important examples and habits given by teachers, particularly older people, are also important. Train the child to always live clean and healthy in line with one of the activities in the parenting program the positive/positive parenting program (Triple P) which aims at developing the capacity of individuals for self-regulation (Sanders, 2008). Self-regulation is the process whereby individuals are taught skills to change their own behavior and be a solution for the problems in the broader social environment that relate to parenting and families (Karoly, in Sanders, 2008). Positive parenting can serve as a model caregiving family to be the solution in maintaining and ensuring that the child has a clean and healthy lifestyle with self-regulation.

3.4 Assisting children in learning

To prevent the chain of transmission of Coronavirus in the school-issued policy the implementation of education in times of emergency the spread of Coronavirus (COVID-19) by the Minister of Education and Culture, through a circular letter on 24 March 2020, about the wisdom of the "learn from home (BDR)" order. This implies that the parents temporarily replace the role of the teacher in assisting children to learn at home.

According to García and Thornton (2014), current research indicates that family involvement in learning helps improve student performance, reduce absenteeism and restore the confidence of parents towards the education of their children. From the results of the review of data , it can be known that the role of parents during the BDR is more on helping the students complete the tasks given by the teachers in the school. These conditions indicate that the activities of the BDR given by the teachers is on the provision of the task as expressed by Nahdi et al. (2020): that the activities provided by the institution of the school in the application of learning at home includes parents giving the task or assignment.

Conditions like this become interesting to study, such as: what is the meaning of learning from that house? Does it have the same meaning with tasks or homework at home? Can these tasks be said to be successful in the role of mentoring in childrens' learning? According to Llamas & Tuazon (2016), parents need to be comfortable when the education system demands involvement in school activities. Parents being involved in the education of their children is a good thing because it improves academic achievement. Learners become more focused in their school work (Kwatubana & Makhalemele, 2015).

...to be an educator for the child while in the home, such as assisting in learning, to help when there are difficulties in a child ...(interview from MN)

...self-study with the parents in the form of a tutorial from the school via mobile or online are to be done by the children in our home, and guided and directed...(excerpt interview SF)

The interview quote above suggests that essentially the efforts of the mentoring can be done through different methods such as giving help when there are difficulties, accompanying the child when online learning takes place, optimizing the things that parents need to improve the knowledge and skills that focus on cognitive, affective, psychomotor in optimizing all aspects of it is development.

3.5 *Doing activities and play activities together*

There are various kinds of activities carried out together between parents and children during this pandemic, such as cleaning the house, cooking, playing, praying, etc. This moment provides an opportunity for parents and children to bond with each other. In line with this UNICEF (2020) reveals that there are several ways that parents can help the parenting process during this pandemic, one of which is by making time quality with the child, for example for kindergarten aged children this is to involve the child in household chores such as cleaning the house or cooking, and this can be suggested as an exciting game. At this point, parents act as developers of various activities that can be done together with children.

The quality of time that parents and children have during the pandemic can be used to build togetherness between family members, Harmaini (Kurniawati, 2020) states states that parental togetherness is needed because parents understand the level of development of their children and the things they need. The busyness of the day often results in limited time with children. Even on holidays, when all the family members gather, each one is busy with his own activities. Thus the role of parents as activity developers can be carried out together by involving children in determining the variety of activities to be carried out so that children avoid feeling bored. According to (Ministry of Education and Culture of the Republic of Indonesia, 2017) there are several activities that can be done together with children to create quality time, namely by praying together, eating together, playing or exercising together, cooking together, cleaning the house together, accompanying children in doing school work, watching television, surfing the internet, and using gadgets.

3.6 *Guiding and motivating children*

The events of the pandemic COVID-19 that occured in almost all the world, is showing the growing importance of the role of the family in the nurturing, care and also education of the child. These events restore the initial function of the family as the center of all activities, a place of education for the child. According to the students ' parents, learning at home is rated as still able to improve the quality of learning as well as with learning in school. Again, it is said that teaching occurs when a more knowledgeable person passes on knowledge to a less knowledgeable individual in order to induce a relatively permanent change in behavior (Lantolf, 2008; Bateman & Waters, 2013; Greenberg, 2005). The aspect of the more knowledgeable person is missing when one is learning alone in the house, especially with concepts they are not familiar with and may need explanation from the teacher or other colleagues to enhance their understanding. Many of the parents also help provide motivation for students required to learn from home because of the government appeal about COVID-19, it also makes deliberately taking the time to do this a significant part of the learning process of children in the home. A lot of parents who agreed that during the learning at home, parents also helped in the tasks given by the teacher. Although more than a few who feel this feel it is an additional activity of the parents in addition to doing household chores.

3.7 *School Support*

The provision of learning materials by the teacher to make the material in the form of videos: TK Budi Mulia 2 Sedayu also makes the program one on one (one teacher one child), but there are still some parents who do not agree with the program. Necessary cooperation of parents with the school to support the ongoing learning activities of the online running smoothly during the pandemic. Based on the description of the class teacher, usually the learners follow the learning activities in school alongside the class teacher a week as much as two times. After the policy of online learning, the learning system changed to giving the task which remains adjusted with study schedules as usual. Tasks assigned according to the Program Plan Daily Lessons (RPPH) given once a week in the beginning of the week. Then, each student sends results of work tasks according to schedule learning by WhatsApp (WA) to each class teacher. The task given at the beginning of the week, usually guardians of pupils go to the school to pick up assignment materials provided by the school, also take thbis opportunity to get a direct explanation from the teacher with regard to how the tasks given to minimize misunderstanding of the task execution. The form of the assignment given by the teachers is processed with the diverse variations in every week so that children do not feel bored. Tasks can be specialized workmanship books magazines, make work, video creating tasks showing children singing, or take the form of other tasks.

4 CONCLUSION

The results show that in general parental support of learning creates a comfortable environment, facilitates the child in learning, maintains and ensures the child has a clean and healthy lifestyle assists children in learning, doing activities and playing together, and guides and motivates the child. TK Budi Mulia 2 Sedayu makes the program one on one. The program was intended to provide learning face-to-face one teacher one child which was made one week at a time.

REFERENCES

Bateman, A., & Waters, J. 2013. Asymmetries Of Knowledge Between Children And Teachers On A New Zealand Bush Walk. *Australian Journal of Communication*, 40(2): 19–31.

Brown, S. M., Doom, J. R., Lechuga-Peña, S., Watamura, S. E., & Koppels, T. 2020. Stress and parenting during the global COVID-19 pandemic. *Child Abuse and Neglect*, 110(June). https://doi.org/10.1016/j.chiabu.2020.104699

Cohen, S. & Mckay, G. 2008. *Social Support, Stress and the Buffering Hypothesis: A Theoretical Analysis.* In Baum, A. Taylor, S., & Singer, J. *Handbook of Psychologyand Health.* New York: Hillsdale.

García, L. E., & Thornton, O. 2014. The Enduring Importance of Parental Involvement. *Nea Today*, Http://Neatoday.Org/2014 / 11 / 18 / The-Enduring-Importanceof Parental involvement-2/.

Greenberg, K. H. 2005. *The Cognitive Enrichment Advantage Family-School Partnership Handbook.* Ox Knoxville: Kcd Harris & Associates Press.

Hasbullah. 2001. *Dasar-Dasar Ilmu Pendidikan.* Jakarta: Rajagrafindo Persada.

Kurniati, E., Nur Alfaeni, D. K., & Andriani, F. 2020. Analisis Peran Orang Tua dalam Mendampingi Anak di Masa Pandemi COVID-19. *Jurnal Obsesi?: Jurnal Pendidikan Anak Usia Dini*, 5(1): 241. https://doi.org/10.31004/obsesi.v5i1.541

Kwatubana, S., & Makhalemele, T. 2015. Parental Involvement In The Process Of Implementation Of The National School Nutrition Programme In Public Schools. *International Journal of Educational Sciences*, 9(3): 315–323. https://doi.org/10.1080/09751122.2015.11890321

Lantolf, J. P. 2008. *Sociocultural Theory and the Teaching Of Second Languages.* London:Equinox Publishing.Brown, S. M., Doom, J. R., Lechuga-Peña, S., Watamura, S. E., & Koppels, T. (2020). Stress and parenting during the global COVID-19 pandemic. *Child Abuse and Neglect*, 110(June). https://doi.org/10.1016/j.chiabu.2020.104699

Lilawati, A. 2020. Peran Orang Tua dalam Mendukung Kegiatan Pembelajaran di Rumah pada Masa Pandemi. *Jurnal Obsesi?: Jurnal Pendidikan Anak Usia Dini*, 5(1): 549. https://doi.org/10.31004/obsesi.v5i1.630

Llamas, A., & Tuazon, A. P. 2016. School Practices in Parental Involvement, Its Expected Results and Barriers in Public Secondary Schools. *International Journal of Educational Science and Research*, 6(1): 69–78.

Nahdi, K., Ramdhani, S., Yuliatin, R., & Hadi, Y. 2020. Implementasi Pembelajaran pada Masa Lockdown bagi Lembaga PAUD di Kabupaten Lombok Timur. *Jurnal Obsesi: Jurnal Pendidikan Anak Usia Dini*, 5(1): 177–186.

Puspitawati, H. 2013. *Konsep Dan Teori Keluarga. Gender Dan Keluarga*, 1–16. Https://Doi.Org/10.1249/01.Mss.0000074580.79648.9d

Sadler, R., & Doo, M. 2016. *Language learning in virtual worlds*. Diambil dari http://www.aupress.ca/books/120254/ebook/99Z_Gregory_et_al_2016-Learning_in_Virtual_Worlds.pdf

Sanders M. R. 2008. Triple P-Positive Parenting Program as a public health approach to strengthening parenting. *Journal of family psychology: journal of the Division of Family Psychology of the American Psychological Association (Division 43)*, 22(4): 506–517.

Sarafino, E.P., & Smith, T.W. 2011. *Health Psychology: Bio Psychosocial Interaction*. New York: John Willey and Sans Inc.

Shonkoff, J. P., Garner, A. S., Siegel, B. S., Dobbins, M. I., Earls, M. F., McGuinn, L., ... & Committee on Early Childhood, Adoption, and Dependent Care. 2012. The lifelong effects of early childhood adversity and toxic stress. *Pediatrics*, 129(1): e232–e246.

Dubey, S., Biswas, P., Ghosh, R., Chatterjee, S., Dubey, M. J., Chatterjee, S., & Lavie, C. J. 2020. Psychosocial impact of COVID-19. Diabetes & Metabolic Syndrome: Clinical Research & Reviews, 14(5): 779–788. https://doi.org/10.1016/j.dsx.2020.05.035

Undang-Undang Republik Indonesia Nomor 35 Tahun 2014 Tentang Perubahan Atas Undang-Undang Nomor 23 Tahun 2002 Tentang Perlindungan Anak, Pub. L. No. 35.

WHO. 2020. Who Director-General's Opening Remarks at the Mission Briefing on COVID-19. 2020. Https://Www.Who.Int/Dg/Speeches/ Detail/Who-Director-General-S-Opening-Remarks-At-The-Missionbriefing-On-COVID-19 (Accessed July 12, 2020).

Wang, G., Zhang, Y., Zhao, J., Zhang, J., & Jiang, F. 2020. Mitigate the effects of home confinement on children during the COVID-19 outbreak. *The Lancet*, 395(10228): 945–947.

Educational Innovation in Society 5.0 Era: Challenges and Opportunities – Purnomo & Herwin (Eds)
© 2021 the authors, ISBN 978-1-032-05392-9

The influence of a gamification platform and learning styles toward student scores in online learning during the Covid-19 pandemic using split-plot design

W.P. Hapsari & Haryanto
Research and Evaluation Education, Graduate School, Universitas Negeri Yogyakarta, Indonesia

U.A. Labib
Biology Education, Faculty of Mathematics and Science, Universitas Negeri Yogyakarta, Indonesia

ABSTRACT: A gamification platform is present as a major complement to the online learning system. Online learning is delivered by various methods that are adapted to the user's environment, especially in the teaching innovations necessary to optimize the COVID-19-influenced learning process. The use of various gamification platforms affects student learning outcomes. This paper explains the difference in the impact of gamification platforms based on the learning styles possessed by students. The subjects of this study consisted of 20 elementary school students, grade six, who were carrying out online learning as an implementation of the COVID-19 pandemic prevention policy. This research was an experimental research with a split-plot design. The results of this study indicate that differences in the use of gamification platforms do not have a significant effect on student learning outcomes with various learning styles, because all gamification platforms are able to accommodate all types of learning styles. This study provides information on the use of the appropriate gamification platform for use in online learning, further analysis is needed to find out the further impact of using the platform.

1 INTRODUCTION

Implementing distance learning in a pandemic situation is a step that must be taken to seek personal safety. Various types of educational institutions have carried this out in almost all regions of the Republic of Indonesia since March 2020. This is according to the Circular of the Minister of Education and Culture Number 3 of 2020 concerning the Prevention of Corona Virus Disease (COVID-19) in the Education Unit. The circular is addressed to the head of the provincial education office, head of the district/city education office, head of higher education service institutions, university leaders, and school principals in Indonesia who are followed up by each educational institution to conduct distance learning or home learning. Besides, the internet turns into a significant mechanism for training conveyance, as an ever-increasing number of courses will be offered in an online arrangement (Diaz & Cartnal, 2010). In spite of the fact that the workforce may endeavor to utilize the very showing strategies in a distance climate that they would utilize in a nearby classroom, the information from the current investigation proposes that faculty will experience essentially extraordinary learning inclinations just as other different understudy attributes.

The implementation of student learning activities in formal and non-formal education units was officially carried out from March 16, 2020 to April 5, 2020 (referring to SE Disdik DKI Jakarta 27/2020) which was then extended according to respective regional policies. It is hoped that this situation will not hamper the education process in areas affected by COVID-19, especially in big cities.

This policy has a real impact on the learning process in the smallest formal educational institutions such as Elementary School (SD), Junior High School (SMP), and Senior High School (SMA). In terms of the basic words, learning is, in a narrow sense, that which can be interpreted as a process or method that is carried out so that someone can carry out learning activities, while learning is a process of behavior change due to individual interactions with the environment and experiences. Learning is interactive and communicative. Interactive means that the learning activity is multidirectional between teachers, students, learning resources, and an environment that affects each other, not dominated by just one component. The interactions that occur in learning are not only at what level and how, but at the why level, the level of seeking meaning, both social and personal meaning. While communicative means that the nature of communication between students and teachers or vice versa, between students themselves, or between teachers that must be able to give, receive and understand each other. The learning process must be endeavored in various conditions, including during the current pandemic, accepting and

132

DOI 10.1201/9781003206019-24

understanding the situation. The learning process must be undertaken in various conditions, including during the current pandemic.

Educational institutions have made various efforts to support the journey of the distance learning process. One of the steps is to conduct distance learning using social media supported by learning applications that can be operated together remotely. Besides the use of social media, some teachers also carry out online learning with the help of online applications that can be processed by students. The learning media is used as a communication medium for teachers and students or vice versa as part of the social interaction process. Media comes from Latin which is the plural form of the word medium, which can literally be interpreted as an intermediary or an observer. (Heinich et al., 1989) revealed that media is a channel of communication, derived from the latin word of between, the term refers to anything that carries information between a source and a receiver. Learning media is a tool that can help students to carry out the learning process. The pandemic period that occurs with a circular from the ministry of education requires students to study at home, therefore the learning process that takes place must be supported by appropriate learning media. The learning can continue with the support of online tools or online media operated by teachers and students.

Courses are generally planned by course originators or coordinators. As a rule, the courses mirror the learning styles of the planners instead of the learning styles of the pupils for whom the courses are implied (Fatt, 2000). Teachers nowadays use many kinds of media or learning platforms to enhance the study practice conducting the distance learning implementation. The platforms used by teacher were vary, depending of the material and the learning instructional necessity. There also many responses in children conducting the assessment score that held in the platform, one of the factors was their learning styles. In recent years, the application of learning styles are also expanded to other areas such as assessment, educational games and media choices (Truong, 2016). There are many ways in delivering learning processes through online methods, one of them was by using a gamification method. Gamification may support an online learning process with many tedious tasks (Rodriguez et al., 2020). Gamification refers to the application of game dynamics, mechanics, and frameworks into a non-game setting (Stott & Neustaedter, 2013). Thus, a gamification learning platform may effect student scores due to their learning styles and attitude responses (Aydın, 2015). This research aims to analyze the difference in gamification platforms' impact based on the learning styles possessed by students. The innovation of the teaching method needs to be sufficient engaging the current issues of the Pandemic situation. The results from this study would inform the reader which steps should be taken to optimize the student learning process.

1.1 *Learning styles*

Everyone has a unique learning style in absorbing, organizing, and processing information received. The manner in which a student responds to the general learning climate makes up the person's learning style. No all-around acknowledged phrasing exists to portray the learning style and its different parts; notwithstanding, how people respond to their learning climate is a central idea (Bluejurnes & Gurdner, 1995). The terms learning style and intellectual style are now and then utilized reciprocally, the term learning style shows up more consistently on paper; it additionally seems, by all accounts, to be the more extensive term. An appropriate learning style is the key to student success in learning. The use of learning styles that are limited to only one style, especially those that are verbal or auditory, of course can cause many differences in absorbing information. Therefore, in learning activities, students must be assisted and directed to identify learning styles that suit themselves so that learning outcomes can be maximized (Bire et al., 2014). The explanation of the various learning styles possessed by these students requires different treatments to develop their potential. Online learning situations also require students to use various online learning platforms. Therefore, students have the characteristics of their respective learning styles according to learning styles, namely Visual, Auditory and Kinesthetic. Visual learners learn by what they see. Auditory learning by listening and kinesthetic learning through motion and touching. In fact, everyone has all three learning styles, but most people tend to use only one of the three styles that are more dominant.

It has been resolved that the learning style inclinations of grade school understudies are huge factors in both foreseeing the mentalities of understudies towards social examination courses both overall and for class levels (Çal & Kılınç, 2012). Students with a visual learning style can be seen from the main characteristics of using learning modalities with the power of the visual senses. Students who have a visual learning style find it easier to remember what they see, such as body language or facial expressions of their teachers, diagrams, pictorial textbooks or videos, so they can understand positions or locales, shapes, numbers, and colors. The characteristics of students who have a visual learning style tend to be neat and organized, speak rather quickly, give priority to appearance in clothes/presentations, are not easily distracted by noise, remember words more by looking at the letter arrangement of words, but they have difficulty receiving verbal instructions. Visual acuity, more pronounced in some people, is still very strong in all people. This is because in the brain there are more components that function to process visual information than all other senses. Meanwhile, according to the object, problems in vision are classified into three groups, namely the first, seeing the shape, the second seeing the inside and the third, seeing the color. This means that students digest more quickly when information is in the form of pictures, colors, and other art forms that are captured by the eye and stored in the brain and will be remembered more often. Visual learning style people are closer to characteristics such as scribbling when talking on the phone, speaking quickly, and prefering to look at a map rather than hear explanations. Generally, people with visual styles in absorbing information explain strong visual strategies with visual images and expressions.

Auditory learning is a learning style that relies on hearing to be able to understand and remember. The characteristics of this kind of learning model actually place as the main means of absorbing information or knowledge. This means that we have to listen first and then be able to remember and understand the information obtained. Students who have this learning style are that all information can only be absorbed through hearing, the second has difficulty absorbing information in the verbal form directly, the third has difficulty in writing or reading. Students who have an auditory learning style can be identified by their characteristics that use more learning modalities with the power of the auditory sense, namely the ear. People with auditory learning styles are closer to characteristics such as preferring to speak themselves, prefering lectures or seminars over reading books, and/or prefer speaking rather than writing.

People with kinesthetic learning styles are closer to characteristics such as better thinking when moving or walking, moving more limbs when talking and finding it difficult to sit still. Generally, people with kinesthetic learning styles in absorbing information apply physical strategies and expressions that are physically characteristic. Students who have a kinesthetic learning style can find reading and listening to be boring activities. Giving instructions that are given in writing or orally are often easy to forget, because they are more likely to understand the task if they try it in person.

Although some research declared that the visual, auditory, and kinesthetic may fail to explain learning and achievement (An & Carr, 2017) they would be still reliable models to pay attention to this context. Student define their own interest to image, sound, or motion would be have the different result of achievement through the different learning platforms. The legitimacy of learning styles has not been logically upheld, but they are accounted for to be valuable by and large (Li et al., 2016). Hence, we ought to be cautious and basic when drawing on learning styles for course planning. Right now, completely denying the significance of learning styles isn't suitable. Notwithstanding, all things considered, over the long haul this more convincing and dependable exploration will happen to permit instructors and strategy producers to arrive at a more authoritative decision about the relevance of student style showing models and their appropriateness in various circumstances

1.2 *Learning platform*

For most students, studying at home is new and requires habituation to be conducive and effective. The success of distance learning cannot be separated from the support and facilities of parents for their children. The facilities that need to be provided by these parents include communication tools, the need for school items, and of course the time needed for children who still need guidance. In fact, according to a preliminary study that has been carried out, not a few students experience problems studying at home in the absence of communication tools, so they still have to go to school to get lesson materials. The creativity of schools and local policy authorities is needed to continue to be able to provide good educational services during the COVID-19 pandemic.

Online learning requires a variety of teacher innovation and creativity. Teachers need to adjust the conditions of each student in order to be able to carry out learning and achieve core and basic competencies as planned. This condition has resulted in efforts from the government to carry out various trainings to realize distance learning efficiently. Software training and making learning videos have been widely implemented. As a result, the teacher has been able to mix learning so that it can be well received by students. One form of mechanism that is carried out is learning that is carried out through online platforms such as google classrooms, and strategic communication through WhatsApp and social media.

Google classroom offers a variety of online classroom features such as class discussion forums, assignment columns and deadlines, as well as simple teacher assessments. The learning evaluation system in the form of daily assessments or formative assessments can be carried out by teachers through the help of applications such as Google Forms and other online platforms. Some of them can be through a game system or what is commonly called gamification. This system makes it possible to form a simple competition between students in working on a quiz or question as a form of formal assessment. This form of gamification can be in the form of badges for the best players, periodic rewards, and points for the team when played in groups.

Several platforms that can be used as a means of gamification in this research which include Google Classroom, Quiziz, Quiz-Maker, and Kahoot. Google Forms is a service from Google that is used to create surveys, questions and answers with online form features that can be tailored to your needs. Meanwhile, Quiziz is a web tool used to create interactive quizzes, which is almost the same as Kahoot and Quiz-Maker.

1.3 *Gamification*

Gamification can be used without being named explicitly, and that the change from the traditional style classroom to the new learning environment (Dicheva et al., 2018). All three gamification platforms have premium features that can only be activated after paying. The use of various platforms is tailored to the needs and facilities of students at home. Gamification is an uprising pattern that applies gaming mechanics as a driver to spur, draw in and improve the client experience (Zainuddin et al., 2020). It is a quickly developing marvel that has appeared to give connecting with and convincing arrangements in the instructive and learning setting. An underlying structure depends on basic mental hypotheses, including and characteristic and outward inspiration. This paper presents a deliberate writing audit to distinguish great exact investigations in the zone of gamification in the instructive field over the 4-year time frame from 2016 to mid-2019. This paper has zeroed in on methodologies embraced in examination, the basic hypothetical models, gamification stages, the most often utilized moving pursuits, the members' inclinations and the most well-known game components utilized in training. The learning results and effect classifications distinguished were student commitment and inspiration, scholarly accomplishment and social availability.

For understudies, gamification effectively minimizes negative feelings that they normally experience in conventional types of instruction. It lets them approach information and abilities, utilizing the learn-by-disappointment procedure that is mainstream in game-like conditions, without the shame factor that generally frames a piece of homeroom instruction (Huang & Soman, 2013). Educators on their part can productively accomplish their set goals and use money based following instruments to get input on their understudies' advancement. In spite of the fact that it is difficult to effectively execute gamification in schooling, a careful methodology, can expand the likelihood of making powerful training gamification technique. It is likewise suggested that educators recollect that gamifying schooling may require significant stretches of calibrating and most certainly ought not supplant the first estimation of human instructing. Gamification in instruction can be an incredible methodology when actualized appropriately, as it can upgrade the training program, and accomplish learning destinations by affecting the conduct of understudies.

The consequences of gamification on learning examinations show that the test based gamification emphatically affected understudy learning contrasted with conventional showing strategies (contrasted with having no treatment and treatment including understanding activities) (Legaki et al., 2020). The investigation, place challenge-based gamification as a helpful instructive device in measurements training in various scholarly majors in specific situations. The method would be recommend to be applied to the available platform to use in the distance learning due to the COVID-19 pandemic.

2 METHOD

The research to be conducted is an experimental research with a quantitative approach. According to Frankel, Wallen, and Hyun (2011), experiment means trying, looking for, and proving. The experimental method is a quantitative research method used to determine the effect of independent variables on depended variables under controlled conditions.

The research design used is a split-plot design. Another term for split-plot design is a divided plot design which is a pattern of factorial designs that have certain characteristics and are fundamentally different from ordinary factorial designs. The difference lies in the arrangement of the experimental plot and in the technical analysis of its variety. In addition, the use of this design is intended to streamline research time and costs.

Table 1. Research design

Learning Platform	Replication 1		
	1	2	3
Google form	a1b1	a2b1	a3b1
Quiziz	a1b2	a2b2	a3b2
Kahoot	a1b3	a2b3	a3b3
Quiz-maker	a1b4	a2b4	a3b4
	Replication 2		
	1	2	3
Google form	a1b1	a2b1	a3b1
Quiziz	a1b2	a2b2	a3b2
Kahoot	a1b3	a2b3	a3b3
Quiz-maker	a1b4	a2b4	a3b4
	Replication 3		
	1	2	3
Google form	a1b1	a2b1	a3b1
Quiziz	a1b2	a2b2	a3b2
Kahoot	a1b3	a2b3	a3b3
Quiz-maker	a1b4	a2b4	a3b4

The split plot design used in this study consists of two independent variables and one dependent variable. The first independent variable of this study is the student learning style which consists of three levels as *whole plot* or main plot. While the independent variables of the two studies are *platform* delivery of learning material which consists of four levels as *sub plot* or sub-plots. The dependent variable of this study is the learning outcomes during the COVID-19 pandemic. The experimental design model is described in a table like the following.

This research was conducted at Gegulu State Elementary School, Kulon Progo. Sampling is done by technique purposive sampling namely by selecting samples with specific aims and objectives, in this case the sample and the intended research subject are students of class VI Gegulu State Elementary School.

The data collection technique used was a survey or questionnaire distributed online. The questionnaire is a collection of data by providing questions or written statements that have been prepared in advance by the researcher and then given to the respondent. Data collection regarding learning styles used an online platform with the assigned method, while student learning outcomes through various platforms used daily tests conducted by class teachers.

3 RESULT AND DISCUSSION

The first research carried out is a classification of student learning styles which is divided into three types, namely visual, auditory, and kinesthetic. This collection produces the following research data.

Table 2. Student learning styles

No	Learning Styles	Students
1.	Visual	14
2.	Auditory	10
3.	Kinesthetic	6

Then, after the test was carried out, the data on the average score of each group of students were obtained as follows.

Table 3. Student score

Learning Platform	Replication 1		
	1	2	3
Google form	75.55	60.21	71.46
Quiziz	82.41	81.89	84.65
Kahoot	74.65	73.52	75.13
Quiz-maker	79.81	78.12	76.34
	Replication 2		
	1	2	3
Google form	91.79	88.92	90.53
Quiziz	84.24	81.34	85.22
Kahoot	81.22	80.98	79.44
Quiz-maker	82.88	83.84	82.37
	Replication 3		
	1	2	3
Google form	89.37	87.88	70.43
Quiziz	80.49	79.45	81.11
Kahoot	80.77	81.38	82.00
Quiz-maker	84.60	83.27	90.25

The analysis of the data obtained produces the following results.

Table 4. ANOVA results

Variance Source	db	JK	KT	F_{count}	F_{table}
Learning Styles	2	451.69	225.84	20.16	5.14
Error (a)	6	67.23	11.2		
Platform	3	77.21	25.7	1.37	3.16
Learning Styles X Platform	6	448.70	74.78	3.99	2.66
Error (b)	18	337.50	18.75		
Total	35				

Because in the interaction AB, $F_{count} = 3.99 > 2.66 = F_{table}$ then Ho is rejected so further tests are carried out. The follow-up test sees the effect of the learning platform factor on the same learning style level.

Based on the research that has been done, it can be seen that the most learning styles by students are visual learning styles with 14 people, then auditory learning styles with 10 people and kinesthetic learning styles with 6 people. This indicates that most students have a learning style that makes it easier to remember what they see, such as body language or facial expressions of the teacher, diagrams, picture textbooks or videos, so that they can understand well about positions or locales, shapes, numbers, and colors. Students learn more easily what he sees than what he hears or does. This can also be a recommendation for learning models that teachers can use either through online or offline learning.

Based on the hypothesis test, the middle quadratic analysis that is processed from the data shows that Ho is rejected, because F count is greater than F table. This provides information that all treatments or all learning platforms carried out on various kinds of learning styles are different. Therefore, it is done by further testing the effect of each variable on the thematic exam results. The first follow-up test provides information that the learning platform used has the same impact on learning styles. Likewise, further testing which informs that the learning style gives the same response as the others.

The student learning process and the environment may take an excellent catalyst to their achievement. Different way of teaching or delivering lesson's material may influence student based on their learning style in the daily basis. Teacher tirelessly should innovate and explore the suitable platform to deliver the perfect way in teaching and make sure that student with vary learning styles can adapt.

So, from the results of this research it can be explained that all the treatments are good platform different learning styles or student learning styles have the same effect on learning outcomes. There are several factors that can be explained from this phenomenon, one of the things that can be implied is that all online learning platforms used offer various online learning features that can accommodate various kinds of learning styles, including images that accommodate students with visual learning styles, video recordings that accommodate students with auditory learning styles or demonstrations that accommodate students with kinesthetic learning styles.

4 CONCLUSION

Based on the research that has been done, it can be concluded that the most common learning style possessed by the student group is a visual learning style. Based on the middle square analysis hypothesis test that is processed from the data, it is found that Ho is rejected, because F count is greater than F table. This provides information that all treatments or all learning platforms carried out on various kinds of learning styles are different. Therefore, it is done by further testing the effect of each variable on the thematic exam results. The first follow-up test provides information that the learning platform used has the same impact on learning styles. Likewise, further tests inform that learning styles give the same response as the others.

So, from the results of this research it can be explained that all the treatments are good *platform* learning or different student learning styles have the same effect on learning outcomes. There are several factors that can be explained from this phenomenon, one of the things that can be implied is that all online learning platforms used offer various online learning features that can accommodate various kinds of learning styles.

REFERENCES

An, D., & Carr, M. 2017. Learning styles theory fails to explain learning and achievement?: Recommendations for alternative approaches. *Personality and Individual Differences*. https://doi.org/10.1016/j.paid.2017.04.050

Aydın, B. 2015. Examination of the Relationship between Eighth Grade Students' Learning Styles and Attitudes towards Mathematics. *Journal of Education and Training Studies*, 4(2): 124–130. https://doi.org/10.11114/jets.v4i2.1162

Bire, A. ., Geradus, U., & Bire, J. 2014. Pengaruh Gaya Belajar Visual, Auditorial, dan Kinestetik Terhadap Prestasi Belajar Siswa. *Jurnal Kependidikan: Penelitian Inovasi Pembelajaran*, 44(2).

Bluejurnes, W., & Gurdner, D. L. 1995. Learning Styles?: Implications for Distance Learning. *New Directions For Adult and Continuing Education*, 67.

Çal, H., & Kılınç, G. 2012. The Relationship Between the Learning Styles of Students and Their Attitudes Towards Social Studies Course. *Procedia – Procedia*

Computer Science, 55: 47–56. https://doi.org/10.1016/j.sbspro.2012.09.476

Diaz, D. P., & Cartnal, R. B. 2010. Students' Learning Styles in Two Classes and Equivalent On-Campus. *College Teaching*, 47(4): 130–135. https://doi.org/10.1080/87567559909595802

Dicheva, D., Dichev, C., Agre, G., & Angelova, G. 2018. Gamification in Education: A Systematic Mappring Study. *Journal of Educational Technology & Society*, 21(1): 152–154.

Fatt, J. 2000. Understanding the Learning Styles of Students: Implications for Educatiors. *International Journal of Sociology and Social Policy*, 20(11/12): 31–45.

Frankel, J. R., Wallen, N. E., & Hyun, H. H. 2011. How to Design and Evaluate Research in Education (8th Edition). In *Boston: McGraw-Hill.*

Heinich, R., Molenda, M., & Russell, J. 1989. Instructional Media and The New Technologies of Instruction. *Macmillan.*

Huang, W. H.-Y., & Soman, D. 2013. A Practitioner's Guide To Gamification Of Education. In *Rotman School of Management University of Toronto*. https://doi.org/10.1007/BF00814831

Legaki, N., Xi, N., Hamari, J., & Karpouzis, K. 2020. International Journal of Human-Computer Studies The e ff ect of challenge-based gami fi cation on learning?: An experiment in the context of statistics education. *International Journal of Human-Computer Studies*, 144(November 2019). https://doi.org/10.1016/j.ijhcs.2020.102496

Li, Y., Medwell, J., Wray, D., Wang, L., & Liu, X. 2016. Learning Styles?: A Review of Validity and Usefulness. *Journal of Education and Training Studies*, 4(10): 90–94. https://doi.org/10.11114/jets.v4i10.1680

Rodriguez, I., Puig, A., Tellols, D., & Samso, K. 2020. Evaluating the Effect of Gamification on the Deployment of Digital Cultural Probes for Children. *International Journal of Human – Computer Studies*, 102395. https://doi.org/10.1016/j.ijhcs.2020.102395

Stott, A., & Neustaedter, C. 2013. Analysis of Gamification in Education. *Carmster.Com*, 1–8. http://carmster.com/clab/uploads/Main/Stott-Gamification.pdf

Surat Edaran Dinas Pendidikan, Pemuda dan Olahraga Daerah Istimewa Yogyakarta Nomor 421/0280 tentang Pembelajaran Jarak Jauh bagi Anak Sekolah dalam Rangka Pencegahan Covid 19. Yogyakarta

Surat Edaran Menteri Pendidikan dan Kebudayaan Nomor 3 Tahun 2020 Tentang Pencegahan Corona Virus Disease (COVID-19) pada Satuan Pendidikan. Jakarta

Truong, H. M. 2016. Computers in Human Behavior Integrating learning styles and adaptive e-learning system?: Current developments , problems and opportunities. *Computers in Human Behavior*, 55: 1185–1193. https://doi.org/10.1016/j.chb.2015.02.014

Zainuddin, Z., Kai, S., Chu, W., Shujahat, M., & Perera, J. 2020. The impact of gamification on learning and instruction: A systematic review of empirical evidence. *Educational Research Review*: 100326. https://doi.org/10.1016/j.edurev.2020.100326

Educational Innovation in Society 5.0 Era: Challenges and
Opportunities – Purnomo & Herwin (Eds)
© 2021 the authors, ISBN 978-1-032-05392-9

Strengthening beginners character education in facing the 2020 general election

D. Purba & B. Juantara
Department of Governmental Science, Social and Political Sciences, University of Lampung, Indonesia

I. Bulan
Department of Language and Arts Education, University of Lampung, Indonesia

ABSTRACT: The Regional Head General Election is an important part of the implementation of democracy in the regions. Unfortunately, the *pilkada* process in post-reform Indonesia has been infiltrated by various interests and has caused problems. This stems from money politics, goods politics, and various other transactional politics. Beginner voters are the new generation of voters who have never been involved in the electoral process before. With the current ease of access, new voter groups are very vulnerable to the impact of political games in the *pilkada*. Character education for young voters is a panacea for protecting their thinking and behavior from pragmatic politics, and teaching them the substantive meaning of elections as a whole. This paper analyzes how character education can protect first-time voters in regional head elections.

1 INTRODUCTION

General elections as a means of democracy have been used in most countries in the world, including Indonesia, in determining the head of government (Suprojo, 2013). The election allows all parties to be accommodated what they want and aspire to in relation to the figure of the desired head of government, with the hope that a better life will be realized.

The community (citizens) is the main component that determines the success of the election. Basically, election is a people's concern, which means that the main actor in this case is the people. In this case, ranyat is not viewed from the person's background. All citizens have the same rights and have the same value in the General Election. Likewise are teenagers who are better known as new voters in the world of politics (Hasyim et al., 2020; Setyawan & Azmi, 2020).

Beginner voters are those who are voting for the first time in the election. The teenagers in question, namely those who are 17 years old or already married, will have the same citizenship responsibilities as citizens in general. In the world of politics, first-time voters are the easiest target to be won over by successful teams to win a candidate's supporters. This is determined because the level of knowledge of first-time voters is still minimally related to the world of politics.

Beginner voters, of course, have to open themselves up to have an insight into the nationality, especially related to the General Election including the Regional Head Election which will soon be held in a number of regions. Insights are important because a sufficient number of first-time voters can contribute to changes

in the next 5 years. An anti-political attitude and abstention will actually harm the state because it loses its support for change. Likewise, on the other hand, first-time voters must also have a protective shield and strong principles in terms of protecting themselves from deviant political behavior. As we know, the success teams in winning a candidate will carry out various strategies to get support from the public, including first-time voters (Hartono & Putri, 2017).

The strategies that are used are of course not always good, but they tend to "spoil" the very innocent personality of the first-time voters. "Money politics" is one of the strategies often used by the success teams to tie the masses, including novice voters, to their preferred form. This kind of strategy could have two effects on first-time voters. First, it will damage the personality of the novice voters so that they will become money oriented and do not vote based on their conscience. Second, it will create a closed attitude and willful ignorance about the General Election or Regional Head Election, so that they choose to abstain.

These two effects are of course very detrimental to our nation which will continue to produce a generation that fails politically. Of course, this must be addressed immediately, namely by planting the values of national character that can protect first-time voters from the above impacts. In accordance with eighteen (18) national character values, there are at least 4 characters that must be absorbed, possessed, and practiced by novice voters (Setiawati Nanda Ayu, 2017). The four character traits referred to are religious, democratic, national spirit, and love for the country traits. It is hoped that the planting and strengthening of national

138

DOI 10.1201/9781003206019-25

character in first-time voters can eradicate bad influences that can endanger the nation's future generations.

2 DISCUSSION

Various studies on the practice of clintelism (money and goods politics) in general elections have been carried out. (Robison & Hadiz 2004,2013; Hadiz 2010; Winters 2011), most of these works see that political and economic life in Indonesia is dominated by the predatorship of the oligarchic elite and their efforts to expand network access and obtain material benefits. One of the efforts to gain power, many politicians partner with oligarchic elites and receive "black money" which will be exchanged for voter support.

However, the practice of clintelism is not always seen as a sign of pathology. This is because not only clientelistic politics can dominate election results and are the key to political success but also determine how state institutions work and the character of government (Aspinall & Berenschot, 2019). From various cases we can see how the government and people's representative institutions do not represent the interests of the community.

Starting from the failure of residents to access their rights, such as the difficulty of village communities in protecting their land from expanding oil palm and mining concessions (Schleifer, 2017), Cramb & Mc Carthy, 2016), the control of large forest enterprises in Indonesia is dominated by large entrepreneurs when compared to the people with a ratio of 96: 4, increasing forest/land conflicts belonging to the people that continue to increase until the disappearance of agricultural land belonging to the people of up to 353 hectares/day due to pressure on the conversion of community production lands (KPK, 2018).

Other problems such as the difficulty of people achieving justice and the right to access to health, education, access to development infrastructure, (Berenschot & van Klinken, 2019) to welfare (Paskarina, Asiah, & Madung, 2016). In addition, the low level of trust, reputation and accountability shown by the government has encouraged pragmatism and public distrust of the state.

Based on the results of the Indo Barometer survey in 2020, it shows that state institutions such as the MPR, DPR, DPD and Polri are at the lowest level in terms of public trust, even intermediary institutions such as political parties have almost no established grassroots power. The tendency of people to abstain (not vote) in elections is getting bigger so that voters are friendly to transactional politics.

Seeing the above problems, of course, must be addressed immediately, so that the consolidation of democracy that seems made due to bad practices that occurs formally and informally does not hinder the flow of change and optimism towards the common ideals of the Indonesian state, namely a just and prosperous society as stated in Pancasila and The 1945 Constitution. Especially for first-time voters, although the number is only 20% to 30% of the total number of voters in the 2019 election, the contribution of these votes can determine the future direction of the nation. Therefore the inculcation of national character values that can protect first-time voters from the above impacts is an important point to develop.

In fact, the cultivation of values and character building has been carried out from childhood to adulthood. Characters that are formed in a person such as attitude, behavior, motivation, and skills are the result of the process of forming or developing every day from the living environment such as in the spheres of family, school, to the community, to digital environments such as social media etc. How far and how well the character values implanted in the environment will affect the success and success of the desired character is another matter.

Some literature, most of whom see the internalization of character education as early as possible up to the level of formal education (Oktarina et al., 2015; Wijanarko, 2011). It is hoped that the cultivation of values carried out in layers can have an impact on knowledge, attitudes and decisions taken in his life.

However, the literature linking character education with novice voters in the context of regional elections has not developed much, even though the space for strengthening character education is very important to produce novice voters who understand the values of truth, exemplary behavior, and daring to advocate for these values in order to realize goodness in the *pilkada* as a whole.

The Ministry of National Education has identified 18 values that come from religion, Pancasila, culture, and national education objectives, namely: (1) religiousness, (2) honesty, (3) tolerance, (4) discipline, (5) hard work, (6) creativity, (7) independence, (8) democracy, (9) curiosity, (10) national spirit, (11) love of the fatherland, (12) respect of achievements, (13) friendliness/communication, (14) love of peace, (15) love of reading, (16) care for the environment, (17) care for social affairs, and (18) responsibility (Kemdiknas: 2011). Of the 18 character education attributes mentioned above, there are 4 characters that have a direct correlation with first-time voters in the context of facing the *pilkada*. The five characters referred to are religiousness, democracy, national spirit, and love for the country.

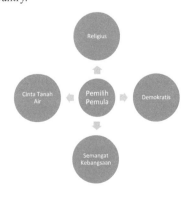

Figure 1. Characters that influence beginner voters.

2.1 Religious value

Religious value is a process of appreciation and implementation of religious teachings in everyday life (Ngaimun Naim, Ch Character Building 2012). Every human attitude and behavior cannot be separated from the internalization of divine values. Theological values and doctrines taught will affect a person's level of belief, especially in making important decisions in his life.

By devoting oneself to God, one is indirectly connected to all the good and forbidden teachings. That way, everyone has a strong foundation in regulating patterns of attitudes and behavior based on something true from religious teachings. In the context of the elections, the practice of exchanging money and goods from politicians to first-time voters is very damaging to the substance of representation. This fast and pragmatic walk is avoided as early as possible by novice voters by practicing and carrying out religious functions and religious teachings that they profess.

The implementation of worship on time and carried out together in houses of worship, praying together, fasting, reading the holy book, and maintaining morality allow novice voters to be awake and aware to always call for the truth at every step. The above activities can be carried out in a family environment, and schools with regular habituation allow the process of religious character education to be strengthened so that the opportunity for tolerance for bad practices of money politics in the implementation of regional elections can be minimized as early as possible by novice voters when choosing candidates.

2.2 Pancasila democratic value

Democratic education is a process to develop all human potentials that appreciate plurality and heterogeneity as a consequence of cultural, ethnic and religious diversity. The main values emphasized in democratic education are equality and tolerance (Rosyad, A. M., & Maarif, M. A. 2020). In the context of Indonesia, democratic values are integrated into the pure values of Pancasila.

In the context of regional elections, channeling the right to vote in regional head elections is a perfect form of manifestation of people's sovereignty. In this arena the people can carry out a process of political evaluation and control of the regional head and the political forces that support him. (Gaffar, 2012).

Especially novice voters who are new to the political process, being silent and apathetic to the electoral process will encourage the low legitimacy of the elected leader, besides deliberately allowing the seat of power to be held by people who have no credibility and do not take the side of the people. Therefore, the choice to participate democratically is the best choice so that the process of consolidating democracy can continue.

2.3 The value of love for the fatherland

Love for the homeland is a feeling of pride and an attitude of self-sacrifice to protect one's territory from various disturbances and threats. In the context of Indonesia, the value of love for the country is very much needed as a driving force for the integration of national diversity. If this value is not maintained, then the opportunity for egocentrism of ethnicity, religion, race, culture, etc. can lead to division and lead to conflict. Unfortunately, the emergence of the dichotomy and excessive fanaticism from the above is not in fact balanced with a spirit of unity and kinship. Instead of the *pilkada* presenting a new space for division, various issues related to Sara and the dominance of identity politics as a political tool for a certain group of entities have heated up the space for public interaction. The negative sentiment that is built continuously will encourage the weakness of the substance of the *pilkada* and lead to abstention.

In anticipating this, the spirit of loving the country must be echoed in a real way to first-time voters. The initial thing is to foster a feeling of pride in one's own homeland, a sense of belonging, respect, high respect and loyalty, being willing to sacrifice for the interests of the nation and country and loving the customs and culture of the people (Nurmantyo, 2016: 9). Internalization of these values can be channeled through self-discipline and tolerance for differences and diversity. In addition, an attitude of prioritizing collective interests is more important than individual interests.

2.4 The value of the national spirit

The practice of elite predators and oligarchy cronies against the domination and monopoly of Indonesia's natural resources and ecumenism has certainly eroded national values. The colonial practice of this new model must be resisted with a high national spirit. Unfortunately, the powerful political actors are unreliable, on the one hand the movement of social activism and freedom has always been hindered in its development.

In the context of first-time voters, the true national spirit can be shown by participating in every activity related to defending the country. This activity can be followed starting from formal learning activities and extracurricular activities, besides building a national community and holding regular discussions is another part of understanding these values well.

In the context of the *pilkada*, first-time voters take part in community activism by overseeing the general election so that it runs according to the rules and regulations, besides that novice voters can actively invite all voters to participate in giving their personal rights in the election, besides that, novice voters can also become a driving force to advocate for alternative actors who care and side with the community.

3 CONCLUSION

Character education for novice voters is a panacea for protecting their thinking and behavior from pragmatic politics and teaching them the substantive meaning of the elections as a whole. This paper tries to link character education with first-time voters in the context of regional elections. Of the 18 character education, there are four characters that have direct correlations with first-time voters in the context of facing the *pilkada*. The five characters referred to are religious, democratic, national spirit, and love for the country. The strengthening of the four character values above greatly influences the characteristics and attitudes to the behavior of novice voters in determining their choices in the *pilkada*.

REFERENCES

Amin, S. et al. 2019. *Citizenship in Indonesia: Struggle for Rights, Identities, and Participation.* Pustaka Obor Indonesia Foundation.

Aspinall, E., & Berenschot, W. 2019. *Democracy for sale: Elections, clientelism, and the state in Indonesia.* Cornell University Press.

Berenschot, W., & van Klinken, G. 2019. *Introduction: Citizenship in Indonesia, the Struggle for Rights, Identity, and Participation. In Citizenship in Indonesia: Struggle for Rights, Identity and Participation.*

Corruption Eradication Commission. 2018. Integrity Assessment Survey. Directorate of Research and Development. Deputy for the Prevention of the Corruption Eradication Commission, Jakarta

Gaffar, J. M. 2012. *Election Law Politics.* Jakarta: Constitution of the Press

Hadiz, V. 2010. *Localising power in post-authoritarian Indonesia: A Southeast Asia perspective.* Stanford University Press.

Hadiz, V. R., & Robison, R. 2013. The political economy of oligarchy and the reorganization of power in Indonesia. *Indonesia*, (96), 35–57.

Hartono, Y., & Putri, E. M. C. 2017. *The views of novice voters on general elections in Indonesia (A study of the views of Pangudi Luhur Van Lith Muntilan High School students on the Legislative and Presidential Elections).* Justitia et Pax.

Hasyim, B., Sartibi, N., & Shiddiq Fauzan, H. 2020. *Political Education for Beginner Voters in Participating in the Implementation of General Elections.* Culture and Society.

Ngaimun Naim, Ch Character Building. 2012. *Optimizing the Role of Education in the Development of Science and National Character Building.* Yogyakarta: Ar Ruzz Media

Nurmantyo, G., 2016. *Understanding Threats, Realizing Self-Identity as Building Capital Towards Indonesia Emas.* Jakarta: Litbang. Indonesian national army

Oktarina, N. et al. 2015. *Character Education Evaluation Model Based on School Culture for Elementary School.* Journal of Research and Method Education, 05 (05), 11

Robison, R., & Hadiz, V. R. 2004. *Reorganising power in Indonesia: The politics of oligarchy in an age of markets* (Vol. 3). Psychology Press.

Rosyad, A. M., & Maarif, M. A. 2020. *Paradigm of Democracy Education and Islamic Education in Facing the Challenges of Globalization in Indonesia.* Nazhruna: Journal of Islamic Education, 3(1): 75–99.

Schleifer, P. 2017. Cramb, Rob A., & John F. McCarthy, eds. 2016. *The Oil Palm Complex: Smallholders, Agribusiness and the State in Indonesia and Malaysia.* Singapore: NUS Press.

Setiawati Nanda Ayu. 2017. *Character Education as a Pillar of National Character Formation.* Proceedings of the Annual National Seminar, Faculty of Social Sciences, State University of Medan.

Setyawan, R., & Azmi, A. 2020. Beginner Voters' Perceptions of Voting Rights in Participating in General Elections in Nagari Tanjung Gadang, Sijunjung Regency. *Journal of Civic Education.* https://doi.org/10.24036/jce.v2i5.289

Suprojo, A. 2013. Analysis of Participation Level of Voter Participation Post-Unum Election Commission Decree About 10 Contesting Parties in 2014 Election in Community Political Development. *Reform: Scientific Journal of Social and Political Sciences, 3*(1).

Paskarina, C., Asiah, M., & Madung, O. G. (Eds.). 2016. *Struggling for control over welfare: Cases of politicization of democracy at the local level.* PolGov.

Wibowo, A. 2012. *Character Education: A Strategy to Build a Civilized Nation Character.*

Wijanarko, W. 2011. *The Influence of the Outbound Method on the Leadership Character Building of Indonesian Natural School Students.* Thesis: UIN Syarif Hidayatullah Jakarta

Winters, J. A. 2011. *Oligarchy.* Cambridge University Press.

Educational Innovation in Society 5.0 Era: Challenges and
Opportunities – Purnomo & Herwin (Eds)
© 2021 the authors, ISBN 978-1-032-05392-9

Revitalization of local wisdom values in strengthening cooperation character towards community civilization 5.0

S.F. Shodiq
Universitas Muhammadiyah Yogyakarta, Indonesia
Universitas Pendidikan Indonesia, Indonesia

D. Budimansyah, E. Suresman, & M. Hidayat
Universitas Pendidikan Indonesia, Indonesia

ABSTRACT: One of the unique characteristics of the Indonesian people is cooperation. However, the rapid flow of globalization causes the values of local wisdom to erode, leading to the decline of the cooperative character. This study aims to analyze strategies for strengthening the character of cooperation from an early age through the revitalization of local wisdom values. This research's approach is descriptive-qualitative, collecting data through direct observation, interviews, and documentation. The data is then analyzed with data reduction, data display, and conclusion drawing/verification. The results of the study have shown (1) the values of local wisdom have a vital role in strengthening the character of cooperation; (2) the values of local wisdom need to be re-packaged to facilitate the value transmission to the students; and (3) the reinforcement of cooperation character through the revitalization of the values of local wisdom can be done through learning activities.

1 INTRODUCTION

Local wisdom is a combination of knowledge and values discovered and acquired by local people in responding to various problems regarding the fulfilment of needs (Meliono, 2011). The local wisdom is a construction of socio-cultural aspects rooted within a particular society which is passed on from generation to generation. In other terminology, this can be called local knowledge, local wisdom, or local genius (Yadi Ruyadi, 2010). Since the local wisdom is a unique characteristic of a particular society, it can be distinctive among societies, depending on the socio-cultural spectrum of each society.

Local wisdom has a significant role in character integration. In other words, local wisdom embraces wise values about the education of character integration. This character is based on the values that are manifested in people's behaviours. In Cirebon society, for example, there is a tradition called *memitu*. According to Basyari, *memitu* is a tradition of *slametan* which means to celebrate the seventh month of pregnancy (Basyari, 2014). This tradition has an educational aspect of character integration in which respecting and loving each other in order to develop sincerity in society. In other words, local wisdom contains great values of character building.

In Indonesian society, mainly, there are many aspects of character integration shown by individuals. The Indonesian Heritage Foundation has listed several character traits, including; the love for Allah and the universe, peace-loving and unity, tolerance, kindness

and humility, a right and just leader, hard work and relentlessness, creativity and confidence, helpfulness and cooperation, love, respectfulness and politeness, honesty, independence, discipline, and responsibility (Raharjo, 2010). Moreover, there are other pillar characters, in which: integrity, diligence, courage, honesty, citizenship, caring, fairness, responsibility, respect, and trustworthiness (Eginton, 1934; Lickona, 1999, 2004; Nucci & Turiel, 2007).

Those pillars of character are born from cooperation. Fajarini assumes that one of these unique characteristics is cooperation (Fajarini, 2014). This is a unique character type that is rooted in Indonesian societies. This character contains several values like caring, citizenship, honesty, and respect. The caring character will be strongly formed among the people in society through cooperation that is implemented by honesty and cooperation. In Javanese society, cooperation, commonly known as *sambatan*, could be implemented in many activities, such as building houses and roads, cleaning the sewer and cemetery, or other collective activities. By implementing this character, many things become more comfortable to handle. No wonder, there is a Javanese saying *"holopis kuntul baris,"* meaning that cooperation and teamwork make people healthy.

Nevertheless, cooperation has begun to disappear from the new generation's characters (Muryanti, 2014). The rapid change of globalization has led to a degradation of this character among the people of Indonesian societies. The development of pragmatism and materialism in modern society has also reduced the culture of "helping others." In line with Marhayani's research, the tradition of sambatan in this globalization era has been substituted by some experts in particular fields (Marhayani, 2016). Furthermore, she gives another example that the tradition of cooperation in the agriculture sector has been substituted by labour intensive with the wage system. If this matter continues without an effort to maintain the character, no wonder if, in the future, the terminology of cooperation can only be found on paper as history.

Seeing into this problem, it is necessary to develop the character of cooperation from an early age as a base for welcoming the era of society 5.0. In other words, the children, as the next generation, should embrace the culture of cooperation in their minds and hearts. They are not only expected to know about the concept but also do implement it in real life. If the children have developed this character in their minds, the cultural transmission from generation to generation will undoubtedly occur.

Cooperation can be implemented as a hidden curriculum by internalizing current local wisdom. It does not mean to deny the formal curriculum in educating the students concerning cooperation. However, formal education tends to focus only on cognitive aspects rather than focusing on affective and psychomotor aspects. By applying the hidden curriculum, it is expected to be an option to develop the character of cooperation at an early age. Therefore, considering the narration above, it is essential to realize how the conception of the hidden curriculum in internalizing the local wisdom works.

2 RESEARCH METHOD

2.1 Methods

This research aims to study the internalization of local wisdom in the character integration of cooperation for children as the base in welcoming the society 5.0. This research applies qualitative approach. Data is collected through direct observation, interviews, and documentation. Data gathered is then analyzed in several stages, in which data reduction, data display, and conclusion drawing or verification (Creswell, 2013).

The subjects of this study are teachers and students of Muhammadiyah Sapen primary school, Keputren 2 Public Primary School, and Muhammadiyah Suronatan Primary School, in Yogyakarta. Data gathered from interviews, observations, and documentation, is then studied and analyzed further through the descriptive-analytic method. The results of the analysis are then formulated as a conception of the internalization of local wisdom in the character integration of cooperation for children in the effort to welcome the era of society 5.0.

2.2 Measurements and procedures

The subjects of the study are interviewed concerning the hidden curriculum, which contains the cooperative character. The questions asked to the interviewees are:

(1) the methods applied in integrating the cooperation character, (2) the effect of reward and punishment on the motivation level of cooperation, and (3) the obstacles dealt during the character integration methods. In the interview, the subjects are asked to explain and answer the questions based on their own experiences. The results of the interview are then used as a conception of cooperation character integration education in universal meaning.

3 FINDING AND DISCUSSION

3.1 Internalization of local wisdom values

Internalization is an effort made to increase knowledge, improve skills and abilities, and integrate values and norms into individuals in a sustainable manner to be absorbed and implemented in social life (Pornpimon, Wallapha, & Prayuth, 2014). By referring to this definition, internalization can be interpreted as a transfer of knowledge, transfer of value, and improvement of skills sustainably from generation to generation so that those values can be understood and implemented in the society (Kusumasari & Alam, 2012).

Bourdieu (1989) reveals that internalization among individuals can be different. The difference can be seen from those individuals' life phases starting from their childhood, adolescence, adulthood, and senior years. Value labels characteristics to things by giving appreciation to them to make them beneficial for human life (Watson, 2019). This definition means that value is the appreciation given by humans to everything beneficial in life.

Schwartz (1992) mentions that value is abstract. The abstractness of value implies that value does not belong to the fact so that there is no discussion on the concept of right or wrong. Instead, it discusses the internalization of something that people like or desire. This opinion shows that value is the reflection of affection from individuals and serves as a united, complete, and mutually influential system which creates a close relationship with a human in which the value system serves as an emotional grip affecting humans' attitudes and personalities (Smaldino, 2019). The close relationship between value and humans covers all aspects, including religiosity or a person's understanding and faith toward religion (Haste & Abrahams, 2008). It signifies that value is life guidance or faith in humanity as a valued creature in presenting assessment or judgment.

Along with the above context, Barni et al. (2014) divides human values into two contexts. First, human values as something objective. The existence of value is absolute even without the presence of the value giver or assessor. Second, humans also view value as something subjective. The existence of value depends on the existence of the value giver and is inherent within the assessors themselves. Based on those two different perspectives, a conclusion can be drawn stating that value is something which has existed and can affect a human's attitude and personality.

Local wisdom is intricate knowledge related to the systems of value, faith, norms, and way of life of a culture in a particular region (Litina, Moriconi, & Zanaj, 2016). Based on the results of interview and observation, local wisdom taught by schools has three dimensions of local wisdom value covering 1) religious values covering the value of a solemn relationship with God, obedience towards religion, good and sincere deeds, payback for good and bad deeds, and feeling of gratitude; 2) independent values covering self-esteem, work ethic, discipline, responsibility, courage, spirit, openness, self-control, positive thinking, and self-potentials; 3) moral values which teach about love and passion, togetherness and working together or cooperation, solidarity, helping each other, tolerance, respecting each other, manners and feeling of shame. Referring to the definition, local wisdom can be understood as a culture of a particular society which contains values functioning as media in shaping and building the character of a nation. From the explanation, internalization of local wisdom value is defined as an effort in understanding values of a culture to be implemented in living in the society, nation, and country, and playing a role as a means of shaping and building the character of the nation

3.2 Character integration for children

The methods which can be applied in integrating good character for children are traditional game and dance, and social agenda as childhood is a time full of games, the internalization used in character integration is a traditional game (Kashima, Bratanova, & Peters, 2017; Ranieri & Barni, 2016; Smaldino, 2019). The traditional game is game inherited from generation to generation by adopting cultural values as a symbolization of knowledge (Kashima et al., 2017). Traditional games not only play roles as entertainment, creativity, recreation, fantasy, and sports media but also as a means of character integration. It also trains solidarity, socializing, skills, hospitality, and other values. Thus, traditional games cannot be held away from their role in shaping character traits since childhood.

Based on the research finding, one of the traditional games taught as an effort in character education of working together or cooperation is the *cublak-cublak suweng* game. This game is begun with *hompimpa* or *gambreng* to decide who will be Mr. Empo or in this particular case refers to the one who should guess the person hiding the stone. After Mr. Empo is chosen, he faces down in the middle of the circle surrounded by other children. Other children sit and put their hands with palms facing up on Mr. Empo's back who has faced down in the middle of the circle. One kid runs the stone from one hand to another while all participants are singing *cublak-cublak suweng* song. When the song is over, Mr. Empo decides which hand hides the stone.

The above traditional game not only serves as entertainment for children but also sharpens their sensitivity, empathy, and focus. Children will be thoughtful towards their environment and able to act fast towards what is needed or expected by their environment. *Cublak-cublak Suweng* also assimilates the value of tolerance for ignorance. The dance also trains solidarity and cooperation.

Other than games, there is also a method for children in their early age which serves as a way of character education integration by doing social activity. Social activities done in schools are usually in the form of working together in cleaning up the field, watering the plants, doing the garbage, and tidying up their toys. Those activities are completed by involving all students so that they understand the meaning of cooperation, although, in its practice, there are usually children who are unwilling to participate. To those passive children, teachers need to approach them.

3.3 The effect of reward and punishment on mutual cooperation motivation

It is very crucial to teach cooperation or to work together (cooperation) which is closely related to reward and punishment method. The two methods are implemented to give more motivation for students to do cooperation program. The reward becomes a trigger for somebody to do something as their actualization as human (Andriani, 2012). Based on the research finding, the reward is given to children who diligently join activities and actively respond to educators and their environment. Reward giving is essential to motivate children during the learning process to meet the learning objective. Beside reward, punishment also plays a role as a form of motivation. However, punishment is given if children commit mistakes or ignore their responsibilities. Punishment will teach children to be responsible, disciplined, and careful.

The reward does not have to be concrete stuff. It can also come in other forms such as a compliment. However, teachers sometimes also give presents as appreciation such as books or stationary to their students (Berkowitz & Hoppe, 2009). As for the punishment method, researchers see that the concept of punishment given by teachers is different from what the majority of people understand. The punishment is in the form of advice and personal approaches to students. As they are still too young, it is impossible to give the type of punishment which is commonly understood by the majority of people as the concept of punishment.

3.4 Obstacles in methods of character integration

Like other educational methods, the implementation of a method frequently meets various constraints. Based on the results of observation and interview, one of those constraints during character integration was less participation from the children during a specific agenda (Litina et al., 2016). To the children who did not participate enough, teachers approached the students by providing advice, understanding, and motivation (Roest, Semon, & Gerris, 2010). However, some children were obedient, while others were not. Teachers were expected to cooperate with parents to continue giving guidance, education, and advice as a follow up from what had been learned (Ranieri & Barni, 2016). Cooperation between the role of teachers and teachers is crucial to uniform the perception between the education children receive at home and school (Barni et al., 2014). Parents' involvement will generate a positive influence if they understand the meaning, form, and purpose of their involvement. On the other hand, the ineffective effect comes from parents who do not understand the form, meaning, and purpose of their involvement.

3.5 *Character education in early childhood*

Children in their early age are a sort of gold investment for Indonesia in the future as successor and trustee for the next generation. Early age is a golden age to develop potential and integrate values in children. Those values form children's characters to be accepted in society (Mei-Ju, Chen-Hsin, & Pin-Chen, 2014). The time when children in their early age are at the right time to be educated have not received many external negative influences so that it will be easier for parents and teachers to teach and integrate character values, and develop their potential (Wang et al., 2015).

The value which requires special attention is the value of character education. According to Lickona & Davidson (2006), some of those good character traits are understanding, caring, and behaving according to fundamental values. Imam Al-Ghazali mentions that character has a close meaning to akhlaq or the spontaneity in behaving or doing actions which becomes part of a human and emerges without any prior thought (Raharjo, 2010).

Empowerment of character education internalization has received special attention from the Indonesian government to improve the education system to anticipate the degradation of students' morals (Pemerintah Indonesia, 2017). Regarding this matter, the government issued a presidential decree of Presidential Decree No. 87 the year 2017 under the following considerations: (1) Indonesia as a cultured nation is a country which upholds decent morals, high values, wisdom, and character. (2) In actualizing a cultured nation through the reinforcement of religious values, honesty, tolerance, discipline, hard work, creativity, independence, democracy, curiosity, national spirit, nationalism, appreciation towards achievement, being communicative, peace loving, love of reading, caring towards the environment and society, and responsibility, character education reinforcement is necessary. (3) Strengthening character education is the responsibility of the family, educational institutions and society.

Considering the above presidential decree, the parties which coherently support character reinforcement are religion, country, and family (García, Heckman, & Ziff, 2019). Rooting from this, all involved elements should be intensified in integrating character

education. The decadence of morals and loss of religious values are not the result of era transformation and rapid progress in science, technology, and art. Instead, it is due to the lack of society's anticipation towards transformation while the society has not yet had a deep understanding on religious teaching which is correctly and comprehensively designed regardless of time and era (Haste & Abrahams, 2008). Therefore, the effort has to be made to overcome the degradation of morals, especially for children in their early age by the reinforcement of religion, family, or society.

Character education for early childhood needs to be adapted to their moral development stages. Piaget (2013) states that moral development is divided into three stages namely (1) pre-moral, (2) moral realism, and (3) moral relativism. Meanwhile, Power & Kohlberg (1989) mention that moral development covers (1) pre-conventional, (2) conventional, and (3) post-conventional. The two types of stages contain the same focus; that is, children have not yet known about regulation, moral, and ethics. It is on the next stage that children become individuals who are aware of regulation, moral, ethics, and who behave according to the values they have understood.

4 CONCLUSION

Local wisdom has an excellent role in character integration education. In other words, in local wisdom, many values which contain character integration education are included within. One of the special characters of Indonesian society is cooperation. The character of cooperation should be developed at an early age so that the children become the right individuals as what is expected by the education itself. The local wisdom should also be revitalized to ease the transmission of tradition towards the students. The strengthening of the cooperative character can be done by revitalizing the local wisdom, which can be implemented in teaching-learning activities, both indoor and outdoor. It can be in the form of traditional games, dances, or doing cooperative activities like cleaning school areas and watering the plants. These ways are considered adequate for early-aged children since education is given unconsciously. All the local wisdom, such as traditional games and dances, will always be in line with the cooperation character education integration for early-aged children.

REFERENCES

Andriani, T. 2012. Permainan Tradisional Dalam Membentuk Karakter Anak Usia Dini. *Jurnal Sosial Budaya*.

Barni, D., Ranieri, S., Scabini, E., Rosnati, R., Barni, D., Ranieri, S., & Scabini, E. 2014. Value transmission in the family: do adolescents accept the values their parents want to transmit Value transmission in the family: do adolescents accept the values their. *Journal of Moral Education*, (November): 37–41. https://doi.org/10.1080/03057240.2011.553797

Basyari, I. W. 2014. Nilai-Nilai Kearifan Lokal (Local Wisdowm) Tradisi Memitu Pada Masyarakat Cirebon. *Edunomic.*

Berkowitz, M. W., & Hoppe, M. A. 2009. Character education and gifted children. *High Ability Studies.* https://doi.org/10.1080/13598130903358493

Bourdieu, P. 1989. Social Space and Symbolic Power. *Sociological Theory.* https://doi.org/10.2307/202060

Creswell, J. W. 2013. *Research Design: Qualitative, Quantitative, and Mixed Method Approaches* (4th ed.). California: SAGE Publications, Inc.

Eginton, D. P. 1934. Principles of Character Education. *Junior-Senior High School Clearing House*, 8(5), 298–305.

Fajarini, U. 2014. Peranan Kearifan Lokal Dalam Pendidikan Karakter. *Sosio Didaktika: Social Science Education Journal.* https://doi.org/10.15408/sd.v1i2.1225

García, J. L., Heckman, J. J., & Ziff, A. L. 2019. Early childhood education and crime. *Infant Mental Health Journal.* https://doi.org/10.1002/imhj.21759

Haste, H., & Abrahams, S. 2008. Morality, culture and the dialogic self: Taking cultural pluralism seriously. *Journal of Moral Education.* https://doi.org/10.1080/0305724080 2227502

Kashima, Y., Bratanova, B., & Peters, K. 2017. Social Transmission and Shared Reality in Cultural Dynamics. *Current Opinion in Psychology*: 1–13. https://doi.org/10.1016/j.copsyc.2017.10.004

Kusumasari, B., & Alam, Q. 2012. Local wisdom-based disaster recovery model in Indonesia. *Disaster Prevention and Management: An International Journal.* https://doi.org/10.1108/09653561211234525

Lickona, T. 1999. Character Education: Seven Crucial Issues. *Action in Teacher Education.* https://doi.org/10.1080/01626620.1999.10462937

Lickona, T. 2004. *Character matters: How to help our children develop good judgment, integrity, and other essential virtues. Touchstone Books.* Carmichael: Touchstone Books.

Lickona, T., & Davidson, M. 2006. Smart & Good High Schools: Integrating Excellence and Ethics for Success in School, Work, and beyond (2005). *Journal of Character Education.*

Litina, A., Moriconi, S., & Zanaj, S. 2016. The Cultural Transmission of Environmental Values: A Comparative Approach. *World Development, xx.* https://doi.org/10.1016/j.worlddev.2016.03.016

Marhayani, D. A. 2016. Development Of Character Education Based On Local Wisdom In Indegenous People Tengahan Sedangagung. *JETL (Journal Of Education, Teaching and Learning).* https://doi.org/10.26737/jetl.v1i2.40

Mei-Ju, C., Chen-Hsin, Y., & Pin-Chen, H. 2014. The Beauty of Character Education on Preschool Children's Parent-child Relationship. *Procedia – Social and Behavioral Sciences.* https://doi.org/10.1016/j.sbspro.2014.07.431

Meliono, I. 2011. Understanding the Nusantara Thought and Local Wisdom as an Aspect of the Indonesian Education. *TAWARIKH: International Journal for Historical Studies.*

Muryanti, M. 2014. Revitalisasi Gotong Royong: Penguat Persaudaraan Masyarakat Muslim di Pedesaan.

Jurnal Sosiologi Reflektif. https://doi.org/10.31227/osf.io/2p4wm

Nucci, L., & Turiel, E. 2007. Development in the moral domain: The role of conflict and relationships in children's and adolescents' welfare and harm judgments. In *Moral development within domain and within context.* Boston.

Pemerintah Indonesia. Perpres no. 87 Tahun 2017 Tentang Penguatan Pendidikan Karakter, Pub. L. No. 87, Sekretariat Negara (2017). Indonesia: Sekretariat Negara. Retrieved from https://setkab.go.id/inilah-materi-perpres-no-87-tahun- 2017 -tentang-penguatan -pendidikan-karakter/

Piaget, J. 2013. *The Moral Judgment Of The Child. The Moral Judgment of the Child.* Routledge. https://doi.org/10.4324/9781315009681

Pornpimon, C., Wallapha, A., & Prayuth, C. 2014. Strategy Challenges the Local Wisdom Applications Sustainability in Schools. *Procedia - Social and Behavioral Sciences*, 112(Iceepsy 2013): 626–634. https://doi.org/10.1016/j.sbspro.2014.01.1210

Power, H., & Kohlberg, L. 1989. Lawrence Kohlberg's approach to moral education. *Choice Reviews Online*, 27(01), 27-0436-27–0436. https: // doi.org/10.5860/CHOICE.27-0436

Raharjo, S. B. 2010. Pendidikan Karakter Sebagai Upaya Menciptakan Akhlak Mulia. *Jurnal Pendidikan Dan Kebudayaan.* https://doi.org/10.24832/jpnk.v16i3.456

Ranieri, S., & Barni, D. 2016. Family and other social contexts in the intergenerational transmission of values Family and other social contexts in the intergenerational transmission of values. *Family Science*, 3(1): 1–3. https://doi.org/10.1080/19424620.2012.714591

Roest, A. M. C., Semon, J., & Gerris, J. R. M. 2010. Value transmissions between parents and children: Gender and developmental phase as transmission belts. *Journal of Adolescence*, 33(1): 21–31. https://doi.org/10.1016/j.adolescence.2009.05.017

Schwartz, S. H. 1992. Universals in the content and structure of values: theoretical advances and empirical tests in 20 countries. *Advances in Experimental Social Psychol- Ogy*, 1–65.

Smaldino, P. E. 2019. Social identity and cooperation in cultural evolution. *Behavioural Processes*, 161(November 2017): 108–116. https://doi.org/10.1016/j.beproc.2017.11.015

Wang, J., Hilliard, L. J., Hershberg, R. M., Bowers, E. P., Chase, P. A., Champine, R. B., …Lerner, R. M. 2015. Character in childhood and early adolescence: models and measurement. *Journal of Moral Education.* https://doi.org/10.1080/03057240.2015.1040381

Watson, L. 2019. Educating for inquisitiveness: A case against exemplarism for intellectual character education. *Journal of Moral Education.* https://doi.org/10.1080/03057240.2019.1589436

Yadi Ruyadi. 2010. Model Pendidikan Karakter Berbasis Kearifan Budaya Lokal (Penelitian terhadap Masyarakat Adat Kampung Benda Kerep Cirebon Provinsi Jawa Barat untuk Pengembangan Pendidikan Karakter di Sekolah). In *Proceedings of The 4th International Conference on Teacher Education; Join Conference UPI & UPSI.*

Educational Innovation in Society 5.0 Era: Challenges and
Opportunities – Purnomo & Herwin (Eds)
© 2021 the authors, ISBN 978-1-032-05392-9

The implementation of project based learning through mind mapping to increased student creativity

U.M. Sadjim & R. Jusuf
Universitas Khairun, Indonesia

ABSTRACT: This study aims to determine the application of the Project Based Learning (PjBL) model assisted with mind mapping to increased creativity and learning result in society 5.0 era. This type of research is Classroom Action Research (CAR) and in each cycle there are steps, namely: planning, observation, and reflection. Data collection techniques used tests, observations, and documentation. Data analysis techniques used in this study are descriptive quantitative and qualitative. The results showed that the first cycle, student creativity reached 57.3%, categorized as not enough, in the second cycle creativity completeness increased 85.4%, with a very good category. The results of learning cycle I, as many as 17 (60.4%) of 28 students who completed and 11 (39.6%) had not yet completed. Cycle II, there was an increase in learning outcomes by 25 (89.6%) which was completed and 3 (10.4%) which were incomplete and categorized as very good, reaching the minimum mastery criteria classically by 70%. Based on the results of the study, it was concluded that the application of the model (PjBL) aided by mind mapping media can remind creativity and learning results in the society 5.0 era.

1 INTRODUCTION

Education is very important in human life as a means to improve, develop, educate, and practice self-skills. Education is one of the efforts and efforts to make the process of habituation which is a guide in determining the resilience and progress of a nation. The education pathway can be obtained through formal, informal and non-formal education channels. Schools is one of the formal education sources that carries out the learning process in space. According to BSNP (Kumala, 2016: 12), learning Natural Sciences (IPA) should be carried out in scientific inquiry to foster the ability to think, work, and be scientific and communicate from various important aspects of life. Therefore, learning science in elementary/MI emphasizes the provision of direct learning experiences through the use and development of scientific process skills and attitudes. The existence of the learning process carried out by the teacher will produce output in the form of learning outcomes by meeting the value of the Minimum Mastery Criteria (MMC). Based on the results of observations, found facts in fourth class students, specifically: (1) in the implementation of teacher learning using drawing media but does not involve students in the teaching and learning process, causing students to feel bored and passive; (2) teachers do not provide opportunities for students to practice the material being learned; and (3) the learning done by the teacher does not foster students' courage to ask questions. The solution used to overcome these problems, namely by using a project-based learning model (project based learning) which is assisted by mind mapping media. Project-based learning is a learning method that uses projects/activities as a medium. Students explore, assess, interpret, synthesize, and provide information to produce various forms of learning results and provide opportunities for teachers to manage learning in class by involving project work and students can also produce a project or real work (Malawi et al. 2019: 109).

2 THEORETICAL REVIEW

2.1 *Learning results and creativity*

Learning results are the results of one's abilities, both regarding cognitive, affective, and psychomotor aspects after students gain experience from learning. Learning results can be seen through evaluations that aim to find out or show the level of student ability to achieve learning goals that have been learned (Utami et al., 2018: 545).

Creativity is the ability of someone who in everyday life is associated with special achievements in creating new things or something that already exists into new concepts, finding ways of solving problems that cannot be found by most people, making ideas new that has never existed, and see the various possibilities that will occur (Fakhriyani, 2016: 195). The conclusion of the above explanation of creativity is the ability to create something new, both in the form of ideas and real work, which is relative to developing ideas through imagination skills that can produce new thought patterns formed from learning experiences or the real world is different from what already exists previous.

DOI 10.1201/9781003206019-27

2.2 Project based learning model

Project-based learning is process-centered learning, relatively timed, focused on problems, using meaningful learning units by integrating concepts from a number of components, be it knowledge, scientific disciplines or the field. Project-based learning is learning that demands student creativity (Pratiwi et al., 2018: 117). Learning activity project based learning takes a long time, which is student-centered and integrated with real-world problems. Project based learning is innovative learning that is centered on students (student centered) and places the teacher as a motivator and facilitator, where students are given the opportunity to work in groups to continue learning (Gunawan et al., 2018: 35). This learning activity provides an opportunity for students to summarize knowledge from various sources and apply it in a work of product. Project based learning is active learning, innovative and fun that uses the project/activity as a goal and ultimately produces real work that can be shown such as reports, essays, and completion of written assignments (Setyawan et al., 2019: 85). The conclusion of the concept of project-based learning is a learning centered on student creativity, process, time, problem solving, learning is more meaningful through teacher guidance as a motivator and facilitator in activities, where students are given the opportunity to work in groups for the continuation of learning in order to produce nana's work or which are real-world topics that can be shown such as reports, essays, and written task completion.

3 METHOD

Whereas classroom action research (CAR) is research conducted by educators in their own classes through self-reflection with the aim of improving the quality and quality of the learning process in the classroom so that learning outcomes are improved (Kurniawan, 2017: 8). In collecting data researchers use three aspects: test, observation and documentation. The technique used to analyze data is a process of processing and interpreting data in order to position various information in accordance with their functions so that they have meaning and meaning that is clearly in accordance with the purpose of the study. The results of data analysis will be used to determine the success of the actions given, the previous action is not successful, then the data analysis will be used as a basis for planning further actions. In the process of data analysis as a result of research using the following steps:

3.1 Calculating completeness

The results of this test know the extent to which the increase in student learning outcomes obtained at the end of the cycle is calculated and then presented. Count students who have completed MMC using the formula:

$$\frac{\text{Number of students completed}}{\text{Number of students}} \times 100\%$$

Calculating the completeness of the scores achieved by each student in solving the problem using the following formula:

$$TP = \frac{\text{Number of students obtained}}{\text{Total score}} \times 100\%$$

Measuring the level of mastery of the material in the study sample, from the test scores obtained are converted into mastery completeness. Then used a five-scale Benchmark Reference Guide (PAP). The completeness in determining the value in PAP is as follows: Maximum score x number of observations. The score results obtained in each aspect are presented and qualified to make conclusions about the level of student learning creativity in learning. To measure the variable of student learning creativity classically can be formulated:

$$P = \frac{f}{N} \times 100\%.$$

The percentage to be achieved at the level of student learning creativity is 60%–79%.

4 RESULTS

This class action research was carried out in two cycles. Before the action of the first cycle meeting, a pre test was conducted elementary school on Monday, January 2020. This initial test was intended to find out the creativity and learning results of students in style material before applying a project-based learning model aided mind mapping media.

4.1 Expose the research process in the first cycle

4.1.1 Student creativity results
Data on creativity in the first cycle got as much as 16 people who got the value of 60 above while 12 people received values below 55. The data in the first cycle has not been good or can be said to be still low.

4.1.2 Student learning result
The results of tests conducted in the first cycle can be seen that the number that reached individual completeness was 17 students while 11 students had not yet reached mastery of learning. In accordance with the MMC that has been determined by Ternate City 46 Elementary School to be classically 70 of the number of students. Therefore the completeness of students learning classically is not yet complete.

4.1.3 Student activities
Student activities in the learning process activities in the first cycle are 75 categories 51–75 (good). During the process of after-learning implementation, the learning activities of students were observed by observers, specifically: (1) student attention to the

teacher/researcher conducting preliminary activities; (2) students can find information about how to use mind mapping; (3) lack of students exploring material; (4) students form groups; (5) student determinations based on the question/questions given by the teacher are still not optimal; (6) students, in designing project completion steps, find there are still obstacles; (7) students complete projects with good facilitation and monitoring; (8) students have not been able to submit reports and present project results as well as the absence of responses from project results from other groups; (9) students do not ask about material that is unclear or incomprehensible; (10) students together with the teacher conclude the activity of learning.

4.1.4 Student creativity in cycle I

The results of class IV creativity research at Elementary school were obtained from the results of tests conducted by students. The creativity of all student indicators get an average value of 1.545 (55.1%). Student learning creativity in learning in the first cycle can pay attention to the results of student creativity tests the value of creativity is 55% with students who scored 60 and above there are 16 (57.3%) people from 28 students, the value is included in the completeness category P = 60% - 79% and there are 1 students get a value of 85 so the completeness category P = 80% -100% while 12 (42.7%) students get grades below 55.7, the value is included in the category of insufficient P = 40% -59% can be seen creativity assessment criteria table 3.3 page 51 in Chapter III.

4.1.5 Learning results in cycle I

After the learning activities in the first cycle lesson plan took place, the teacher gave post-test questions that were followed by 28 students in class and can be seen in the graph below.

4.1.6 Observation results of student activities in cycle I

The results of observing scores obtained from observations of student activities during the learning process, it can be seen that student activities during the learning process are included in both criteria (51-75) for (replace words) scores obtained 75%. There are a number of activities that must be improved between, specifically; (1) students are exploring the material described using mind mapping media less; (2) students do not respond to questions given by the teacher; (3) students lack confidence when presenting/presenting their projects; (5) students do not ask enough questions about the material.

4.2 Expose the process and results of research on cycle II

4.2.1 Learning results in cycle II

After conducting learning activities with a project-based learning model (project-based learning) assisted with mind mapping media, then students are given an evaluation test to determine the extent of student learning outcomes in understanding the benefits of style material.

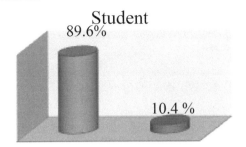

Figure 1. Complete classification of learning outcomes graphs.

The results of the completeness classification of learning outcomes in the graph above can be seen that 25 (89.6%) students have completed their studies, 3 (10.4%) students have not completed. These results indicate the classical value achieved by students can already meet the expected completeness criteria.

4.2.2 Student activities in the second cycle

Table 1. Comparison results of teacher and student activities

No	Aspect	Cycle I	Cycle II	Improvement
1	Observation of teacher activity	82.5%	92.5%	10
2	Observation of Student activities	75%	90 %	15

The data from the analysis of the observations of student activities during the learning process in class explained that during the learning process students were very active and creative in making a project and focus on learning. In accordance with the results of observations of student activities, the scores obtained were 36 out of a maximum score of 40. The overall calculation of student observations shows that the value of student activities in learning activities in the second cycle is classified as very good (70% -100%) with the acquisition of a score of 90 %.

4.2.3 Data on student learning outcomes and creativity in cycle I and cycle II

Table 2. Comparison of learning outcomes and creativity

Aspect	Cycle I	Cycle II	Information			
Creativity	57.3%	85.4 %	42.7%	14.6%	12	4
Learning result	60.4%	89.6%	39.6%	10.4%	11	3

Based on this data, it can be said that there has been an increase in creativity and significant learning outcomes from cycle I to cycle II by applying a mind mapping media-based learning model. The value achieved by students has reached the criteria and learning completeness that have been set on the indicators of success.

5 DISCUSSION AND CONCLUSION

This research was conducted in an elementary school with 28 students in carrying out an action in the first cycle. The researcher first conducted a pre-test to determine the level of initial knowledge of students before carrying out the action in cycle I. This creativity has 4 indicators used for research that's fluency includes two aspects, the first is asking lots of questions, the second is answering with a number of answers if there are questions, delicacy includes one aspect, one of which is in discussing/discussing a situation always has a different or conflicting position from the majority of groups, elaboration includes one aspect of which is more like synthesizing than analysing something. The first cycle got 57.3% (16) students who completed and 12 students who did not complete 42.7%. Therefore, the results of the creativity of students who scored 60 above there were 16 students from a total of 28 students but the grades were included in the good creativity category (60% – 79%). Learning outcomes can be seen from the completeness of students in the first cycle increased to 60.4% (17) students who completed and 11 (39.6%) students who did not complete. From the test in the first cycle creativity and learning outcomes have still not reached the classical completeness of 70%. With that, the study continued the actions in the second cycle so that it can be seen the value of creativity from 4 indicators increased from 85.4% (24) students who completed with an average of 67.8 and 14.6% (4) students did not complete. Project-based learning models (mind-based learning) assisted with mind mapping media can increase creativity and learning outcomes in fourth class elementary school.

REFERENCES

Fakhriyani, D. V. (2016). Pengembangan kreativitas anak usia dini. Wacana Didaktika, 4(2): 193–200.

Gunawan, B., & Hardini, A. A. T. (2018). Penerapan Model Pembelajaran Project Based Learning Untuk Meningkatkan Hasil Belajar IPA dan Kemampuan Berfikir Kreatif Siswa Kelas V SD. JTIEE (Journal of Teaching in Elementary Education), 2(1): 32–46.

Kumala F.N. 2016. Pembelajaran IPA Sekolah Dasar Malang: Ediide infografika.

Kurniawan, N. 2017. Penelitian Tindakan Kelas (PTK). Yogyakarta: Deepublish.

Malawi, I., Kadarwati, A., & Dayu, D. P. K. 2019. Teori dan Aplikasi Pembelajaran Terpadu. Yogyakarta: AE. Mediagrafika.

Pratiwi, C. D., Karistin, F. & Anugrahani, I. 2018. Penerapan Model Project Based Learning (PJBL) Berbantuan Media Mind Map untuk Meningkatkan Keaktifan dan Hasil Belajar Siswa Kelas 4 SD. Jurnal Guru Kita (JGK), 2(3): 116–125.

Setyawan, R.I., Purwanto, A., & Sari, N.K. 2019. Model Pembelajaran Berbasis Proyek untuk Meningkatkan Hasil Belajar. Jurnal Dikdas Bantara, 2(2): 81–93.

Utami, T., Kristin, F., & Anugraheni, I. 2018. Penerapan Model Pembelajaran Project Based Learning (PjBL) untuk Meningkatkan Kreativitas dan Hasil Belajar IPA Siswa Kelas 3 SD. e-Jurnal Mitra Pendidikan, 2(6): 541–552.

Educational Innovation in Society 5.0 Era: Challenges and
Opportunities – Purnomo & Herwin (Eds)
© 2021 the authors, ISBN 978-1-032-05392-9

The dilemma of Timorese education in the COVID-19 pandemic

Syahrul, Arifin & A. Datuk
Universitas Muhammadiyah Kupang, Kupang, East Nusa Tenggara, Indonesia

ABSTRACT: This study focused on the learning process during the COVID-19 pandemic in Timor. The research used a qualitative method and purposive sampling to select subjects such as teachers, students, and parents. We collected the data through interviews and observations. We analyzed the data using the Creswell method. The results of the study showed three dilemmas of the learning process in Timor Island, namely, (1) most parents found difficulty in teaching their children at home because they did not have the ability in science. As a result, many parents in Timor cannot teach their children at home. (2) Learning by social distancing with the teachers requires technology and an internet connection, while most Timorese have a low economic income, and they have difficulty in accessing an internet connection. (3) Door-to-door learning is very effective for students, but the teachers are exhausted because there are many students who should be visited.

1 INTRODUCTION

Most of Timorese do not have computers and cellphones, and they are also far from technological systems because infrastructure and natural resources are poor. Therefore, many students and teachers find it difficult to access transportation, information, and the internet. Several schools also do not have standardization to carry out the learning process because the building of the schools does not have facilities (Bexley & Tchailoro, 2013; Nygaard-Christensen, 2013; Quinn, 2013). Although several schools are good in the building and facilities, they are only in Kupang City as the capital, while if we visited several villages, we will find many schools which are under government standardization (Kent & Kinsella, 2015; Roja, 2020). Moreover, during COVID-19, education in Timor is more difficult than before the pandemic because most Timorese people do not have a computer, cellphone, and internet to carry out online learning.

Based on the research, the condition of Timorese education before the COVID-19 pandemic has many problems because of poverty and low technological systems (King et al., 2018). It affected many children who cannot access formal education and let to them dropping out. Timor occupies a low educational rank in Indonesia (Quinn, 2013). Furthermore, COVID-19 makes the difficult for Timorese people to access formal education because, during the pandemic in Timor, they did not hold the learning process at school, instead studying from home and online. This is due to the parents never studying at school, therefore they cannot teach mathematics, sociology, physics, biology, geography, and language to their children. On the other hand, poverty makes it difficult for parents to buy online learning devices such as computers and

cellphones (Brinke, 2018; King et al., 2018). Although they have a computer and a cellphone, they cannot download the subject matter from the teacher because there is no internet connection.

Moreover, both offline and online learning is difficult to conduct due to the condition of Timorese education, which is like a free day, namely, there is no study from home, work from home or a learning process at school. If the student came to school, they only fulfilled the administration, this is especially true for freshmen. Thus, the condition of Timorese students is that they are only playing a game with their friends in their houses and school, while they do not wear masks like before the COVID-19 pandemic.

The study from home also caused problems in several provinces in Indonesia as Hutapea's research showed that he found several problems in the study from home in several regencies, namely, (1) many teachers, lecturers, and students found it difficult to access the internet because there was no internet in their village; (2) packaged internet was very expensive to use in online learning; (3) the jobs of teachers and parents were not practical to conduct studying from home; (4) the teachers had no skill and ability to use the platform of online learning, and (5) the parents found it difficult to teach and guide their children to study from home into online learning (Hutapea, 2020).

According to Gunawan, online learning has many problems, namely, there were 40% of students did not know how to use the platforms of a learning management system like Moodle, Google Classroom, Zoom Meetings, and Act (Gunawan et al., 2020). On the other hand, teachers and students had many troubles in online learning such as less packaged internet services and poor internet connections (Bayham and Fenichel, 2020; Lancker and Parolin, 2020; Rundle

DOI 10.1201/9781003206019-28

151

et al., 2020). Therefore, there were 78% of students who did not join online learning because the packaged internet was running out, and 53% of lecturers did not access the internet to carry out online learning. Thus, many students cannot access the learning process, subject matter, or interact with teachers and their friends (Gunawan et al., 2020).

On the other hand, the survey of KPAI shows that studying from home has many problems. It found 76.7% of students disliked studying from home because they disliked many assignments from the teacher during online learning. Students also argued that the assignments were very difficult, while the teacher had not explained how to do them. Several students chose to study at school than at home because it was easier than at home, whereas there were 23.3% of students were interested in studying from home because they must wake up early morning, and they did not have to wear their uniforms (Rahman, 2019).

Moreover, Purwanto and Viner found that the learning process during the COVID-19 pandemic has several characteristics, namely, (1) students were pushed to online learning without facilities such as a computer, cellphone, and internet, (2) students found it difficult to study through online learning because they usually study through face-to-face in the classroom, (3) many teachers and lecturers were unable to use the platforms of online learning, (4) many parents, students, teachers, and lecturers refused online learning because packaged internet was very expensive, (5) students had lost their social lives because they did not interact with their teachers and friends (Purwanto et al., 2020; Viner et al., 2020).

The policy of studying from home for obeying social distancing during the COVID-19 pandemic is the only solution to carry out the learning process, but online learning activities must use technological devices. Therefore, teachers and students are required to have technological devices such as a computers, cellphones, and internet. Meanwhile, these are difficult to own by Timorese people because the internet in Timor can only be accessed in Kupang City and a few regencies. Many students complained about poor internet connections because online learning was difficult to hold in several regencies. Because most Timorese people live in a rural place (villages), if they want to carry out online learning, they should come to the city and climb the mountain to find the internet. This is unreasonable to conduct the learning process because this makes the learning process abnormal. As a result, this study elaborated on the condition of Timorese education during the COVID-19 pandemic.

2 METHOD

The research used a qualitative method to reveal the condition of Timorese education during the COVID-19 pandemic. We used purposive sampling to select subjects such as teachers, students, and parents. We collected the data through interviews about the experiences of teachers and students to carry out online learning during the COVID-19 pandemic. We also interviewed the parents about how to guide their children during the study from home period. Moreover, we used observations when we visited several villages and schools to find information about the condition of villages and schools. Thus, we analyzed the data into one circle among data collection, data categories, reading, memo, description, classification, interpretation, and visualization (Creswell, 2013).

3 RESULTS AND DISCUSSION

3.1 The dilemma of Timorese education

During the pandemic, the Indonesian Government conducted social distancing in the learning process. This policy shocked the Timorese people because the Government applied it suddenly. Moreover, it also shocked students, teachers, and lecturers in several schools and high education in Timor. Although online learning is held in several universities in Indonesia, it is different from NTT Province, particularly Timor, because they are limited in technological systems. These problems are increasing during the COVID-19 pandemic. Meanwhile, the Government pushes them to use technological devices to carry out studies from home. Moreover, the condition of Timorese education during the COVID-19 pandemic can be seen in Figure 1.

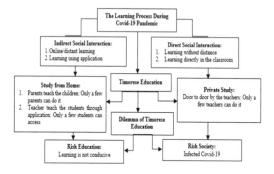

Figure 1. Scheme of education taking place in Timor.

Figure 1 shows that being between two policies that are not relevant to social conditions creates the dilemma of Timorese students and teachers during the COVID-19 pandemic. There are many learning models, methods, and strategies to carry out the learning process during the COVID-19 pandemic, but this is just discourse for Timorese. Several problems hinder the effectiveness of learning for the Timorese during the COVID-19 pandemic, namely, (1) lack of ability to use information technology by teachers and students. Most teachers in Timor do not know how to use technological devices, therefore this prevents them from using information technology as a medium for online learning. As a result, students also have the same conditions as their teachers because it is certain that if the teacher cannot use online learning, it will certainly have an impact on their students.

(2) Inadequate facilities and infrastructure in Timor. The expensive online learning devices have made the Timorese, who are mostly poor, unable to have them. In reality, many teachers in Timor are still in poor economic conditions. The poor condition of teachers and students has prevented them from using the information technology and infrastructure needed during the COVID-19 pandemic as online learning. (3) An unstable internet connection in Timor. The internet network is uneven in Timor, especially in the villages. Almost all people in Timor villages cannot access the internet, although there is an internet connection it cannot be used for online learning.

On the other hand, some places can do online learning in Timor such as Kupang City and other central regencies, but they face obstacles that have created many risks, namely, (1) "Risk education" will grow when the learning process is forced on the online learning. Thus, during the COVID-19 pandemic, the learning outcomes of students are not optimal because students never used engineering learning optimally. The learning process is never effective because students do not face-to-face with their teachers directly, therefore students cannot express their abilities because they cannot discuss and interact directly with their teachers and friends. Students find it difficult to understand the subject matter because they just sit in front of the computer and cellphones.

(2) "Risk society" will grow if the learning process is forced through direct social interaction between teacher and student, they will be infected by COVID-19. Even though this learning process is risky, teachers in Timor continue to use this strategy. They carry it out by following health protocols of the government such as wearing masks and avoiding crowds, but this learning process is no longer because many teachers have complained. Based on these two risks, the Timorese occupy a dilemma in the implementation of education during the COVID-19 pandemic. Schools in several regions in Indonesia choose online learning, but for them, this is very unlikely due to unfavourable social conditions. Furthermore, if they do door-to-door learning, it is very burdensome for teachers and also still allows the transmission of COVID-19. As a result, they can only be between two educational policies during the COVID-19 pandemic.

Based on figure 1, there were two paradigms sociologically of the educational policy during the COVID-19 pandemic in Timor.

3.1.1 *Indirect social interaction: Study from home*

The pedagogical strategy of indirect social interaction in the learning process carried out without face-to-face between students and teachers, but the learning process was conducted by online learning through technological devices. This has two characteristics, namely, (1) online/distant learning is the learning process using social distancing and (2) learning using an application is the learning process through Zoom, Google Meeting, and other platforms. These helped many students and teachers during the learning process in the COVID-19 pandemic, while Firman and Rahayu said that students were undoubted to ask a question for their teachers and to express their opinions in online learning because social distancing has reduced students shame to discuss and debate with the teachers and friends (Firman and Rahayu, 2020). Moreover, the learning process without face-to-face with the teacher made the students not afraid of showing their opinion. As a result, a study from home can grow independent learning in students because they were independently researching subject matter and information, and they did homework without guidance directly by a teacher.

This research was different from what we found in Timor during the learning process in the COVID-19 pandemic. It supported by an interview with the teachers who said that:

> When I tried using the new learning processes for the COVID-19 pandemic for the first time, I taught over WhatsApp and Zoom, but I only did this for a few meetings because, for the next time, it was not effective, and I got many problems, namely, many parents complained that online learning was very expensive and they have no money to buy an internet package. On the other hand, my students were too lazy to study through online learning because they found it difficult to understand the subjects. I also find it difficult to explain the subject details because communication in online learning was very limited. Therefore, until now, I only gave assignments to my students for what they did in their homes (Marten, 2020).

This interview shows that online learning is not similar to imagination because it is difficult to reach the standardization of education, while essentially education is part of human social relationships such as interaction of individuals to groups, group to individuals, individuals to individuals, and groups to groups. On the other hand, teachers cannot control their students directly in online learning. Therefore, online learning has little effectiveness because the emotional relationships of people were not possible. The reality of the learning process is that teachers should understand the social condition of their students, whereas the students should interact directly with their teachers to understand the subject.

Many schools in Timor faced these problems, namely, although teachers can carry out online learning, several students can access and join in online learning because they cannot connect to the internet. This is also supported by the research of Pujilestari, namely, technological devices and the internet had many benefits during the COVID-19 pandemic, but these were not used to the extent of their potential in Indonesia because of a lack of technological maturity (Pujilestari, 2020). The Indonesian Government is not ready to face digital life because of the ocean of human resources, technological infrastructure, technological transformation, and the constitution of technological empowering that must be addressed. Therefore, the goodness of technological information is only found in the capital city, while if we visited several regencies such Timor, Flores, Alor, Sumba, and others, we consider that Indonesia is very far from the technological system. Thus, online learning has not held again in Timor because students and teachers do not have technological devices and an internet connection to carry out online learning.

The distance of the village to other villages in Timor made it difficult for teachers to hold door-to-door learning. There was a teacher who gave an assignment to their students, but he cannot control directly the activities of his students during the study from home. In several places in Indonesia, the teacher used a smartphone to control their students such as using video calls and platforms, but in Timor, it is very difficult to use because students and teachers cannot access the internet. Although the teachers can hold online learning, most students cannot focus to study because they were more fantasy in online learning than face-to-face learning, and also they spent more time playing games. Therefore, online learning is held in a short time because students were difficult to engage for a long time (students cannot study for more than one hour) in front of a computer and smartphone.

Firman and Rahayu have found cases as same as in Timor that many students found it difficult to understand the subject matter which was given by their teachers through online learning (Firman and Rahayu, 2020). The subject matter such as reading papers and counting were difficult to understand by the students. It is caused by the learning process of only reading papers and completing assignments, but essentially students need an explanation directly by a teacher to explain some more complicated subject matter. Therefore, if the teachers only held the learning process by online learning through a platform and messenger, it is not enough to explain the details of the whole the subject matter from the teachers and textbooks.

Although the study from home did not carry out in Timor, they have obeyed the educational constitution of Government during the COVID-19 pandemic, namely, teachers and parents conducted the learning process based on the social condition. It is supported by an interview with the teacher who said that,

during the COVID-19 pandemic, I have been conducting the learning process using two strategies such as (1) I only gave homework to students who don't have a smartphone and internet connection in their home, while the parents should guide them to do their assignments. On the other hand, I did door-to-door to my students to control the homework. (2) Students who have a smartphone and internet connection in their home were taught in online learning (Arifin, 2020).

This interview shows that study from home can be held in two strategies such as students can be taught directly by their parents and students can be taught indirectly by a teacher through online learning.

3.1.1.1 *The study from home by the parents*
The study from home obligated that the parent should have academic skills, while in Timor, most of the parents never studied at schools (formal education). Based on demand, many parents complained because most of them do not have the academic skills to teach their children at home. Therefore they cannot read and write. Thus, study from home for Timorese people cannot be done.

Moreover, the study from home by parents needs an internet connection due to the fact that they should communicate with a teacher about the subject matter which they want to teach. The poor internet connection in Timor obstructed study from home/online learning by parents because they cannot communicate with the teachers. Therefore, there were a few parents who can carry out online learning because they live in a central city of regency, but they complained because packaged internet was very expensive, and it added their burden during COVID-19. An internet connection is not free, and it requires more money when we access services such as Zoom, WhatsApp, Google, and others (Rahman, 2019).

3.1.1.2 *The study from home by the teacher*
The study from home by teachers requires an internet connection and technological devices. We knew that most of Timorese did not have the ability to buy all online learning devices, while we cannot also access the internet in Timor. Therefore, there were a few students can join in online learning held by a teacher. On the other hand, teachers also had similar problems with the students because, during the COVID-19 pandemic, many teachers and students return to their hometown/village to avoid the COVID-19 pandemic in the capital city. When they came to the village, they found it difficult to carry out online learning because most of the villages in Timor did not have an internet connection (Marten, 2020; Seran, 2020).

The impact of online learning for the teacher in Timor during COVID-19 pandemic was (1) many senior teachers who did not know how to use computer and platform for online learning, (2) the infrastructure of online learning such as a computer, internet, and smartphone which was not enough, therefore many teachers were difficult to carry out online learning, (3) Timor which is backward regency, therefore online learning is a new strategy because, before the COVID-19, they only knew face-to-face learning in the classroom. These points show that many teachers were shocked by the new policy in the learning process because they were not able to use a technological device as a learning process. They had pedagogical skills, but it was only direct learning in the classroom. As a result, teachers in Timor need more times how to use the technological device and how to conduct online learning.

3.1.1.3 *Direct social interaction: Private study*

The learning process that was carried out by the teachers and students was basic learning to need and develop social relation between teachers and students without social distance among them. The learning process that used social distancing during the COVID-19 pandemic has made the difficult for Timorese to reach maximal education because sociologically education is part of social interaction. If the learning process used social distancing between teachers and students, it will remove the essence of education. On the other hand, the learning process that used online learning devices was not optimal because the emotional relationship between students and teachers cannot grow. Precisely, the learning process needs an emotional relationship between teachers and students to keep social and to make it easier for teachers to understand their students whereas to make it easier for students to understand the subject from a teacher.

The learning process without social distancing between teachers and students was very effective because it grew social relations between students and teachers, but now this is banned during the COVID-19 pandemic because it can lead to the transmission of COVID-19. Therefore, direct learning in the classroom did not carry out, but the teachers changed with other strategies, namely, door-to-door learning. Meanwhile, teachers who cannot hold online learning can use alternative learning such as textbooks, assignments, homework, and others which were easily obtained by students. They can also study to follow the schedule from the teachers. As a result, this obligated the teachers using a door-to-door learning.

The strategy of the door-to-door learning was arranged by the teacher for a week, namely, he divided his times to visit three students for a day, and arranging the subject matter which will be given to students, namely, he visited directly his students to give a paper and exercise. Meanwhile, the teachers provide 40 minutes to explain the subject matter, after that students did the exercise independently, but sometimes the students were guided by their parents. For the next week, the teacher revisits the student to check on his assignment, while there were a few parents who check on their children's assignment then they report to the teacher via message. In Timor, many parents did not study at school, therefore only 5% of parents can check the assignment and communicate with the teachers about the learning progress of their children.

The door-to-door learning is followed by private study, but this is the only strategy that can be done by teachers in Timor during the COVID-19 pandemic. Door-to-door learning is a harmful learning strategy for teachers because many teachers were exhausted when they visited their students every day, while the villages are far from other villages. It will be proved by the interview with the teachers who said that,

> for every day, I only visited three students because the distance from student home to other homes was very far, even if I should visit from one village to other villages. I also found several

problems such as fatigue, damaged roads, and expensive transportation costs (Arifin, 2020).

This interview shows that the door-to-door learning is beneficial for students because they stay at home without transport fees to the school, whereas the teacher came to visit them. On the contrary, it was detrimental to teachers because they spent a lot of money on transport, and they were also very tired. Therefore, the door-to-door learning was not effective because many teachers did not want to do it. According to the teachers, the door-to-door learning spends much money on transport and drains a lot of energy, while they did not receive an additional salary and transport costs. As a result, the learning process in Timor during the COVID-19 pandemic was not carried out because the door-to-door learning or the private study causes many losses to teachers such as loss of labour and economy.

4 CONCLUSION

The learning process by indirect social interaction makes it very difficult to reach optimal education because education is part of human social interaction such as interaction of individuals to group, groups to individuals, individuals to individuals, and groups to groups. Moreover, the learning process through an application is difficult to use optimally because human social interaction emotionally cannot be done. Meanwhile, in the learning process, the teacher should know the characteristics and social conditions of their students, whereas the students should directly interact with their teacher to understand the subject. On the other hand, the learning process by direct social interaction is effective because it built social relationships between teachers and students, but during the COVID-19 pandemic, this learning strategy is banned because direct social interaction makes the transmission of COVID-19. Although direct learning in the classroom did not apply again, the teacher chose another strategy, namely, door-to-door learning.

REFERENCES

Arifin, 2020. Study From Home by a Teacher.

Bayham, J., Fenichel, E.P., 2020. Impact of school closures for COVID-19 on the US health-care workforce and net mortality: a modelling study. The Lancet Public Health 5, e271–e278.

Bexley, A., Tchailoro, N.R., 2013. Consuming Youth: Timorese in the Resistance Against Indonesian Occupation. The Asia Pacific Journal of Anthropology 14, 405–422.

Brinke, S. ten, 2018. Citizens by waiting: Timorese young adults between state politics and customary authority. Citizenship Studies 22, 882–896.

Creswell, J.W., 2013. Qualitative Inquiry and Researc Design. SAGE Publication, New Delhi.

Firman, F., Rahayu, S., 2020. Pembelajaran Online di Tengah Pandemi Covid-19. Indonesian Journal of Educational Science (IJES) 2, 81–89.

Gunawan, G., Suranti, N.M.Y., Fathoroni, F., 2020. Variations of Models and Learning Platforms for Prospective Teachers During the COVID-19 Pandemic Period. IJTE 1, 61–70.

Hutapea, R.H., 2020. Kreativitas Mengajar Guru Pendidikan Agama Kristen Di Masa Covid-19. Didache: Journal of Christian Education 1, 1–12.

Illich, I., 2002. Deschooling Society, reprint, reissue, revised. ed. Marion Boyars.

Kent, L., Kinsella, N., 2015. A Luta Kontinua (The Struggle Continues). International Feminist Journal of Politics 17, 473–494.

King, M., Luan, B., Lopes, E., 2018. Experiences of Timorese language teachers in a blended Massive Open Online Course (MOOC) for Continuing Professional Development (CPD). Open Praxis 10, 279–287.

Lancker, W.V., Parolin, Z., 2020. COVID-19, school closures, and child poverty: a social crisis in the making. The Lancet Public Health 5, e243–e244.

Marten, 2020. Study From Home by a Teacher.

Menteri Pendidikan dan Kebudayaan, 2020. Pelaksanaan Kebijakan Pendidikan dalam Masa Darurat Penyebaran Covid-19.

Nygaard-Christensen, M., 2013. Negotiating Indonesia: Political Genealogies of Timorese Democracy. The Asia Pacific Journal of Anthropology 14, 423–437.

Pujilestari, Y., 2020. Dampak Positif Pembelajaran Online Dalam Sistem Pendidikan Indonesia Pasca Pandemi Covid-19. 'ADALAH 4.

Purwanto, A., Pramono, R., Asbari, M., Hyun, C.C., Wijayanti, L.M., Putri, R.S., Santoso, priyono B., 2020. Studi Eksploratif Dampak Pandemi COVID-19 Terhadap Proses Pembelajaran Online di Sekolah Dasar. EduPsyCouns: Journal of Education, Psychology and Counseling 2, 1–12.

Quinn, M., 2013. Talking to learn in Timorese classrooms. Language, Culture and Curriculum 26, 179–196.

Rahman, K., 2019. Dewan Pendidikan di Tengah Pusaran Covid-19. AL MURABBI 5, 69–81.

Roja, M.L., 2020. Kebijakan Pendidikan Anak Terlantar di Panti Asuhan St. Louis De Monfort Kota Kupang. Sociological Education 1, 1–10.

Rundle, A.G., Park, Y., Herbstman, J.B., Kinsey, E.W., Wang, Y.C., 2020. COVID-19–Related School Closings and Risk of Weight Gain Among Children. Obesity 28, 1008–1009.

Seran, 2020. Study From Home by a Parent.

Viner, R.M., Russell, S.J., Croker, H., Packer, J., Ward, J., Stansfield, C., Mytton, O., Bonell, C., Booy, R., 2020. School closure and management practices during coronavirus outbreaks including COVID-19: a rapid systematic review. The Lancet Child & Adolescent Health 4, 397–404.

Educational Innovation in Society 5.0 Era: Challenges and Opportunities – Purnomo & Herwin (Eds)
© 2021 the authors, ISBN 978-1-032-05392-9

Model of project based learning in online learning during and after the Covid-19 pandemic

E. Wijaya, Nopriansah & M. Susanti
Universitas Dehasen Bengkulu, Province Bengkulu, Indonesia

ABSTRACT: This research is focused on providing experience to teachers specifically in developing teaching materials by using *project based learning*. The purpose of this study is to assist teachers in improving the effectiveness and creativity of students in developing student confidence during and after the Covid-19 pandemic. Research and development continued with quasi-experiment in collaboration with classroom teachers. Stages of research to be carried out, namely: (1) preparation, (2) development, (3) experimental phase, (4) evaluation. Furthermore, at the implementation stage a collaborative experiment was carried out where the implementation of learning collaborates with the teacher and was designed online. The results of this study with the learning of the project model can develop students' confidence and can improve students' self-creativity. Students were required to study independently at home during and after the Covid-19 pandemic.

1 INTRODUCTION

Efforts to improve student mastery of economics at the school level really need to be improved. This can provide sufficient provisions in life and the field of work so as to solve problems relating to the economy. One of the efforts is the implementation of the 2013 curriculum which emphasizes students to think with the aim of making Indonesian individuals who are productive, creative, innovative, and affective through strengthening attitudes. Economic lessons are very important in equipping students in real life. The importance of economic lessons requires all parties to make improvements, especially those directly related to learning activities. In addition, economic subjects listed in economics are used as a benchmark for graduation in high school. Furthermore, in the selection of an economic college one of the subjects that determine the graduation prerequisites for the social and humanities choice. This shows that economics is important for students to master. However, the reality shows that one of the student learning outcomes in economic subjects needs to be improved. Data from the Ministry of Education and Culture (2019) shows that the average score of the Computer-Based National Examination (UNBK) at the Social Sciences Department level in the 2018/2019 academic year is 46.86 from the 0–100 scale. This shows that the subjects tested were economics that were classified as low.

The results of the initial survey in one of the high schools in the city of Bengkulu through an interview with one of the high school teachers obtained several findings. These findings include: (1) students were not accustomed to finding their own concepts, (2) students still have difficulty in conveying the results of group discussions, (3) students were not accustomed to using real cases in learning, (4) learning materials were not widely found economics that can encourage students' thinking abilities and student skills. One effort that can be done to improve the quality of learning by designing learning so as to facilitate students in developing their abilities and confidence in learning. These efforts using learning media in the form of teaching materials specifically were designed to develop students' abilities in understanding the concept of the material. However, the reality in schools shows that learning tools are rarely found that can be used by teachers directly for learning, especially in developing students' abilities and confidence.

In this view, *Project Based Learning* (PBL) is one of the recommended methods to use. PBL refers to a method that allows "students to design, plan, and implement extended projects that produce outputs that are publicly exhibited such as products, publications, or presentations" (Patton 2012: 13). Through PBL, students engage in communication aimed at completing authentic activities (project-work), so that they have the opportunity to use language in a relatively natural context (Haines 1989, as cited in Fragoulis 2009) and participate in meaningful activities which requires the use of native languages (Fragoulis 2009). The successful implementation of PBL has been reported by Gaer (1998) who taught speaking skills to the Southeast Asian refugee population who were already in their early grade ESOL (Economy for Speakers of Other Languages) classes. Their speaking skills were enhanced through PBL economic students. This study seeks to find out whether PBL can improve students

DOI 10.1201/9781003206019-29

'speaking skills or not, what aspects of speech are improved, and what speaking activities are used to improve students' speaking skills. The scope of this study reveals the use of PBL in developing student confidence and improving student creativity during and after the Covid-19 pandemic. During and after the pandemic, students used the online method of communicating with teachers. The transition period from face to face method changes using online, this is not easily accepted by students. Especially in remote areas in Bengkulu province, where access to the internet still faces constraints on weak networks or signals.

The researchers of this study analyzed how the *project-based learning* model could be accepted and could improve students' creativity in applying learning theory both during and after the Covid-19 pandemic.

2 BACKGROUND

The efforts to improve student mastery of economics at the school level really need to be improved. This can provide sufficient provisions in life and the field of work so as to solve problems relating to the economy. One of the efforts is the implementation of the 2013 curriculum which emphasizes students to think with the aim of making Indonesian individuals who are productive, creative, innovative, and affective through strengthening attitudes. Economic lessons are very important in equipping students in real life. The importance of economic lessons requires all parties to make improvements, especially those directly related to learning activities.

Project-based learning is centered on the learners and gives students the opportunity for in-depth investigation of viable topics. Students are more independent because they build artifacts that are personally meaningful which are representations of their learning, Grant (2002). The results of research by Astrini et al. (2014) that the implementation of a *project-based learning* model is proven to improve the learning process and the learning model of *project-based learning* will be more easily implemented. (Susanti et al. 2020), Project Based Learning can improve students' confidence. By learning project models, it can improve participants' speaking skills, Darini Bilqis Maulany (2013). *Project-based learning* provides project practitioners with the tools to implement *project-based learning* effectively in the context of foreign languages, (Tsiplakides & Fragoulis 2009).

WHO establishes the Corona virus as a pandemic. Covid-19 stands for Corona Virus Disease 2019, which means the Covid-19 corona virus first appeared in 2019 and has already spread to almost all countries in the world. Corona virus is a large family of viruses that can cause diseases, ranging from the common cold to the most severe respiratory illnesses, such as the Middle East Respiratory Syndrome (MERS) and Severe Acute Respiratory Syndrome (SARS). Since it was first detected in Wuhan, China, in December 2019, this outbreak has been growing very fast. Considering this, the Minister of Education and Culture of the Republic of Indonesia issued a Circular no. 4 of 2020 concerning methods of learning during the pandemic. The physical and mental health of students, teachers, school principals and all school residents are a major consideration in implementing policies to keep distance so that the distribution chain can be broken. Implications of the Circular no.4 / 2020 make schools do learning from home for students, to be able to produce meaningful learning according to point 2a, the teacher must choose the right learning model in order to become meaningful learning. Home learning continues until May 2, 2020 which is the National Education Day where the Minister of Education and Culture gives the mandate as a guide for ceremonial activities commemorating the Education Day, (Yuliana 2020).

In designing learning model selection or learning approach is the main key to the implementation of learning. One learning model that can be used is that learning can be done by involving students directly to develop abilities, one of which is the *project based learning* (PjBL) model. The PjBL model facilitates students to make products in order to solve real life problems. Projects for making products can be done individually or in groups. The problem that is the focus of this research is to produce teaching material products. The product produced is economic teaching materials for high school students in grade X. The research design used is in the form of quasi-experimental research. Specifically the purpose of this study is the level of practicality of the learning model provided by teachers during and after the Covid-19 pandemic.

3 LITERATURE REVIEW

3.1 *Teaching materials*

Barab et al. (2007) stated that students bring their own ideas into the classroom and organize them. These organized ideas are not normally used in order to generate a school science activity, but rather an attempt is made to transmit to the students "extracts" from the scientific consensus model and to contrast and highlight the differences between these and the students' alternative ideas. In this way, school tends towards the teaching of a "standard" science, a "true" science that is more or less related to the scientific consensus models. Teaching and learning science seen as a modeling process is different from the transmission of a "scientific consensus model," which involves a didactic transposition adapted to the pupil's age. The various models that can be generated in early school years are provisional representations that explain aspects of reality which are gradually interrelated, leading to the evolution of these models. Teaching materials can be produced quickly with the proposed approach, (Shih et al. 2008). Analysis, modification, and application of the curriculum are core components of teacher practice.

Beginner teachers rely heavily on curriculum materials that often have poor quality to guide their practice, (Schwarz et al. 2008). In order for similar high-quality material to see the light of day, we need communication between publishers, researchers, teachers, and textbook authors to be improved, (Harwood 2005). Two types of material evaluations: predictive evaluations designed to make decisions about what material to use, and retrospective evaluations designed to examine the material to be used, and retrospective evaluations designed to examine the material that has actually been used, (Ellis 1997). Training may be needed if students must be productively involved in managing their learning, (Allwright 1981). The creation of a situational curriculum that supports the learning of specific discipline formalism, (Barab et al. 2007).

3.2 Model project-based learning

Project-based learning is centered on the learner and gives students the opportunity for an in-depth investigation of viable topics (Grant 2002). Many advantages are said by academics who are dismissed by students, who feel that material posted online gives them extra work; represent revocation of lecturer teaching assignments and shift the cost of printing from the institution to students, (Williams 2002). Tomlinson (2012) reviewed the literature on the field of developing material that is relatively new for language learning and teaching. Susanti et al. (2020), learning with the PBL model can improve students' abilities and confidence. Based on a theoretical study of the steps of its application, it can be concluded that another advantage of the *project based learning* model is that it can enhance collaboration. The importance of group work in projects causes students to be able to develop and practice communication skills and scientific performance of students. Based on this explanation, it can be concluded that group work is beneficial to train students' social attitudes. Students can help each other to complete the project, they can be good at helping the less clever, and remind each other to do their respective tasks well.

3.3 The concept of self confidence

Darch & Carnine (1986) Probes and posttests have significant positive effects for groups that are taught with visual displays. There was no significant difference in the transfer test. Eggen & Kauchak (2010) stated that self-confidence is a statement that describes a belief, a cognitive idea is accepted if true without the need to consider other things that support it. In this case we see different ways in each of our self-beliefs that influence our motivation in learning such as: (a) Confidence in things to come, (b) Confidence in terms of intelligence of thought, (c) Confidence in skills, (d) Confidence in content (context), (e) Confidence in achievement.

Furthermore, Willis (Gufron 2010) believes that self-confidence is the belief that someone is able to overcome a problem with the best situation and can provide something that is pleasant for others. With confidence, someone when faced with problems can be resolved properly if they have self-confidence and can provide something valuable for others.

Lauster (Gufron 2010) stated that aspects of self-confidence are as follows:

a. Self-confidence
   A person's positive attitude about himself is a belief in one's ability. Someone is really capable of what he will do
b. Optimistic
   The positive attitude that exists in a person, always has a positive outlook in dealing with everything about his abilities and himself
c. Objective
   Someone who views something or problem not according to himself, but according to the correct truth
d. To be responsible
   Everything that is borne by someone who has become a consequence is someone's responsibility for something
e. To be rational and realistic

Rational and realistic are thoughts that are used to analyze something, an event, and a problem where the thought can be accepted by reason and in accordance with reality.

Based on some of the opinions above, aspects of self-confidence adopted by researchers are: confidence in the ability of self, optimistic, responsible, not influenced by others, and able to overcome problems. These five aspects are used as a reference in measuring confidence in this study

3.4 Research hypothesis development framework

The application of *project-based learning* models in economic learning was assumed to be able facilitate students in developing students' abilities and confidence. Some empirical studies show that PjBL has a role in improving learning outcomes in schools. One of the results of Anita's research (2015) shows that the application of the PjBL learning model improved student creativity in the concept of economic problems. Furthermore, Santoso (2017) stated that this learning model involved the active role of students to produce products or projects that are able to encourage students' ability to understand knowledge through systematic syntax.

Based on the findings of previous researches and the formulation of the problem in this study, the hypotheses of this study are as follows.

a. The application of teaching materials based on valid and reliable project based learning models is effective in increasing the economic abilities of students in class XI of high schools in Bengkulu city.
b. The application of teaching materials based on valid and reliable project based learning models

effectively develops the confidence of students in class XI of high schools in Bengkulu city.

4 METHOD

The research design used was in the form of quasi-experimental research. The purpose of this study is to test the practicality of the learning model given by the teacher, namely the *project based learning* model, with this PBL model the teachers can improve the quality of learning of senior high school students in social sciences (IPS) department. In this case, the researchers got involved directly into the classroom from diagnosing difficulties / obstacles encountered in the learning process then formulating an action plan, implementing learning, monitoring the action process, reflecting and refining the action process, and evaluating the results of the actions or effectiveness of the model. This research activity were carried out through the following stages:

a. Preparation (conducting material and curriculum studies)
b. Development, At this stage two main activities were carried out, namely: (1) Validation, (2) practicality, (3) Field Trial Activities
c. Stage of Collaborative Experiments Conduct learning trials using teaching materials that were designed in the ways of face-to-face and online.

4.1 Data collecting technique

Data collection techniques were divided into two stages, namely development research carried out by observation and distributing validity and practicality evaluation sheets. Validity data collection was done online by contacting experts, namely economics lecturers. While the practicalism test was conducted with a small group conducted by visiting students and teachers of high schools in Bengkulu city while applying the health protocol. In the experimental stage the data collection was done by giving tests to students after the treatment was given. In addition, observations were made to observe the implementation of the learning stages in accordance with the project based learning model.

4.2 Data analysis

Development Research
1. Validity Analysis
The validity estimation used in this study applies the item validity index proposed by Aiken with the following formula:

$$V = \frac{\sum s}{n(c-1)}, \text{ with } s = r - I_0 \qquad (1)$$

Information:
V = validity index
s = the score set for each rater minus the lowest score
r = rater choice category score
I_0 = lowest score in the scoring category
c = the number of categories rater can choose from
n = the number of raters (Retnawati 2014: 3)

4.3 Practical analysis

The trial data that had been obtained was then converted into qualitative data on a five scale. The conversion on a scale of five was adapted from Widoyoko (2009) as in Table 1.

Table 1. Criteria for practicality of learning devices.

Score Interval	Category
$X > \overline{X}_i + 1,8sb_i$	Very practical
$\overline{X}_i + 0,6sb_i < X \leq \overline{X}_i + 1,8sb_i$	Practical
$\overline{X}_i - 0,6sb_i < X \leq \overline{X}_i + 0,6sb_i$	Moderate
$\overline{X}_i - 1,8sb_i < X \leq \overline{X}_i - 0,6sb_i$	Less practical
$X \leq \overline{X}_i - 1,8sb_i$	Not practical

4.4 Collaborative experiment research

Average Learning Outcomes

$$\overline{X} = \frac{\sum X}{N}$$

Information :
$\overline{X}$ = Students' Average Score
$\sum X$ = Total number of students' score
N = Number of students (Sudjana 2009)

4.5 Hypothesis test

In the collaborative experiment stage, data analysis was performed to test the following hypotheses.

$H_0 : \mu_1 = \mu_2$

$H_1 : \mu_1 \neq \mu_2$

μ_1 = average learning outcomes with teaching using teaching materials
μ_2 = average learning outcomes without teaching materials
The statistical hypothesis was tested using the t-test with formulas:

$$t = \frac{\overline{x}_1 - \overline{x}_2}{\sqrt{s^2 gab\left\{\left(\frac{1}{n_1}\right) + \left(\frac{1}{n_2}\right)\right\}}} \qquad (2)$$

With s_2 gab $= \dfrac{(n_1 - 1)s_1^2 + (n_2 - 1)s_1^2}{n_1 + n_2 - 2}$ $\qquad (3)$

5 RESULTS

5.1 Development research results

Project based learning based learning materials that have been prepared were assessed by experts who aim to see the quality of the product in terms of content. The trial results show that teaching materials in the form of Student Activity Sheets (LKS) met valid criteria. The validity test results of teaching materials are as follows (Table 2).

Table 2. Teaching material validation results.

No	Rated Aspect	Aiken Index	Criteria
1	Formulation and Purpose	0.70	Valid
2	Conformity of Content and Material	0.75	Valid
3	Learning Activities	0.73	Valid
4	Language accuracy	0.74	Valid
5	Learning Resources	0.71	Valid
6	Penerapan PjBL	0.72	Valid

The results of the validator's assessment of the teaching material above show a valid category. This shows that in theory economic teaching materials based on *project based learning* models met valid criteria.

5.2 Results of collaborative experiments

The schools selected in the implementation of collaborative experiments consisted of three schools, namely: (1) SMA Negeri 1 Bengkulu City, (2) SMA Negeri 3 Bengkulu City, and (3) SMA Negeri 6 Bengkulu City. In each school one class was chosen as the experimental class, namely class XI majoring in social studies. Description of student learning outcomes after being given learning by using project-based learning based learning materials is shown in Table 3.

Table 3. Student learning outcomes data.

Names of School	Number of Students	Average Scores	Percentage (%)
SMAN 1 Kota Bengkulu	30	83.13	83.00
SMAN 3 Kota Bengkulu	31	82.00	80.02
SMAN 6 Kota Bengkulu	32	81.24	80.00

Based on the table above, it can be seen that the percentage of students mastery learning classically that reaches KKM is more than 65%. In addition, the average value of the two trial classes has reached the KKM value. This shows that the learning tools developed have met the effective criteria.

6 DISCUSSION

Project-based economics teaching materials have been assessed by experts with criteria to assess the suitability of content and materials, learning activities, language accuracy, educator practicality, and student practicality testing (Table 2). The teaching materials used have been declared valid by the assessment team. Using this economic teaching material can increase student grades.

Evidence for improvement is in the pretest and posttest results. The scores of students who used teaching materials were higher than students who did not use teaching materials (Table 3).

The purpose of assessing the validity of teaching materials is to see the quality of the product in terms of content. The test results show that the teaching materials in the form of Student Activity Sheets (LKS) meet the valid criteria (Table 2). Learning with this project model can increase students' self-confidence

7 CONCLUSION

After applying the model of learning i.e. *project based learning*, the result is the ability to learn and the confidence of students of SMAN 1, SMAN 3 and SMAN 6 Bengkulu City can be improved. This can be seen in the results of the *t*-test, where the results of *t* count are greater than *t* table. The *t*-test table shows the students' knowledge after the use of teaching materials with *t*-count > *t*-table and significant <0.005. The research by developing teaching materials using *project based learning* could also improve students' independence in the process of accepting economic subject materials. The students were even more creative in developing innovation in the field of science being studied.

So the results of this study are accepting HO and rejecting Ha, meaning that after the *project based learning* model was applied, there was an improvement in knowledge and confidence of students of SMAN 1, SMAN 3 and SMAN 6 Bengkulu City, this study is in line with the research of Susanti et al. (2020) which shows that the learning model with the *project based learning* model is effective in terms of the conceptual knowledge aspects.

This research has a weakness that is the lack of tools used in the development of teaching materials such as student worksheets (LKS), internet networks in schools were still very low connected, along with props used in developing *project-based learning* models to improve the abilities of high school students in Bengkulu city.

REFERENCES

Allwright, R. L. 1981. What do we want teaching materials for? *ELT Journal*, 36(1): 5–18. https://doi.org/10.1093/elt/36.1.5

Astrini, K. S., Suwendra, I. W., & Suwarna, I. K. 2014. Pengaruh CAR, LDR dan Bank Size Terhadap NPL pada Lembaga Perbankan yang Terdaftar di Bursa Efek Indonesia. *E-Journal Bisma Universitas Pendidikan Ganesha Jurusan Manajemen*, 2(1), 1–8.

Barab, S., Zuiker, S., Warren, S., Hickey, D. A. N., Ingram-goble, A., Kwon, E., Kouper, I., & Herring, S. C. 2007. Curriculum: Relating Formalisms and Contexts. *Science Education*, 91: 750–782. https://doi.org/10.1002/sce

Darch, C., & Carnine, D. 1986. Teaching Content Area Material to Learning Disabled Students. *Exceptional Children*, 53(3): 240–246. https://doi.org/10.1177/001440298605300307

Darini Bilqis Maulany. 2013. The Use Of Project-Based Learning In Improving The Students' Speaking Skill (A Classroom Action Research at One of Primary Schools in Bandung). *Journal of English and Education*, 1(June): 52–58.

Ellis, R. 1997. The empirical evaluation of language teaching materials. *ELT Journal*, 51(1): 36–42. https://doi.org/10.1093/elt/51.1.36

Grant, M. M. 2002. Getting a grip on project-based learning: Theory, cases and recommendations. In *Meridian* (Vol. 5, Issue 1). https://www.scopus.com/inward/record.uri?partnerID=HzOxMe3b&scp = 3042553637&origin= inward

Harwood, N. 2005. What do we want EAP teaching materials for? *Journal of English for Academic Purposes*, 4(2): 149–161. https://doi.org/10.1016/j.jeap.2004.07.008

Schwarz, C. V., Gunckel, K. L., Smith, E. L., Covitt, B. A., Bae, M., Enfield, M., & Tsurusaki, B. K. 2008. Helping elementary preservice teachers learn to use curriculum materials for effective science teaching. *Science Education*, 92(2): 345–377. https://doi.org/10.1002/sce.20243

Shih, W. C., Tseng, S. S., & Yang, C. T. 2008. Wiki-based rapid prototyping for teaching-material design in e-Learning grids. *Computers and Education*, 51(3): 1037–1057. https://doi.org/10.1016/j.compedu.2007.10.007

Susanti, M., Herfianti, M., Damarsiwi, E. P. M., Perdim, F. E., & Joniswan. 2020. Project-based learning model to improve students' ability. *International Journal of Psychosocial Rehabilitation*, 24(2). https://doi.org/10.37200/IJPR/V24I2/PR200437

Tomlinson, B. 2012. Materials development for language learning and teaching. In *Language Teaching*, 45(2), 143–179. https://doi.org/10.1017/S0261444811000528

Tsiplakides, I., & Fragoulis, I. 2009. Project-based learning in the teaching of English as a foreign language in Greek primary schools: from theory to practice. *English Language Teaching*, 2(3): 113–119. https://doi.org/10.5539/elt.v2n3p113

Williams, P. 2002. The Learning Web. *Active Learning in Higher Education*, 3(1): 40–53. https://doi.org/10.1177/1469787402003001004

Yuliana, C. 2020. *Pengembang Teknologi Pembelajaran.* https://lpmplampung.kemdikbud.go.id/detailpost/project-based-learning-model-pembelajaran-bermakna-di-masa-pandemi-covid-19

*Educational Innovation in Society 5.0 Era: Challenges and
Opportunities – Purnomo & Herwin (Eds)
© 2021 the authors, ISBN 978-1-032-05392-9*

Communal learning model as blended learning strategy in primary school during the Covid-19 pandemic

K.I. Sujati, A. Syamsudin, Haryanto & W.P. Hapsari
Universitas Negeri Yogyakarta, Yogyakarta, Indonesia

ABSTRACT: Covid-19 pandemic caused changes including the development of various models and strategies in the learning process to preserve its continuity. This study is aimed to determine the implementation process of the communal learning model as a blended learning strategy in primary school during the Covid-19 pandemic. A mixed-methods sequential explanatory research design used to examine the communal learning model implementation. The results of this study indicate the communal learning model could be used as a blended learning strategy in primary school during the Covid-19 pandemic.

1 INTRODUCTION

The coronavirus has been already spread across the world as pandemic (WHO 2020) includes in Indonesia. Covid-19 pandemic is giving many impacts for almost daily live aspects. Economic, health, education, policy, trades, tourism, and business enterprise are some impacted aspects caused by the coronavirus (Capano, Howlett, Jarvis, Ramesh, & Goyal 2020; Centre for Economic Policy Research 2020; Dwivedi et al. 2020; Gössling, Scott, & Hall 2020; Macartney et al. 2020; Maliszewska, Mattoo, & van der Mensbrugghe 2020; Mann, Krueger, & Vohs 2020; Panovska-Griffiths et al. 2020; Stang, Standl, & Jöckel 2020). Those impacts are being perceived in Indonesia.

The Minister of Education and Culture of Republic Indonesia responded to the emergencies spread situation of the coronavirus disease by issuing a circular letter for the implementation of the education policies in the Covid-19 pandemic since March 2020 (Menteri Pendidikan dan Kebudayaan RI 2020b). The online learning process held in Indonesia brought some learning process evaluation especially for the primary students since March 2020. The students could not attend to school. Despite the online learning process keep it running until this new academic year (General Secretary of Education and Culture Ministry of Republic Indonesia 2020), few teachers finally choose to held blended learning used the communal learning model in the learning process while still following health protocols. Communal learning model had chosen as the most fit learning model that could be used in the blended learning process. Otherwise, this study would determine the implementation process of the communal learning model as a blended learning strategy in primary school during the Covid-19 pandemic.

2 COMMUNAL LEARNING MODEL AS BLENDED LEARNING STRATEGY

2.1 *Learning model*

A learning process is a process that showed students increase their competencies by achieving the learning goals (Cebrián & Junyent 2015; Rožman & Koren 2013). Therefore, the teachers should facilitate the students by maximizing the learning design and the learning implementation (Education and Cultural Human Resources Development Agency and Education Quality Assurance 2014). The learning design is the essential thing that should be prepared by the teachers. The learning design related to the components of learning involves curriculum, learning goals, learning model, learning media, teaching system, and the assessment (Ramadhani & Ayriza 2019).

The learning process happens effectively if the teachers have well-prepared for the learning design (Davidson, Prahalad, & Harwood 2020; Nilsson 2009). Teachers could give more attention when setting the learning materials, choose the learning strategies, the learning model, learning media, and sources, also choose the instruments for the students' assessment. For those reasons, teachers could give more attention to choose the learning model for increasing the learning process effectively (Amrullah & Suwarjo 2018; Wahyuningsih & Kiswaga 2019), includes during in the coronavirus pandemic. The teachers should consider the suit one of the learning models to be used in the learning process.

2.2 *Communal learning model*

Communal learning is the one kind of learning model used by the person to share their knowledge integratively through their organization or community

DOI 10.1201/9781003206019-30

(Alina Ali Zani, Anir Norman, & Abdul Ghani 2018). In this study, the term communal learning model refers to a learning model that implemented by grouping the students based on the nearest area of their residency. This is the part of classroom management doing by the teachers to solve the learning problem during the Covid-19 pandemic. For those points, the communal learning model gives a chance for the students to engage with each other in their learning group (Burnard & Dragovic 2014). Communal learning model was used to engage the blended learning

2.3 Blended learning

Blended learning is one of the learning strategies that could be used in the learning process (Bonk & Graham 2012; Wang 2010). Blended learning commonly used to define the learning combination between face to face with computer technology (Sharma 2010). Blended learning is given by formal education program that provides the students to learn the content and instruction through online delivery and supervise brick-and-mortar home location's away (Staker & Horn 2012).

Some blended learning models could be used in the learning process. Staker & Horn (2012) mentioned four models of blended learning: (1) rotation model, (2) flex model, (3) self-blended model, and (4) enriched-virtual model. The respondents of this study used the enriched-virtual model in this study. The enriched-virtual model is the blended learning model that gives a whole-school experience to provide the content or materials and learning instruction using online delivery and the students experience their divide time to attend a brick-and-mortar campus (Staker & Horn 2012).

2.4 Communal learning model: Learning strategy during the Covid-19 pandemic

World Health Organization (WHO) directly responded to the pandemic situation through the action guideline and control in school to protect the students and children from the coronavirus disease. The guideline gives key messages and actions which could be aware and applicate in the learning process during this pandemic. This guideline provides the information ensuing: (1) the school administrators, teachers, and staff, (2) parents or caregivers and community members, (3) children and the students. World Health Organization (WHO) also gives the detail of age-specific health education during this pandemic for the primary students (Unicef, WHO, & IFRC 2020).

Indonesia government through Ministry of Education and Culture also responded the pandemic situation through the education policy in March (Menteri Pendidikan dan Kebudayaan RI 2020b) which is emphasized in May with the General Secretary of Education and Culture Ministry's circular letter (General Secretary of Education and Culture Ministry of Republic Indonesia 2020). The Covid-19 pandemic situation also responded by the region government, education local authorities, until the school headmasters and the dorp administrator.

Since March until July, the students accomplished the second semester by using a full online learning method. Each school has its evaluation of the students learning achievements during online learning. Some of the teachers felt the online learning not completely effective (Bahasoan, Ayuandiani, & Mukhram 2020; Fauzi & Sastra Khusuma 2020; Wajdi et al. 2020). So, the teachers made the learning model innovation to increase the learning effectiveness during the Covid-19 pandemic.

The teachers who given supports by some stakeholders to held the blended learning strategies on the learning process start in the new academic year. Which some of the teachers choose the communal learning model as the blended learning strategies.

3 METHOD

3.1 Design

The mixed-methods sequential explanatory research design used to examine the communal learning model implementation. This method adopted from Gray et al. (2020) and developed according to this study's needs. The mixed-methods sequential explanatory design figures out in Figure 1.

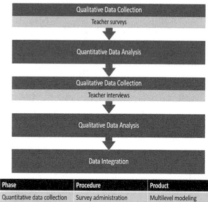

Figure 1. The sequence of mixed-methods design.

The data which collected before is structured and analyzed in two phases: quantitative and then qualitative (De Graaf, Van Klinken, Zweers, & Teunissen 2020; Draucker, Rawl, Vode, & Carter-Harris 2020; Glover, Shah, Bennett, Wilson, & Barnes 2020). The quantitative phase was prioritized use to assess the data and allowed us to examine furthermore.

In this study, the quantitative methods used to assess the learning condition. These data collected through the survey which fulfilled by teachers. These data analyzed to find out the communal learning model used in their blended learning process. The qualitative phase necessaries needed to provide descriptive data of the communal learning model. The triangulation used to culminate the data presentation which obtained before.

The methodological approaches used integration to interpret and explanate the study results. The study results collected from the survey and interviews analyzed and integrated into the qualitative phase, so found out the determination of how the communal learning model as the blended learning strategy implementation. In the specific data analysis, we connected the survey items to teacher interview question development and get the framework about the quantitative data. The probing areas were used to inform the conversations of the interview. The decisions of the overall study were integrated by emphasized and emerged the intersection of the quantitative and qualitative findings.

3.2 *Participants and setting*

The research period was done at the end of July 2020. We observed six teachers from four different regions of Indonesia. The detailed profile from all participant describes in Table 1. All participants have the same teaching experience in the range of 1-5 years and also the same age average at under 30. Two of six is teaching in nonprivate school. Most of them are teaching in a rural area. All of them is the homeroom teacher who has the same educational background. Whole participants school enrolled in the National Primary School Curriculum program.

Table 1. The participants' profile.

Participants	Gender	Teach of Grade	Student	School
Teacher 1	Women	V	12	Private School
Teacher 2	Men	VI	27	Private School
Teacher 3	Men	V	28	Nonprivate School
Teacher 4	Women	VI	5	Private School
Teacher 5	Men	V	21	Private School
Teacher 6	Men	III	23	Nonprivate School

The lead researcher gave the survey to be fulfilled by participants. Then interviewing them for the

learning process implementation during the pandemic situation and giving the participants explained the communal learning model which used as the blended learning strategies. Before the researcher began, the lead researcher has informed the research stages and all the participants disposed of in this study.

3.3 *Procedure*

As shown in Figure 1, our research activities included six teachers, one survey administrations, and a follow up with interviews with teachers. The data collected in the quantitative phase informed the data collection of qualitative phases was conducted and analyzed to further data integration which illustrates the communal learning model implementation as blended learning strategies.

4 RESULTS AND FINDINGS

4.1 *Communal learning model implementation*

Based on the survey results, the communal learning model was implemented of this study used two method application designs: home visitation and grouping the students based on the nearest regency. One teacher used the home visitation method and five teachers used the grouping method. All over the learning process handled by doing the learning-rotation process based on the students' group. The teachers held home-learning group visitation in two methods. First, making the student groups in the early semester based on the nearest residency and teacher will hold the home-learning group visitation based on the learning group schedule. Second, the home-learning visitation will be held by the teacher if one student or more has difficulty in their learning period, the home-learning visitation also based on the communal group. The difference is the first method hold in regular schedule, but the second method just hold when the students faced learning problems.

The communal learning model used to be held offline learning, concurrently teachers give the instruction and content or materials learning through online method. Teachers used the face-to-face method on the communal learning model when held the offline learning. Teachers explained the learning materials, answer the students' problems, and many other learning activities. So, the implementation of blended learning used combination of the communal learning model in the face-to-face supplementation and the online instruction and content.

4.2 *Learning policy brief of communal learning model*

There are two vigorous reasons: rational learning reasons and stakeholder engagement. First, the rational learning reasons disclosed by the participants to some point. The communal learning model held because the students have obstacles when understanding the learning material without the teacher's explanations. Some teachers informed that their students in the previous semester could not fully understand the learning material although have explained using some accessible learning media. It caused the learning competencies were not maximally achieved if only using the online learning method. The other problems are the internet limit access, the unsupported learning device, and bad signal. Second, the communal learning model held by the related stakeholder decision. The teachers bravely held the blended learning of course the related stakeholders' support which engaged the headmaster, the homeroom teacher, the parents, the school committee, the school supervisor, region education authorities, even the nearest village administrator.

4.3 Learning process experience of communal learning model

The communal learning model held in various duration which customized with the unit of learning duration for Indonesian student primary school (Minister of Education and Culture of the Republic of Indonesia 2013). The communal learning model could make the learning process more effective, so the students could more understand the learning material. Moreover, the communal learning model increased the achievement of competencies through various assessment models like the quiz test, student action video, and portfolio.

The teachers were experiencing various kinds of learning obstacles when used the communal learning model. Those obstacles are the limited learning facilities, limited learning duration, the communal group distance, and student attention. The teacher and student have limited access to some learning facilities and learning media. Some teachers have a solution, for example, bring their speaker and a personal laptop to give learning experience for their students. The limited learning duration mostly felt by the students and teachers when need more time to explain or learn rather difficult of the learning materials. The communal group distance mostly felt by the nonprivate school teacher where the students' communal group area has far distance from the school or the teacher's house. So, the teachers have offered more strength and stamina to held communal learning according to the communal schedule. When the communal learning model held outdoor, some students could not focus at some moment, so the teacher used some learning model to support the learning process.

5 DISCUSSION

Designing the learning process in the coronavirus pandemic needs more considerations. In the new academic year, the teachers faced new challenges in the learning process. Physical and psychological problems appeared in the online learning process. The teachers and parents need to be aware about the anxiety and stress in the learning that could be felt by the students in the learning process during the Covid-19 pandemic. Wong (2020) found that the students' stress and anxiety affected the online learning basic needs; the competence, autonomy, relatedness, and arousal. This case found out in the learning process of the previous term, so the communal learning model had chosen to solve the learning problem.

The Minister of Education and Culture of Republic Indonesia showed the situation awareness by issued the policies of the learning process implementation during the coronavirus disease pandemic (Menteri Pendidikan dan Kebudayaan RI 2020b). These policies are giving relaxation for the teachers to reformulate the learning design that will be implemented on theirs each class (Menteri Pendidikan dan Kebudayaan RI 2020a) based on their online learning evaluation from the previous term, especially the online learning when held during March, the first online learning implemented, until June 2020. The teachers need to make meaningful learning, focusing on the life skills education, the learning media access, and the meaningful learning feedback (Menteri Pendidikan dan Kebudayaan RI 2020b).

On the other hand, the teachers faced another learning problem: the online learning accessibility. During the online learning implementation period, mostly the learning process relied on the smartphone supports as the main learning tools (Basilaia & Kvavadze 2020; Iivari, Sharma, & Ventä-Olkkonen 2020; Iyengar, Upadhyaya, Vaishya, & Jain 2020). This has an impact on the students who their family just have a smartphone which used by the parents for main work tools, so the students couldn't access the learning materials and learning process maximally. The lack broadband smartphone signal also impacted the learning process. So, the teachers supported by the parents, the headmaster, and the related stakeholders decided together to held the blended learning to reduce the problems, especially for the students' competencies achievements and learning accessibility.

Blended learning was chosen as the learning strategy to increase the learning effectiveness during the Covid-19 pandemic (Darras et al. 2020; Wargadinata, Maimunah, Dewi, & Rofiq 2020). This learning model also increased the students' awareness about what happen nowadays, build the environmental responsibility, and cooperative skills (Gray et al. 2020). The students increased their awareness about the coronavirus disease and the health protocol which should be applicated in daily life, especially when used the face to face method.

The communal learning model used the enriched-virtual learning design considered of the learning modalities as the common consideration when held the blended learning (Dziuban, Graham, Moskal, Norberg, & Sicilia 2018; Staker & Horn 2012). Accommodating the contemporary condition of the learning process during the pandemic, the communal learning model could be implemented in the blended condition.

6 CONCLUSION

The communal learning model could be used as a blended learning strategy in primary school during the Covid-19 pandemic. The communal learning model could be implemented in the learning process with various considerations that should be understood by the teacher and related stakeholders. The main consideration is how to protect the students while they still attend the learning process during the coronavirus pandemic and make them aware the health education along the learning process. Communal learning model could increase the effectiveness of the learning process, also increase the learning achievement.

REFERENCES

Alina Ali Zani, A., Anir Norman, A., & Abdul Ghani, N. 2018. A Review of Security Awareness Approach: Ensuring Communal Learning. *PACIS 2018 Proceedings*, 278. Retrieved from https://aisel.aisnet.org/pacis2018/278

Amrullah, K., & Suwarjo, S. 2018. The effectiveness of the cooperative problem-based learning in improving the elementary school students' critical thinking skills and interpersonal intelligence. *Jurnal Prima Edukasia*, 6(1): 66. https://doi.org/10.21831/jpe.v6i1.11253

Bahasoan, A., Ayuandiani, W., & Mukhram, M. 2020. *Effectiveness of Online Learning In Pandemic Covid-19*. 100–106.

Basilaia, G., & Kvavadze, D. 2020. Transition to Online Education in Schools during a SARS-CoV-2 Coronavirus (Covid-19) Pandemic in Georgia. *Pedagogical Research*, 5(4). https://doi.org/10.29333/pr/7937

Bonk, C. J., & Graham, C. R. 2012. *The Handbook of Blended Learning: Global Perspectives, Local Designs*. New Jersey, USA: John Wiley & Sons, Inc.

Burnard, P., & Dragovic, T. 2014. Characterizing Communal Creativity in Instrumental Group Learning. *Departures in Critical Qualitative Research*, 3(3): 336–362. https://doi.org/10.1525/dcqr.2014.3.3.336

Capano, G., Howlett, M., Jarvis, D. S. L., Ramesh, M., & Goyal, N. 2020. Mobilizing Policy (In)Capacity to Fight Covid-19: Understanding Variations in State Responses. *Policy and Society*, 39(3): 1–24. https://doi.org/10.1080/14494035.2020.1787628

Cebrián, G., & Junyent, M. 2015. Competencies in Education for Sustainable Development. *Sustainability*, 7(3): 2768–2786. https://doi.org/10.3390/su7032768 T4 - Exploring the Student Teachers' Views M4 - Citavi

Centre for Economic Policy Research. 2020. Economics in the Time of COVID-19. In R. Baldwin & B. W. di Mauro (Eds.), *Economics in the Time of Covid-19*. Retrieved from www.csepr.org

Darras, K. E., Spouge, R. J., de Bruin, A. B. H., Sedlic, A., Hague, C., & Forster, B. B. 2020. Undergraduate Radiology Education During the Covid-19 Pandemic: A Review of Teaching and Learning Strategies. *Canadian Association of Radiologists Journal*. https://doi.org/10.1177/0846537120944821

Davidson, J., Prahalad, V., & Harwood, A. 2020. Design precepts for online experiential learning programs to address wicked sustainability problems. *Journal of Geography in Higher Education*, 00(00): 1–23. https://doi.org/10.1080/03098265.2020.1849061

De Graaf, E., Van Klinken, M., Zweers, D., & Teunissen, S. 2020. From concept to practice, is multidimensional care the leading principle in hospice care? An exploratory mixed method study. *BMJ Supportive and Palliative Care*, 10(1): 1–9. https://doi.org/10.1136/bmjspcare-2016-001200

Draucker, C. B., Rawl, S. M., Vode, E., & Carter-Harris, L. 2020. Integration Through Connecting in Explanatory Sequential Mixed Method Studies. *Western Journal of Nursing Research*. https://doi.org/10.1177/0193945920914647

Dwivedi, Y. K., Hughes, D. L., Coombs, C., Constantiou, I., Duan, Y., Edwards, J. S., …Upadhyay, N. 2020. Impact of Covid-19 pandemic on information management research and practice: Transforming education, work and life. *International Journal of Information Management*, (July), 102211. https://doi.org/10.1016/J.IJINFOMGT.2020.102211

Dziuban, C., Graham, C. R., Moskal, P. D., Norberg, A., & Sicilia, N. 2018. Blended learning: the new normal and emerging technologies. *International Journal of Educational Technology in Higher Education*, 15(1): 1–16. https://doi.org/10.1186/s41239-017-0087-5

Education and Cultural Human Resources Development Agency and Education Quality Assurance. 2014. *Teacher Training Material of 2013 Curriculum Implementation*. Indonesia: Ministry of Education and Culture of Republic Indonesia.

Fauzi, I., & Sastra Khusuma, I. H. 2020. Teachers' Elementary School in Online Learning of COVID-19 Pandemic Conditions. *Jurnal Iqra': Kajian Ilmu Pendidikan*, 5(1): 58–70. https://doi.org/10.25217/ji.v5i1.914

General Secretary of Education and Culture Ministry of Republic Indonesia. *Guidelines for Managing Learning from Home in an Emergency of The Spread of Coronavirus Disease (Covid-19).* , Pub. L. No. 15 (2020).

Glover, C. M., Shah, R. C., Bennett, D. A., Wilson, R. S., & Barnes, L. L. 2020. The Health Equity Through Aging Research And Discussion (HEARD) Study: A Proposed Two-Phase Sequential Mixed-Methods Research Design To Understand Barriers And Facilitators Of Brain Donation Among Diverse Older Adults: Brain donation decision making among. *Experimental Aging Research*, 46(4): 311–322. https://doi.org/10.1080/0361073X.2020.1747266

Gössling, S., Scott, D., & Hall, C. M. 2020. Pandemics, tourism and global change: a rapid assessment of Covid-19. *Journal of Sustainable Tourism*, 0(0): 1–20. https://doi.org/10.1080/09669582.2020.1758708

Gray, D. L. L., McElveen, T. L., Green, B. P., & Bryant, L. H. 2020. Engaging Black and Latinx students through communal learning opportunities: A relevance intervention for middle schoolers in STEM elective classrooms. *Contemporary Educational Psychology*, 60(December 2019), 101833. https://doi.org/10.1016/j.cedpsych.2019.101833

Iivari, N., Sharma, S., & Ventä-Olkkonen, L. 2020. Digital transformation of everyday life – How Covid-19 pandemic transformed the basic education of the young generation and why information management research should care? *International Journal of Information Management*, (June), 102183. https://doi.org/10.1016/j.ijinfomgt.2020.102183

Iyengar, K., Upadhyaya, G. K., Vaishya, R., & Jain, V. 2020. Covid-19 and applications of smartphone technology in the current pandemic. *Diabetes and Metabolic Syndrome: Clinical Research and Reviews*, 14(5): 733–737. https://doi.org/10.1016/j.dsx.2020.05.033

Macartney, K., Quinn, H. E., Pillsbury, A. J., Koirala, A., Deng, L., Winkler, N., …Sullivan, M. V. N. O. 2020. *Transmission of SARS-CoV-2 in Australian educational settings?: a prospective cohort study*. 4642(20): 1–10. https://doi.org/10.1016/S2352-4642(20)30251-0

Maliszewska, M., Mattoo, A., & van der Mensbrugghe, D. 2020. *The Potential Impact of Covid-19 on GDP and Trade: A Preliminary Assessment*. https://doi.org/10.1596/1813-9450-9211

Mann, F. D., Krueger, R. F., & Vohs, K. D. 2020. Personal economic anxiety in response to Covid-19. *Personality and Individual Differences*, 167(July), 110233. https://doi.org/10.1016/j.paid.2020.110233

Menteri Pendidikan dan Kebudayaan RI. *Pedoman Pelaksanaan Kurikulum pada Satuan Pendidikan dalam Kondisi Khusus.* , Pub. L. No. 719/P/2020 (2020).

Menteri Pendidikan dan Kebudayaan RI. *Pelaksanaan Kebijakan Pendidikan dalam Masa Darurat Penyebaran Coronavirus Disease (Covid-19).* , Pub. L. No. 4 (2020).

Minister of Education and Culture of Republic of Indonesia. *The Regulation of Education and Culture Minister of Republic of Indonesia Number 81A Year 2013 about The Implementation of Curriculum.* , Pub. L. No. 81A (2013).

Nilsson, P. 2009. From lesson plan to new comprehension: Exploring student teachers' pedagogical reasoning in learning about teaching. *European Journal of Teacher Education*, 32(3): 239–258. https://doi.org/10.1080/02619760802553048

Panovska-griffiths, J., Kerr, C. C., Stuart, R. M., Mistry, D., Klein, D. J., Viner, R. M., & Bonell, C. 2020. Determining the optimal strategy for reopening schools, the impact of test and trace interventions , and the risk of occurrence of a second Covid-19 epidemic wave in the UK: a modelling study. *The Lancet Child and Adolescent Health*, 4642(20). https://doi.org/10.1016/S2352-4642(20)30250-9

Ramadhani, M. I., & Ayriza, Y. 2019. The effectiveness of quantum teaching learning model on improving the critical thinking skills and the social science concept understanding of the elementary school students. *Jurnal Prima Edukasia*, 7(1): 47–57. https://doi.org/10.21831/jpe.v7i1.11291

Rožman, L., & Koren, A. 2013. Learning To Learn As a Key Competence and Setting Learning Goals. *Management, Knowledge and Learning*, 1211–1218.

Sharma, P. 2010. Blended learning. *ELT Journal*, 64(4): 456–458. https://doi.org/10.1093/elt/ccq043

Staker, H., & Horn, M. B. 2012. Classifying K-12 Blended Learning. *Innosight Institute*, (May), 22. https://doi.org/10.1007/s10639-007-9037-5

Stang, A., Standl, F., & Jöckel, K. H. 2020. Characteristics of Covid-19 pandemic and public health consequences. *Herz*, 45(4): 313–315. https://doi.org/10.1007/s00059-020-04932-0

Unicef, WHO, & IFRC. 2020. Key Messages and Actions for Prevention and Control in Schools. *Unicef*, p. 2. Retrieved from who.int

Wahyuningsih, A., & Kiswaga, G. E. 2019. The effectiveness of CIRC learning model and PQ4R learning model on reading comprehension skills of elementary school students. *Jurnal Prima Edukasia*, 7(1): 82–93. https://doi.org/10.21831/jpe.v7i1.9701

Wajdi, M. B. N., Iwan Kuswandi, Umar Al Faruq, Zulhijra, Z., Khairudin, K., & Khoiriyah, K. 2020. Education Policy Overcome Coronavirus, A Study of Indonesians. *Edutec?: Journal of Education And Technology*, 3(2): 96–106. https://doi.org/10.29062/edu.v3i2.42

Wang, M. J. 2010. Online collaboration and offline interaction between students using asynchronous tools in blended learning. *Australasian Journal of Educational Technology*, 26(6): 830–846. https://doi.org/10.14742/ajet.1045

Wargadinata, W., Maimunah, I., Dewi, E., & Rofiq, Z. 2020. Student's Responses on Learning in the Early Covid-19 Pandemic. *Tadris: Jurnal Keguruan Dan Ilmu Tarbiyah*, 5(1): 141–153. https://doi.org/10.24042/tadris.v5i1.6153

WHO. 2020. Coronavirus Disease Situation Report World Health Organization. In *World Health Organization* (Vol. 19).

Wong, R., & Wong, R. 2020. When no one can go to school: does online learning meet students' basic learning needs? students' basic learning? needs *Interactive Learning Environments*, 0(0): 1–17. https://doi.org/10.1080/10494820.2020.1789672

Educational Innovation in Society 5.0 Era: Challenges and Opportunities – Purnomo & Herwin (Eds)
© 2021 the authors, ISBN 978-1-032-05392-9

Online learning in the medicinal education during the pandemic era: How effective are the platforms?

Y. Febriani, H. Haritani, P. Hariadi, T.P. Yuliana, A. Rafsanjani, M. Azim & E.E. Oktresia
Universitas Hamzanwadi, Lombok, Indonesia

ABSTRACT: The coronavirus pandemic has significant impacts on a whole life, including the learning process. Social distancing is applied in every part of activities as one way to reduce the spread of this disease and hence urge the community to support learning activities from distant. Thereby, the use of information technology and various kinds of platforms in online learning are immediately needed. This research aimed to analyze the effectiveness of multiple platforms in online learning in medicinal education during a pandemic and the purposive sampling method used in this research. The students of health faculty fullfilled the questionnaires by using Google Forms. The survey was categorized into three major questions: (1) the various platforms used in online learning during Pandemic; (2) tools used in online learning; and (3) internet data capacity and network stability. There were 118 responses to the survey. The research showed that the platform used in online learning during Pandemic is more effective if it has fewer internet data capacity and network stability. It is supported in mobile devices and contains audio-text-visual aspects. Based on this situation, we must continue to develop technologies in the online learning process in this pandemic era to augment the new world of medicinal education.

1 INTRODUCTION

COVID-19 has a significant effect in a whole life, including medicine and education. It was also dramatically impacting the learning process. It has changed how we learn and train our students. Some of our annual meetings and programs have been postponed until uncertainty or canceled (Nicola et al. 2020; Plancher, Shanmugam, & Petterson 2020). All efforts to minimize the spreading of this disease have been made. Physical distancing, social distancing, and healthy life were continuously implemented around the world. Some of the public policies, including working from home as consequences to face this Pandemic, have been implicated. Indeed, the Learning process also significantly changed (Allcott et al. 2020; Nicola et al. 2020). This condition forced us rapidly to adapt and made us become fast learners to decide the best way of educational delivery. Therefore, using information technology in instructional delivery was important in this pandemic era.

Information technology (ICT) knowledge is one of the excellent communication skills prepared for every university or higher education to their students, including critical thinking, problem-solving, creative thinking, and cooperation (Deacon & Hajek 2011;

Setiawan, Malik, Suhandi, & Permanasari 2018). In pharmacies, universities or higher educations must give their student critically thinking for solving the complex problem of patients, analyzing and producing pharmacy's toll for practice, and solving environmental issues considering pharmacist skill (Education 2015; Maudsley & Strivens 2000). For encouraging those skills, the learning process must be supported by instructional methods for theory and practice. Educators used multiple teaching approaches while educators used critical thinking in practice for correlating course research experiences. Those various teaching approaches were supported by information technology, an e-learning system, and a new pedagogic or curriculum to enhance education quality (Holland 2019; Martin, Donohoe, & Holdford 2016; Persky, Medina, & Castleberry 2019). Using critical thinking skills to the course research experiences in the context of laboratory research shows a positive correlation to improving practice skills, particularly in pharmacy students (Haritani, Febriani, Puspita, & Arviana 2019). Presently, educators have difficulties in the commonly educational delivery process because of this Pandemic. A learning process in every level of education held online to minimize the effect of COVID-19 nor the theory and practice

DOI 10.1201/9781003206019-31

169

(Crawford et al. 2020; Dwivedi et al. 2020). Thereby, information, technology, and using technology in the learning process become significantly implemented in this pandemic era. We must adopt, adapt, and integrated technology in online learning. The extremely and challenges condition made the new world in the pharmacy education process.

Nowadays, the utilization of technology platforms became an essential aspect of the online learning process. There are many various platforms, including virtual meetings, e-learning, and mobile-devices applications. For virtual encounters, platforms as Skype, Google meet, and zoom was used rapidly. Webinar and podcasts also used to survive in the new world for the learning process (Chen et al. 2020; Huang, Liu, Tlili, Yang, & Wang 2020). Every platform has its characteristics, also a positive and negative side in a learning process. But the important thing that we must realize was network stability and internet data capacity. All the platforms used excellent network stability and high internet data capacity. Even though our research shows that students of Universitas Hamzanwadi used the high intensity and frequency of internet usage (Doni Septu Marsa & Yuyun 2018), this data becomes a primary data to developing and improving an e-learning system. But now in Pandemic, students allowed to come back to their hometown. Thereby, students took access to online learning from their hometown. The problems based on this condition were network stability and internet data capacity. We must remember that not all of our students live in places with excellent network stability and high data capacity and the tools they use for online learning. Not all of the students live in the center of the city. Some students live in a remote area that has an unstable network. So, as educators, we need a highly selective for choosing the best platform to achieve our goal as a standard education process (Goldie 2016).

This research focused on analyzing the various platforms' effectiveness in the online learning process during pandemic situations. Observation and spreading the questionnaires in all students of the pharmacy were emphasized during Pandemic. We also controlled and evaluated the learning process continuously. There are various researches to evaluate, assess, and improve technology in the online learning process during Pandemic, but none of our best knowledge focused on analyzing the multiple platforms used in online learning during this Pandemic. This research objective was analyzing the effectiveness of the various platforms used in online learning. Knowing the platform's effectiveness made us more comfortable choosing and considered the best platform in the online process based on our condition, as unique as COVID-19 Pandemic. Further, our researches aimed to answer the following questions: (1) what are the various platforms used in the educational delivery process during the pandemic era? (2) How effective the multiple platforms used in the online learning process in medicinal education during this Pandemic?

2 METHODS

2.1 Design

An inductive approach of qualitative descriptive was used in this research (Kemparaj & Chavan 2013; Liu 2016; Van Wyk 2012). For collecting the data, we used observation during online learning and questionnaires (Fathy & Morsy n.d.; Morgan, Pullon, Macdonald, McKinlay, & Gray 2017). Surveys were spread in Google form format (Rayhan et al. 2013) to all undergraduate medicinal faculty students at a specific time.

2.2 Participants

The purposive sampling was used in this research. All the medicinal faculty students in each grade at Universitas Hamzanwadi, east Lombok, participated in fulfilling the questionnaires. One hundred eighteen responses followed on this survey. They were 25 man and 93 women with 18–21 years old as interval age. They also have a different home town which spread in the various provinces in Indonesia, such as Sumatera, West Nusa Tenggara (NTB), and East Nusa Tenggara (NTT).

2.3 Data collection

Data collection were collected by observation and survey of questionnaires. The representation was investigated during online learning in this pandemic time. For inquiries, they were divided into three essential questions: (1) the various platforms were used in online learning during Pandemic; (2) tools were used in online learning; and (3) internet data capacity and network stability during education delivery.

2.4 Data analysis

Descriptive analysis and the percentage of the total samples (118 responses) used to analyze the data. The studies were included three significant questions of the questionnaires and observation during the learning process in pandemic time.

3 RESULTS AND DISCUSSION

3.1 The qualitative purpose description of observing the online learning process during pandemic

The learning process in the pharmacy faculty of Universitas Hamzanwadi has started on the ninth of March 2020. According to COVID-19 disease, Indonesia gave officially announcement of this disease on the first of March 2020 (Organization 2020). Thereby, Universitas Hamzanwadi has started to perform online learning as work from home on the sixteenth of March. Two weeks later, on the twenty-fourth of March 2020, West

Nusa Tenggara province gave officially announced that the first of their civil society positively infected with this virus (Ayuningtyas, Haq, & Utami 2020). Therefore, online learning from home continuously held until the end of the even semester. The uncertainty time at the end of this Pandemic made us give our most substantial and best effort to provide educational delivery and a normal process.

Based on this situation, the pharmacy faculty gave the form for monitoring the online learning process during Pandemic to their lectures. The lectures must fulfill the structure, in each subject after providing online learning in their classes. The form included the identity of the item and performing online learning. To identify the question, there were a name of lecture, a name of a subject, class/semester, date and time, a name of faculty and department, and time of evaluation. The dean and head of the department evaluated the process of online learning in each month. This evaluation's results were used to improve the quality of the online learning process in this pandemic time. Furthermore, performing online learning format included of a resume of subject matter, online platform learning, teaching materials, the number of participated students, learning assessment, and the problem or obstacle of online learning during Pandemic.

For the learning process as work from home, at least 12 meetings of 16 meetings each semester, including middle and final exams held in this Pandemic. There was no change as usual condition. There was also no regulation about this process of the education ministry. So, the sum of the meeting was the same as usual. The change was only the implementation that used learning from home. The results of the monitoring from the first meeting during Pandemic to the fifth meeting were:

a. The lectures and students still gave their best efforts to adapt in a new situations during a pandemic,
b. Educators even tried and errored the platform that used in learning from home,
c. Student problem in learning from home, including tools or hardware that they used, internet data capacity, network stability, especially students in the remote area, and the learning habit of students that rapidly changed.

Further the sixth meeting to the eighth meeting (before the middle exam), the lectures and students started to adapt in this challenging time. The problems were teaching materials, teaching methods, and approaches with various platforms, internet-data capacity, and network stability. The lectures must choose the best teaching methods and approaches for delivering their teaching materials to students (Crosby Joy 2000; Roehl, Reddy, & Shannon 2013). They had difficulties in evaluating their process and giving the students understanding equivalence with standard conditions. For the last meeting during the middle exam to the final exam, the lectures started to know which the best platform they used based on their situation. However, the kind of platform still made the problem based on internet data capacity and network stabilization. Thereby, we need more and more struggle for finding the best way to achieve our goal in the learning process during this Pandemic. Developing and improving delivery education were required at this time.

3.2 Number of the effectiveness of the various platforms in online learning during pandemic

There were three major questions of questionnaires that spread on Google form format. All responses of the students of pharmacy faculty participated in this survey. The survey held on 13th to 25th of July 2020 or two weeks after the final exam had finished. The results of the study were:

3.3 The various platforms were used in online learning during pandemic

In the pharmacy department of health faculty, delivery education in learning from home used various platforms, including virtual meetings (Zoom and Google meet), Google classroom, YouTube, WhatsApp, and Open Broadcast Software (OBS). Students can choose more than one platform in this survey. The descriptive analysis and percentage of the various platforms show in Table 1.

Table 1. The descriptive analysis and percentage of the various platforms during the pandemic.

The criteria of platforms	Sum and Percentage				
	WhatsApp	YouTube	Virtual Meeting	Google Classroom	OBS
Platforms online learning	123 (97.6%)	59 (46.8%)	47 (37.3%)	92 (73%)	–
The most frequently platforms online learning	120 (95.2%)	1 (0.8%)	–	5 (4%)	–
The most effective platforms online learning (students perception)	106 (84.8%)	7 (5.6%)	9 (7.2%)	3 (2.4%)	–

3.4 Tools were used in online learning

The students in the pharmacy department used the various tools in learning from home during this condition, such as laptop, smartphone, and notebook. The content of education delivery divided into text, photo or picture, sound, video, and PowerPoint slides combined with audio. The descriptive analysis and percentage of the various platforms show in the Figure 1.

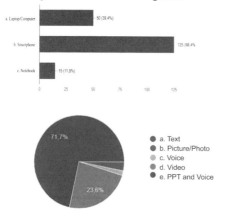

Figure 1. The descriptive analysis and percentage of tools and content were used in online learning during Pandemic.

3.5 The Internet data capacity and network stability of education delivery during pandemic

Pharmacy department students used internet capacity and network stability to access the platform used in this learning from home. With this data, we are able to know how internet data capacity and network stability performance in the different hometown of the students, the survey included internet data capacity, network stability, and accessibility, trust the influence of network stability and availability toward the learning comprehensive, and trust the impact of network stability and accessibility for the learning motivations. Descriptive analysis and percentage of internet data capacity used by students described in the Figure 2.

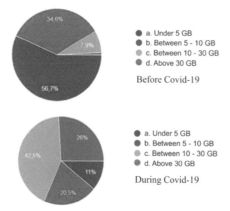

Figure 2. The internet data capacity before and after COVID-19.

Further, the descriptive analysis and percentage of network stability and accessibility show in Table 2.

Table 2. The descriptive analysis and percentage of network stability and accessibility during Pandemic

The criteria of network stability and accessibility	Sum and Percentage			
	Low	Middle	High	Very High
Network stability and accessibility in your hometown/place	36 (28.8%)	83 (66.4%)	5 (4.0%)	1 (0.8%)
Trust the influence of network stability and accessibility toward the learning comprehensive	5 (4%)	44 (35.2%)	48 (38.4%)	28 (22.4%)
Trust the influence of network stability and accessibility toward the learning motivations	6 (4.8%)	11 (8.7%)	55 (43.7%)	54 (42.9%)

Based on the data above, we explained the tool used by students in accessing online learning and the content of education delivery during this pandemic condition. Students used dominantly their mobile phone than a laptop or notebook in their learning process. This learning habit for using tools show equally as a reasonable condition or before COVID-19 Pandemic. Further, the content in education delivery, the students choose PowerPoint slides combined with sound with 71.4% and video with 23.8%. The data mean a consideration to lecture in arranging the learning material. A learning material should be in the PowerPoint combined sound. This learning habit shows that students wished online learning in a policy of learning from home the same as usual conditions. They still needed their lecture to guide them in a learning process. Also, the teacher must consider that their learning material should be accessed easier with a mobile phone. Mobile phone devices gave students the practice and convenience to access and participate in online learning during Pandemic. Therefore, mobile phone devices were dominantly used in education. This data also shows that developing knowledge and skill in information technology becomes necessary and improves continuously (Brewer & Brewer 2010).

There were various platforms in the pharmacy department, including virtual meetings (Zoom and Google meet), Google classroom, YouTube, WhatsApp, and Open Broadcast Software (OBS). Using the various platforms must consider the condition during the Pandemic. Every place has a different situation, so we need more effort to achieve the goal of learning eve in this pandemic time. It is not an easy thing for students to improve and increase their comprehensive knowledge. They also need a long time, and more struggling and continuously improving were required to achieve the learning process (Schmoker 2018). Based on various platforms above, students dominantly used WhatsApp platforms than other platforms in online learning. It can be explained below: (1) WhatsApp was more comfortable downloading in every device like a smartphone or laptop.;2) All of the students in the pharmacy department used this application to communicate in their daily life. So, the students were more familiar with this application and no need effort to operate it. (3) The students used this application as one of the learning media to share everything, including learning material before the Pandemic. So they feel more comfortable using this application in online learning in this pandemic time. (4) This application needed not high internet data capacity, network stability, and accessibility in running it. The student can download and saved the learning material when they have excellent network stability and availability. Further, they can learn the content without an internet connection. This condition was very appropriate for students in a remote area.

The data show that 28.8% of students have low network stability and accessibility to an internet connection in their hometown. Further, 66.4% of students have network stability and availability in the middle range. Moreover, only 4% of students have a good internet connection. This percentage described us why WhatsApp was dominantly used. The other platforms, such as Google class or virtual meeting (Zoom and Google Meet), need high internet data capacity and a stable network to the internet connection. For the internet data capacity, the data show a significant increase in using internet data capacity before and during Pandemic. This enhancement can be explained that students need high internet data capacity because all the subjects used online learning during this pandemic time. It is a painful condition because the higher internet data capacity used the higher cost financed of students. Moreover, network stability and accessibility influenced the student's trust in their learning comprehension and motivation.

Based on the explanation above, concerning the COVID-19 Pandemic, we measured that choosing platforms used in this pandemic time must consider the condition, especially the student's habit in learning, network stability, and accessibility to the internet connection. Thereby, using platforms such as WhatsApp was an appropriate method in delivering learning material during this pandemic time, especially for the students in a very remote area. Combining Power-Point slides with sound and uploaded in the WhatsApp platform can be an option in this challenging time.

Further research will be started in the odd semester by applying this measure to the pharmacy department in online learning during the Pandemic. We do not even know when this Pandemic is over. The uncertain time of this Pandemic, made us must continue developing and improving the quality of the learning process.

4 CONCLUSION

The pandemic situation made us become the faster learner. We must adapt to this rapidly changing in every whole life, including education. The online learning process must consider the condition, the habit learning of students, and network stability and accessibility to the internet connection. By knowing this condition, the lectures were expected to give their best effort to the learning process and achieve it. Furthermore, we must continuously develop the information technologies in the online learning process in this pandemic era to augment the new world of medicinal education.

ACKNOWLEDGEMENT

Our deeply grateful gave to all the pharmacy department students, health faculty of Universitas Hamzanwadi, who participated in this survey and research. This research did not receive any grant of public and commercial funding.

REFERENCES

Allcott, H., Boxell, L., Conway, J., Gentzkow, M., Thaler, M., & Yang, D. Y. 2020. Polarization and public health: Partisan differences in social distancing during the Coronavirus pandemic. NBER Working Paper, (w26946).

Ayuningtyas, D., Haq, H. U., & Utami, R. R. M. 2020. Initiating Global Civil Society as a Strategy for Handling the COVID-19 Public Health Threat: A Policy Review. Kesmas: National Public Health Journal, 15(2).

Brewer, P. D., & Brewer, K. L. 2010. Knowledge management, human resource management, and higher education: A theoretical model. Journal of Education for Business, 85(6), 330–335.

Chen, T., Peng, L., Yin, X., Rong, J., Yang, J., & Cong, G. 2020. Analysis of User Satisfaction with Online Education Platforms in China during the COVID-19 Pandemic. Healthcare, 8(3), 200. Multidisciplinary Digital Publishing Institute.

Crawford, J., Butler-Henderson, K., Rudolph, J., Malkawi, B., Glowatz, M., Burton, R., …Lam, S. 2020. COVID-19: 20 countries' higher education intra-period digital pedagogy responses. Journal of Applied Learning & Teaching, 3(1), 1–20.

Crosby Joy, R. M. H. 2000. AMEE Guide No 20: The good teacher is more than a lecturer-the twelve roles of the teacher. Medical Teacher, 22(4), 334–347.

Deacon, C., & Hajek, A. 2011. Student perceptions of the value of physics laboratories. International Journal of Science Education, 33(7), 943–977.

Doni Septu Marsa, I., & Yuyun, F. 2018. Preliminary Studies: The Influences of Internet Usage by Student in Developing E-Content of E-Learning System.

Dwivedi, Y. K., Hughes, D. L., Coombs, C., Constantiou, I., Duan, Y., Edwards, J. S., …Prashant, P. 2020. Impact of COVID-19 Pandemic on information management research and practice: Transforming education, work and life. International Journal of Information Management, 102211.

Education, A. C. for P. 2015. Accreditation standards and key elements for the professional program in pharmacy leading to the doctor of pharmacy degree. (Standards 2016). Accreditation Council for Pharmacy Education Chicago, IL.

Fathy, N., & Morsy, S. (n.d.). Chemical Structure, Quality Chemical Quality Indices and Bioactivity of Essential Oil Constituents Essential Oil Constituents.

Goldie, J. G. S. 2016. Connectivism: A knowledge learning theory for the digital age? Medical Teacher, 38(10), 1064–1069.

Haritani, H., Febriani, Y., Puspita, T., & Arviana, E. 2019. The Correlation of Undergraduate Course Research Experience and Critical Thinking Skills. 5(6), 336–347.

Holland, B. 2019. Factors and strategies that influence faculty involvement in public service. Building the Field of Higher Education Engagement: Foundational Ideas and Future Directions.

Huang, R. H., Liu, D. J., Tlili, A., Yang, J. F., & Wang, H. H. 2020. Handbook on facilitating flexible learning during educational disruption: The Chinese experience in maintaining undisrupted learning in COVID-19 Outbreak. Beijing: Smart Learning Institute of Beijing Normal University.

Kemparaj, U., & Chavan, S. 2013. Qualitative research: a brief description. Indian Journal of Medical Sciences, 67.

Liu, L. 2016. Using Generic Inductive Approach in Qualitative Educational Research: A Case Study Analysis. Journal of Education and Learning, 5(2), 129–135.

Martin, L. C., Donohoe, K. L., & Holdford, D. A. 2016. Decision-making and problem-solving approaches in pharmacy education. American Journal of Pharmaceutical Education, 80(3).

Maudsley, G., & Strivens, J. 2000. Promoting professional knowledge, experiential learning and critical thinking for medical students. Medical Education, 34(7), 535–544.

Morgan, S. J., Pullon, S. R. H., Macdonald, L. M., McKinlay, E. M., & Gray, B. V. 2017. Case study observational research: A framework for conducting case study research where observation data are the focus. Qualitative Health Research, 27(7), 1060–1068.

Nicola, M., Alsafi, Z., Sohrabi, C., Kerwan, A., Al-Jabir, A., Iosifidis, C., …Agha, R. 2020. The socio-economic implications of the coronavirus pandemic (COVID-19): A review. International Journal of Surgery (London, England), 78, 185.

Organization, W. H. 2020. Coronavirus disease 2019 (COVID-19)ý: situation report, 88.

Persky, A. M., Medina, M. S., & Castleberry, A. N. 2019. Developing critical thinking skills in pharmacy students. American Journal of Pharmaceutical Education, 83(2).

Plancher, K.D., Shanmugam, J.P., & Petterson, S.C. 2020. The Changing Face of Orthopedic Education: Searching for the New Reality After COVID-19. Arthroscopy, Sports Medicine, and Rehabilitation. https://doi.org/10.1016/j.asmr.2020.04.007

Rayhan, R.U., Zheng, Y., Uddin, E., Timbol, C., Adewuyi, O., & Baraniuk, J.N. 2013. Administer and collect medical questionnaires with Google documents: a simple, safe, and free system. Applied Medical Informatics, 33(3): 12.

Roehl, A., Reddy, S. L., & Shannon, G. J. 2013. The flipped classroom: An opportunity to engage millennial students through active learning strategies. Journal of Family & Consumer Sciences, 105(2): 44–49.

Schmoker, M. 2018. Focus: Elevating the essentials to radically improve student learning. Ascd.

Setiawan, A., Malik, A., Suhandi, A., & Permanasari, A. 2018. Effect of higher order thinking laboratory on the improvement of critical and creative thinking skills. IOP Conference Series: Materials Science and Engineering, 306(1), 12008. IOP Publishing.

Van Wyk, B. 2012. Research design and methods Part I. University of Western Cape.

Educational Innovation in Society 5.0 Era: Challenges and
Opportunities – Purnomo & Herwin (Eds)
© 2021 the authors, ISBN 978-1-032-05392-9

The effectiveness of the storybooks on the love of the homeland character trait for elementary schools

W. Wuryandani, Fathurrohman, E.K.E. Sartono, Suparlan & H. Prasetia
Universitas Negeri Yogyakarta, Sleman, Yogyakarta, Indonesia

ABSTRACT: The success of strengthening character values is supported by the message of learning messages to students. Learning media plays a role in whether or not the learning message reaches students. One of the mediums that can be used to convey character messages is storybooks. This study aims to examine the effectiveness of storybooks on the love of homeland character trait. This research is a quasi-experimental type with a pre-test post-test control group design. The subjects of this study were fourth-grade elementary school students. Data was collected using tests. The research instrument was a questionnaire on the scale of the love of homeland characters. The hypothesis test used is Wilcoxon. The results showed that storybooks' use was influential in the love of homeland character. There is a difference between groups of students who use storybooks and groups of students who do not use storybooks. The difference obtained from the significance test was 0.022, which means less than 0.05. From this it can be concluded that there is a difference between the pre-test and post-test results in the experimental class group.

1 INTRODUCTION

Character education is one aspect that must receive attention in the education process in schools, including elementary schools. Internalization of character values needs to give good values to students as a basis for them to behave in the future. Armed with good character values will lower their deviant behavior expectations within the family, school, and community. Character education in Indonesia is a program launched by the government that one implementation does mandate to school. Through the education process in schools, students are equipped with the moral values of good (Kamaruddin, S.A., 2012).

In line with these opinions, Suyitno et al. (2019) Character education is the effort made by the state (government), community, family, and education units to make the people of Indonesia a nation of noble character. Through this character education, fair values are internalized to form students with a noble character as a basis or guide them to behave. The values of good character are expected to be inherent in a person in daily life.

One of the character values that need to be developed today is the love of the homeland. Love for the homeland is a way of thinking, behaving, and acting that shows loyalty, concern, and high respect for language, the physical environment, socio-culture, economy, and national politics (Wibowo, 2012). The character of love for the country grown in the student's

condition begins with a love for the local culture in their area. Then, the students will get to know the components of love for students' homeland in their country.

The arrival of the globalization era has opened access for all people, including students, to access the world. All information is available, whenever and wherever they will be able to receive it. The information obtained will undoubtedly affect the daily behavior patterns of students. It is hoped that students' understanding of the outside world in the current era of globalization will not reduce their love for their homeland. Therefore, it is necessary to instill the character values of love for the homeland since school age.

Elementary school is a school level that takes longer than most other education levels. So that elementary school is the most extended place for students to study. The long duration of time to take learning in elementary school can internalize students' good character values. Internalization of character values to students can do through the various learning mediums used. Students' information through learning media becomes a provision for students' moral knowledge, which later becomes a guide to behavior.

Learning media plays an essential role in delivering information messages during the learning process. The role of learning media is as an intermediary for delivering messages from the message source to the message recipient, in this case, students.

DOI 10.1201/9781003206019-32

175

2 METHOD

This research is an experimental study with a non-equivalent pre-test design post-test control group design. The study conducted in class IV Demakijo public elementary school using one control class and one experimental class). The class given by the storybook is experimental. Data collection has carried out using a student self-assessment questionnaire related to the character of love for the homeland. Data analysis was performed using the Wilcoxon non-parametric statistical test because the data obtained did not meet the requirements to test using parametric statistics.

3 RESULT

This research conducted using two classes; the first is the control, and the second is the experiment group. The research begins by giving a pre-test for both the control and the experimental. The pre-test conduct to measure the love for the homeland of students in both groups.

The next step is to provide treatment for the experimental class using storybooks after the two classes gave a pre-test. This treatment gave in two lessons. After that, the students' love of their homeland character measured using a self-assessment questionnaire.

The pre-test and post-test results of the experimental class using storybooks show that there are differences in the questionnaire results on homeland character's love before and after the treatment. Researchers tested the difference between pre-test and post-test in the experimental class and got a significance test value of 0.022. The significance test value was 0.022, which means it was smaller than 0.05. The researchers concluded a difference in results between the experimental class before and after treatment based on the different test data.

Meanwhile, the pre and post-test results given to the control class that did not use the storybooks did not differ. It means that the homeland student character's love did not show any differences between before and after treatment. The results of the pre-test and post-test control class produced a significance value of 0.413 > 0.05. The researcher concluded no difference in the pre-test results for class 4B given X with the post-test result based on this value.

Based on these explanations, the researchers concluded that there are differences in measurement results homeland love the character of students between control and experimental classes. The control class shows no difference in the character of love for the students' homeland before and after the treatment. Meanwhile, the experimental class obtained data on the difference between before and after treatment using picture storybooks. This condition becomes the basis for the conclusion that picture stories are effective against the character of love for the homeland fourth-grade students of SDN. Demakijo, Yogyakarta City.

4 DISCUSSION

Character education is an important thing that needs to be emphasized by the teacher by implementing good character values since students take learning in schools, including elementary schools. Character education has the task of developing the moral knowledge, moral feelings, and moral behavior of students. Kamarudin (2012: 224) explains that character strengthening is an essential part of the education process's performance. Student characteristics form with the hope of being able to stick to him and become a guide in behavior. Therefore schools should not override the character education process in school programs.

The achievement of character education goals influences educators' messages about students' character values. Knowledge of character values is also part of the process of forming a student's character. For that, the delivery of learning information must use appropriate and diverse learning media. Through instructional media, communication between teachers and students becomes more effective (Naz & Akbar, 2008: 35).

Hope and homeland love character formation in instructional media storybooks will be based on a student's behavior to reflect the character values love for the homeland. Hudi (2017), in his research, explained that there is a correlation between the knowledge possessed by students and the moral behavior shown. Based on this research, students need to know their values when expecting students to love the country.

Berkowitz & Bier (2007) explain that one of the things in character education is choosing a pedagogical strategy to implement character education. In this study, the chosen pedagogical strategy involved the use of storybooks to implement learning. At the time, the storybook learning media is part of classroom activities, hopeful that strengthening the character of love for students will be more optimal than those who do not use storybooks.

The use of storybooks in character education reinforce by Turan & Ulutas's (2016) research findings. In their research, Turan and Ulutas revealed that picture storybooks effectively taught students to understand character education. This study stated that picture storybooks could be a medium for student literacy activities. The information received through the material in the illustrated storybooks that students read will provide moral knowledge to guide students' behavior.

The presence of storybooks is intended to attract students' interest in learning. Elementary school students will love the exciting pictures and stories. Therefore, the visual storybook media make by paying attention to these elementary school students' characteristics. Adipta, Hasanah, and Maryaeni (2016: 989) explained that the use of picture storybooks was sufficient enough to attract students' interest in learning. Furthermore, Suryani (2017) stated that reading interest significantly influences understanding the reading content in textbooks.

Regarding the picture book media used in the character education of the country's love, the picture book's material is also related to its love values. Students need to get reinforcement first by learning about local cultural values to love the homeland. Aspin Wasino & Yusuf (20210) explained that the value of love for the country is a form of cultivating character values in a person through language, culture, and education, which will produce ideas or ways of thinking to individuals, attitudes or behavior in respect, respect, love for the environment and the country.

Students find it easier to understand learning on broader material by utilizing the environment around students according to elementary school-age children's characteristics. Given that elementary school students find it easier to understand the material more broadly if starting from their immediate environment. The story is packaged in an exciting way to arouse students' curiosity about the environment in the area where they live to arouse students' curiosity about the storybook's material. Mustadi et al. (2017) explained in their research that a good understanding of students' concepts begins with good curiosity.

Based on the discussion of various kinds of literature above, picture storybooks are influential for learning the character of love for the homeland. Character strengthening through picture storybooks begins with students' understanding of local cultural diversity, so they have a sense of love for their region. After that, it is hoped that they will get to know other more expansive areas to get to know their country's full cultural diversity to provide them with a sense of love for their homeland.

5 CONCLUSION

Based on the results of the research findings and data analysis, it can be concluded that the picture book is effective for fostering the character of the love of the land for fifth-grade students. The use of picture storybooks provides knowledge about homeland love for students to underscore their behavior that reflects the values of love for the homeland in their daily lives. The introduction of the values of love for the country starts from the students' closest environment, then extends until finally getting to know their country as a whole. The picture storybooks are arranged according to the characteristics of elementary school students in order to be able to arouse students' interest in learning so that the love of the country contained in the book can reach students properly.

REFERENCES

Adipta H., Maryaeni, Hasanah, M. 2016. Pemanfaatan Buku Cerita Bergambar Sebagai Sumber Bacaan Siswa SD. *Jurnal Pendidikan: Teori, Penelitian, dan Pengembangan*, *1*(5): 989–992.

Aspin, P.Y., Wasino, & Yusuf, A., 2021. Values of Nationalism in the Extracurricular Activity of Boys Scout at Public Elementary School 3 Palu. *Journal of Primary Education*, *10*(1): 93–104.

Berkowitz, M.V., & Bier, M. C. 2007. What Works In Character Education. *Journal of Research in Character Education*, *5*(1): 29–48.

Hudi, I. 2017. Pengaruh Pengetahuan Moral Terhadap Perilaku Moral Pada Siswa SMP Negeri Kota Pekan Baru Berdasarkan Pendidikan Orangtua. *Jurnal Moral Kemasyarakatan*, *2*(1): 30–44.

Kamaruddin, S. A., 2012. Character Education and Students Social Behavior. *Journal of Education and Learning*, *6*(4): 223–230.

Mustadi, A., et al. 2017. Character-Based Reflective Picture Storybook: Improving Student's Social Self-Concept In Elementary School. *Cakrawala Pendidikan*, XXXVI (3): 369–381.

Naz, A.A. & Akbar, R.A., 2008. Use of Media for Effective Instruction its Importance: Some Consideration. *Journal of Elementary Education, 18*(1–2): 35–40.

Suryani, S.N. 2017. Pengaruh penguasaan sintaksis, tingkat pengetahuan dongeng, dan minat baca terhadap pemahaman bacaan dalam BSE. *LingTera*, *4*(1): 98–111.

Suyitno, et., all. 2019. Nationalism and integrity values in the teaching-learning process of mathematics at elementary school of Japan. *Journal of Physics: Conference*, 1321(2): 1–6.

Turan, F. & Ulutas, I. 2016. Using Storybooks as a Character Education Tools. *Journal of Education and Practice*, 7(15): 169–176.

Educational Innovation in Society 5.0 Era: Challenges and Opportunities – Purnomo & Herwin (Eds)
© 2021 the authors, ISBN 978-1-032-05392-9

Strengthening a student's character in the era of society 5.0 in primary school

Jamilah, T. Sukitman, & M. Ridwan
Primary School Teacher Education Study Program, STKIP PGRI Sumenep, East Java, Indonesia

ABSTRACT: Basically, education is one of the ways to improve the abilities of human beings. Therefore, in the long run, these people are expected to be characterized as adults and be able to live better independently. The use of technology and communication is identical to the social life of a community in the era of revolution 4.0 and the needs of society 5.0. This requires preventive actions against the negative impact of these developments so that individuals respond more wisely in dealing with changes. The negative impact of technological development in the form of television, printed media and social media leads to moral degradation in Indonesian people, such as increasing cases of corruption, harassment, humiliation, and so on. Therefore, character education is one of the methods used by the government through the new educational policies stated in the national education curriculum. Furthermore, the government has established primary education as the foundation for character building. This study reveals how to explore the strengthening of character education for students in the era of society 5.0 in primary schools through a literature review based on various sources.

1 INTRODUCTION

Strengthening character education according to the Presidential Regulation Number 87 of 2017 and the Regulations of the Ministry of Education and Culture Number 20 of 2018 is of the utmost importance and is intended to be implemented in schools to overcome the moral decadence of the younger generation; moreover, information along with the influence of information and communication technology changes rapidly (Dalyono, B dan Lestariningsih, 2016). Education is a crucial need of every individual.

At present, several technological developments that have occurred in Indonesia have affected moral decline. Both online and offline media have spread a lot of news that has affected student behavior towards learning, the ways to interact with friends, educators, and surroundings by going against social norms. This, of course requires everyone to think more critically than ever. One of the means for someone to develop a self-critical attitude is through character education in school (Gleason, 2018)

The Japanese government has initiated the concept of society 5.0, namely applying aspects of technology to facilitate the activities of human life. According to Skobelev, P and Borovik (2017), the concept of society 5.0 is not limited to solving problems in the manufacturing sector but also related to social problems with assistance integrating with physical spaces and virtual spaces. Society 5.0 impacts all sectors of life including education and technology.

According to Ahmad (2018), education in Indonesia currently uses online learning. One of the factors is the presence of strong and excellent technical ability

and professionals' abilities in their respective fields to carry out their professions. The behavior presented in mass-media news is in the form of insulting teachers, disrespecting friends, bullying others, student conflict, and even criminal acts.

Society 5.0 will have an impact on all aspects of human life, especially in the domain of education. This is in line with Risdianto (2019) who states that in the era of 5.0 in the learning process, teachers or educators not only focus on transferring knowledge but also on strengthening character education because both soft and hard skills cannot be replaced with any sophisticated technological tools (Judiani, 2010).

The results of research conducted by previous researchers show that most of them were still focused on the learning of character values in the form of a lesson plan therefore, the study was only at the level of concept and implementation as stated in the lesson plan. Based on the research results it can be seen that there were gaps in understanding among the concepts of character education, implementation, and its usefulness in facing technological developments as a form of readiness to face changes and developments in technology and social life.

Based on the explanation above, this study aims to explore character education and education as the initial foundation for character building in the era of society 5.0 by focusing on how characterbuilding techniques in primary schools are in line with the goals of the national education, character education, development of information technology, and society so that character building can be a natural consequence following the national goals on an ongoing basis.

178

DOI 10.1201/9781003206019-33

2 RESEARCH METHOD

This is a qualitative research paper using a character education strengthening approach in the era of society 5.0. This qualitative research aims to comprehend social phenomena through a holistic model and create deeper understanding (Sugiyono, 2013). The data in this study used a literature search method derived from various reference sources both from online and offline sources, which was done by following the focus of this study. The data sources of this study are textbooks, laws, curriculum 2013 (K-13), and other sources that back up the achievement of the objectives of this study. Based on the data found, we then carried out a deeper examination to find conclusions from statements and questions in this study.

3 RESULT

3.1 *Character education in basic education*

The application of character values must be adapted to the values of Pancasila, including religious values, honesty, tolerance, discipline, hardwork, creativity, independence, democracy, curiosity, a spirit of nationality, loving the motherland, appreciating achievement, communication, loving peace, loving to read, caring for the environment, social care, and being responsible.

The main values of those eighteen values of character education are religiousness, nationalism, independence, integrity, and mutual assistance. The students' character building and strengthening can be done appropriately if the main character assessment objective is not to assess the character of students, but to obtain whole and comprehensive information regarding the development of the students' character.

This is in line with the opinion of Lickona (1991) which states that character habits coming from thinking, habituation in the heart, and habituation in action forms good character from a knowledge of goodness, a desire for good, and from doing good. In other words, character building can be done through habituation. Character habituation is done from childhood to adulthood.

Schools can build characters as the focus of character building or strengthening at schools. This selected character becomes one of the school's missions in a certain period of time. For example Luqman Al-hakim Integral Primary School Sumenep which has a school motto, i.e. *a character with integrity* by setting honesty and independence as the character emphasis. The characters chosen are then used as an integrated character education program of the school. Luqman Alhakim Integral Primary School organizes an honesty canteen for its students when buying food and drinks.

At the school canteen, when the students buy food or drinks, the money must be put in the box provided according to the price of the food or drinks the students buy. In this way, students are given the freedom to buy and students are given the trust and trained to be honest in buying and selling transactions. If the students are found to be dishonest, then the students will receive punishment, such as being reproved individually in a neutral atmosphere to admit the mistakes they made.

Strengthening and character-building can be implemented as school culture. An effective process for building a school culture is to involve and invite all parties or stakeholders to jointly commit to strengthening the character learning process at schools. The main belief of the school must be focused on the effort to build character, manners, morals, values, and norms. There are many values that must be habituated at schools, such as caring and creative values, honesty, responsibility, discipline, health and hygiene, and caring for each other among school members. School is like a garden or fertile land where the seeds of these values are seeded and planted. For this reason, principals, teachers, and employees must focus on organizational efforts that lead to the above expectations.

3.2 *Society era 5.0*

Japan has initiated the launch of the concept of society 5.0, which is defined as a human-centered society that balances economic progress with solving social problems through a system that deeply integrates virtual and physical space. Furthermore, a significant difference between the era of industrial revolution 4.0 and society 5.0 is as follows: industrial revolution 4.0 demands connectivity in all aspects using the internet of things. Meanwhile, the concept of the society 5.0 era is new innovation from society 1.0 to society 5.0 in the history of human civilization.

Figure 1 shows that the 5.0 era focuses on the human component of society, while still using artificial intelligence as a tool and as media. The internet is not only used for information but also for living life. Society 5.0 is also an era where all technology is part of humans themselves. The development of technology can minimize the gap in humans and economic problems in the future, where this era offers a society centered on balance.

Figure 1. The learning model of society era 5.0.

The 4.0 industrial revolution is followed by the development of society 5.0, which is characterized by the following situations 1. Humans and everything else is connected to the internet, knowledge and information are quickly received from all corners, and finally new values in society will emerge 2. Social problems will arise and people will be freed from various problems 3. Artificial intelligence frees humans from the burden of processing large amounts of information 4. The use of robots and automatic machines can ease human work.

4 DISCUSSION

4.1 *Strengthening character education in society era 5.0*

Basic education character formation techniques, which are used as the basic foundation for character education in the era of revolution 4.0 and society 5.0 can be carried out through concrete actions, namely through activities carried out outside involving individuals (children) directly and/or through storytelling techniques. This technique requires measuring indicators that can be designed by educators according to the character that they want to put forward. The curriculum in this case determines the indicators to be developed by educators by following the specific circumstances and conditions in the field (Zurqoni; Retnawati, H; Arlinwibowo, J; and Apino, 2018).

This is in line with Salgues, (2018) who states that society 5.0 has the characteristics of the strength of science and where the community has various values: sustainability, inclusion, effectiveness, and the power of knowledge in a very integrated way. Therefore, every human being must have a strong character such as being able to adapt and making changes, creativity, good communication, lifelong learning, collaboration, innovation, and skillfullness.

In the era of society 5.0 students must have the highest skills needed to solve complex problems, think critically, and be creative. Strengthening these three core abilities is a big responsibility for practitioners of the world of education, especially educators in the learning process at schools. Based on this social phenomenon, the 2013 curiculum has revised curriculum emphasis learning on the integration of four abilities, namely: strengthening character education (PPK), literacy, creativity, critical thinking, communication, and collaboration (4C) and higher order thinking skills (HOTS). These four skills are important for students who live in the 21st century. Students are equipped with these skills to help them respond to various real problems in social life (Salgues, 2018)

To get to the high-level thinking stage students need to be equipped with literacy skills to make them easy to understand and process the information obtained to solve problems. Through this ability students can collect information by reading data from printed media and online as well as on websites via internet access. Providing students with this ability will greatly assist the output of educational institutions in understanding the various problems that exist in the era of society 5.0 by collecting, understanding and processing information to find the right solution.

By following the demands of the society 5.0 era in the 2013 revised curriculum learning, students are also trained in mastery of creative, critical, communicative, and collaborative thinking abilities (creativity, critical thinking, communication, and collaboration/4C). Mastery of these abilities can be realized by applying learning methods and models that provide opportunities for students to find knowledge concepts through activity-based learning.

This kind of learning process will encourage students to think creatively and critically. Learning can not only be done in the classroom but also outside the classroom, students are introduced to the complexity of problems in the real world, especially problems in the era of society 5.0. The problems raised are also not only focused on the problems in the surrounding environment but also in the universal environment that can be obtained through internet website facilities.This is so that they can learn to critically analyze existing problems and to find solutions for the problems by applying the learning concepts.

Educators as facilitators have a role to offer direction for students in finding solutions. The solutions offered are not outdated or old solutions, but solutions that have a novel value according to the context of the situation at hand. In the process this activity can take advantage of the development of industrial technology 4.0 to collect information and procurement of teaching materials and discussion material, such as using laptops, cellphones, and an internet connection to access learning videos. Digital transformation has changed and been implicated in various habits and ways of life in humans of various industries.

The habituation of HOTS can be obtained by introducing various real-world situations to students. By recognizing the real world, students can find out the various complexities of character education problems at schools.

By following the opinion of Yusnaini (2019), increasing the competence of students in building student character includes: critical thinking, creative thinking, interpersonal skills and communication, teamwork and collaboration, and confidence. The internet and computers are elements of the learning media used by students in the learning process. Learning can be done by online learning or through other platforms to support the online learning process. One of the obstacles is that the number of students is too large, causing various difficulties in the learning process which are still not evenly distributed in every society.

5 CONCLUSION

Parents, schools, and communities, as well as the government and all components of the nation are responsible for strengthening the character values of students in this era of 5.0 where information is flowing so fast and cannot be blocked. Therefore, strengthening character education is one of the answers. Students need such a strengthened character to face various novel aspects of life in the era of society 5.0. Strengthening character education will integrate the abilities of students so that they have the ability to solve problems and find the right solutions, and in the long run, bring prosperity to all.

REFERENCES

Ahmad, I. 2018. *Proses Pembelajaran Digital Dalam era Revolusi Industri 4.0*. Direktur Jenderal Pembelajaran dan Kemahasiswaan, Kemenristek Dikti.

Dalyono, B & Lestariningsih, E. D. 2016. Implementasi Penguatan Pendidikan Karakter Di Sekolah. *Bangun Reka Prima Majalah Ilmiah Pengembangan Rekayasa Sosial Dan Humaniora*, 3(2): 33–42.

Gleason, W. C. 2018. *Higher Education in The Fourth Industrial Revolution*. McMillan.

Judiani, S. 2010. Implementasi Pendidikan Karakter di Sekolah Dasar melalui Penguatan Pelaksanaan Kurikulum. *Jurnal Pendidikan Dan Kebudayaan*, 16(9): 280–289. https://doi.org/https://doi.org/10.24832/jpnk.v16i9.519

Lickona, T. 1991. *Education For Character*. Bantam Books.

Risdianto, E. 2019. Analisis Pendidikan Indonesia Di Era Revolusi Industri 4.0. *Akademia*, 07. https://www.academia.edu/38553914/Analisis_Pendidikan_Indonesia_Di-Era_Revolusi_Industri_4.0.pdf

Salgues, B. 2018. *Society 5.0: Industry of the Future, Technologies, Methods and ToolsNo Title*. ISTE, Ltd.

Skobelev, P., & Borovik, Y. 2017. On The Way From Industry 4.0 to Industry 5.0: From Digital Manufacturing To Digital Society. *International Scientific Research Journal Industry*, 307–3011.

Sugiyono. 2013. *Metode Penelitian Pendidikan, Pendekatan Kuantitatif, Pendekatan Kualitatif, Dan R & D*. Alfabeta.

Yusnaini, Y. 2019. Era Revolusi Industri 4.0: Tantangan dan Peluang Dalam Upaya Meningkatkan LiterasiPendidikano Title. *Seminar Nasional Program Pascasarjana Universitas PGRI Palembang*.

Zurqoni, Retnawati, H., Arlinwibowo, J. & Apino, E. 2018. Strategy and Implementation of Character Education in Senior High School and Vocational High School. *Journal of Social Studies Education Research*, 9(3): 370–397.

Educational Innovation in Society 5.0 Era: Challenges and
Opportunities – Purnomo & Herwin (Eds)
© 2021 the authors, ISBN 978-1-032-05392-9

Healthy school behaviour in public elementary school in Sanden Bantul

S.N. Isvandari & C.S.A. Jabar
Educational Management, Postgraduate Program, Yogyakarta State University, Yogyakarta, Indonesia

ABSTRACT: Healthy behaviour is the outcome of developing a healthy school behaviour that is consistently implemented for a school. School behaviour in elementary schools assists develops students' potential to be healthy individuals, both physically, psychologically and socially. Several schools have implemented healthy school behaviour, but only some of them have not been able to develop it. This is proven by the number of children who have unhealthy behaviour. This objective of the study is to explore healthy behaviour in elementary schools that may be sustainable be implemented for other schools. Data were collected through indepth interviews and observations. The data were analyzed by using qualitative analysis of Miles, Huberman, and Saldana. This study reveals that SD 2 Sanden and SD 2 Gadingharjo have been able to develop a clean and healthy life according to healthy school standards, they were able to provide facilities according to standards, providing nutritious healthy food, creating a healthy, clean, and conducive environment, providing counselling, and having effective waste management. This behaviour is formed through clean and healthy living habits with the implementation of the School Health Unit (UKS) TRIAS code of conduction. This implementation is supported by facilities that lead to habitual hygiene and healthy living behaviour for students and is supported by commitment and coordination between school as well as collaboration with related institutions. It may create a high sense of mutual trust and belonging to the school and has an aim in creating feelings as a family. This feeling will lead to an intimate, conducive and more responsible collaboration.

1 INTRODUCTION

The era of globalization made challenges for children that can threaten their physical and mental health if they are not developed of positive characters from an earlier age. Schools as educational institutions are responsible for educating, developing potential, and creating the positive characters of children (Wiyani, 2013: 21). Character is essential to build strong human beings. This character needs to be created as early as possible through family education and also developed through basic education. The basis for the implementation of character education is stated in Law Number 20 of 2003 concerning the national education system in article 3. One of the objectives of character education is to develop the potential of students to become healthy humans, both physically, psychologically, and socially. As a follow-up, it takes a strategic and planned role from schools for character building that can instil understanding and healthy living behaviour through healthy school behaviour.

Healthy school behaviour is behaviour from the government to put into functions of improving children's knowledge, abilities, attitudes, and health as a provision for improving the standard of living in a future life (Warwick et al., 2005; Lee, Tsang, Lee, & To, 2003; Passmore & Donovan, 2014). In its implementation, the development of healthy schools must include the provision of information, the formation of values and attitudes, the creation of activities that

support civilization, and the acquisition of skills that enable it to support behaviour change. So that, cultivating healthy schools will train people to develop and empower themselves so that they can improve their quality (Passmore & Donovan, 2014; Lee et al., 2003). Increasing the quality of self will have an impact on improving the social-economic performance of modern society (Lee et al., 2003). Therefore, this behaviour needs to be implemented as early as possible and carried out comprehensively through basic education (Passmore & Donovan 2014). Basic education is a place or foundation to introduce and develop a healthy life both physically, mentally and socially (Warwick et al., 2005; Passmore & Donovan, 2014).

This healthy school behaviour has been implemented in many countries for example in England, Australia, Poland, Czechoslovakia, New Zealand, Canada, and Indonesia (Warwick et al. 2005; Lee et al., 2003). However, in the implementation of this healthy school behaviour, there are several obstacles encountered. They are lack of infrastructure, funds, time, knowledge, lack of self-confidence from schools, and policies that are still weak and incomprehensive to develop this culture (Eggert et al., 2018). In supporting the behaviour of healthy schools, schools need curriculum adjustments, complete infrastructure, and collaboration with external partners (Passmore & Donovan 2014; Warwick et al., 2005; Lee et al., 2003; Mensink, Schwinghammer, & Smeets, 2012; Yun et al., 2017). Also, in implementing this behaviour,

182

DOI 10.1201/9781003206019-34

classroom teachers are the main character to creating learning opportunities internally and externally of the classroom and developing children's future capacities to grow (Lee et al., 2003). Teachers will assist children to develop positive self-esteem and personal and social skills. Therefore, teacher training is essential.

Indonesia, as one of the countries that has implemented this healthy school behaviour, has also experienced several similar obstacles. Some schools can implement and develop this behaviour well, but many of them are not yet able to develop it. Also, in this era of globalization, there are many challenges for children that can threaten their physical and mental health. Not a few children have unhealthy behaviour (Zubaidah, Ismanto, & Sulasmono 2017). Departemen Kesehatan (2010) also states that many health problems in elementary school-age children are usually caused by inadequate personal hygiene, a slum environment, and an unhealthy lifestyle. This statement is reinforced by the results of research by the Food and Drug Supervisory Agency in 2014 which also showed that 60% of elementary school-age children snack carelessly and do not meet safety quality standards. Research data also found the emergence of various diseases that often attack school-age children, among them is diarrhea (Maryuani, 2013). This is an indication of the importance of exploring a clean and healthy culture in elementary schools. SDN 2 Gadingharjo (1st place in a healthy school competition in 2017) and SDN 2 Sanden (1st place in a healthy school competition in 2016) can be used as a pilot of public elementary schools to give impact to other schools in Sanden.

2 METHOD

The study uses qualitative research methods with a phenomenological approach. This qualitative study was conducted to explore the meaning of the experience of independent healthy schools at SD Negeri 2 Sanden and SD Negeri 2 Gadingharjo. These two schools are schools that have successfully implemented the behaviour of healthy schools in Sanden District, Bantul Regency. They got awards for winning the regency level healthy school competition and being able to develop and maintain a culture that has been embedded. The participants of this study were the principal, teachers, school committee, parents of students, students, and health centres at SD Negeri 2 Sanden and SD Negeri 2 Gadingharjo. Data was collected using in-depth interviews and observation techniques. The instruments used were interview guidelines and observation guidelines. Data analysis was performed using the data analysis model Miles, Huberman, and Saldana which consisted of data reduction, data presentation, and conclusion.

3 RESULTS AND DISCUSSION

Generally, the form of school behaviour can be analyzed by the behaviour, behaviour habits, and physical form that is in school. So, schools need to prepare rules, activities and supporting facilities that lead to the behaviour of students' healthy hygiene behaviour. Based on the results of interviews and observations that have been carried out, it appears that the school has been able to develop a healthy school culture. School members must have healthy behavior, physically, psychologically and socially. Schools have established rules and supporting facilities. Moreover, the most important thing is that there is a strong commitment from the school in cultivating a healthy school. All the parties I interviewed felt that they had many positive things that they got from this culture. They also feel that this civilization has made schools more active, productive, and increased school performance. Further, the manifestations of the school culture at SD 2 Sanden and SD 2 Gadingharjo will be explained below.

3.1 Healthy school rules

SD 2 Sanden and SD 2 Gadingharjo are two schools in Sanden Sub-District that have been able to develop a healthy school behaviour. These two schools have their motto and slogan in developing a clean and healthy life. As for SD 2 Sanden's motto, "tesamatesa," find trash, put it in the trash can. Meanwhile, SD 2 Gadingharjo sparked the school's slogan, namely: "Clean my school, green my environment, healthy my body and soul, succeed in my achievement." This can remind and encourage children to always get used to a clean and healthy life.

Furthermore through mottos and slogans, the school also made a rules in written. These healthy school rules are made by the school together with the school committee and approved by the parents of students. These healthy school rules are not only applied at school but home as well. The school coordinates with students' parents through class teachers. So, the communication between the class teacher and the parents of the students is quite good. This communication is usually done using a book or direct communication. Parents feel it's very helpful in directing their children. Similarly, the teachers, through this cooperation the teacher feels that it is not in vain because the directions that are given at school are also controlled at the home.

3.2 Healthy school physical environment

Based on the results of the assessment of the healthy school competition, there is no doubt that the school has facilitated its citizens to live healthily, following existing standards. The observations also reveal that schools have a physical environment that supports a healthy school culture. The environment looks clean, comfortable, and conducive to teaching and learning activities. Schools have air circulation, adequate lighting and there are abundant sources of clean water. This school is also provided with UKS, healthy canteen, counseling room, life shop, living pharmacy, play area, hand washing area, sanitation, sortable trashcans and processing. Its supportive physical environment cannot be obtained instantly but through the hard work of all parties. The school collaborates with parents, health centers, alumni and other related parties to provide additional material, training, and other donations.

3.3 *Healthy school culture activities*

Activities in health culture at SD 2 Sanden and SD 2 Gadingharjo are carried out through the School Health Unit (UKS) culture (TRIAS School Health Unit (UKS), through health education, health services, and environmental development. Health education is carried out in learning physical education, sports and health. Healthy learning activities are allocated as 1 lesson hour per week for each class. Health services are implemented in schools to assist dealing with students who experience illness at school, however, if the School Health Unit (UKS) is not capable, they are referred to the public health center accompanied by the teacher's behaviour of healthy living habits and socialization are conveyed in the class involved in the subject and outside the classroom. Fostering a healthy school environment is carried out through environmental maintenance and maintenance of plants around the school which is carried out in collaboration with BPPT Sanden and the Women's Farmers Group (KWT). The form of cooperation is socialization and assistance in plant maintenance both from BPPT Sanden and from KWT. More specifically, habituation is carried out through the following activities.

1. Mosquito Nest Eradication Activities by SICAN-TIK (student larvae search and eradication)
   This activity is carried out by a little doctor who has been assigned to monitor larvae and their nesting sites, as well as to mobilize their friends to carry out the eradication of mosquito nests. This activity was motivated by the large number of people affected by dengue fever around the school.
2. PKS (School Security Patrol)
   The School Security Patrol is assigned to regulate the course of traffic near the school before coming and after school. The two schools are directly adjacent to the main road, therefore this activity is helpful.
3. Canteen "AMANAH"
   AMANAH canteen (safe, tasty, and halal) provides healthy and safe and halal snacks. It has been tested by the Bantul District Health Office and gained a certificate of good hygiene. The canteen manager has obtained a health snack training certificate from the Bantul Health Office. The provision of a healthy canteen in schools is supportive for children learning to choose healthy snacks.
4. Checking nails and hair lice
   It is carried out by little doctors, sports teachers, and UKS teachers on Mondays for grades 1 to 3, and Wednesdays from grade 4 to grade 6.
5. Pen Gayur
   It is an acronym for herbal and Vegetable Growers, created to provide knowledge and habituation to students about agriculture. This is following the

area around the school where the majority of the population is farmers.

6. Making Healthy Snacks for DONU Schools (purple sweet potato doughnuts)
   This activity is an activity to familiarize students with processing healthy snack foods according to food ingredients around the school.
7. Manufacture of Healthy Drinks DALIYA (DAWET ALOE VERA)
   This activity is an activity to familiarize students with processing healthy snack foods according to food ingredients around the school.
8. Making DANGSERA Healthy Drink (Red Lemongrass Drinking)
   This activity is an activity to familiarize students with processing healthy snack foods according to the food ingredients that are around the school.

All activities can be carried out well with the commitment, coordination and cooperation of school members. Furthermore, to support the achievement and smooth running of activities, the school also maintains a harmonious relationship with student guardians and school committees, the community around the school, and related government agencies. As has been implemented by SD 2 Sanden and SD 2 Gadingharjo, they have collaborated with community health centers, health offices, BB POM, BMT, PMI, waste managers, police, military, agricultural services, farmer groups, BLH, and local health colleges.

Table 1. Indication of the success of the habit of healthy behaviour in SD 2 Sanden

SD 2 Sanden

1. Teachers and employees are actively involved in assisting health services.
2. Decreasing Dengue Fever sufferers in the environment around SD 2 Sanden.
3. The implementation of the civilization of the doctors who monitor school children's snacks and the high interest of students to become cadres of "little" doctors.
4. The civilization of little dentists was implemented as evidenced by the decrease in dental problems of students at SD 2 Sanden.
5. Physical and spiritual cultivation starting with a handshake has been cultured.
6. The culture of 3R waste management independently by students and school residents has become a culture.
7. Students of SD 2 Sanden are ready, alert, and responsive to disasters.
8. A clean and healthy lifestyle has become a habit.
9. UKS services are carried out excellently.
10. The health status of SD 2 Sanden students increases.
11. Increased concern for the environment.
12. The use of medicinal plants is increasing.
13. The existence of a healthy canteen which provides nutritious and healthy food.
14. Clean and shady environment.
15. The realization of a clean and well-maintained bathroom.
16. Utilization of school land maximally by healthy school teams.

Table 2. Indication of the success of the habit of healthy behaviour in SD 2 Gadingharjo

SD 2 Gadingharjo

1. Decreasing Dengue Fever sufferers in SD 2 Gadingharjo.
2. The waste sorting and disposal process has been carried out at SD 2 Gadingharjo.
3. Checking hair and nail lice is a routine at SD 2 Gadingharjo
4. Implementation of PKS becomes routine every day
5. All students have a sense of belonging to the plants in the school, so as not to destroy them.
6. The percentage of students who were admitted due to illness decreased.
7. Students are accustomed to snacks in the school canteen rather than buying snacks outside of school.
8. Bathrooms are clean and well-maintained.
9. The environment is shady and comfortable.
10. Teachers and employees are actively involved in health services.

All parties said they were happy and lucky with the behaviour of healthy living in this school. There are so many benefits they get from this culture. They admit that students rarely get sick, students understand the importance of health and hygiene, student skills improve, the school environment becomes more conducive and comfortable for learning, creates a culture of healthy and clean living at home, and communication between parties becomes more intensive.

4 CONCLUSION

The result of developing healthy behaviour is a consistent increase in healthy behaviour. The development of healthy behaviour is carried out by making behavioral rules, supporting physical facilities, and habituating healthy living habits through activities that support health. These activities need to be supported by commitment, coordination and cooperation from school members in realizing a healthy culture. Also, there is a need for cooperation with related government agencies. In creating a healthy culture, it needs to

be accompanied by a high sense of mutual trust and belonging to the school and to have a goal of creating feelings as a family. With this feeling, an intimate, conducive and more responsible collaboration will emerge. All parties said they were happy and lucky with the culture of healthy living in this school.

REFERENCES

Departemen Kesehatan. 2010. *Riset Kesehatan Dasar.* Jakarta: Badan Penelitian dan Pengembangan Kesehatan Kementerian Kesehatan RI.

Eggert, E., Overby, H., McCormack, L., & Meendering, J. 2018. Use of a Model Wellness Policy May Not Increase the Strength and Comprehensiveness of Written School Wellness Policies. *The Journal of school health*, *88*(7): 516–523. https://doi.org/10.1111/josh.12635

Lee, A., Tsang, C., Lee, S. H., & To, C. Y. 2003. A comprehensive "Healthy Schools Programme" to promote school health: the Hong Kong experience in joining the efforts of health and education sectors. *Journal of epidemiology and community health*, *57*(3): 174–177. https://doi.org/10.1136/jech.57.3.174

Mensink, F., Schwinghammer, S. A., & Smeets, A. 2012. The Healthy School Canteen programme: a promising intervention to make the school food environment healthier. *Journal of environmental and public health*, *2012*, 415746. https://doi.org/10.1155/2012/415746

Maryuani. 2013. *Perilaku Hidup Bersih dan Sehat.* Yogyakarta: Nuha Medika.

Passmore, S., & Donovan, M. (2014). Health for Life in Primary Schools Program, United Kingdom: a Program Impact Pathways (PIP) analysis. *Food and nutrition bulletin*, *35*(3 Suppl), S154–S162. https://doi.org/10.1177/15648265140353S209

Warwick, I., Aggleton, P., Chase, E., Schagen, S., Blenkinsop, S., Schagen, I., Scott, E., & Eggers, M. 2005. Evaluating healthy schools: perceptions of impact among school-based respondents. *Health Education Research, 20*(6): 697–708.

Wiyani, N. A. 2013. *Membumikan Pendidikan Karakter di SD.* Yogyakarta: Ar-Ruzz Media.

Yun, Y. H., Kim, Y., Sim, J. A., Choi, S. H., Lim, C., & Kang, J. H. (2018). Development and validity testing of the school health score card. *Journal of School Health*, *88*(8), 569–575.

Zubaidah, S., Ismanto, B., & Sulasmono, B. S. 2017. Evaluasi program sekolah sehat di Sekolah Dasar Negeri. *Kelola: Jurnal Manajemen Pendidikan*, *4*(1): 72–82.

Educational Innovation in Society 5.0 Era: Challenges and Opportunities – Purnomo & Herwin (Eds)
© 2021 the authors, ISBN 978-1-032-05392-9

Cooperative learning model talking stick type: To improve speaking skills?

A.J. Verrawati, A. Mustadi & W. Wuryandani
Universitas Negeri Yogyakarta, Yogyakarta, Indonesia

ABSTRACT: This research is motivated by problems that occur in IA class SD Negeri 2 Percobaan, namely the average speaking ability of students included in the low category. The research objective is to improve students' speaking skills through the cooperative learning model of Talking Stick types. This research uses Classroom Action Research (CAR) using the Kemmis & Mc Taggart model. The subject of this research was class 1A of SD Negeri 2 Percobaan Yogyakarta. The results show that the average pre-cycle of speaking skills is 57, including skilled enough. After the action in the first cycle, speaking skills increase to 69.5 in the skill category. After the action in the second cycle, speaking skills increase again to 81.0 and are categorized as very successful. So, the conclusion of this research is that cooperative learning model of the Talking Stick type in each learning process can improve students' speaking skills.

1 INTRODUCTION

Speaking skills are important for elementary school students, because by speaking students will be able to convey ideas, ideas, and responses to other students, teachers, and other people. Talking has meaning, which is an act of delivering messages from participants (communicators and communicants). The purpose of speaking is so that the message from the speaker reaches the communicant. Speaking skill is a person's ability to transfer an idea, idea, or opinion to other people. Talking is a means of verbal communication to convey ideas, ideas, and opinions to others (Efrizal, 2012); (Robert E. Owens, 2012). Speaking requires language for its means of communication (Efrizal, 2012). Based on these opinions, speaking is the ability that a person has to convey, express, and express information in the form of ideas, ideas, opinions, and feelings to others using verbal communication.

Conditions in the field prove that at the elementary school level (SD), speaking skills are often difficult skills for students to master. This is supported by the results of observations at SD N 2 Percobaan Yogyakarta class IA with a total of 32 students carried out on 6–10 February 2018. The results of the observations resulted in data that the value of speaking skills and the level of student participation in learning speaking skills were in the fairly skilled category. Based on the observation data obtained, 44% of students who passed were scored ≥ 65 (complete) and 66% got a value <66 (not completed).

Learning conditions at SD N 2 Percobaan at the time of initial observation showed a lack of variation in learning that facilitates students to practice their speaking skills and a lack of habituation in the learning process which gives students the opportunity to speak. As a result, the speaking skills of class IA students at SD N 2 Percobaan Yogyakarta are still experiencing problems in improving them. The facts found in the field prove that teachers still play a central role in learning, this event is not very good, because students' thinking patterns will be very limited, whereas in the current era of globalization, various changes have occurred according to the demands and needs of society, and have challenges. in the form of local problems as well as international (global) problems that are happening so rapidly.

Based on the results of interviews with class I teachers, information was obtained that in the learning process, teachers had difficulty finding the right model or method to teach speaking skills well, so that the learning process was not optimal. Efforts to improve speaking skills are using the Talking Stick type of cooperative learning model. The Talking Stick cooperative learning model is expected to be able to overcome the problems that exist in students' speaking skills. This is in accordance with the research conducted by (Widiyanti, 2018) the results showed that the application of the talking stick model could improve the speaking skills of students in class V-3 at SD N Suryakencana CBM Sukabumi in the 2017/2018 academic year. Similar research was also conducted by (Antara, Kristiantari, & Suadnyana, 2019). The results showed that the Talking Stick learning model assisted with newspaper rubrics had an effect on the speaking skills of fifth grade students of SD Negeri cluster IX Kintamani for the 2017/2018 school year.

Cooperative learning is a learning model that prioritizes learning that focuses on students as a learning center and is interactive. This is supported by the opinion of (Barkley, 2012) which states that cooperative

186

DOI 10.1201/9781003206019-35

learning requires students to cooperate, interact by sharing information, and giving to one another. Cooperative learning models include cooperative learning type Talking Stick, think pair share model (exchanging pairs in pairs), round robin model (responding in turn), talking chips model (talking pieces), and others.

The Talking Stick type of cooperative learning model is a learning model that can encourage students to have the courage to express their opinions and speak in public. This is supported based on research by (Fajri, Yoesef, & Nur, 2016) which states that cooperative learning with the Talking Stick type is very appropriate for students from SD, SMP, and SMA/SMK. There are several basic elements in cooperative learning which is the Talking Stick type, which is to make students able to work together in groups and work independently and understand the material quickly (Shoimin, 2014). Students are trained in speaking skills when students have to answer questions from the teacher while holding a stick. Apart from practicing speaking, this learning model also creates a pleasant atmosphere and keeps students active.

The Talking Stick type of cooperative learning model is appropriate for the PAIKEM learning process (participatory, active, innovative, creative, effective, and fun). The cooperative learning model with the Talking Stick type makes students more courageous to speak and express opinions (Suprijono, 2016).

Some of the advantages of the Talking Stick type of cooperative learning model by Shoimin (Shoimin, 2014) is that it a) can test students preparation when learning, b) train students' abilities to be able to master the subject matter quickly, c) motivate students to learn before learning begins (students learn more actively) d) allow students to become confident to put forward ideas, information, ideas and responses. The Talking Stick type of cooperative learning model in the implementation of learning is that the teacher presents the material and the material can be studied in groups which makes it easier for children to exchange ideas to understand the subject matter. After the material given by the teacher has been completed by the students, the students will be asked questions by the teacher with the help of sticks. Students are trained in speaking skills when students must answer questions from the teacher when they get a chance to hold the stick. Apart from practicing speaking skills, this learning model also creates a pleasant atmosphere and keeps students active.

The cooperative learning model of the Talking Stick type makes students confident and courageous to express ideas, ideas, and opinions (Suprijono, 2015). The Talking Stick type of learning model trains students to dare to speak, namely by giving students the opportunity to answer questions given by the teacher. According to opinion (Shoimin, 2014), regarding the advantages of the Talking Stick type of cooperative learning model in addition to training students to speak up, it is also a model that can make students more active in learning.

Based on the facts in the field, it shows that the speaking skills of the IA class students of SD N 2 Percobaan still need a variety of learning models. A variety of learning models must be chosen according to the needs, lessons, and also the student's environment. Therefore, to overcome the problem of speaking skills of the IA class students of SD N 2 Percobaan Yogyakarta, which was still lacking, a study was conducted using a cooperative learning model with the Talking Stick type to improve the speaking skills of the IA class students of SD N 2 Percobaan Yogyakarta. Based on the facts obtained in the field, research was carried out using a cooperative learning model with the Talking Stick type in an effort to improve the speaking skills of the IA class students of SD N 2 Percobaan Yogyakarta.

2 METHODS

This research is a classroom action research using the research model developed by Stephen Kemmis and Robin Mc Taggart, which uses four research components in each step, namely planning, acting, observing, and reflecting (Wiriaatmadja, 2014). Kemmis and Mc Taggart's model. If visualized, Kemmis and Mc Taggart's model will look like Figure 1 below.

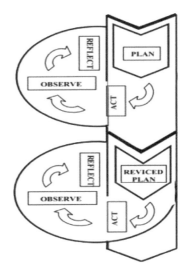

Figure 1. Classroom action research model Kemmis and Mc. Taggart.

The research was conducted from February to June 2018. The research site was SD N 2 Percobaan Yogyakarta. The research subjects were class IA SD N Percobaan 2 Yogyakarta. The number of subjects in this study were 32 students consisting of 18 male students and 14 female students with various student diversity.

This research is classroom action research. This study was designed using a spiral model cycle developed by Kemmis and Mc. Taggart (Wiriaatmadja, 2014) in each cycle will use four components of action namely (1) Plan, (2) Act, (3) Observe and (4) Reflect in an interlocking spiral. The use of the classroom action research model Kemmis and Mc Taggart's model in research is described as follows.

The first stage of planning, at this stage starting from finding problems that occur in the implementation of learning, then identifying and analyzing their feasibility to be overcome with classroom action research. The second stage is acting, the stage of implementing the action is implementing the plan that has been prepared at the planning stage. The third stage of observing, at this stage the teacher makes observations simultaneously with the implementation of learning carried out by researchers and students. The fourth stage of reflection reflection, the reflection stage is the stage where the researcher analyzes or evaluates the actions and the results are associated with the goals to be achieved (Suharsaputra, 2014). Reflection reads the implications of follow-up in the next cycle if the goal has not been achieved or stops the research for later reporting. Reflection must be carried out openly and fairly and by means of team discussions.

Percentage of students' speaking skills:

$$1 + \frac{n}{N} \times 100\%$$

Information:
n = scores obtained by students
N = the total number of students

Table 1. Determine the value in the assessment category according to (Arikunto, 2016) below.

Category	Value Range
Very skilled	80–100
Skilled	66–79
Quite skilled	56–65
Less Skilled	40–55
Very less skilled	30–39

3 RESULT AND DISCUSSION

3.1 Result

The implementation of learning speaking skills of students was surveyed during the pre-cycle. The initial survey activity was carried out to determine the initial conditions in the field regarding learning speaking skills. The results of this initial survey form the basis for relevant actions in an effort to overcome existing problems experienced by teachers and students. Observations and interviews were carried out in this initial survey. The survey was conducted on 6–10 February 2018 to see the learning process of speaking skills for class IA students at SD N Percobaan 2 Yogyakarta. The results of the initial survey activities of this study are as follows.

The survey results show several problems that occur during the learning process, namely students are still shy in expressing opinions, students still point to each other when they are required to come forward to convey opinions/discussion results, students still convey arguments/opinions that are very much in accordance with what is recorded in the book and have not been able to express opinions using the students 'own language, the students' voices when presenting the results of the discussion were also not loud enough.

Based on the results of observations and interviews, it was found that the students' speaking skills were still classified as low and had not achieved the goals of learning. The learning process activity is assessed based on attention, courage to appear in front of the class, cooperation, and the abilities of students related to speaking skills using Indonesian that are coherent, good, and correct. Based on all aspects of the assessment, it was obtained data that the percentage of success was 43.7% or 14 students who passed out of a total of 32 students. Efforts to overcome students' low speaking skills are by using a cooperative learning model with the Talking Stick type.

The research was conducted in 2 cycles, each of which consisted of 4 stages, namely: planning, doing, observation, and reflection. The results of the study proved that there was an increase in speaking skills of class IA students after the use of cooperative learning model Talking Stick type during the learning process.

Details of the comparison of student scores on the aspect of speaking skills in pre-cycle, cycle I, and also cycle II class IA SDN 2 Percobaan Yogyakarta can be seen in Figure 2 below.

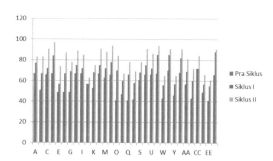

Figure 2. Comparison of student scores in the speaking skills aspect of students in pre-cycle, cycle I, and also cycle II.

Speaking skills from pre-cycle, then in cycle I, and finally in cycle II have increased as expected as an indicator of success, both in terms of pronunciation, intonation, diction and word choice, sentence structure, a calm and natural attitude, gesturing/mimicking correctly, volume, fluency, and mastery of the topic. The class average speaking skills in the nine aspects of students' speaking skills.

The use of a cooperative learning model with the Talking Stick type in thematic learning activities has a positive effect so that it can improve speaking skills

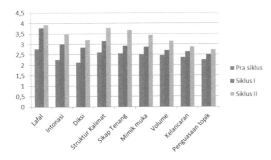

Figure 3. Improved speaking skills on nine indicators.

and increase speaking activities for class IA students of SD N Percobaan 2 Yogyakarta. Students' ability in speaking activities. The increase in the average value in the aspect of speaking skills was 43.8 (the average initial condition was 57.2, the increase in the first cycle was 69.5, increased to 81.0 in the second cycle).

The initial condition shows that the average score of speaking skills is still low, that is, 14 students are categorized as skilled and 18 students are categorized as less skilled. Some of the reasons are that the majority of students still have low motivation and interest in speaking activities, the atmosphere of learning activities is less fun and less conducive, and the use of models in the teaching and learning process carried out by the teacher does not direct or provide opportunities for students to be able to practice speaking skills.

After the first cycle of action, the average speaking skill score began to increase by 17.7 (initially 57.2 to 69.5). Based on existing value data from cycle I at the first, second, and third meetings combined, the average value is 69.5. In the first cycle it is known that 21 students have reached the skilled and highly skilled category with a value of ≥66. In this first cycle, from 5 aspects of speaking skills assessment, aspects of sentence structure and calm and natural attitudes have increased.

Some of the factors that increase the ability to speak are psychological factors of students (motivation and interest). In cycle I, students are more motivated to learn and have more opportunities to develop and practice speaking skills. These factors are closely related to aspects of fluency and accuracy, mastery of material during the learning process because each student must be able to master all learning topics by discussing with groups.

Although in cycle I there was an increase in student speaking ability, the success index in speaking skills was 75% of the total number of students that had not been fulfilled, and there were still some problems, namely mastery of the topic and optimal use of sentence structures. When discussing, several students discussed a topic that did not fit the context. So a revision is needed to continue in the second cycle with a time limit when discussing.

In cycle II, speaking skills increased by 16.8 (mean in cycle I 69.5 increased to 81.0). It is known that 28 students have completed their studies or 87.5% of the 75% indicators of success. In cycle II, the time to discuss the material from the teacher is more effective, the topics discussed are focused and optimal, most students have been able to follow the steps of the Talking Stick learning model well, and students have shown active speaking that meets the criteria of being very skilled. These results prove that Talking Sticks are a cooperative and fun learning model. The Talking Stick learning model is carried out with the assistance of a stick, the student who gets or holds the stick must be ready to speak to answer questions from the teacher after the students learn the subject matter.

3.2 *Discussion*

Based on the explanation of the research results above, the use of the Talking Stick type cooperative model in thematic learning activities has a positive effect so that it can improve students' speaking skills and increase speaking activities of class IA students of SD N 2 Percobaan Yogyakarta. Students' ability in speaking activities. The increase in speaking skills was 43.8 (the initial condition, mean 57.2, increased in the first cycle, namely 69.5, increased to 81.0).

In the initial conditions, the mean score of students' speaking skills was in the fairly skilled category. The reason is that the use of learning models carried out by the teacher has not yet referred to or provided opportunities for students to practice speaking skills.

After the first cycle of action, the students' speaking skills increased by 17.7 (initially 57.2 to 69.5). Based on the existing value data from cycle I, it is known that 21 students have finished learning or 65% of the 75% success indicator. In this first cycle, from 5 aspects of speaking skills assessment, aspects of sentence structure and calm and reasonable attitude have increased.

Some of the factors that cause students' speaking skills to increase are students' psychological factors (motivation and interest). In cycle I, students are more motivated to learn and have more opportunities to develop and practice speaking skills. These factors are closely related to aspects of fluency and accuracy, mastery of learning materials because each student must be able to master all learning topics by discussing with his group.

Even though in cycle I there was an increase in students' speaking skills, the success index of speaking skills, which is 75% of the total number of students, has not been fulfilled, and there are still several problems, namely mastery of the topic and the use of optimal sentence structure. When discussing, some students discuss topics that are out of context. So to overcome this, it is necessary to revise it to be continued in cycle II with time restrictions when discussing.

In cycle II, students' speaking skills increased by 16.8 (mean in cycle I 69.5 increased to 81.0). It is known that 28 students have finished learning or

87.5% of the 75% success indicator. In cycle II, the time to discuss the material provided by the teacher was more effective, the topics discussed were focused and optimal, most students were able to follow the steps of the Talking Stick learning model well, and students had shown activeness to speak that met the criteria of being very skilled. This is in accordance with the theory that the talking stick is a cooperative learning model. This learning model is carried out with the help of a stick, who is holding the stick must answer questions from the teacher after students learn the subject matter. Talking Stick learning is very suitable for elementary, junior high, and high school/ vocational students. In addition to practicing speaking, this learning will create a fun atmosphere and keep students active (Fajri, Yoesef, & Nur, 2016).

Overall, most of the students were categorized as skilled and highly skilled at speaking according to the criteria set in the speaking skills assessment. There are 4 students who have not achieved the success criteria because of several factors, including two students who are often sickly resulting in students being less enthusiastic in learning, and two students who are less enthusiastic and less interested when the discussion takes place.

Although in cycle I there was an increase in students' speaking ability, the success index in speaking skills was 75% of the total number of students that had not been fulfilled, and there were still some problems, namely mastery of the topic and optimal use of sentence structures. When discussing, some students discuss topics that do not fit the context. The effort to follow up on this problem is by carrying out a revision process to continue the second cycle with a time limit during discussion.

In the second cycle, speaking skills increased 16.8 (average in the first cycle 68.9 increased to 85.7). It is known that 28 students have completed their studies or 87.5% of the 75% indicators of success. In cycle II, the time to discuss the material provided by the teacher is more effective, the topics discussed are focused and optimal, most students have been able to follow the steps of the Talking Stick learning model well, and students have shown speaking activeness that meets the criteria of being very skilled. This is in accordance with the theory that the cooperative learning model with the Talking Stick type can improve speaking skills. The Talking Stick learning model uses a stick as a medium of assistance during the learning process, students who hold the stick are required to speak to present ideas or responses that are in accordance with the questions from the teacher after students learn the subject matter.

The use of a cooperative learning model with the Talking Stick type is very suitable for students of various school ages ranging from elementary school, junior high school and SMA/SMK. Learning the Talking Stick model trains students to speak, and learning is able to make students active and learning becomes more fun (Fajri, Yoesef, & Nur, 2016). Overall, most of the students were categorized as skilled and highly

skilled in speaking according to the success criteria set out in the speaking skills assessment, as many as 28 out of 32 students. There are four students who have not achieved the success criteria due to several factors, including two students who often get sick and cause students to be less enthusiastic in learning, and two students who are less enthusiastic and less interested when the discussion takes place.

4 CONCLUSION

Based on the results of this study, it was concluded that the effort to improve speaking skills on sticks of grade I A students at SD N 2 Percobaan Yogyakarta academic year 2017/2018 was successful by using a cooperative learning model with the talking type. This is evidenced by the results of the research which showed that there was an increase in speaking skills at the beginning before the cycle (pre-cycle), then in the first cycle that increased from the pre-cycle, and finally in the second cycle the speaking skills increased again in class IA and had met the indicators. success, that is, 75% of students are categorized as skilled/highly skilled. In this second cycle the research was stopped because it had met the indicators of research success.

Researchers provide suggestions to educators that it is important to apply and develop various learning models. The learning model makes the learning process more meaningful and can also develop students' speaking skills. Therefore, it is hoped that this research can become a reference for carrying out further research so that it can further optimize the application of creative and innovative learning models for the creation of effective learning.

The implication is based on the research results, namely that cooperative learning will further train students' abilities if learning uses a learning model that is able to make students more confident and fun. The Talking Stick is one type of cooperative learning model that is able to give students the motivation to dare to express opinions, ideas, and arguments because of the steps in the Talking Stick learning model which requires students to speak after getting sticks and questions from the teacher during the learning process. Based on the results of this study, it is expected to be able to become a reference for further research so that it can optimize the application of cooperative learning models that are able to train students' abilities and be able to improve students' abilities.

REFERENCES

Antara, I. N. P., Kristiantari, M. G. R., & Suadnyana, I. N. (2019). *Pengaruh Model Pembelajaran Talking Stick Berbantuan Rubrik Surat Kabar Terhadap Keterampilan Berbicara, 3*(4), 423–430.

Arikunto, S. (2016). *Prosedur Penelitian Suatu Pendekatan Praktek.* Jakarta: Rineka Cipta.

Barkley, E. E. (2012). *Collaborative Learning Techniques Teknik-teknik Pembelajaran Kolaboratif.* Bandung: Penerbit Nusa Media.

Efrizal, D. (2012). *Improving students ' speaking through communicative language teaching method at Mts Jaalhaq, Sentot Ali Basa Islamic Boarding School of Bengkulu , Indonesia.* International Journal of Hhmanities and Social Science, *2*(20), 127–134.

Fajri, N., Yoesef, A., & Nur, M. (2016). *Pengaruh Model Pembelajaran Kooperatif Tipe Talking Stick Dengan Strategi Joyful Learning Terhadap Prestasi Belajar Siswa Pada Mata Pelajaran Ips Kelas VII MTSN Meuraxa Banda Aceh.* Pendidikan Sejarah, *1*(1), 98–109.

Robert E. Owens, J. (2012). *Language development: an introduction (8rd ed)* (8rd ed). New York: Pearson.

Shoimin, A. (2014). *68 Model Pembelajaran Inovatif dalam Kurikulum 2013.* Yogyakarta: Ar-Ruzz Media.

Suharsaputra, U. (2014). *Metode Penelitian: Kuantitatif, Kualitatif dan Tindakan.* Bandung: PT. Refika Aditama.

Suprijono, A. (2015). *Cooperative Learning.* Yogyakarta: Pustaka Pelajar.

Widiyanti, D. (2018). *Penerapan model pembelajaran talking stick untuk meningkatkan keterampilan berbicara menyampaikan informasi di kelas tinggi.* Skripsi thesis: Universitas Muhammadiyah Sukabumi.

Wiriaatmadja, R. (2014). *Metode Penelitian Tindakan Kelas untuk Meningkatkan Kinerja Guru dan Dosen.* Bandung: PT. Remaja Rosdakarya.

Educational Innovation in Society 5.0 Era: Challenges and
Opportunities – Purnomo & Herwin (Eds)
© 2021 the authors, ISBN 978-1-032-05392-9

Building the critical thinking skills of elementary students through science thematic learning using a guided inquiry model

P. Pujiastuti & D. Rahmawati
Yogyakarta State University, Indonesia

ABSTRACT: The critical thinking skills of elementary students requires more comprehensive research. This research aimed to develop the critical thinking ability of students through thematic learning charged natural science use with a guided inquiry model at elementary school. The research subjects were 30 students from five classes at the Elementary School of Giwangan. The research design used a classroom action research cycle model that repeatedly and sustainable. This activity consists of (1) identification of problems, (2) planning, (3) learning implementation, (4) observation and reflection, and (5) conclusion. The data collection methods used to test, observation, and field notes. Data were analyzed descriptively quantitatively and qualitatively. The results showed that there was an increase in pre-cycle critical thinking skills reaching 53.87, increasing in the first cycle to 62.23, then in the second cycle reaching 78.50. The second cycle has met the success criteria determined by the class average at ≥ 75, so it was stopped.

1 INTRODUCTION

Implementation of the 2013 Curriculum is expected to realize the productive, creative, innovative, effective Indonesians through strengthening integrated attitudes, skills, and knowledge. Various studies, crystallization of ideas, and ideal concepts about education, curriculum development, preparation, and assignment of educators are intended to support their success. Currently, the 2013 curriculum has got into the implementation stage at the national level, and then the development level of the quality of learning is to apply thematic learning through a scientific approach. Thematic learning is an effort to combine knowledge in a comprehensive and integrated manner, integrating several subjects into one unit bound by themes (Fogarti, 1991). Thematic learning integrates various subject content, knowledge competencies, attitudes, and skills at the implementation of the 2013 curriculum. Therefore, thematic learning with a scientific approach can develop aspects of knowledge, skills, and attitudes in an integrated manner.

The teachers are required to be able to change the learning paradigm from teacher-centered to student-centered, rote learning to discover science concepts, teachers are expected to develop curriculae in each educational institution and apply learning models adapted to local environmental conditions, referring to the development of science and technology in the future. Learning models that can accommodate this paradigm shift are constructivist and collaborative learning models, namely guided inquiry learning. Critical thinking skills at students can be developed

through discussion, problem-solving, and activity that requires cognitive strategies and interactive classes, as well as active student learning. The critical thinking indicator (Costa, 1985) consists of 5 aspects, that elementary clarification, basic support, inference, advanced clarification, and strategic and tactics.

The three levels of inquiry were based on how the role of teachers and student's involvement. Level I inquiry is a basic "cookbook," that called guided inquiry or discovery because the students are carefully guided through the investigation up to the discovery process. An inquiry is level II, that the students decide and plan the investigation process to be carried out. Inquiry level III, the students recognized, problem identification, designing problem-solving processes, and conclude. The Inquiry model able to develop thinking skills. The Inquiry process (Schmidt, 2003) consists of problem identification, hypothesis, verification, report, and compare to other groups. Guided inquiry or discovery is a learning activity that emphasizes the process, which begins with the teacher asking the problem and solved by the student with instruction from the teacher.

The advantages of inquiry model are (1) intellectual potential is well developed, (2) intrinsic motive as a consequence of inquiry model, (3) the opportunity of guided inquiry, (4) guided inquiry can help memory retention (Carin and Sund, 1995), and (5) students centered. Inquiry learning emphasizes the cognitive process. The results of Wartono's research (2006) show that inquiry learning models more effectively develop the skills at Elementary School in the city. Starting from the description, the researcher felt the need to apply a guided inquiry learning model in science-based thematic learning to improve student's critical thinking skills. The aim of this research to develop critical thinking skills at thematic learning science use the inquiry learning model at elementary students.

2 METHODS

This research was conducted in the odd semester of the 2018/2019 academic year, at the elementary school of Giwangan Yogyakarta. The research subjects were 30 students of five class, and the research object is the improvement critical thinking skill of students. This research is a classroom action research that consists of planning, acting, observation, and reflection. The data collection methods used to test, observation, and field notes. The test method was measuring critical thinking skills, using multiple-choice test questions. Filed notes are carried to based on observations of events during the learning process. The data analysis uses a descriptive quantitative percentage to illustrate that the actions able to improve, increase, and change for the better when compared to the previous situation. The analysis of critical thinking skills using guided inquiry models. A formula of the analysis:

$$M = \frac{\Sigma fX}{N} \tag{1}$$

where M = average; $\sum$ fX = total value of all students; and N = total students.

The success criteria are if the students get an average critical thinking skills of at least 75. The value range table, the value of 80–100 in the very good category, 66–79 in the good category, the value of 56–65 in the sufficient category, the value for 40–45 in the poor category and 30–39 for failed category (Arikunto, 2006).

3 RESULT AND DISCUSSION

The learning carried out in 2 cycles, which is theme 2 "Clean air for health", consists of sub-theme 1 "how to body treats clean air" (cycle 1) and sub-theme 2 "the importance of clean air for breathing" (cycle 2).

3.1 Cycle 1

The action of one cycle consists of three meetings. The first meeting, students reading the textbook about clean air and answer the question from the mind map with the questions what, where, when, who, and how according to the worksheets. Then, followed by group activities observing the movements of the mouth and gill cover of the fish in the pond and then observation of video movements of the fish. The results of this observation were presented in groups. Then, experiments to find concepts the breathing requires oxygen using an animal respirometer. Results are reported in

writing and presentations. This experiment expected to provide experience to students that it makes it easier to learn so that learning becomes meaningful, gain concrete experiences (Twining, 1991), and conceptual knowledge helps students to connect fact with theory, additional meaning (Barrow, 2012). An inquiry familiar environment is effective at developing skills, training students' psychomotor skills (Wartono 2006).

In the second meeting, students are to do experiments by blowing clear lime water in a glass. This activity shows that breathing gives off carbon dioxide and water vapor. The activity was continued to calculate the respiratory rate in humans. Students compare it before the activity and after the activity. The next study examines the breathing regulation with singing. The third meeting, students to do observation use breathing apparatus model, identification of breathing organs in human beings, and breathing arrangements when singing. The results of students' critical thinking skill tests in pre-cycle and cycle 1 activities are in Table 1.

Table 1. Average critical thinking skills of students at pre-cycle and cycle 1

Aspect	Pre-Cycle	Cycle 1
Critical thinking skills	53.87	62.23

Based on Table 1, there was an increase in the achievement of critical thinking skills from pre-cycle (53.87) to cycle 1 (62.23). It shows that from the discussion, students provide explanations of their findings, analyze data, and make a conclusion, which is expected to train students' critical thinking skills. Costa (1985) states that the development of critical thinking skills requires interactive classes. Adding Naghl et al. (2012) that the students participate actively in learning to find gains, it is easier for students to understand the material being studied. For more details, the category of achievement of the results is presented in Table 2.

Table 2. Achievement category results test of students' critical thinking skills at pre-cycle and cycle 1

		Pre-cycle		Cycle 1	
Score	Category	Total Students	Percentage	Total Students	Percentage
80–100	Very good	1	3,33%	3	10,00%
66–79	Good	6	20,00%	11	36,57%
56–65	Sufficient	7	23,33%	8	26,67%

3.1.1 Observation and reflection

Based on learning at cycle 1, students enthusiastically participate in learning, do experiments, observe, have a group discussion, and working on group assignments. However, there are still some students who are less active and not participate. Teacher guidance is needed to guide students, especially those related to skills to analyze and work on worksheets. The results of the thinking ability test did not meet the success criteria, so the learning cycle 2 was continued with some improvements. The results of the evaluation and reflection in cycle 1 and improvement in cycle 2 are presented in Table 3.

Table 3. Results of evaluation and reflection cycle 1 and improvement in cycle 2

No	Results of evaluation and reflection cycle 1	Improvement in cycle 2
1	There are still some students who do not focus on learning	The number of a group member is reduced
2	There are still some students who have difficulty concluding the material	Guidance is carried out for groups that are experiencing difficulties
3	There are still some missing instructions that are not filled in working on worksheets	The instructions on the worksheets that must be filled in more emphasized

3.2 Cycle 2

Learning is carried out according to the material that has been prepared. Students read material about the causes of respiratory disorders in humans, then the discussion on the group about the topic. The results of group work are presented, then the students reading the text, making a question, and answer it. In the second meeting, the students conduct group discussions to answer the questions about respiratory disorders in humans, answer the questions, and then making a chart of human respiratory organ diseases and their symptoms. A critical thinking skills test was conducted at the end of cycle 2.

Table 4. The average critical thinking skills of students at pre-cycle, cycle 1 and cycle 2

Aspect that are measured	Pre-cycle	Cycle 1	Cycle 2
Critical thinking skills	53,87	62,23	78,50

For more details, the categories of the achievement of students's critical thinking skills in pre-cycle, cycle 1, and cycle 2 are presented in Figure 1.

Figure 1 shows that the highest number of students in the pre-cycle is in the poor category (16 students), cycle 1 is in a good category (11 students), as well as in cycle 2 (17 students). This shown an increase in critical thinking skills of students.

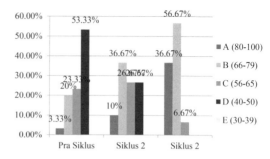

Figure 1. The achievement of students critical thinking skills at pre-cycle, cycle 1, and cycle 2.

3.2.1 Evaluation and reflection

Based on the learning observation and reflection at cycle 2, the students excited about participating in elementary school learning. The student's group work activities are carried out under teacher guidance. It's especially related to data analysis and concluding the experiment results that have finished. The results of the critical thinking skills test have met the success criteria ≥75 so that learning ends in cycle 2. The stages of the guided inquiry learning models consist of identification and determination if the problem scope, formulating a hypothesis, collecting data, analyzing data, interpreting data, hypothesis, and determining conclusions (Kinsvatter, 1996). Guided inquiry activities able to train critical thinking skills of students, so that learning becomes meaningful and it can be built to be improved. The teacher has the responsibility to encourage and facilitate students in developing critical thinking skills (Duruk & Akgun 2016).

4 CONCLUSION

The development of a student's critical thinking skills is indicated by an increase after followed learning using a guided inquiry model. Increase at pre-cycle reached 53.87, increased at cycle 1 to 62.23, and then in cycle 2 reached 78.50. the increase in cycle 1 to cycle 2 is due to improvements in learning, that reduced member group, intensive guidance for groups experiencing difficulties, and reinforce the instructions on the worksheets. The cycle was stopped because the success criteria fulfilled, it is the average of critical thinking skills and the value of the responsibility character ≥75.

REFERENCES

Arikunto, S. 2006. *Dasar-dasar Evaluasi Pendidikan*. Jakarta: Bumi Aksara.

Barrow, L.H. 2012. Helping Students Construct Understanding about Shadows. *Canadian Center Of Science and Education*, 1(2): 1888–191.

Carin, A.A dan Sund, B.R. 1995. *Teaching Science Through Discovery*. Columbus: Charles E, Meril Publishing Co. A Bell & Howel Company.

Challahan. J.R., Clark., L.H., Kellough. R., D., 1982. *Teaching in the Middle and Secondary Schools.* Macmillan Pubishing Company. Newyork.

Duruk, U & Akgun, A (2016). The Investigation of Pre-service Science Teacher Critical Thinking Dispositions In The Context Of Personal and Social Factors. *Science Educational International.* 27(1): 3–15.

Fogarti R. 2009. *How to Integrated the Curricula.* Corwin ASAGE Company: USA

Jufri, W. 2009. Peranan Perangkat Pembelajaran Berbasis Inkuiri Dan Implementasinya Dengan Strategi Kooperatif Terhadap Perkembangan Keterampilan Berpikir Kritis. *Jurnal Pendidikan Biologi,* 1(1): 87–92.

Kellough, R. D., Kellough, N. G. Dan Hough, D. L. 1993. *Middle School Teaching: Method and Resources.* New York: Macmillan Publishing Company.

Kemmis, S and Mc Taggart, R. 1990. The Action Reserch Planner. Deakin University.

Kinsvatter, R.D., Wilen, W. dan Ishler, M. 1996. *Dynamic Off Effective Teaching.* London: Longman Publisher USA.

Maleong, L.J. 2000. *Metodologi Penelitian Kuantitatif.* Bandung: PT Remaja Rosdakarya.

Naghl, M.G, Obadovic, D.Z. & Segedinac, M. 2012. Effective Teaching of Physics and Scientific Method. *TEM Journal,* 1(2): 88

Schmidt, S.M. 2003. Learning by Doing: Teaching the Process of Inquiry. *Science Scope.* 27(1): 27–30.

Slamet, Y. 2007. *Alternatif Pengembangan Kemampuan Berpikir Secara Nalar dan Kreatif Dalam Pembelajaran Bahasa Indonesia,* (online), http://www.uns.ac.id/cp/penelitian.php?act=det&idA=264, diakses 16 November 2009).

Slavin, R. E. 1995. *Cooperative Learning Theory, Research, and Practice.* London: Allyn and Bacon.

Educational Innovation in Society 5.0 Era: Challenges and
Opportunities – Purnomo & Herwin (Eds)
© 2021 the authors, ISBN 978-1-032-05392-9

Analysis of improving student's statistics thinking mathematic education

T.H. Nio & B. Manullang
Universitas Kristen Indonesia, Jakarta, Indonesia

H. Suyitno, Kartono & Sc. Maryani
Universitas Negeri Semarang, Semarang, Indonesia

ABSTRACT: All this time, there is not much information about the level of student statistical thinking and the increase in student learning achievement in class. Therefore, a study was conducted with the aim of explaining students' statistical thinking skills at each level of statistical thinking, and the achievement of the improvement during one semester of learning. The results will be used to develop lesson plans for teaching needs. In addition, there is an urgency to reform conventional learning into online classes that teach introductory statistics, how students learn statistics, and how to assess learning outcomes.

1   INTRODUCTION

In statistics learning, it is important to build high-level statistical thinking as recommended by statisticians and statistics educators in the American Statistics Association report (2016), so that through educational statistics all citizens are aware and able to make decisions (Vallecillos & Moreno 2002).

Statistics learning usually emphasizes computations, memorizing procedures, applying formulas and neglects the development of views that are integrated with statistical problem solving. Pfannkuch & Wild (2002) state that traditional teaching approaches, which focus on skills development, fail to produce statistical thinking skills.

The concept of statistics taught on campus rarely derives directly from the real problems at hand. Visible environment-oriented learning stiff, does not prioritize the recognition of real problems that arise on the surface. Students are less involved in finding learning materials that are appropriate to problems that are familiar with life. For example, describing congestion problems on the road, identifying variations that appear in student learning outcomes scores.

Apart from that, there are weaknesses in describing data, presenting data in the form of multiple tables and graphs, transitioning quantitative data into graphical data, quantifying data, making interpretations and conclusions in accordance with relevant facts. Even recognizing the procedure for making tables is not easy for students (Pfannkuch, Rubick, & Yoon 2002).

Furthermore, it is said that making conclusions is not based on existing data, incomplete conclusions, lack of understanding of the use of observational statistics with standard statistics so that it has implications for inaccurate and inconsistent conclusions with the

data, many students have not been able to identify samples and populations correctly, and 50 percent did not provide reasons for selecting samples.

A number of observational phenomena above appear repeatedly in introductory statistics learning every semester. It means that student statistics are generally low. This is in line with the findings of Martadiputra (2009) in his research on basic education mathematics teachers of PLPG certification participants in Lembang that the statistical thinking ability of teachers is classified as low.

Various definitions have been put forward regarding statistical thinking. Snee (1990) states that statistical thinking is the process of looking at the variation around and that variation is present in all work done. The statistical processes are interrelated when identifying, characterizing, quantifying, and reducing data that provide opportunities for improvement.

Wild & Plannkuch (1999) state thinking of statistics as art and touchstones in a statistician. This shows that statistical thinking has beauty, harmony and attracts attention like a work of art by experts, whose results can be seen by many people, but at the same time it is a big challenge that must be mastered to be used as a tool to solve real problems at hand.

According to Langrall & Mooney (2002) have compiled the characteristics of statistical thinking in a framework consisting of 4 (four) statistical processes to be performed in data handling, namely: describing data, organizing and reducing data, representing data, and analyzing and interpreting data.

Steel, Liermann & Guttorp (2019) made three recommendations for changes in the curricular of statistics as identified by a group of statistical activists which emphasized statistical thinking, more data and concepts, less theory, and encouraging active learning.

196

DOI 10.1201/9781003206019-37

From some of the definitions and recommendations stated above, the researcher concludes that statistical thinking is a mental activity to organize and manage data and variations in data that appear in work towards increasing better work results. On that basis, this research is directed to answer research questions: (1) what is the statistical thinking ability of students who take introductory statistics course. (2) how is the improvement and achievement of students' statistical thinking in the place where the research was carried out? The above problem is interesting to research, the results of which are used to develop lesson plans on teaching needs. In addition, there is an urgency to reform conventional learning into online classes to teach certain topics of introductory statistics.

A number of the problems above are interesting to research. However, in this study, it is necessary to disclose how the statistical thinking skills of students who take the introductory statistics course. Research findings will be used to develop lesson plans on teaching needs. In addition, there is an urgency to reform conventional learning into online classes to teach certain topics of introductory statistics.

The results of the research are used to improve the learning program for introductory statistics courses (lesson plan). The scope of the research is focused on what Shaughnessy, Garfield, & Greer (1996) argues: explaining data, organizing, presenting, analyzing and interpreting data, on simple linear correlation and regression material.

Four processes are the dimensions of compiling a statistical thinking ability data collection test. The domain of operational tasks from delMas (2002) is used as assessment items (tests) of statistical thinking, namely applying, criticizing, evaluating, and making generalizations from numerical or categorical data (Sudjana 1992).

2 METHOD

This type of research is expose facto with the subject of 60 mathematics education students in 2020. Because the number of students is limited, it purposively becomes the subject of research. The dimensions of the test instrument capture data on statistical thinking skills consisting of: explaining data, organizing data, presenting, analyzing and interpreting data, in simple linear correlation and regression material. The domain of the scope of tasks that characterizes statistical thinking by delMas (2002) is used and adjusted for statistical thinking test items, namely applying, criticizing, evaluating, and generalizing in handling data.

The quantitative data obtained through the test were analyzed to see the increase in pretest and posttest scores and the average achievement with the gain normality rule $<g>$, where $<n\text{-gain}> = $ posttest $-$pretest$/(SI -$ pretest), with SI (maximum ideal score) $= 100$. Then the posttest and pretest mean comparison test was carried out with the t statistic to see whether the average score increase was statistically significant (Hake 1998).

3 RESULTS AND DISCUSSION

The results showed that the increase in student statistical thinking was as follows There are as many as 5 percent have high abilities; 82 percent moderate; and 13 percent low. The depiction of statistical thinking improvement in students is shown in Table 1.

Table 1. Improved statistical thinking scores.

N-gain Interval	n	Category
$g \geq (\text{Mg} + \text{Sg}) = 30.802$	3 (5%)	high
$(\text{Mg} - \text{Sg} = 0.032) < g$ $< (\text{Mg} + \text{Sg}) = 30.802$	49 (82%)	moderate
$(\text{Mg} - \text{Sg} = 0.032) \leq g$	8 (13%)	low

Mg: mean gain; Sg: Standard deviation of the gain data.

In addition to the increase in thinking skills above, the achievement of learning outcomes shows an average normalized increase $<n\text{-gain}>$, see Table 2.

Table 2. Normalized average increase.

N-gain interval	n	Category
$0.70 < \text{n-gain} \leq 1$	1 (2 %)	high
$0.30 < \text{n-gain} \leq 0.70$	40 (67%)	moderate
$\text{n-gain} \leq 0.30$	19 (31 %)	low

Tables 2 and 3 show the average statistical thinking skills at the pretest and posttest, respectively 56.42 and 70.33, with a standard deviation of 14.76 and 14.23, respectively. These results indicate that the normalized gain average $<n\text{-g}>$ of 0.21, which means that the statistical thinking ability of the studied students is low. If we look closely, Kaizen thinks that statistics show as much as 2 percent high performance, 67 percent medium achievement, and 31 percent low achievement. If we look closely, Kaizen thinks that statistics show as much as 2 percent high performance, 67 percent medium achievement, and 31 percent low achievement. Based on these results it can be said that most of the students studied had the ability to think in the moderate category.

Table 3. Normalized average gain.

Group	N	Min	Max	Mean	SD	$<n\text{-g}>$
Pre test	60	25	95	56.42	14.76	0.21
Posttest	60	25	90	70.33	14.23	

Furthermore, analyzed the difference between the average posttest and pretest at a confidence level of 0.95. With the 2-tail t-test, the results obtained with the

SPSS output are the probability sig. $0.00 < 0.05$. With a comparative test between the pretest and posttest scores, the observation statistic is obtained $= 5.258$ for df: 58. The results show that the t-observation value is greater than the t-critical value $= 1.671$. It is concluded that the average difference between the posttest and pretest is statistically significant. This result confirms that the increase in statistical thinking skills of the students studied increases statistically significant.

Based on the statistical results obtained in Table 1, the increase in the score from pretest to posttest is generally classified as moderate. Based on the statistical results obtained in Table 1, the increase in the score from the pretest to posttest is generally classified as moderate. More specifically, the 5 percent finding was high; 82 percent moderate; and 13 percent low. This result strengthens the research findings of Pfannkuch et al. (2002) which state that the statistical thinking category of students is low.

The student's statistical thinking achievement is classified as low with $<n\text{-}g> = 0.21$. These results also support the finding that students' statistical thinking skills are low. This illustrates that weak statistical thinking skills have implications for making conclusions that are incompatible with the data or the contextual problems at hand.

Among the results put forward, there are several interesting things, namely there is an increase in the average statistical thinking ability of students from 56.42 to 70.33. This increase in results was statistically significant. Research findings can be used as input for lecturers who teach statistics courses to complete lesson plans in statistical introductory classes by developing statistical thinking for students.

Research findings can be used as input for lecturers who teach statistics courses to complete lesson plans in statistical introductory classes by developing statistical thinking for students.

4   CONCLUSION

Based on the analysis and discussion of the results of the research, it is generally concluded that the increase in statistical thinking skills of the students studied is moderate with low achievement; The average statistical thinking ability of students increased statistically

significantly, and the findings of this study are in accordance with the findings of previous researchers which confirms the low statistical thinking ability of students and mathematics teachers. It is recommended that introductory statistics teachers complement their lesson plans with the domain of statistical thinking.

REFERENCES

DelMas, R. C. 2002. Statistical literacy, reasoning, and thinking: A commentary. *Journal of Statistics Education, 10*(2): 1–11.

Hake, R. R. 1998. Interactive-engagement versus traditional methods: A six-thousand-student survey of mechanics test data for introductory physics courses. *American journal of Physics, 66*(1): 64–74.

Langrall, C. W., & Mooney, E. S. 2002. The development of a framework characterizing middle school students' statistical thinking. In *Sixth International Conference on Teaching Statistics (ICOTS6), Cape Town.*

Martadiputra, B.A.P. 2009. Kajian Tentang Kemampuan Melek Statistis (Statistical Literacy), Penalaran Statistis (Statistical Reasoning), Dan Berpikir Statistis (Statistical Thinking) Guru Smp/Sma (Studi Terhadap Guru Smp/Sma) yang mengikut kegiatan PPM Dosen Jurdikamat UPI di Kab. Subang dan peserta PLPG Sertifikasi Guru Guru Matematika SMP di BMI Lembang.

Pfannkuch, M., & Wild, C. 2002. Statistical thinking models. *The University of Auckland. NewZealand. ICOTS6.*

Pfannkuch, M., Rubick, A., & Yoon, C. 2002. Statistical thinking and transnumeration. *Mathematics Education in the South Pacific*, 567–574.

Shaughnessy, J. M., Garfield, J., & Greer, B. 1996. Data handling. In *International handbook of mathematics education* (pp. 205–237). Springer, Dordrecht.

Snee, R. D. 1990. Statistical thinking and its contribution to total quality. *The American Statistician, 44*(2): 116–121.

Steel, E. A., Liermann, M., & Guttorp, P. 2019. Beyond calculations: A course in statistical thinking. *The American Statistician, 73*(sup1): 392–401.

Sudjana. 1992. Metoda Statistik. Bandung: Penerbit Tarsito.

Wild, C. J., & Pfannkuch, M. 1999. Statistical thinking in empirical enquiry. *International statistical review, 67*(3): 223–248.

Vallecillos, A., & Moreno, A. 2002. Framework for instruction and assessment on elementary inferential statistics thinking. *Presentation at the Second International Conference on the Teaching of Mathematics, Crete, Greece, July. 2002.*

Educational Innovation in Society 5.0 Era: Challenges and Opportunities – Purnomo & Herwin (Eds)
© 2021 the authors, ISBN 978-1-032-05392-9

The effect of PORPE strategy to improve the understanding of the concept of social sciences in online learning

L.R. Hidayah, M.N. Wangid & S.P. Kawuryan
Graduate School, Yogyakarta State University, Yogyakarta, Indonesia

ABSTRACT: This study aims to examine how the effect of implementing the PORPE strategy improve understanding of the concept of social sciences in elementary schools through online learning during the Covid-19 pandemic. It involved 42 subjects of undergraduate students from elementary school teachers (PGSD) of Yogyakarta State University. This study used a quantitative approach with a quasi-experimental method of one group pre-test/post-test design. The data was collected through tests, observations, and documentation. The obtained data was analyzed using descriptive statistics which was preceded by a prerequisite analysis test in the form of a normality test. While the hypothesis was tested using the T-test. The result showed that the PORPE strategy had a significant effect on the understanding of the concept of social sciences in elementary school materials for PGSD students with the t-count value of 11.955 > t-table value of 2.021 and a significance value of $0.000 < 0.05$.

1 INTRODUCTION

Currently, reading skills have become the most frequently discussed issue along with the paradigm shift towards a literate society. Reading is the narrowest definition of literacy, but the measure of literacy is highly determined by a person's reading skills. Based on the 2018 PISA survey, Indonesia obtained a score of 371 and was ranked 6th from the bottom position. It indicates that the public's reading skills are still very low. In response to this, the implementation of education is required to consider the improvement of students' reading abilities and skills.

Reading is not merely to read from the start to the end of the text and it is not only to play with words. Reading is something complex and involves many aspects, not only reciting writing but also involving visual, thinking, psycholinguistic, and metacognitive activities (Crawley & Montain, in Rahim, 2007). In line with this, Leonhardt (in Natadjumena, 2006) states that reading activities should be fun without coercion, pressure, and formal assignment.

Concerning the demand for achievement, elementary school teacher (PGSD) students are expected to have a good conceptual understanding of what they read. It is because understanding a concept is an indicator of successful learning in the context of everyday life (Utami & Rohaeti, 2019). In this case, understanding is knowing the meaning, paraphrasing a concept (Bloom, in Widodo, 2006).

The indicators of conceptual understanding are interpreting, exemplifying, classifying, summarizing, inferring, comparing, and explaining (Anderson & Krathwohl, 2001). Therefore, understanding is not just remembering (Jacobsen et al., 2009) and it involves doing complex activities to understand factual knowledge such as recognizing, remembering, interpreting, summarizing, and explaining information (Anderson, in Muehleck et al., 2014).

Learning in social sciences cannot be separated from reading books or reading texts. The textbooks of social sciences have been known to be difficult to read due to their high reading skill requirements, long sentences, technical vocabulary, broad content (Apriyanti et al., 2017, Tyas, 2018), and their function as a major learning resource (Johnson, in Fry & Gosky, 2007). The PGSD students also felt the same way when they had to study social sciences teaching materials for elementary school levels to understand the content and prepare themselves to take part in learning evaluations by lecturers.

In this case, the field condition shows that some students' reading activities are less effective in which they only waste time to reading the text without really understanding the content. It can be seen from their response to the lecturer's questions. In this case, the use of learning strategies is needed for the acquisition of conceptual understanding (Anderson, in Alao & Guthrie, 1999). For example, students use learning strategies to learn concepts, theories, abilities, and skills, also to remember and understand different academic concepts (Pressley, et al. in Alao & Guthrie, 1999). By using the appropriate strategy, it is expected that students can achieve optimal results

DOI 10.1201/9781003206019-38

in their learning, especially related to understanding reading sources.

Lecturers do not need to teach their students how to read, because students can certainly read. However, lecturers need to encourage their students to be able to read effectively and to understand the content of the reading. One of the strategies that can be applied is PORPE (Predict, Organize, Rehearse, Practice, and Evaluate). Simpson & Hynd (1988) stated that PORPE is an independent learning strategy that operationalizes cognitive and metacognitive processes in which readers are effectively involved to understand and study the content. Moreover, PORPE can be a strong, durable, and efficient independent learning strategy for students to learn the concept of materials (Simpson & Stahl, 1987).

The PORPE strategy does not only focus on reading activities but also involves writing activities. It is intended to make the reader understand the content of the reading more properly. Besides, writing can be used to see a person's reading skills, for example in the essay test. This strategy can facilitate students who have concerns in writing their essays. Therefore, it can be said that reading and writing are closely related. Being unable to read means being unable to write (Barth, 1931).

The PORPE strategy is designed to help students in (1) actively designing, monitoring, and evaluating the reading material being studied; (2) studying the process involved in preparing for an essay test; and (3) using the writing process as a means to study materials of certain subjects. Furthermore, Simpson (in Tierney R. J. and Readence J., 2005) stated that PORPE was structured as a response to (1) the desire to see whether writers can be used as an independent learning strategy for various types of reading, and (2) students' anxiety and lack of knowledge to prepare for and face the essay test.

The implementation of PORPE is expected to form readers' reading habits indirectly. The five-step of the PORPE strategy requires students to behave like effective readers who have awareness and control over their cognitive activities as they read and learn. Baker and Brown (in Simpson, 1986) explained that effective readers are individuals who follow metacognitive abilities. The criteria for effective readers can be seen below.

1. Explain the purpose of reading
2. Identify important aspects of the reading message
3. Focus on the main content (idea) and not on trivial aspects
4. Monitor activities to investigate the understanding
5. Use your own questions to determine whether the reading goal has been achieved
6. Improve activities when finding disturbances in understanding

The PORPE strategy has five stages of learning that need to be organized. Referring to the experts, Simpson & Stahl (1987) formulates the stages of learning below.

1. Predict
   Predict potential essay questions to guide future research. A person is required to clarify the purpose of the reading activities being performed and focus on important points.
2. Organize
   Organize key ideas using their own words, structure, and methods. The ideas are manifested in the form of an outline such as a mind map or concept map.
3. Rehearse
   Practice with key ideas. At this stage, a person practices to memorize keywords that have been organized before. After practicing, the next stage is testing by answering essay questions orally. The focus at this stage is transferring information to brain memory.
4. Practice
   Practice the retrieval of key ideas in independent writing assignments that require analytical thinking skills. At this stage, a person is required to rewrite the predictive answer for the essay that has been compiled.
5. Evaluate
   Evaluate the completeness, accuracy, and suitability of their written product for the essay questions and self-predicted answers.

Those five stages of the PORPE strategy are synergistic as they support each other and guide students through the necessary processes to read and study the content of the material (Simpson & Hynd, 1988).

Learning during the Covid-19 pandemic has changed the learning system that has been used so far. There is a shift from traditional learning to online learning, also known as e-learning, and web-based learning (Cappel & Hayen, 2004). The concept of e-learning relies on an independent learning approach, but it is monitored according to the PORPE strategy criteria as an independent learning strategy. Thus, e-learning can also be said to be learning that supports the concept of lifelong learning. Through this learning, students get their learning materials through electronic media such as the internet, intranet, extranet, satellite, audio/video equipment, and CD (Sulčlč & Lesjak, 2009). Particularly, e-learning is intended as an effort to transform the learning process at schools or colleges into a digital form bridged by internet access (Munir, 2012).

The presence of online learning cannot be denied and it has various advantages (Karwati, 2014), namely:

1. For students
   E-learning enables the development of a high degree of learning flexibility. Teaching materials can be accessed at any time and repeatedly. So can the interaction between the student and the lecturer. Therefore, students can further strengthen their understanding or mastery of the material.
2. For lecturers
   E-learning provides benefits to lecturers, particularly in updating teaching materials, as a means

of self-development or research to broaden horizons, controlling student learning activities, checking assignment fulfillment, checking answers, and informing the results to the students.

Various research on the PORPE strategies has been carried out, particularly regarding reading and writing activities. M. L. Simpson (1986), the initiator of this strategy, found that the application of PORPE provides significant changes to the results of essay tests or even multiple-choice tests and can be implemented in various fields of scientific studies. It is also in accordance with the situation in the field where the implementation of learning evaluation is presented in the form of multiple-choice questions and essays.

Based on the description above, this present study aims to examine the effect of implementing the PORPE strategy on understanding the concept of social sciences in online learning for PGSD undergraduate students of Yogyakarta State University.

The research hypothesis can be seen below.

Ho: There is no significant effect between the implementation of the PORPE strategy and students' conceptual understanding.

Ha: There is a significant effect between the implementation of the PORPE strategy and the students' conceptual understanding.

2 METHOD

This quantitative research used an experimental research method. According to (Sugiyono, 2012), experimental research is research to find the effect of certain treatments on others under controllable conditions. It used one group pre-test/post-test design in which it only involved students from one class without a control class. The treatment was carried out between the pretest and posttest implementation.

This study was conducted from the 15th–30th of April 2020 with two treatments. Considering that the strategy has many stages and requires a pause between stages, the implementation of each treatment was divided into two meetings. Online learning was carried out with the help of the Google Classroom and Quizziz platforms.

This study involved 42 subjects of undergraduate students from Class 2B PGSD of Yogyakarta State University. The reading material consisted of two types. First, it deals with the use of maps and globes in elementary schools. Second, it relates to literacy and strategies for reading social sciences materials.

The data was collected through tests, observations, and documentation. The pre-test and post-test had a total of 30 questions consisting of 20 multiple choice questions and 10 essay questions. The questions were developed based on the indicators of conceptual understanding as proposed by Anderson & Krathwohl (2001). The observation was to identify whether the learning process is in accordance with the instrument developed from the stages of the PORPE strategy as

proposed by M. L. Simpson, (1986). Meanwhile, the documentation was to support the findings.

The obtained data were analyzed using the normality test of the Kolmogorov test by comparing the significance value of 0.05. Meanwhile, the hypothesis test and conclusion used the Paired Sample T-test.

3 RESULTS AND DISCUSSION

3.1 *Results*

Based on the results of observations, all students actively participated in learning. Each stage of the PORPE strategy was carried out properly and thoroughly as evidenced by the submission of each activity from each stage. The results of the pre-test showed that the students' average score reached 49.67. Meanwhile, the post-test which was carried out after the treatment showed that the students' average score was 72.48. The results of the pre-test and post-test indicate an increased value. However, no student manages to achieve the maximum score.

Table 1. Descriptive statistics

Component	Obtained Data	
	Pretest	Posttest
N	42	42
Mean	49.67	72.68
Standard Deviation	9.343	72.48
Maximum	74	98
Minimum	30	43

The score obtained from the pre-test and post-test was then tested to identify the normality of the data and the results showed that the data were normally distributed with the normality test results > 0.05, with the pre-test value of 0.077 and post-test value of 0.200. After finding that the data were normally distributed, the Parametric Test was carried out through the Paired sample t-test to test the hypothesis.

The paired sample t-test found that the t-count value of 11.955 is higher than the t-table value of 2.021 with a significant value of 0.000 < 0.05. It means that Ho is rejected and Ha is accepted. In other words, the PORPE strategy has a significant effect on the students' understanding of the concept of social science material.

3.2 *Discussion*

Based on the results of the study, the use of strategies is necessary to improve the quality of learning. Schools need to ensure that students can come up with their learning strategies appropriately (Wegner et al., 2013). Therefore, the application of learning strategies can positively affect learning.

Table 2. Results of paired sample t-test

Pre-test-						95% Confidence Interval of the Difference	
Post-test	t	df	Sig. (2-tailed)	Mean	Std. Error Mean	Lower	Upper
	−11.955	41	0.000	−22.810	1.908	−26.663	−18.956

*Processed data using SPSS 23.0

Reading activities are only understood as spelling word by word thus far. Thus, it is not surprising that it cannot make readers understand the contents of the reading text. Ideally, reading comprehension depends on several cognitive and linguistic processes (Muijselaar et al., 2017). Moreover, Pressley (2002) revealed that good text comprehension arises if the reader can predict what the text is about, relate information in the text to their background knowledge, ask questions, monitor the understanding of the text, and summarize the text being read. Generally, all of these principles have been implemented in each PORPE stage in this study. Therefore, the increase in understanding of the concept of the materials using the PORPE strategy is due to the implementation of the treatment in which previously reading activities were only in the form of appeals or assignments and now turns to structured cognitive and metacognitive activities.

The use of the PORPE strategy shows that there is an increase in the post-test score which is presented in multiple-choice questions and essays. M. L. Simpson et al., (1989) stated that PORPE is prepared to answer multiple-choice questions well. Moreover, M. L. Simpson (1986) added that PORPE can be an answer to students' concerns in answering essay questions. Therefore, it can be said that the PORPE strategy is an appropriate learning strategy to prepare students to answer multiple-choice questions and essays.

The implementation of the PORPE strategy in online learning can be said appropriate and suitable considering that both highlight the concepts of lifelong learning and independent learning (M. L. Simpson et al., 1988). Therefore, this strategy is highly recommended for students to academically succeed (M. L. Simpson & Hynd, 1988).

4 CONCLUSION

The PORPE strategy is one of the reading strategies focusing on independent learning and it is applied as preparation for various learning evaluations. This strategy has the same characteristics as online learning and college students. Based on the result of the analysis, it can be concluded that the PORPE strategy in online learning has a significant effect to increase understanding of the concept of social sciences material for undergraduate students. Implementing this strategy

requires habituation for optimal academic achievement considering that this strategy is used in reading activities which are the most basic learning activities before learning others.

REFERENCES

Alao, S., & Guthrie, J. T. 1999. Predicting conceptual understanding with cognitive and motivational variables. *Journal of Educational Research*, *92*(4): 243–254.

Anderson, L. W., & Krathwohl, D. R. 2001. *A Taxonomy for Learning, Teaching, and Assessing*. New York: Longman.

Apriyanti, L., Pargito, & Pujiati. 2017. Pengembangan Buku Bacaan pada Mata Pelajaran IPS untuk Meningkatkan Aktivitas Belajar. *Jurnal Studi Sosial*, *5*(2): 3–10.

Barth, J. L. 1931. *Methods of Instruction in Social Studies Education* (3 ed.). Lanham: University Press of America.

Cappel, J. J., & Hayen, R. L. 2004. Evaluating e-learning: A case study. *Journal of Computer Information Systems*, *44*(4): 49–56.

Fry, S. W., & Gosky, R. 2007. Supporting social studies reading comprehension with an electronic pop-up dictionary. *Journal of Research on Technology in Education*, *40*(2): 127–139.

Jacobsen, D. A., Eggen, P., & Kauchak, D. 2009. *Methods for teahcing: Metode-metode pengajaran meningkatkan belajarn siswa TK-SMA*. Yogyakarta: Pustaka Pelajar.

Karwati, E. 2014. Pengaruh Pembelajaran Elektronik (E-Learning) terhadap Mutu Belajar Mahasiswa. *Jurnal Penelitian Komunikasi*.

Muehleck, J. K., Smith, C. L., & Allen, J. M. 2014. Understanding the Advising Learning Process Using Learning Taxonomies. *NACADA Journal*, *34*(2): 63–74.

Muijselaar, M. M. ., Swart, N. M., Planting, E. G. S., Droop, M., Verhoeven, L., & De Jong, P. F. 2017. Developmental Relations Between Reading Comprehension and Reading Strategies. *Scientific Studies of Reading*, *21*(3): 1–16.

Munir. 2012. Multimedia Konsep & Aplikasi Dalam Pendidikan. Bandung: Alfabeta.

Natadjumena, R. 2006. *Perpustakaan Sekolah-Lahan Tidur Pustakawan*.

Pressley, M. 2002. Metacognition and self-regulated comprehension. In *What Reserach has to say about reading instruction* (3 ed., hal. 291–309). International Reading Association.

Rahim, F. 2007. *Pengajaran Membaca di Sekolah Dasar*. Jakarta: Bumi Aksara.

Simpson, M. L. 1986. PORPE?: A Writing Strategy for Studying and Learning in the Content Areas. *Wiley on behalf of the International Reading Association Stable*, *29*(5): 407–414.

Simpson, M. L., Hayes, C. G., Stahl, N. A., Connor, R. T., & Weaver, D. 1988. An initial validation of a study strategy system. *Journal of reading behavior*, *20*(2): 149–180.

Simpson, M. L., & Hynd, C. 1988. An initial validation of a strategy for studying narrative text. *Journal of College Reading and Learning*, *21*(1): 41–47.

Simpson, M. L., Stahl, N. A., & Hayes, C. G. 1989. PORPE: A Research Validation. *Journal of Reading*, *33*(1): 22–28.

Simpson, M., & Stahl, N. A. 1987. Porpe: A comprehensive study strategy utilizing self-assigned writing. *Journal of College Reading and Learning*, *20*(1): 51–57.

Sugiyono. 2012. Metode Penelitian Kuantitatif, Kualitatif dan R & D. Bandung: Alfabeta. *Metode Penelitian Kuantitatif, Kualitatif dan R & D.Bandung:Alfabeta.*

Sulčlč, V., & Lesjak, D. 2009. E-learning and study effectiveness. *Journal of Computer Information Systems*, *49*(3): 40–47.

Tierney R. J. and Readence J. 2005. *Reading Strategies And Practices* (6 ed.). Pearson Education.

Tyas, K.K. M. 2018. Penggunaan Novel Sejarah sebagai Sumber Belajar Sejarah. *Pendidikan Sejarah*, *7*(2): 85–103.

Utami, R.P., & Rohaeti, E. 2019. Students' Concept Understanding in Chemistry Learning Using Macromedia Flash Based Inquiry Learning. *International Journal on New Trends in Education and Their*, *10*, 1–12.

Wegner, C., Minnaert, L., & Strehlke, F. 2013. The importance of learning strategies and how the project "Kolombus-Kids" promotes the successfully. *European Journal of Science and Mathematics Education*, *1*(3): 137–143.

Widodo, A. 2006. *Revisi Taksonomi Bloom dan Pengembangan Butir Soal* (hal. 1–14). Universitas Pendidikan Indonesia.

Educational Innovation in Society 5.0 Era: Challenges and Opportunities – Purnomo & Herwin (Eds)
© 2021 the authors, ISBN 978-1-032-05392-9

A case study on the implementation of marketing competencies for deaf children in entrepreneurship education

E. Bunyanuddin & N. Azizah

Special Needs Educational Program, Graduate School, Yogyakarta State University, Indonesia

ABSTRACT: This study aims to determine the aspects related to marketing competence developed in deaf children in entrepreneurship education. It uses the case study approach applying in-depth interview techniques in its data collection. The results of this study indicate that teachers and deaf children discuss and conduct surveys as a need for research in society. The teacher invites the deaf children to analyze existing competitors' products. Deaf children are also encouraged to use digital applications to expand their marketing range. The teacher also guides them on how to serve customers so they are satisfied.

1 INTRODUCTION

Entrepreneurship is a modern phenomenon. Globally, a country that is developed and has good competitiveness is determined by its economic strength (Lake, 2018). Moreover, free trade creates a growing need for product trade between countries. Of course, the conditions for each of these products also need to compete fiercely with economic actors in the destination country.

The intense competition creates a competitive culture. Countries that have not yet prepared for the development of entrepreneurs can fall into an abundance of imported goods. This can happen to such an extent that unhealthy imports and exports can be, in the aggregate, risks to the country's financial balance (Xu, Lian & Zheng, 2017) not only for imported products, but also with the absorption of employment. The manufacturing of products can actually open up significant job opportunities compared to selling imported products. Therefore, there is a need to prepare the community to become entrepreneurs.

Currently, the program to prepare people to become entrepreneurs is known as entrepreneurship education. Entrepreneurship education is often carried out in community groups, universities and even elementary schools. The development of entrepreneurship education is quite good. It is supported by the government, which issued regulations that encourage entrepreneurship education. Although in practice there are several concepts, from apprenticeships and incubator concepts to project-based concepts that are also used (Purnomo, 2015).

The target of entrepreneurship education in schools is all students, including deaf children. Deaf children who experience communication problems due to their hearing impairment have the potential to become

entrepreneurs. Visual capital, communication between communities, and being able to think logically provides opportunities for deaf children develop into entrepreneurs.

In Indonesia, the goal of entrepreneurship education is the ability to manage and develop small and medium enterprises (Nasution & Lubis, 2018). A realistic target is many markets in which to buy products, supported by current technology which makes it easier for people to shop. In the midst of a large market and technological advances, a marketing strategy is needed so that products can be accepted by consumers (Indriani, 2006). For example, a producer will stop the production if the product does not sell because the marketing does not reach potential consumers. So, in entrepreneurship education, it is necessary to teach marketing to all students, including deaf children.

2 LITERATURE REVIEW

2.1 *Marketing competence*

Marketing is the key to doing business. A company's product is already good, but if it cannot be marketed or sold, it becomes a waste. Likewise, a system that has been built in the business will have difficulty working if there are no sales due to bad marketing. So, it is not uncommon for the marketing department in a business to sometimes require large funding in order to achieve sales targets.

The scope of marketing capabilities includes market research, assessing markets, marketing products and services, persuasion, making people excited about the business ideas that are established, and serving customers (Lakceus, 2015). Meanwhile, the marketing ability according to Rentz, Shepherd, Tashcian,

204

DOI 10.1201/9781003206019-39

Dabholkar and Ladd (2013) is divided into several indicators, two of which are sales skills and technical knowledge.

Sales ability is described as the ability to attract customers, fulfill customer requirements, create relationships with customers, close sales, create sales messages, and improve marketing. Meanwhile, marketing technical knowledge skills include knowledge of customer and product markets, company procedures, competitors' products, services, marketing policies, and product knowledge, including features and benefits, product arrangement, employee training, imagination in supplying products, and services that meet customer needs.

Marketing competence also has an important role in product development. Sales market research will provide buyers with the choices they need. Products that do not meet the needs of the buyers will not be sold. Systematically, marketing capability also spurs business people to compete according to market expectations and compete against existing competitors.

In addition, marketing education must use an educational model that is able to accommodate the rapidly changing world, such as social media (Faulds & Mangold in Crittenden & Crittenden, 2015). An educational approach that uses technology will make it easier for students to be skilled and innovative so that they can enter the market gaps that exist in marketing in atypical and productive ways. Of course, how to market in the real and virtual world is different. So, marketing education requires the involvement of technology, because the use of educational technology has a positive impact on the learning process (Dowell & Small, 2011).

2.2 Deaf children

Limited language development contributes to behavioral problems in the deaf, and research shows that poor sign language and verbal skills are linked to social difficulties. When the level of sign language and oral language skills is high, social difficulties are recorded as no greater than children with normal hearing (Fellinger, Holzinger, & Pollard, 2012).

This is because the biggest problem faced by deaf people is not only the inability to hear, but also the acquisition of language in the way the majority of people do, even though this method is an efficient choice (Brinkley, 2011). Linguistic obstruction also affects cognitive development, world knowledge, and social function, all of which affect each other cumulatively over time (Knoors & Marschark, 2014).

The inability to hear also results in the inability to produce sounds unlike other children who do not have hearing impairments. Losing the ability to hear in deaf children, in addition to limiting the perception of speech, will also result in an inability to monitor the production of language and speech (Sunardi & Sunaryo, 2007).

This can be seen in the language development of deaf children that stops at the spelling stage. The spelling stage is the stage of repeating syllable-like sounds, such as *ma ma ma*. This repetition is prompted by environmental sounds around the child. For ordinary children, there will be a process of repeating the pronunciations they hear, which is also followed by the listening process. Hearing the sound, they will try repeatedly to improve sound production.

In deaf children, this process is different because the inability to hear results in the inability to improve the way of producing sound repeatedly. So, it becomes natural that the language skills and voice production in deaf children will not be the same as that of the ordinary children due to difficulties in the process of correcting the sounds produced. The ability to produce sound is good enough for deaf children when it comes to producing sounds.

However, for all educational backgrounds, the main challenge in educating deaf children is meeting their communication needs. More than 95% of children with a hearing impairment have parents who speak and hear, but because of their hearing loss, children with a hearing impairment have limited access to spoken language. Thus, a large proportion of deaf children arrive at school with a significant delay in language development compared to their hearing peers (Knoors & Marschark, 2014).

It is understood that deaf children will experience language deficiency and difficulty producing sounds. Even so, the language skills of deaf children can be compensated for with oral and sign language, which is the language that relies on the modal sense of sight. Close communication between deaf children communities also provides added value for deaf children to reduce the impact of language deficiency. The language deficiency of children with hearing impairment does not rule out the possibility of becoming entrepreneurs.

3 MATERIAL AND METHODOLOGY

This research is qualitative in nature with a case study research approach. The case is in the form of entrepreneurship education, especially marketing competence. The research was carried out at SLB N Cilacap, Jawa Tengah, Indonesia. The consideration of the research setting is that the implementation of entrepreneurship education has been going on for more than a decade and is one of the largest special needs schools in the region. The research subjects were five teachers who taught entrepreneurship education to deaf children. The criteria for selecting research subjects are that they have taught entrepreneurship for at least five years so that they can provide complete information. The data collection was carried out through interviews with the teachers who taught entrepreneurship education. The data analysis was performed using qualitative descriptive techniques.

4 FINDINGS AND DISCUSSION

4.1 *Findings*

4.1.1 *Market research*

Market research is the first finding revealed in entrepreneurship education for deaf children. It serves as the stage for determining the prospects for the product to be released to the public.

> "We then read the opportunity too; if the local community is still mostly traditional, then we will also follow the traditional needs of the community. If the skill orientation is art, which is an elementary school orientation and we are in high school, the skill is oriented towards independence" (Teacher-5)
> "In the past, we tried everything. There was wood. There was a workshop. There was carpentry. These things were done in accordance with the needs of the community and adapted to the abilities of the child" (Teacher-4).

Marketing material begins by observing the needs or desires of the community for a product. The results of the observations are then analyzed so that they have the potential to become ready-to-sell products. Market research is also carried out by experimenting with various types of businesses according to the ability of the children to those that are acceptable to the community.

4.1.2 *Use of digital applications*

The ease of technology that is currently available is also used as teaching material to expand marketing coverage.

> "I have an IG that I share with you. Now, that's for marketing too. These are the batiks. There are continuous deliveries. We gathered yesterday for vocational, for entrepreneurial skills. This CSE Production, Cilacap Special Education, is done to accommodate all vocational results" (Teacher-1).
> "Yes, there is lots of media. There is Tokopedia. Yes, we introduce it and shopping-based applications. There is also Shopee. At most, we will do this so that the promotion doesn't have to be from outside because if you leave, the children are difficult. This is because we understand that the direct technique is difficult" (Teacher-2).

The use of digital applications is through the teacher's stage of creating an account on Instagram. Then, deaf children are guided to use and optimize the account as an online marketing medium to complement the limitations of face-to-face communication.

4.1.3 *Competitor product analysis*

The efforts to maintain competitiveness are also necessary to test product strengths and weaknesses, including knowing and analyzing competitors' products.

> "With the existing educational media, let's say we want to make ginger, you know, for example, it tastes like this. Now they know what this is like. So, children with hearing impairment must know what it looks like. We take the children to the stalls. We first bring the product. The ginger there (which is in the shop) is cheaper, actually cheaper, 1,000 per sachet. But we try to wrap it more beautifully. We also provide a tester. We also explain how we use the heater, and we provide the tester, we say 'please, if anyone wants to try, just try it.' To overcome competitors which are cheaper and are packed in plastic sachets, the teachers of the Cilacap State special school also teach children with disabilities strategies to stay superior, saying their price is 1,000, cheaper, and our price is 1,500. Maybe they are large in scale, and maybe they have economical packaging. Yes, they have been dredged, neatly, without being damaged. Even though there is competition, we are still running even though the sales are not that big. But we can still sell it, because our customers are not the people who don't know, but those who have tried our products" (Teacher-4).

This analysis finds the advantages and disadvantages of the product, and then it can be improved with the products that will be made by children with a hearing impairment so as to add their value and be able to compete in the market.

4.1.4 *Good service to customers*

Service to customers needs to be done in various ways, such as using the right language. Especially for children with hearing impairment, sign language and oral language can be taught. This service is important to maintain customer satisfaction and it aims to make customers want to shop again.

> "Both of them, if you are here, between friends, prefer a gesture. But for me, sometimes it's oral signals too. On a small level, he was taught orally, looking at lips. Not accustomed to total cues. But we also give them the knowledge that gestures are also important, and if they meet deaf people outside, they can communicate with them" (Teacher-1).
> "Both, when a high school student is older. We emphasize, 'you are already in touch with the community. So you have to talk too; they can understand. Between friends, it's a gesture, and it's okay. But if you talk to a teacher, lips must speak. But sometimes there are children who get sick if forced to speak. But there are also voices coming out, and we can follow the expression and read lips. Even if they can't, they can communicate through writing if they have a cellphone now. If they wanted to write, they have to look for paper first; otherwise they have to write on the ground. If there is no writing tools at all, they write on the ground.

Now, if you bring your own cellphone, you can write it down. Even if the language is reversed" (Teacher-4).

"Teaching convinces the children that they have to be brave first. What is your name? Your product is good. You have to explain it. Try to tell them this because they won't know. Who knows, teachers might know, but outsiders do not necessarily know. So, we are educated, yes, in combination, in a gesture and implicitly" (Teacher-2).

Sign language is mostly used for children with a hearing impairment or for customers who are able to use sign language. Meanwhile, oral language or those that prioritize facial expressions is intended for customers who do not understand sign language. Written language is used when the customer does not understand sign language or oral language, usually by typing on a cellphone.

4.1.5 *Take persuasive action*
Persuasive action can be interpreted as an effort to persuade customers to believe in buying or using services from deaf children.

"Oh, for example, yesterday I got the handover order, right? That's what I didn't ask to accept. I kept making notes for the children, but I told you beforehand. For example, the cost of six delivery boxes. How many details do you have? Later, you will explain it, including yesterday there were those, for example the price of one box of service was 50,000. The mother was given a discount of 10,000 per box. This is how we stay rich. Yes, the point is that children are sometimes just not confident in themselves; just accompanied by their mother. 'Say! You talk. Just wait. Just wait and feel confident. That means, if later you don't know, you can be left behind like that. But if you order it yourself, sometimes you won't be'" (Teacher-3).

"Usually, call the teacher. If there is no teacher, say, 'I don't know.' They are not confident yet. They want to talk to that person. But if that person doesn't know his language, he ends up being the rich one. Fine. Like yesterday during the Cilacap Expo, there was a fashion show, and they bought it from traders. They were not ashamed. It's only possible that if you explain the details to other people and other people don't know his language, it's like he feels his efforts were in vain. 'Yes, it is difficult for me to explain; I let my teacher do it,' they say, but if it's really hard and no one is around, the child asks for a cellphone, keeps writing, explains via cellphone like that. If the teacher's language is complicated, then it is simplified, for example, 'I want the batik to be red,' the teacher says, 'Is this the color replaced with red? Is it good or not?' Later, if he does not understand, he will

say, 'Repeat. Repeat one or two times, God willing, understand'" (Teacher-1).

Deaf children were invited to directly serve the buyers, then given encouragement to try to persuade the buyers. This service is in the form of explaining the product, discussing the product the buyer wanted, and receiving orders from the buyers.

4.2 *Discussion*

Based on the research findings, the implementation of marketing skills includes four indicators, namely market research, marketing optimization through digital applications, understanding competitor products, and trying to be persuasive to potential buyers. Market research is taught to deaf children in the form of paying attention to community needs first, then it is carried out in schools.

The implementation of entrepreneurship education is based on the needs of the community, not only for children to make products but also for sale. If the implementation and results of market research are not satisfactory, then changing the type of marketing business has also been carried out using digital applications as an effort to reach a wider market. This finding is in line with the opinion of Lackeus (2015) that the implementation of entrepreneurship education is based on the results of market research on the development of community needs, so that children can learn to create value from real entrepreneurial experiences.

The teacher invites deaf children to do research on competitor's products by showing them in class. Then, together they analyze the advantages and disadvantages of each product. Then the product is evaluated and developed again. Deaf children not only receive knowledge in the classroom, but also practice directly recognizing their competitors in the community.

This research finding is in line with the finding of the research by Rentz et al (2013) which states that skilled marketing requires the ability to attract customers to buy, the knowledge of the customer market, and the knowledge of competitor products.

5 CONCLUSION

Marketing competence of deaf children can be developed through five aspects, namely market research, use of digital applications, analysis of competitor products, persuasive action, and customer service. Teachers can invite deaf children together in class or individually. This study supports optimizing the role of teachers to become facilitators of deaf children in business development. The results of this study can be used as a guide in designing entrepreneurship education for teachers. They can be used also as a basis for further research with a broader subject so that the findings can be generalized.

REFERENCES

Brinkley, D. 2011. *Supporting Deaf Children and Young People*. New York: Continuum International Publishing Group.

Crittenden, V & Crittenden, W. 2015. Digital and Social Media Marketing in Business Education: Implications for the Marketing Curriculum. *Journal of Marketing Education*. 37(2): 71–75.

Dowell, D.J., & Small, F.A. 2011. What is the impact of online resource materials on student self-learning strategies? *Journal of Marketing Education*, 33: 140–148.

Fellinger, J., Holzinger, D., Pollard, R. 2012. *Mental Health of Deaf People*. Austria: Institute of Neurology of Senses and Language

Indriani, F. 2006. Studies Regarding Innovation Orientation, Product Development And Promotion Effectiveness As A Strategy To Improve Product Performance. *Jurnal Studi Manajemen & Organisasi*, 3(2): 82–92.

Knoors, H., & Marschark, M. 2014. *Teaching Deaf Learners: Psychological and Development Foundations*. Oxford: Oxford University Press.

Lackeus, M. 2015. *Entrepreneurship in Education: What, Why, When, How*. Paris Perancis: OECD.

Lake, D.A. 2018. Economic Openness and Great Power Competition: Lessons for China and the United States. *The Chinese Journal of International Politics*. 237–270.

Nasution, D.P. & Lubis, A.I.F.. 2018. The Role of MSMEs on Economic Growth in Indonesia. *Jurnal Kajian Ekonomi Dan Kebijakan Publik*. 58–66.

Purnomo, M. 2015. Dynamics of Entrepreneurship Education: A Systematic Mapping of Education, Teaching, and Learning. *Jurnal Dinamika Manajemen*, 6(1): 97–120.

Rentz, J. O., Shepherd, C., David., Tashchian., Araien., Dabholkar, P. A., and Ladd, R.T. 2013. A Measure of Selling Skill: Scale Development and Validation. *Journal of Personal Selling and Sales Management*. Edition October 2013.

Sunardi & Sunaryo. 2007. *Early Intervention of Children with Special Needs*. Jakarta: Indonesian Ministry of National Education.

Xu, Y., Liu, D., & Zheng, L. 2017. Econometric Model Analysis of Influencing Factors of Import and Export in Zhanjiang City. *Advances in Economics, Business and Management Research*, 3: 810–813.

Educational Innovation in Society 5.0 Era: Challenges and
Opportunities – Purnomo & Herwin (Eds)
© 2021 the authors, ISBN 978-1-032-05392-9

The strategy of the principal of the elementary school of Jogja Green School in facing the Covid-19 pandemic

B.D. Jaswanti & E. Purwanta
Yogyakarta State University, Yogyakarta, Indonesia

ABSTRACT: The purpose of this research is to describe the management strategy of the principal of elementary school of Jogja Green School in facing the Covid-19 Pandemic. This was a qualitative research with a phenomenological approach. The research data was collected through interviews, observation and document analysis. The results show: 1) environmental scanning: the principal filters the internal and external environment in the Jogja Green School; 2) formulating: in accordance with the vision and mission of the school as an inclusive school the principal formulates long-term plans for opportunities and threats effectively; 3) implementing includes: students: student learning is conducted through online, curriculum: modified curriculum, educators: creative in providing learning to students, facilities and infrastructure: facilities to support the health protocols, finance: budget efficiency, 4) evaluation: continue to conduct evaluations both in human resources and online to maintain the quality of education.

1 INTRODUCTION

At the end of 2019, the world was shocked by the outbreak caused by Covid-19 (Corona Virus Disease-2019). This virus was first detected in Wuhan, China in November of 2019. This virus outbreak spread very fast in a short time, from person to person and from country to country in the world.

In accordance with the Policy on Implementation of Education in Emergency for Covid-19 which was issued by the Minister of Education and Culture in Circular Number 4 of 2020, the learning process is conducted from home. Distance learning from various online methods to stop the spreads of Covid-19 face various problems. Further research in Purwanto (2020: 8–9) reveals that online distance learning during the pandemic causes students, parents and teachers to work to master the technology required. Besides this factor, additional internet packages, additional parents' jobs for assisting children in the learning process, the decreasing of communication and socialization and the increasing working hours of teachers because they have to communicate and coordinate with parents, other teachers and the principal are also new obstacles.

Meanwhile, Zaharah's research (2020: 279–280) shows that the obstacles to adjusting online learning include unfulfilled internet networks, unfamiliar understanding of students and teachers with online learning and unreadiness of parents to give assistance at home. Both studies show that it takes readiness of all parties to support the learning process, in terms of facilities, budget, and mastery of technology, parents support and environment support. Another related obstacle found is the learning of special needs students who find difficulties to learn without an accompanying shadows teacher.

In this case, the principal should take responsibility for the comfort, the school environment and the school residents including teachers, students and parents, such as the safety and comfort during the Covid-19 emergency response. Therefore, it is possible for the principal to adopt a strategic approach that integrates several management aspects to manage and achieve the educational institution or school goals effectively and efficiently.

Poister (2010) suggests that strategic management consists of (a) strategic planning; (b) budgeting, performance measurement and management and evaluation (how to implement); and (c) feedback between these elements to enhance mission fulfillment, mandate meetings and sustainable public value creation through strategic learning. The strategic management approach is intended to assist the strategic efforts of public leaders and managers to coordinate important decisions across levels and functions within an organization and across organizations. These approaches vary in completeness, formality, and tight control over the planning and implementation process. This definition includes an important part of strategic management that often receives less attention (thus triggering further criticism), namely continuous strategic

DOI 10.1201/9781003206019-40

learning. By incorporating strategic learning, strategic management can ensure that an organization or other entity periodically and continuously assesses the relevance of its strategies to determine whether they are functioning or emerging or needed (Van Dooren, Bouckaert, & Halligan, 2015).

Strategic management is based on managers' understanding of their organization, markets, prices, raw material suppliers, distributors, governments, creditors, shareholders, and rival customers around the world. It is a critical factor for business success in today's fast changing world. Strategic thinking enables organizations to deal with and adapt their management to future conditions. The strategic plan describes the process of achieving from an existing situation (organizational mission) to a desired situation that describes the future of the organization. Because environmental recognition is so good, they can help an organization to improve its performance. (Ilembe & Were, 2014).

Strategic planning is a strategic tool that must be used in comprehensive quality schools to provide a clear vision to meet the needs of schools and the community (Latorre & Blanco, 2013). The aim of comprehensive quality schools is to provide process-oriented quality management and comprehensive quality management. In this case, one of the fourteen general principles of management to improve the quality of schools is to instill strategic planning in schools because we cannot manage changes in the school process, unless the school has made a strategic plan for itself (Toorani, 2012).

Wheelen & Hungler explain that strategic management is a set of managerial decisions and actions that determine the long-term performance in an organization. This includes environmental scans (both external and internal), strategy formulation (strategic or long-term planning), strategic implementation and evaluation and control. Strategic management studies, therefore, emphasize monitoring and evaluation of external opportunities and threats given the strengths and weaknesses of the organization. The topic of strategic management of business policy include strategic planning, environmental scanning and industrial analysis.

The rational planning model predicts that when environmental uncertainty increases, an organization that works more diligently to analyze and predict more accurately situational change in which they operate and will outperform those who do not. Strategic management consists of four basic elements, those are:

1.1 Scanning the environment

Environmental scanning is monitoring, evaluation and dissemination of information from the external and internal environment as a key aspect in organization. It aims to identify the strategic factors through external and internal elements that will determine the future cooperation used by organization to gain a competitive advantage.

1.2 Formulation

Strategy formulation is the development of a long-term plan for effective management of environmental opportunities and threats, taking into account organizational strengths and weaknesses (SWOT). It includes defining the organization's mission, determining objectives that can be achieved, developing strategies and setting policy guidelines (Sindhu, 2004: 40–45).

1.3 Strategy implementation

Strategy implementation is the process by which strategies and policies are implemented through the development of programs, budgets and procedures. This process might involve changes in culture, structure or management system of an organization. Excepting, when drastic changes throughout the organization are needed, however, strategy implementation is usually carried out by middle and lower-level managers and reviewed by top management. Sometimes referred to as operational planning, strategy implementation often involves daily decisions in resource allocation.

1.4 Evaluation and control

Evaluation and control are the process by which organizational activities and performance results are monitored, so actual performance can be compared with desired performance. Managers in all levels use the generated information to take corrective action and solve problems. Although evaluation and control are the final main elements of strategic management, they can also show weaknesses in the strategic plans that have been implemented previously and thus stimulate the whole process. Performance is the final result of the activity. This includes the actual results of the strategy management process.

In Veranti and Sasongko's research (2017) regarding principals in educational services with inclusive educational system in regular schools must adapt to the needs and potentials of each student, especially special needs students, so strategic management skills in inclusive schools are implemented. Based on the finding, it shows the need for a special strategy for principals in inclusive schools to provide educational services for all students. On the other hand, Hidayat (2018) reveals that the principal's commitment through the strategies implementation to improve the quality of education, including: 1. increasing teacher resources, 2. facilities and infrastructure, 3. principal management, 4. curriculum, 5. students and 6. public relations. It shows that the principal is the determinant of the strategy in an effort to improve the educational quality in schools. The importance of managerial ability to be able to achieve the expected education in the conditions during the pandemic is not sufficient only with general managerial abilities, but it needs to apply strategic management skills especially in inclusive schools to achieve school goals in line with vision and mission.

Strategic management according to Jauch & Glueck as quoted by Akdon (2011: 17) is a number of decisions and actions leading to the formulation of an effective strategy to help achieve organizational goals. Elementary school is childhood to instill the basic knowledge that is important for the success of further education, and the principal is very instrumental in implementing effective and efficient policies in managing the conditions of the Covid-19 outbreak.

Elementary School of Jogja Green School, is a school implementing the concept of contextual learning (nature-based). School of nature (Maryati 2007: 180) is a school based on the natural environment as the learning object. The profile of this school is different from schools in general, but its existence is able to accommodate if education is liberating, fun and having the best inclusive school reputation.

Based on obstacle found, the researchers are interested to conduct research related to the context of the principal's strategic management in Elementary School of Jogja Green School dealing with the Covid-19 spread.

2 METHOD

This was a qualitative research project. It was used to match empirical reality with the prevailing theory using descriptive methods. In this qualitative method, researcher tries to understand phenomena through data collection techniques including observation, interviews and documentation.

This data was collected in the form of primary and secondary data. According to Sugiyono (2014: 39), primary data is a data source that directly provides data to data collectors. This primary data is the main data containing interview result with informants and direct observation notes in the field. Sugiyono (2013: 309) argues that secondary data is data sources that do not directly provide data to data collectors, for example through other people or through documents. Secondary data obtained in this research were student attitudes records, photo files of teacher and student activities in the implementation of learning, school culture and literature study results.

Data collection techniques used in this study include interviews, observation and documentation. This interview aims to obtain information related to how strategies for inclusion principals deal with the pandemic what are the supporting and inhibiting factors of the principal's strategy and how to measure success in the principal's managerial strategies which need to be further developed for learning during the pandemic.

Observations were carried out to find out how the principal's strategy of exclusion in dealing with the Covid-19 pandemic. Observations were made by observing and recording the characteristics of the environment. In the documentation, the researchers were looking for data from written data, in the form of an archive of attitude score data and a number of incident records in inclusive-based school namely Jogja Green School.

Researchers used the validity test of credibility, including 1) increasing perseverance through making careful and continuous observations by re-examining the data obtained after observation and interviews. If the data was deemed lacking, the researcher conducted another interview until the required data could be fulfilled 2) Triangulation in the forms of source triangulation, technical triangulation, and time triangulation. Triangulation was done by using data validity test using source triangulation by checking the obtained data through several data sources that have been determined by principals, assistant teachers, several subject teachers and administrative staff at Jogja Green School including NN, RN, WD, DS, EF, EN, FH and other students. The researcher also used triangulation techniques through various techniques such as observation and interviews with the same informant and using triangulation for several weeks 3) Used reference materials using research aids in the form of Nikon D3100 cameras as image documentation tools and used Samsung as a voice recorder. Therefore, data can be better trusted for reliability.

3 RESULTS AND DISCUSSION

Wheelen & Hungler explain that strategic management is a set of managerial decisions and actions that determine the long-term performance of an organization. It includes environmental scans (both external and internal), strategy formulation (strategic or long-term planning), strategy implementation and evaluation and control. Strategic management studies, therefore, emphasize monitoring and evaluation of external opportunities and threats given the strengths and weaknesses of the organization. (Wheelen & Hungler, 2012: 7).

The first stage is environmental screening that is strategy management undertaken by the principal of Elementary School of Jogja Green School to conduct an environmental screening in the planning scope. The scan includes description of the region's background or the planning of the principal's strategies carried out in dealing with the Covid-19 outbreak. The principal decides on a plan by observing and considering from political and economic context. In this case, the principal identifies the functions of the education component in achieving school goals to be well conducted amid the pandemic. The components analyzed by the principal of Jogja Green School include internal and external components by considering several standards, including distance learning, student management with standard government procedures during pandemic, a modified curriculum amid distance learning, development of technological needs of teachers and education staff, funding adjusted in pandemic conditions efficiently and relations both with society and parents to cooperate. Then, the principal of the Elementary School of Jogja Green School conducts a SWOT analysis to see the strengths, opportunities, weaknesses and obstacles that might be faced in the future. Principal of Elementary School of Jogja Green School analysis is strengthened by internal and external factors, how ready the school is to proceed to the next stage.

The second stage is formulating. Formulating of strategy is the development of long-term plans to manage external opportunities and threats as well as measuring internal strengths and weaknesses, so the vision of the school can be implemented effectively. The stages of strategy formulation include determining the vision and mission of the organization or school, determining the reach of objects, developing strategies and making organizational policy direction. In this case, the Principal of Elementary School of Jogja Green School understands the vision and mission as one of stake holder of inclusive education and character educational school which needs to organize schools by determining the direction of implementing the program structure. One of the visions and missions of the school is how to shape students' characters even though the schooling is carried out remotely. Therefore, students' home assignments are related to general school learning material and the rest is character lessons that take the form of helping in the home environment, including cleaning the house, watering plants or planting, cleaning the house as a form of love for the environment as taught in school. Within a week the learning task is not only general material but also character learning monitored by parents.

The policy that the of distance learning activities for normal students and school learning activities for children with special needs is weighed based on the situation and conditions of the school environment. However, the principal reduces the portion of learning activities for special needs students and focuses on daily activities. In addition, learning materials for special needs students are in the form of pictures and videos, so they are easy to understand.

The third stage is implementing. The principal understands that as an inclusive school even in the circumstances and conditions amid the pandemic. The principal provides education with multicultural insights that help students understand and accept conditions during the pandemic through life value, character education, physical and psychological abilities. Among the implementation activities are developing a culture that supports the strategy dealing with Covid-19, creating an effective organizational structure, directing marketing patterns and linking performance between staff. The implementation of the strategic plan consists of several aspects, including programs, budgets, procedures, evaluation and supervision. The budget includes two aspects, namely the income and expenditure aspect.

The implementation of distance learning for the Elementary School of Jogja Green School uses the previous curriculum, only for grades 4, 5, 6. There will be change from thematic curriculum to use module. For new students, the implementation of the curriculum during this pandemic is in accordance with the

development based on the teachers' input. The curriculum adjustment strategy is carried out by adjusting the duties of teachers and programs implemented in school.

During the distance learning period, the teacher adjusts the learning method and makes interesting learning media. On the other hand, the teachers' work time has also been adjusted. At the beginning of the pandemic, teachers were more intense in carrying out learning at home. Only in June the picket was held with a schedule of once a week. To support distance learning activities, the principal provides a policy by providing facilities to teachers in the form of websites, videos, wi-fi and internet packages

Besides, the principal also provides a policy for students to use existing facilities in their homes. If it is still not sufficient, students can borrow existing facilities at school. Meanwhile, learning infrastructure is not carried out. The principal makes a budget to provide procurement operational costs for internet package assistance to teachers. Schools make use of existing learning media in schools such as laptops, wi-fi and stationery. The use of facilities in the form of laptops must be used with the principal's permission, while other facilities can be used without prior permission.

The pandemic outbreak affects the financial condition of schools and parents of students, so the principal makes a financing adjustment strategy. Especially for parents of students affected by the Covid-19 pandemic. The principal provides a policy in adjusting the education development donation by providing a discount. The policy is also based on the condition of parents who have experienced the impact of Covid-19.

The principal's strategy in adjusting school finances is by coordinating school members to develop an entrepreneurial spirit through making salted eggs, selling fish, selling fruits and selling breadfruit and chicken. Apart from overcoming financial problems, this activity is also an effort to maintain food security. The school only receives funding from the foundation and the students' parents. During the pandemic, schools make financial adjustments, so the existing financing system is not affected. The obstacle in school is the time. Previously, the school was implemented only until noon, but in distance learning the work time is longer. Meanwhile, the obstacle for students and parents at home is their honesty and responsibility. This requires the school to contact them through online or sending letters to parents to collect assignments. The level of learning effectiveness at home is lower than at school. This is influenced by the condition of each student's home. Home learning is parents' responsibility. During implementing learning, not all parents are able to focus on it. There are only adjustments related to program submissions that are necessary or not to be made during this pandemic.

The final stage is evaluating. Basically, monitoring and evaluating are two interrelated activities. Supervision of the principal is needed to compare the performance achieved with the expected performance.

In addition, it also needs to assess and measure the extent to which the principal's goals dealing with pandemic related to the availability of schools, so the education can be conducted as the stated goals. The principal's evaluation alludes to the SWOT analysis which may be less sharp or incomplete and needs improvement. Evaluation can be done as a whole. If the actual performance is far from expectations, then one by one the previous activities need to be evaluated starting from the objectives, strategies, budgets and procedures. The evaluation of the principal of the Elementary School of Yogyakarta Green School is routinely carried out every week by evaluating teacher constraints during distance teaching and learning activities through personal contact via telephone or meetings using Zoom and WhatsApp.

The principal and the teacher evaluate the learning for students by working on questions. However, the National Exam (UN) score was taken from the report scores. In addition, the principal instructs the teachers to keep making a mid-semester report containing student progress reports and final semester report cards containing reports on students learning progress. Evaluation of student learning is usually carried out weekly with the results of deliberation at work meetings with teachers. However, amid the pandemic condition, the evaluation is carried out online. In addition, the school also conducts non-formal supervision of educators and education personnel both online and face-to-face. Furthermore, reporting to the education office has not been carried out during the pandemic. Meanwhile, reporting to foundations is carried out online to foundation leaders, institutional leaders, coordinators or finance and administration departments.

4 CONCLUSION

Based on the explanation above, it can be concluded that in general, the principal of Elementary School of Jogja Green School has carried out the basic model of strategic management, namely: 1) scanning the environment done by the principal of SD Green School who conducts environmental screening and decides plans for curriculum modification by distance learning, developing the technology facilities for teachers, efficient funding, and making collaboration with students' parents. 2) Formulating, in accordance with the vision and mission of the school as an inclusive school and character education, distance learning with general materials and character learning which can be done by helping and loving the home environment, for special needs students are provided pictures and videos to make them able to learn independently. 3) Implementing some aspects, such as, students: student learning is conducted through online; curriculum: curriculum implementation with modifications amid the pandemic; educators: creative in providing learning to students, and collaborating with parents. Facilities and infrastructure: there are additional facilities to support the protocol of health; finance: budget efficiency due to the impact of school finances in a pandemic situation; homework: optimizing social media to facilitate learning, introducing schools to prospective students in new normal learning; special services: providing maximum service to all students. 4) Evaluation: continuing to evaluate both in-person and online to maintain the quality of education, in terms of infrastructure, learning, teaching staff, students and the readiness of parents in providing assistance in distance learning.

From this research, it can be concluded that there are obstacles faced by the principal in implementing management strategies including lack of awareness of parents in accompanying their children in distance learning, an increase in funding to provide equipment that supports health during the pandemic and the ineffective distance learning for special needs students

REFERENCES

Akdon. 2006. *Strategic Management for Educational Management*. Bandung: Alfabeta

CNN Indonesia. 2020 CNN Indonesia. *Corona dan Dilema Guru Mengajar Siswa Berkebutuhan Khusus*. Retrieved from: https://www.cnnindonesia.com/nasional/20200421 163541-20-495770/corona-dan-dilema-guru-mengajar-siswa-berkebutuhan-khusus

Ilembe, W., & Were, S. 2014. . International Journal of Social Sciences Management & Entrepreneurship, *Challenges affecting implementation of strategic planning in management of secondary schools in Kimbu Country* (2), 121–140.

Hopkins, Johns CSSE. 2020. Coronavirus COVID-19 Global Cases by the Center for Systems Science and Engineering (CSSE) at Johns Hopkins University (JHU). From: ArcGIS: https://gisanddata.maps.arcgis.com/apps/opsdashboard/index.html#/bda7594740fd40299423467b48e9ecf6

Maryati. 2007. Sekolah Alam. Yogyakarta: FMIPA UNY. Seminar Nasional Penelitian, Pendidikan dan Penerapan., MIPA. *Alternatif Pendidikan Sains yang Membebaskan dan Menyenangkan*.

Poister, T. H. 2010. The future of strategic planning in the public sector*: Linking strategic management and performance. Public Administration Review*, 70 (Suppl. 1), s246–s254.

Toorani, H. 2012. Introduction to process-based management schools. Tehran: Tazkieh Publication.

Wendell, L., Bell, C. H., & Zawacki, R. A. 2011. Managing effective change (6th ed). Organization development and transformation. Boston: McGraw-Hill.

Sugiyono. 2014. *Metode Penelitian Kuantitatif, Kualitatif, dan R&D*. Bandung. Alfabeta.

Van Dooren, W., Bouckaert, G., & Halligan, J. 2015. *Performance management in the public sector (2nd Ed.)*. New York, NY: Routledge.

Wheelen, T. L. & Hunger, J. D. 2012 *Strategic Management and Business Policy (13th Ed.)*. New York: Pearson.

Educational Innovation in Society 5.0 Era: Challenges and
Opportunities – Purnomo & Herwin (Eds)
© 2021 the authors, ISBN 978-1-032-05392-9

Development of Star Book Media to influence writing skills and carrying attitude in grade IV elementary school students

E. Zubaidah, S. Sugiarsih & A. Mustadi
Faculty of Education, Universitas Negeri Yogyakarta, Yogyakarta

ABSTRACT: The purpose of carrying out this research and development is to produce Star Book Media that can streamline story writing skills and caring characters for fourth-grade elementary school students who are declared worthy of media expertise and material expertise. This research and development refers to Borg & Gall's research and development steps. The material expert gave a final average score of 5 in the excellent category. The media expert gave a final average score of 4.69 in the excellent category. The results of operational trials show that Star Book media is effective in improving story writing learning. Based on the average score of the experiment, 1 class is 82, experiment class 2 has an average of 77, and the control class is 70. This shows that star book media is useful for improving the writing skill of fourth-grade elementary school students.

1 INTRODUCTION

The rapid development of science and technology has caused the emergence of several problems that occur today. Muslich (2011) states that the recent moral crisis can be overcome with character education in every school environment, namely in education and teaching carried out by teachers and their students. Character education (Kemendiknas, 2010; Kemdikbud, 2017; Fadlillah and Khorida, 2013) is a growing educational initiative supporting students' social, emotional, and ethical development. The best way to teach character is to allow students to apply concepts in real-life situations (Anderson & Glover, 2017).

One of the characters that are starting to look faded with the current development is the caring character. That is not only caused by changes and advances in technology and times, but also from within a person. The caring character itself attempts to think about one's own will, other's needs, or trying to help others (Raatma, 2000). Furthermore, caring is defined as an activity to help solve others' problems with individual or collective good (Octavia, Syatibi, Ali, Gunawan, & Hilmi, 2014). Yulianti & Hartatik (2014) explained that the character of caring is fundamental in order to foster a sense of brotherhood and kinship and to distance oneself from arrogant, selfish, and individualistic traits. These various meanings show that caring is the attitude of paying attention to something, both inside the person and outside himself. In social life, everyone needs a caring attitude.

There are many opportunities that teachers can use in developing character which one does find in

learning the Indonesian Language and Literature in every educational institution (Zubaidah, 2013). Learning the Indonesian language and literature through thematic learning in schools should achieve a caring character education through listening, speaking, reading, and writing skills. It is said (Huck, Helper, & Hickman, 1987; Stewig 1980) that children can develop their personalities through literature. Elementary school (SD) students will find it easier to improve their caring character and writing skills if done through the material in daily life. The material in everyday life is good through listening, speaking, reading, and writing.

Tompkins and Hoskisson (1991) stated that language skills in schools, including listening, speaking, reading, and writing, are closely related so that each skill underlies another skill. This understanding shows that writing is a skill acquired after students can read, speak, and listen beforehand. Students can write if they are used to reading or seeing anything. Through reading or being seen, students can easily develop their ideas or ideas into a story that can be told or discussed and or written. Thus, it is very logical not to write because they cannot read or even have never known writing. This problem requires teachers to learn to write, use, and show students media.

According to its type, the media divided into visual media (visual), listening media (audio), and listening media (audiovisual). For learning to read and write, Jalongo (2007) and Fellowes and Oakley (2010) mentioned more than ten media types. However, from these types of media, there are no Star Book Media books mentioned. Therefore, in this study, the Media

214   DOI 10.1201/9781003206019-41

Star Book was chosen to be studied. Through this, story writing skills and caring characters can be improved.

The picture story series in the Star Book that is presented to students must provide an example. Abbas (2006) states that series pictures are media images in chronological events that can help students find ideas and tell stories. Series pictures are usually in the form of a pictorial story. A pictorial story is a form of moving art that uses images that are not arranged in such a way as to form a storyline (Susanto, 2011). In line with this opinion, Arsyad (2002) explains that a picture is a series of activities or stories presented sequentially. Matulka (2008) describes several components that need to be considered in making a storybook series, including the outside of the book, shape, and size, other supporting parts, illustrations, and typography.

The development of fourth-grade elementary students, apart from having fun playing and making friends, also contains rapid motor development and rapid social development. Language development is also present. Through these developments, it is hoped that the character of the individual can be improved. Therefore, through the various characteristics of student development, it is necessary to learn to write stories and develop caring characters for fourth-grade elementary school students. However, based on observations and interviews with students and teachers in several schools where this research was conducted, the media to improve story writing skills in Star Books does not yet exist.

Based on several reasons behind the problem and some media needs in the fourth grade, it is quite reasonable to believe that the feasibility of research on Star Book Media Development can be tested on material experts and media experts. Their effectiveness can be tested in the school where this research is conducted.

The purpose of this research and development is twofold. (1) Produce Star Book Media appropriate for the effectiveness of story writing skills and caring character for fourth-grade elementary school students. (2) Find out the effectiveness of Star Book Media in making story writing skills effective and character caring for fourth-grade elementary school students.

2 METHOD

The research was conducted in Yogyakarta, namely SD Kintelan, Keputran, and SD Surokarsan. The instruments used in this study were sheets: interviews, observations, questionnaires, validation of material experts and media experts, and writing skills test sheets. Researchers need an instrument developed based on book media's quality aspects to produce a suitable star book media product. From this instrument, we can see the quality of the product produced. Table 1 below is the aspects used to evaluate the Star Book media by media experts and material experts.

Table 1. Aspects of the assessment of material experts and media experts

No	Material aspects	Aspects of the media
1	Language	Outside layer
2	Presentation	Shape and size
3	The suitability of the image to the story	Support layer
4	Format and content of the material	Illustration
5	Usage for teachers	Typograph
6	Usage for student	-

The data obtained through an assessment questionnaire by material experts and media experts was analyzed using descriptive statistics. The results are used to assess the quality of the product developed in the form of scores. The scores obtained were then summed and averaged, then converted into a value with a scale of 5. Data in the form of interviews and observations was analyzed using descriptive techniques.

The Media Star Book effectiveness test was tested on a limited scale and an expanded scale. Limited trials (initial field trials) were conducted at SD Kintelan, while expanded trials (main field trials) were conducted at SD Surokarsan and SD Keputran.

3 RESULT AND DISCUSSION

3.1 Result

The interviews with teachers in the initial study showed that some students still had difficulty writing stories. Teachers already know a lot about media in improving story writing skills and increasing students' caring character, especially Star Book Media. Schools only provide storybooks stored in the library. Students are usually asked to come to the library to read storybooks. Students are then asked to summarize the stories they have read. The teacher rarely uses the storybook media stored in the library for use in the classroom. Usually, the teacher asks students to read thematic books that students already have.

Another activity that can improve students' writing story skills is that the teacher provides examples of stories. The teacher gives a reading to the students, and then the students are asked to summarize the reading content. Another activity is to make a synopsis and make a story based on the teacher's pictures.

The development of story writing skills and caring characters has been done in schools using pictures, posters, and video clips. The teacher stated that in learning, there were difficulties in linking caring characters in learning. Therefore, writing stories and caring characters was carried out using pictures, posters, and video shows prepared by the teacher. However, this is only done occasionally, considering that schools still have limited school facilities.

Other actions to improve the teacher's story writing skills and caring character are giving students gifts and warning students. Students who have good story writing skills have their work posted on the classroom wall magazine with a grade. However, those whose stories are not good are not posted on the classroom wall magazine. Likewise, students who have a caring character, for example, are diligent in picking up to clean the class as an example. Conversely, the teacher does not set an example of lazy students.

Information gathering is obtained by conducting a literature study related to development research using other references as additional information related to the media to be developed. The researcher also made observations on several storybook media following student development and the media of storybooks that attracted students' attention. Some of the information obtained from literature studies is used to consider the product to be developed. It is hoped that this will become the basis for product development which will suit needed.

Analysis of the results of gathering information in the initial study contained several problems. These problems are the following. Students have difficulty in developing story writing skills. In schools, there are no learning media available to improve story writing skills. The unavailability of book media teaches students how to write stories that support story writing skills and enhance caring characters in classroom learning.

Based on the results of observations, interviews, and the distribution of questionnaires to the teachers and students, the researcher tries to design a media for a series pictorial storybook using an exciting form, like a star (Star Book).

3.1.1 *Star book concept*
The concept used in the development of the Star Book media is a story of everyday experiences. The experiences that are told are familiar with the daily lives of students at the research site. The Media Star Book was made contextually by taking into account the school environment in both elementary schools. Between SD Surokarsan and SD Kintelan, a river is quite famous, namely the Code river. In the river, waste and garbage have been polluted, so the river looks dirty and not beautiful. Most of the students come from the river environment. Thus, the theme of the river environment is close to the world of the students.

A mandate is an application of caring characters, especially caring for the environment, which students can apply in everyday life through the story's contents. Therefore, with the development of this Star Book, students can develop their writing skills and can be used to recognize the values of caring characters presented in the story.

3.1.2 *Format and size of the book*
This book is made in a large size so that it can be used as a group. This book measures 38 cm x 48 cm with a total of 12 pages. Coupled with the front and back covers so that it becomes 14 pages. A Media Star Book is bound by using the star binding technique. A Media Star Book is equipped with presenting the elements in the story, and the outline of an essay as an example of a written story developed independently by students. The learning is done in groups.

3.1.3 *Title and content of star book media*
The title of the story is placed on the Media Star Book page. The story is entitled Trio of Environmental Heroes. The Trio of Environmental Heroes is about the friendship of three elementary school children who have experience fishing in rivers and cleaning dirty environments. Caring values are shown through the three characters' behavior to clean up the river environment, polluted with garbage and cloudy water. Through their efforts to clean up the river environment, they deserve to be called the Village Heads and the community as the trio of environmental heroes. The trio who cares about the environment.

3.1.4 *Figures and characterizations*
Character selection refers to a character who has a caring character. This is done so that the images used attract students' attention and encourage students to become like characters in the story. The main character in the Star Book media is a boy named Andi. Andi is a good boy, friendly, and has a leadership spirit. Besides, the character Andi also has a caring character, as shown in the story in the picture. Andi and his friends clean the river, which is littered with waste and garbage. They saw and felt sad while fishing in that murky and dirty river. The depiction of this character or characterization, friendly Andi, is shown through Andi's togetherness with his friends named Budi and Tono. Andi's choice as the main character is intended to have awareness and concern for their surroundings. Apart from that, the researcher hopes that elementary students will read this story and hope that they will become pioneers of clean life and care for their environment.

3.1.5 *Selection of material and themes*
The material presented in the Star Book is entitled "Trio Pahlawan Lingkungan." This material is a material that has met the level of conformity. It has compatibility with the environment, age, and language of students, mainly grade IV elementary school students. The language is made simple to attract students' attention to read and understand the story.

The presentation of the material is made according to the stages of developing the story correctly and adequately. First, students are exemplified by good stories. Students then learn about the building blocks of stories and are given examples of how to outline a story before the story is made. Students can then write stories independently.

Examples of developed stories are stories that contain caring characters that can be applied in everyday life. For this reason, it is based on intimate experience with the children's environment in the two primary schools where the initial research was carried out, namely SD Surokarsan and SD Kintelan Yogyakarta. The theme chosen is environmental. The selection of environmental themes makes it easier for children to apply caring characters in students' daily lives in their environment.

3.1.6   *Selection of image types and colors*
The image selected is an image that matches the story title. This image can imagine the child about the atmosphere and place when fishing in the river and cleaning the river environment. The color of the image is chosen following the development of fourth-grade elementary school students. They like bright colors, but the color of the t-shirts they wear represents a brave boy. Black pants and a t-shirt in blue, dark blue and green. The river environment is described by growing large trees accompanied by natural colors.

3.1.7   *Making star book media designs*
The first stage in developing a Star Book design is creating a book design template. Selection of images that match the material and stories in Star Book media. Initially, the image design used images from the internet, then based on media experts' advice, the final image was created manually. Images that have been created manually are then processed digitally and given coloring through the Photoshop CS 6 application. After the image is finished, it is then laid out with material that has been developed using Corel Draw X5.

3.1.8   *Typography selection*
The typography used in the Star Book development used a choice of fonts, namely the letters "Linnote." Linnote letters were chosen to take into account the typeface that students could easily imitate in handwritten form.

3.1.9   *Print technique*
The researcher's type of paper to print the cover and content pages was ivory 210 paper. In the finishing process, the researcher bound it manually using the star binding technique, beginning with cutting paperboard paper to size. The paperboard paper is then affixed with an ivory paper print, which is the title page and the story material page with castol glue, page by page. Therefore, a paperboard with a thickness of 300 mg is used as a base for sticking the printout so that it is strong and not easily torn or damaged. When the mold's sticking has been completed, it is then assembled and bound with tape to form a book that resembles a star. After finishing, the bookbinding process can be used for learning.

Media development in the form of a series pictorial storybook in the form of a Star Book refers to fourth-grade elementary school students' media and material requirements. This is done so that the developed Star Book media suits students' needs in grade IV SD. The color selection used in this study uses bright colors tailored to the story's color needs.

3.1.10   *Validation*
The product designs that have been developed by researchers are then tested on media and material experts to assess the feasibility of the Star Book Media products that have been developed. Validating the Star Book Media was carried out by media expert lecturers twice, and the material experts were tested twice. The first validation test resulted in a total score of 68, with an average of 4.25. This shows that the value obtained in the media validation test falls into the "perfect" category. The media expert lecturer gave several suggestions, including images, preferably using original images (not from the internet). Besides, the use of images needs to adjust the story text, and the use of colors should use colors according to student characteristics with colors that attract students' attention.

Based on the material expert validation test, it is known that the total score is 64, with an average of 4. This shows that the value obtained in this media's validation test is in the "good" category. The material expert lecturer gave several suggestions, including writing a few words in the story that should be adjusted to the level of language development of grade IV elementary school students. Sentences are not too long, and appropriate punctuation is used. The story text is adapted to the image. The picture must show the three characters of the story while cleaning the environment to match the theme and learning objectives.

The second stage of the material expert test obtained a total score of 80, with 5. This shows that the value obtained in the media validation test is in the "perfect" category. In this second test, the media has experienced an increase in the score without any revision from the material expert lecturer. Table 2 The following summarizes the total score from the Validation of media experts and material experts.

Table 2.   Summary of total scores of media and material experts

Validation stage	Media expert scoring	Category	Material expert scoring	Category
Step 1	4,25	excellent	4	good
Step 2	4,69	excellent	5	excellent

The average score of SD Kintelan students in the initial field trial was 78.33. The ideas and content aspects showed excellent results. This shows that Star Book Media can solve difficulties in writing students' ideas into stories. The writing aspect still shows low grades. Therefore, teachers can improve students' writing story skills through teacher revision activities in student writing. In the main field trial, the average grade IV story writing score at SD Surokarsan was 85.5. This is because the Star Book media has undergone revisions based on teacher input in the initial field test. Thus, students find it easier to write stories.

Based on these results, it can be seen that the total score of 29 students in the experimental class 1 is 2341, with an average value of 82. This score shows that the students' writing skills are included in the perfect category. The main field trial results, Keputran Elementary School students had an average story writing score of 82, while in other experimental classes, the students' writing scores already exceeded the minimum competency in learning to write stories, namely getting a score of 77. This shows that students can develop story writing skills using Star media. Book. The score for writing stories of control class students reached a minimum score of 70.2. This shows that learning to write stories with the teacher's explanation without media in front of the class has not made learning writing stories useful for fourth-grade elementary school students.

3.2 Discussion

The teacher reveals that storybooks are quite sufficient to use in the classroom during learning. It is just that published storybooks cannot be used as a medium to improve story writing skills. Media Star Book contains sample stories, analysis of elements in the story, and the essay's framework can be used. This can help teachers to teach story writing skills to fourth-grade elementary school students. Besides, picture storybooks can be a medium for teachers to show character values based on the activities carried out by the characters in the story.

Through the story text that shows the application of caring values, students are given an overview of problem-solving related to an event around them. Star Book Media development needs to be developed to improve story writing skills in classroom learning. Thus, choosing a picture book with a unique shape (star shape) can stimulate students' creativity in writing stories. This is following Semi's (2007) opinion, which explains that writing or composing is a creative process of transferring ideas into writing symbols so that the Star Book media is very appropriate to bring out students' creativity in developing ideas into story texts.

The Media Star Book presents the story building elements. The teacher can explain various elements in the story to help students develop their essays. At the end of the Star Book media page, a written outline based on the title of the story "Trio Pahlawan Lingkungan" is presented. The teacher can provide explanations for students to modify the essay's outline into an essay framework that students will develop. The teacher does this by changing the story elements such as themes, messages, characters, characterizations, settings, and storylines. Themes are replaceable elements so that students can implement caring characters in their stories. By developing stories based on stories made by students themselves, story writing skills will increase and make it easier for students to develop their skills to write stories (Tompkins and Hoskisson 1995).

Based on the analysis of the fourth-grade teacher's responses at SD Kintelan, Star Book Media got an average score of 3.1 with a perfect category. The assessment is based on a rating scale of 4. The results of the average score indicate that the Star Book Media is beneficial for teachers in learning. This is because this media presents pictorial stories and is also accompanied by explanations of the elements of the story and the framework of the essays. An essay outline will be beneficial in teaching story writing skills to students. Students can more easily develop their ideas by making an essay outline first.

The implementation of learning to write stories for fourth-grade students of SD Kintelan using Star Book media received a positive response from students. This is shown through the activities of students who are more active in expressing their ideas. Based on the teacher's assessment of story writing skills, students got an average value of 78.33. Besides, based on the student's assessment of the Media Star Book, the student response questionnaire got 115 out of a maximum score of 132. Based on this assessment, Star Book media was proven to facilitate the learning process and attract students' attention at SD Kintelan.

When assessing the writing skills of the fourth-grade students of SD Surokarsan, the students' mean score was 85.5. This shows that students can develop story writing skills using Star Book Media materials. Meanwhile, based on the student response score, it is 206 out of a maximum score of 220. Students are interested in using this media because Star Book Media is unique and provides exciting stories. According to students, the Star Book Media pictures make it easier for students to imagine the events experienced by the "Trio Pahlawan Lingkungan."

The analysis of the responses of experimental class students related to the use of Star Book Media in classroom learning has a positive response. Based on the questionnaire distributed to students, 293 scores were obtained from a maximum score of 308. This shows that Star Book Media is easy to use for learning to write stories for students. In addition, students can learn caring character values based on stories presented in Star Book Media. The Star Book Media pictures are quite exciting, and the material presented is evident for students. Students can find it easier to find ideas in developing essays.

Star Books contain caring characters. This can be seen from the mandate, which is used to message students to care for their surroundings. Lickona (2004) said that school could help students form an attachment to attention to adults and each other. This caring relationship will foster a desire to learn and a desire to be a good person. Caring characters can be developed through modeling (example). In addition to using examples, the teacher can provide descriptions of how to develop caring characters through stories. Raatma (2000) states several aspects of caring character, keeping the environment clean. Also, teachers can develop their own based on the aspects that arise related to caring characters.

4 CONCLUSION

Based on the research results, it can be concluded that Star Book media is declared feasible to streamline story writing skills and caring character development for fourth-grade elementary school students based on the assessment of media experts and material experts. The material expert gave a final average score of 5 in the perfect category. Meanwhile, media experts gave a final average score of 4.69 in the perfect category. Based on the results of operational trials conducted, there is a significant influence related to Star Book media's use in learning to write stories. Based on the mean score between the experimental class and the control class. The experimental class 1 got a mean of 82, the experimental class 2 got a mean of 77, and the control class got a mean of 70.

REFERENCES

Abbas, S. 2006. Pembelajaran Bahasa Indonesia yang Efektif di Sekolah Dasar. Jakarta: Depdiknas.

Anderson, L., & Glover, D. R., 2017. Building character, community, and a growth mindset in physical. Champaign: Human Kinetics.

Arsyad, A., 2002. Media Pembelajaran. Jakarta: Raja Grafindo Persada

Borg, W.R. & Gall, M.D., 1983. Educational Research: An Introduction (4th ed). London: Longman, Inc.

Matulka, D.I., 2008. A Picture Book Primer. London: Libraries Unlimited.

Fadlillah, M. & Khorida, L. M., 2013. Pendidikan Karakter Anak Usia Dini. Yogyakarta: Ar-Ruzz Media.

Fellowes, J. & Oakley, G., 2010. Language Literacy and Early Childhood Education. Melbourne: Oxford University Press

Huck, C.S.; Helper, S.; Hickman, J., 1987. *Children Literature in The Elementary School*. New York: Holt Rinehart and Winston.

Jalongo, M. R., 2007. Early Childhood Language Art. Boston: Pearson Education.

Kementrian Pendidikan dan Kebudayaan., 2017. Konsep dan Pedoman Penguatan Pendidikan Karakter. Jakarta: Kementrian Pendidikan dan Kebudayaan.

Kementrian Pendidikan Nasional., 2010. Buku Induk Pengembangan Karakter Bangsa. Jakarta: Kementrian Pendidikan Nasional.

Kosasih, E., 2008. Apresiasi Sastra Indonesia. Jakarta: Nobel Edumedia

Lickona, T., 2004. Character matters: how to help our children develop good judgment, integrity, and other essential virtues. New York: Simon & Schuster

Muslich, M., 2011. Pendidikan Karakter Menjawab Tantangan Krisis Multidimensional. Jakarta: Presindo.

Octavia, L., Syatibi, I., Ali, M., Gunawan, R., & Hilmi, A., 2014. Pendidikan karakter berbasis tradisi pesantren. Jakarta: Rene Book.

Raatma, L. 2000. *Caring*. Minnesota: Capstonepress.

Semi, M.A., 2007. Dasar-Dasar Keterampilan Menulis. Bandung: Angkasa.

Stewig. J.W., 1980. Children and Literature. Chicago: Rand Mc. Nallu Collage Publishing Company.

Susanto, M., 2011. Diksi Rupa. Yogyakarta: DictiArt Lab & Djagad Art House

Tompkins, G. E. dan Hoskinson, K., 1991. Language Arts, Content and Teaching.New York: Maxwell Macmilan.

Tompkins, G. E. dan Hoskinson, K., 1995. Language Arts, Content and Teaching. New Jersey. Prentice Hall, Inc.

Yulianti, & Hartatik., 2014. Implementasi pendidikan karakter di kantin kejujuran. Malang: Gunung Samudera.

Zubaidah, Enny., 2012. Peningkatan kemampuan menulis cerita anak melalui strategi menulis terbimbing. Pascasarjana Universitas Negeri Jakarta: Jakarta.

Zubaidah, Enny., 2013. Pendidikan untuk pencerahan kemandirian bangsa. Yogyakarta: Ash-Shaff.

Educational Innovation in Society 5.0 Era: Challenges and
Opportunities – Purnomo & Herwin (Eds)
© 2021 the authors, ISBN 978-1-032-05392-9

Study motivation and students' participation in distance learning during the Covid-19 pandemic

H. Nuryani & Haryanto
Universitas Negeri Yogyakarta, Yogyakarta, Indonesia

ABSTRACT: This research aims to discover the relationship between study motivation and participation of Elementary School Teacher Education students during the course of distance learning as a consequence of the Covid-19 pandemic. This was a quantitative research that used the ex-post facto type technique. The total number of samples in this research were 250 students of 2nd semester of total population of Elementary School Teacher Education students in Yogyakarta. This research implemented the technique of data collection in the form of an online survey. Then, the data was analyzed by using the moment product correlation test. The result of the research shows that there is a relationship marked by positive and significant results. It can be seen through the significant number of study motivation and participation is 0.000 < 0.05 and r-calculator of study motivation and participation is 0.62 > r-table, 0.124 as well as has a relationship in the category of "moderate."

1 INTRODUCTION

Starting from the end of 2019, the world faced the global health crisis, namely the Coronavirus outbreak or Covid-19 pandemic. This phenomenon initially comes from China and has had a profound impact on the alteration in the educational sector globally. Therefore, the educational system has significantly changed in some aspects, especially in which the educational system condition before the Coronavirus outbreak was designed with face-to-face interaction in the class. Otherwise, it is suddenly amid the pandemic turned to distance learning assisted by learning applications (Luthra & Mackenzie, 2020). The transformation of the educational system starts with the closure of schools and universities in many countries, then it is designed to be Distance Learning (DL).

UNESCO supports the nations in terms of e-learning application as an alternative caused Covid-19 outbreak. The data published by UNESCO (2020) mentions that there are 153 schools closed all over the world and almost 70% of the students population in the world suffer the impact of this policy nationally. One of them, for instance, there is a policy designed by Indian government regarding the closure of educational institutions followed by national rules in combating the Coronavirus by implementing lockdown for 21 days (Gupta & Goplani, 2020). This is conducted as an alternative for educational institutions in response to stop Covid-19 transmission.

The Coronavirus outbrea has affected the implementation of education in Indonesia. According to the circular letter by the Minister of Culture and Education Number 4 of 2020 concerning the

implementation of education in the emergency period of the Coronavirus is with distance learning, life skills education, and varied task activities (Kementerian Pendidikan dan Kebudayaan, 2020). The policy of DL has been implemented by colleges in Indonesia in the form of e-learning, utilizing Learning Management System (LMS), learning application or other supporting features.

Fast revolution in the educational paradigm as a consequence of Covid-19 outbreak becomes a challenge for all parties in implementing education in Indonesia because of the different situation between the long-distance and face-to-face class. Distance learning unites physically the distance among students, learning facilitators and other students through various two-way communication media such as video, social media and discussion forum (Saykili, 2018). Furthermore, the quality of Long Distance educational services can be measured by six dimensions, including substantive, reliable, comprehensive, transmissive, guaranteed and participa-tive (Nsamba & Makoe, 2017). Nsamba dan Makoe mention that the supporting service through the facility of technology usage is highly essential to students' academic success, because of the physical distance and the interaction between both.

Therefore, the implementation of DL needs to be evaluated comprehensively and there are still some obstacles to be faced. These obstacles are the access of the technology resources and internet, anxiety of expression and a perceived lack of writing skill, lack of topic for discussion and time, shy personalities, less confidence and incorrect answers which become obstacles to create a more participative online learning

220

DOI 10.1201/9781003206019-42

environment (Bardakci et al., 2018). It is different from the implementation of learning activities that involve interactive online technologies, the activities not only provides the online access in the form of books (e-book) and videos but also involving effectively the active pedagogy to motivate in enhancing the learning outcome (Holmes et al., 2019). It can be estimated that the impact of the Covid-19 outbreak lead to almost all universities in Indonesia to set up the policy to implement DL or online educational process or e-learning as the strategic planning in upgrading the students' achievement.

Along with the government policy about the DL program during this pandemic, it needs the readiness and full support from both educators and students. This can be in the form of highly necessary tutor support that can motivate college students to be more enthusiastic and to be able to influence the students' satisfaction through participation (Herman, 2017). In university level, the Elementary School Teacher Education students are also relevant. The success of DL implementation can be seen in the study motivation levels and student participation. In any condition, an individual that has a high motivation will have an adaptive attitude and behavior which they will use to achieve their purposes. In the educator group, the motivation appears through being the participation in quality improvement in the program to make a practice change more innovative through online forums (Trust, 2017). The physical distance between students and teachers in online learning causes a lack of communication, interaction, and low confidence in the study group, as a consequence, it affects the motivation, participation and enthusiasm of students (Chaiprasurt & Esichaiku, 2013).

A key factor causing less motivation of students in online learning is the geography factor of each student and the low internet network speed to support online learning during the pandemic. The internet becomes the main access method for online learning activities during the pandemic. Study motivation inside of every individual needs to be stimulated first, especially in DL, which prioritizes studying independently. Besides, the success of DL can be seen through assessing the measure of students' participation during learning process activities. The factor that determines the participation is also prioritized in three main elements, such as participation frequency, participation consistency and attendance (Precourt & Gainor, 2018). The consistency of student involvement is important in learning. This is related to the active learning process which will leads to reciprocity, both students-lecturers and students-students. There is a mutual cooperation between lectures and students or among students that can facilitate to reach the purpose of education desired. In the terms of online learning, students' participation is also triggered by intrinsic motivation. Therefore, this research focuses on "is there any relationship between study motivation and student participation in DL as an impact of Covid-19?" Study motivation and student participation in DL will assist the process course of online learning or e-learning.

2 METHOD

2.1 Research type

This was a quantitative research that used an ex-post facto type technique. The research type of ex-post facto was chosen to examine the relationship between independent and dependent variables and make this relationship clear when contrasted to the real experiment research (Cohen et al., 2007). This research was conducted to discover the relationship between study motivation and participation of the Elementary School Teacher Education students in distance learning as the impact of Covid-19.

2.2 Participant

The population of this research was the 2nd-semester of Elementary School Teacher Education students in Yogyakarta. The total of subjects in this research were 250 students from Universitas PGRI Yogyakarta, Universitas Negeri Yogyakarta, Universitas Ahmad Dahlan.

2.3 Data collection and research instrument

The data collection was collected through online survey with the questionnaire instrument of study motivation and participation. The instrument of study motivation and participation refers to the relevant source by adjusting the learning model used (e-learning). The instrument in this research was tested firstly by searching for reliability and validity. The reliability test of the study motivation instrument shows the value of 0.928. The trial result of the study motivation instrument is 0.334, the validity test result of the study motivation instrument in all points is more than significant, which means that the study motivation instrument is valid. Then, the reliability test of the study participation instrument shows the value of 0.904. Meanwhile, the participation test result is 0.334 for all points, so the instrument of participation can be said valid.

2.4 Data analysis

The collected data was analyzed by using SPSS 22 software with the statistical significance of 0.05. to discover the relationship between the study motivation and student participation of the elementary school teacher study program in DL amid the Covid-19 outbreak, the moment product correlation test was conducted with the normally assumed data.

2.5 Research procedure

Research data in this research was collected through online survey results regarding study motivation and participation. From the online survey, it can be obtained the results from fulfilling the questionnaire letter of the study motivation and participation of the 2nd-semester of Elementary School Teacher Education students during DL as the impact of Coronavirus outbreak through a google form.

3 RESULT AND DISCUSSION

3.1 Result

Before summing up the relationship between study motivation and participation of Elementary School Teacher Education students in Distance Learning as the pandemic impact, those need to be tested the normality of data firstly. The Normality test is the precondition for the statistic analysis test. Significance (Sig.) shows that the result is 0.200 which means that Sig. value > 0.05, then it is distributed normally.

Furthermore, the simple correlation test is run and the presented data results on the following table.

Table 1. The correlation test result

		Study Motivation	Participation
Study Motivation	Pearson Correlation	1	.622**
	Sig. (2-tailed) N	250	.000 250
Participation	Pearson Correlation	.662**	
	Sig. (2-tailed) N	.000 250	250

**. Correlation is significant at the 0.01 level (2-tailed).

The correlation test was conducted to detect the relationship between study motivation and participation of Elementary School Teacher Education students in DL during the pandemic. Data collected in this research is the form of interval data, so it was conducted the moment product correlation test. Based on the table output above the (Sig.) of study motivation and students' participation is 0.000 and the output of the Pearson correlation (r) of study motivation and students' participation is 0.662 that both have the positive value.

The conclusion of the data processing results is by comparing (Sig.)-value with the (Sig.)-level of 5% (0.05) as well as the comparison between (r) with (r)-table at the moment product correlation test. It is known that Table 1 output (Sig.)-value of study motivation, in the score of 0.000 means (Sig.) < 0.05 and output of (Sig.)-value of participation that is 0.000 means the (Sig.)-value is <0.05, this can be summed up in that H_0 is rejected and H_1 is accepted. Furthermore, it can be seen that r-table is 0.124 that is obtained the value of (r) for study motivation and students' participation accounts for 0.662 > 0.124, so (r) > r-table. Consequently, H_0 is rejected. According to the (Sig.)-value and (r) that has compared, it can be concluded that there is a significant relationship between study motivation and participation of Elementary School Teacher Education students in DL process during the pandemic. To discover the level of relationship between both variables.

The value output of the Pearson Correlation (r-calculation) can be interpreted based on the category of the correlation coefficient (Jackson, 2009). The Pearson Correlation (r) output of both variables in this research is the amount of 0.662 that is in the scale ± 0.30 – 0.69, so it has a relationship in the category of "moderate." Then, study motivation and participation of Elementary School Teacher Education students have a positive relationship with the correlation coefficient level of "moderate" in DL during the Covid-19 pandemic that can be clarified with the graphic below:

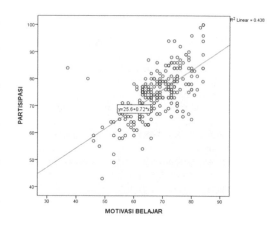

Figure 1. The graph results of the relationship between study motivation and student participation.

The above graphic explains an increase of one variable followed by other variables. The output of data on the graphic above is that there is an increase of study motivation as well as the condition of participation that experiences an increase. This means that the Elementary School Teacher Education students with the high motivation, tend to have high participation as well. Moreover, the graphic above also describes the positive and linear relationship between study motivation and participation that represents the relationship among variables. This tends to lean more to the right and the point of all data was close enough to the line.

3.2 Discussion

Distance learning (DL) is an alternative methodology for the education institutions to respond to the impact of the Covid-19 pandemic. As social distancing, it has changed the educational system of face-to-face interactions switched to online learning. Universities of the United States also decide to make the policy of online learning in response to the issue of the university closure of the Covid-19 (Murphy, 2020). The first education course program of Hongkong University also applies the face-to-face learning online model as the impact of the Covid-19 pandemic (Moorhouse,

2020). Some nations all over the world make the policy to close the schools and swift the methodology of learning that is arguably effective to minimize the spread of Coronavirus.

Basically, the success of Long Distance learning in individuals is affected by some factors, such as infrastructure technique, pedagogy and psychology (Cak$ır$ et al., 2018). The study motivation is the predictor of the three factors of DL program's success. The main motivation cames from the students' behaviour or intrinsic motivation. However, it does not disappear to be the effect of extrinsic motivation. Intrinsic motivation points to the individual's desire referring to convenience and curiosity. But it is not only the assumptions that can be made to the knowledge level content (Sansone & Harackieicz, 2000). Furthermore, the impact of intrinsic motivation is possibly caused by the persistence and success of the students (Simons et al., 2019). Moreover, in the terms of the extrinsic motivation, it can be in the forms of the support tutor gives. The functions of the virtual or distance learning include 1) pedagogy: planning, management, counseling, being counselor, evaluator and process moderator; 2) social: interpersonal skill, being able to create the cooperative and collaborative learning environment, being well-mannered or polite and responsible to each other in the cognitive behaviors and interaction that the student obtains in the different virtual learning; 3) technique: having the knowledge and abilities to apply the technology for the process of education; 4) administrative: applying the rules and norms of the behaviors about the purpose of debate, the theoretical and practical scheme that relates logically to the instructive performances (Espinoza-Freire & Rojas-Garcia, 2019). Therefore, either the good intrinsic motivation or the good extrinsic motivation is essential to shape the students' behavior of the Elementary School Teacher Education program. This behavior pattern aims to achieve the success of the online learning process during the pandemic.

If the study counseling support is inadequate to worry about the students' motivation, it will be lost and step back from virtual learning. The study motivation is related to the aim and hope to follow the learning process in the class, consequently, leading to unacceptable behaviors compared to the level of each students' motivation (Anas & Aryani, 2014). Study motivation will also foster active students' participation. Active participation is one of the behavior manifestations that the Elementary School Teacher Education students do in the process of learning in the class. This research focuses on the relationship between learning motivation and par-ticipation of Elementary School Teacher Education students in the implementation of DL due to Covid-19 spread.

Being compatible with the research results that has been conducted, this research shows the positive relationship between study motivation and participation of Elementary School Teacher Education students that accounts for Sig. of $0.000 < 0.05$ dan $r = 0.662 >$ r-table $= 0.124$. This is corresponding to the research by Mendari & Kewal (2015) that states that the study

students' motivation is in the level of moderate with the average value of 3.21 and there is the relation between study motivation toward the academic achievement of the students with the Sig. $= 0.007 < 0.05$. The research about the study motivation of the students is conducted by Hadiyanti (2017) with the result of r-value is 0.69 marked positive and Sig. $= 0.000 < 0.05$ which means that there is the relationship between study motivation and students' achievement.

On the other hand, turning to the research regarding with the students' participation that correlates with Sayidiman & Lambogo (2016) shows that the learning process that involves the student comprehensively physically, mentality and emotionally can encourage the emergence of the maximum potential. The research Widuroyekti (2006) shows that an increase of students' participation in the tutorial process in 3 cycles of learning with the actively engaging efforts on the subject of Indonesian and Mathematics. This is part of the most dominant factors of students' participation that is actively giving responses, understanding of learners, actively delivering answers, clarification skill, recapitulation ability, confident to give questions (Ginanjar et al., 2019). Study motivation and participation of Elementary School Teacher Education students affect to the process of DL amid the pandemic. The difference of this research from previous researches is that there is the relationship between study motivation and participation of the 2nd-semester of Elementary School Teacher Education students in the application of DL program during the pandemic. The research focusing the relationship of both the variables shows that there is a positive relationship with student abilities in the implementation of DL program.

The relationship between the students' motivation and participation is also linear with some previous studies. The motivation factor for achievement has an important role in the result of implementing the concept, which is clear and different findings among the group that has a weak and high motivation with the average score of 12.6234 (Sumarno et al., 2017). Meanwhile, the level of satisfaction seems high enough for students' participation and active learning with the score (> 4.0) for over 6 months online learning conducted in faculty (Walters et al., 2017). The finding from Bardakci et al. (2018) regarding with the online learning with the 30 respondences shows that 28 students take the advantages from the online discussion, 23 students do not experience the difficulties or have problems in giving expressions clearly and pleasurably, as well as 27 students have a positive perception toward teacher participation and think of that as motivation to be more active in the online discussion process. Puspitasari & Oetoyo (2018) states that with distance learning, the achieving students in the average age of 29 have motivation to study to support their career progress by learning for 3–4 hours for each subject a day.

On the other hand, the student participation phenomenon that fewer support in the process of DL can come as the problems in the implementation or operation of online learning media. It is same as in the research of Zou et al. (2019), which states that the platform that assists analyzing the online activities affects positively to students' participation motivation through the interaction of students on the multimedia platform including development, capability, achievement, and acknowledgment. The capability to operate the learning platform will have a significant impact to the students' desire to be able to follow the online lecturing activities. Therefore, between study motivation and participation are interconnected. The students with the high study motivation and participation will tend to be active to participate in the intention to reach the purpose. Therefore, the study motivation has a positive relationship with students' participation during DL process amid the pandemic.

4 CONCLUSION

Based on the discussion, it can be concluded that the study motivation of students has a relationship with study participation in the activities during the Distance Learning (DL) program. Amid pandemic outbreak, the study motivation has given a positive impact to study participation of the 2nd-semester of Elementary School Teacher Education students. The students' motivation becomes one of the influences toward students' participation in the process of DL program. This is because of the relation between study motivation and students' participation, namely the relationship of both research variables. The higher study motivation is the higher students' participation in the course of DL as an impact of the Covid-19 outbreak. The study motivation and participation of the 2nd-semester of Elementary School Teacher Education students as the variables in this research have a positive relationship in DL as the result of the disasterous impact caused by Coronavirus outbreak. Based on the analysis of the data and hypothesis test on both variables. There is a positive relationship with the category of moderate proven by the analysis results with the value of Sig. (study motivation) $= 0.000 < 0.05$ and Sig. (participation) $= 0.000 < 0.05$, so H_0 is rejected and H_1 is accepted. Then, the result of N = with df = 248 which is 0.124 resulted value of r (study motivation) = 0.662 > r-table = 0.124 and r (participation) = 0.662 > r-table = 0.124, so H_0 is rejected H_1 is accepted.

REFERENCES

Anas, M., & Aryani, F. 2014. Motivasi Belajar Mahasiswa. *Penelitian Pendidikan INSANI*, 16(1), 41–46.

Bardakci, S., Arslan, O., & Can, Y. 2018. Online Learning and High School Student: A Cultural Perspective. *Turkish Online Journal of Distance Education*, 19 (4)(October), 126–146. https://doi.org/https://doi.org/10.17718/tojde.471909

Cakır, O., Karademir, T., & Erdogdu, F. 2018. Psychological Variables Of Estimating Distance Learners' Motivation. *Turkish Online Journal of Distance Education*, 19(1), 163–182. https://doi.org/https://doi.org/10.17718/tojde.382795

Chaiprasurt, C., & Esichaiku, V. 2013. Enhancing Motivation in Online Courses With Mobile Communication Tool Support: A Comparative Study. *International Review of Research in Open and Distance Learning*, 14(3), 377–401. https://doi.org/10.19173/irrodl.v14i3.1416

Cohen, L., Manion, L., & Marrison, K. 2007. *Research Methods in Education. The American Biology Teacher* (Sixth, Vol. 63). New York: Routledge Taylor & Francis Group.

Espinoza-Freire, E. E., & Rojas-Garcia, C. R. 2019. The Tutoring Influences in Distance Education at El Oro Province Ecuador. *European Journal of Educational Research*, 8(4), 1093–1099. https://doi.org/10.12973/eu-jer.8.4.1093

Ginanjar, E., Darmawan, B., & Sriyono. 2019. Faktor-Faktor yang Mempengaruhi Rendahnya Partisipasi Belajar Peserta Didik SMK. *Journal of Mechanical Engineering Education*, 6(2), 206–219. https://doi.org/10.17509/jmee.v6i2.21797

Gupta, A., & Goplani, M. M. 2020. Impact of Covid-19 on Educational Institutions in India. *UGC Care Journal*, 31(21) (May). https://doi.org/10.13140/RG.2.2.32141.36321

Hadiyanti, A. H. D. 2017. Hubungan Motivasi Belajar Dengan Prestasi Belajar Mahasiswa PGSD Universitas Sanata Dharma Yogyakarta Pada Mata Kuliah IPA Biologi. *Jurnal Ilmiah PGSD*, 12(2), 1–10. https://doi.org/10.1017/CBO9781107415324.004

Herman. 2017. Loyality, Trust, Satisfaction and Participation in Universitas Terbuka Ambiance: Students' Perception. *Turkish Online Journal of Distance Education*, 18(3)(July), 84–95. https://doi.org/https://doi.org/10.17718/tojde.328937

Holmes, W., Nguyen, Q., Zhang, J., Mavrikis, M., & Rienties, B. 2019. Learning Analytics for Learning Design in Online Distance Learning. *Distance Education*, 40(3), 309–329. https://doi.org/10.1080/01587919.2019.1637716

Jackson, S. L. 2009. *Research Methods and Statistics A Critical Thinking Approach*. Diambil dari www.ichapters.com

Kementerian Pendidikan dan Kebudayaan. 2020. Surat Edaran Kemdikbud No 4 Tahun 2020 mengenai Pelaksanaan Pendidikan Dalam Masa Darurat Coronavirus Disease (Covid-19). Taken 22 Mei 2020, from kemdikbud.go.id / main/blog/2020/03/mendikbud-terbitkan-se-tentang-pelaksanaan-pendidikan-dalam-masa-darurat-covid19

Luthra, P., & Mackenzie, S. 2020. 4 Ways COVID-19 Could Change How We Educate Future Generations. Taken 21 Mei 2020, from https://www.weforum.org/agenda/2020/03/4-ways-covid-19-education-future-generations/

Mendari, A. S., & Kewal, S. S. 2015. Motivasi Belajar Pada Mahasiswa. *Jurnal Pendidikan Akuntansi Indonesia*, XIII13(1), 1–13. https://doi.org/https://doi.org/10.21831/jpai.v13i2.10304

Moorhouse, B. L. 2020. Adaptations to A Face-to-face Initial Teacher Education Course ' Forced ' Online Due to The COVID-19 Pandemic. *Journal of Education for Teaching*, 1–3. https://doi.org/10.1080 /02607476.2020.1755205

Murphy, M. P. A. 2020. COVID-19 and Emergency eLearning?: Consequences of The Securitization of Higher Education for Post-Pandemic Pedagogy. *Contemporary Security Policy*, 1–14. https://doi.org/10.1080/13523260.2020.1761749

Nsamba, A., & Makoe, M. 2017. Evaluating Quality of Students' Support Service in Open Distance Learning. *Turkish Online Journal of Distance Education*, 18(4)(October), 91–103. https://doi.org/https://doi.org/10.17718/tojde.340391

Precourt, E., & Gainor, M. 2018. Factors Affecting Classroom Participation and How Participation Leads to A Better Learning. *Accounting Education*, 1–19. https://doi.org/10.1080/09639284.2018.1505530

Puspitasari, K. A., & Oetoyo, B. 2018. Successful Students In An Open and Distance Learning System. *Turkish Online Journal of Distance Education*, 19(2)(April), 189–200. https://doi.org/https://doi.org/10.17718/tojde.415837

Sansone, C., & Harackieicz, J. M. 2000. *Intrinsic and Extrinsic Motivation*. London: Academic Press.

Sayidiman, S., & Lambogo, A. 2016. Partisipasi Belajar Mahasiswa Dalam Pembelajaran Berbasis Andragogi. *Publikasi Pendidikan*, 6(3). https://doi.org/10.26858/publikan.v6i3.2278

Saykili, A. 2018. Distance Education: Definitions, Generations, Key Concepts and Future Directions. *International Journal of Contemporary Educational Research*, 5(1), 2–17. Taken from http://search.ebscohost.com/login.aspx?direct=true&db=eric&AN=EJ1207516&site=ehost-live&scope=site

Simons, J., Leverett, S., & Beaumont, K. 2019. Success of Distance Learning Graduates and The Role of Intrinsic Motivation Intrinsic Motivation. *Open Learning: The Journal of Open, Distance and e-Learning*, 1–17. https://doi.org/10.1080/02680513.2019.1696183

Sumarno, Setyosari, P., & Haryono. 2017. Effect of Feedback Strategy and Motivation of Achievement to Improving Learning Results Concept in Learning Civic Education in Vocational High School. *European Journal of Educational Research*, 6(4), 441–453. https://doi.org/10.12973/eu-jer.6.4.441

Trust, T. 2017. Motivation, Empowerment, and Innovation: Teachers' Beliefs About How Participating in the Edmodo Math Subject Community Shapes Teaching and Learning. *Journal of Research on Technology in Education*, 49(1–2), 16–30. https://doi.org/10.1080/15391523.2017.1291317

UNESCO. 2020. COVID-19 Educational Disruption and Response. Taken 21 Mei 2020, from https://en.unesco.org/covid19/educationresponse

Walters, S., Grover, K. S., Turner, R. C., & Alexander, J. C. 2017. Faculty Perceptions Related to Teaching Online: A Starting Point for Designing Faculty Development Initiatives. *Turkish Online Journal of Distance Education*, 18 (4) (October), 4–19. https://doi.org/https://doi.org/10.17718/tojde.340365

Widuroyekti, B. 2006. Pendekatan Belajar Aktif Dan Peningkatan Partisipasi Mahasiswa Dalam Proses Tutorial Tatap Muka. *Jurnal Pendidikan*, 7(1), 55–65.

Zou, J., Liu, K., & Han, L. 2019. The Impact of A Rich Media Platform to Table Tennis Learners ' Performance and Participation Motivation. *Interactive Learning Environments*, 1–13. https://doi.org/10.1080/10494820.2019.1619087

Educational Innovation in Society 5.0 Era: Challenges and
Opportunities – Purnomo & Herwin (Eds)
© 2021 the authors, ISBN 978-1-032-05392-9

Increasing ecological intelligence through strengthening social studies education

R.A. Basit, A. Yuliyanto & B. Maftuh
Elementary Education Study Program, School of Postgraduate Studies, Universitas Pendidikan Indonesia, Bandung, Indonesia

H.E. Putri
Elementary School Teacher Education, Universitas Pendidikan Indonesia in Purwakarta, Indonesia

ABSTRACT: This literature study aims to reveal the increase in ecological intelligence through strengthening social studies education against world climate change. The perverse way of life of humans causes the earth's temperature to rise, now every part of the earth is experiencing global warming or climate change. This perverse way of life is caused by (1) minimal human knowledge regarding climate change, such as knowledge of the causes of climate change, the adverse effects of climate change, and ways of overcoming climate change; (2) the lack of empathy and social attitudes towards climate change; (3) and our lack of efforts in preventing climate change and cope with climate change that is now occurring. Social studies education has an important role in developing students' ecological intelligence, with good ecological intelligence it is hoped that they will become responsible citizens of the world, namely by not destroying and repairing the damaged natural environment to return nature to a more sustainable state and to rid of climate change problem.

1 INTRODUCTION

God has created the universe we live in, and throughout all the planets in the universe, Earth is the only known inhabitable planet. Unfortunately, our planet's current condition is not well. Even though COVID-19 has successfully reduced human consumptions and emissions, it has not succeeded in restoring Earth's natural sustainability. The reason is that humans as inhabitants of the earth are not aware of the way to preserve nature.

History has proven that changes that occur on Earth are related to human activities. Previously, there were no problems regarding Earth and its inhabitants, but over time the requirements became more complicated, making the humans' way of life a threat to nature, the damage to nature increased the Earth's temperature. Currently, all parts of the earth are undergoing global warming or climate change. Earth's climate changes are a response to a series of anthropogenic disturbances, especially the release of greenhouse gases (Zalasiewicz & Williams 2009).

Scientists assume this situation is not entirely due to natural disasters but is a consequence of human activity. Unfortunately, there are many gaps between the understanding of scientists and the public, there are still many people who do not care about climate change. The study states that people's understanding of climate change is limited to increasing sea levels, and those who live outside of Indonesia cannot see the influence of climate change around them, even though they have felt it. (LIPI 2015).

Climate change is defined as a shift in climate patterns caused by greenhouse gas emissions. Greenhouse gases cause energy to be trapped in the earth's atmosphere and are the main factor in climate change. The main sources of these emissions are the system of nature itself and human activities (Fawzy, Osman, Doran, & Rooney 2020). The negative impacts of climate change include: (1) health threat, every year there are 150.000 deaths due to climate change, and human injuries due to extreme weather; (2) impact on food production; (3) pattern changes on disease transmission (Arora 2019; Ashalatha, Munisamy, & Bhat 2012; World Health Organization 2002).

Greenhouse gases, which are the cause of climate changes, consisting of carbon dioxide, nitrogen oxides, and methane. The sun's energy enters the earth's surface, then part of the energy is absorbed, the residue is returned to space. The energy that does not return to space and causing the earth to warm is known as the greenhouse effect. In a normal context, the greenhouse effect is good for the earth, because without the greenhouse effect the energy will disappear into space and the earth's temperature will be colder, namely -18 ° C. But if the amount of greenhouse gases that are unreleased to earth is considered excessive, it may cause major disasters because there is too much energy left in the atmosphere. The study said the consequences of the greenhouse effect could be dangerous for humans,

226

DOI 10.1201/9781003206019-43

such as melting polar ice which caused sea levels to increase. As a result, the coastal fertile land is submerged. If flooding occurs with high intensity, it can threaten agriculture, crops to sink, and also causes freshwater to disappear (Mikhaylov, Moiseev, Aleshin, & Burkhardt 2020).

The excess greenhouse effect is caused by human activities, namely burning fossil fuels, deforestation, livestock manure, and garbage (Global Carbon Report 2019). Besides, there are fewer plants that function as pollutants from carbon emissions. Furthermore, the number of humans has increased, causing the need for consumption to increase, one type of food favored by the majority of humans is meat, so many animal farms are working to meet these needs. Whereas livestock produces large amounts of livestock manure, they produced toxic gases such as hydrogen sulfide, anomia, and methane which fly into the air, as methane traps more energy than carbon dioxide. (Humane Society International 2011). And lastly is garbage, landfills are the largest methane gas pollutant in Indonesia, the production of waste in Indonesia in 2018 was recorded at around 66 million tons, dominated by organic waste (60%) and plastics (15%) (Kementerian Lingkungan Hidup dan Kehutanan 2019). Other findings showed that 87 cities on the coast of Indonesia contribute about 1.27 million tons of waste to the sea and around 9 million tons of plastic (World Bank Group 2018). This condition causes Indonesia to become the 2nd largest country to waste the sea after China.

In Indonesia, climate change causes flooding in various regions, continual drought, increasing air temperatures due to energy waves, and rain accompanied by strong winds often occur. As a result of climate change, Indonesia's coast is threatened with drowning. By 2050 sea floods are expected to threaten 23 million people, they will have to face annual tidal floods due to climate change. (Climate Central 2019).

The findings suggest that the threat of sinking Indonesia's coast is very worrying. The increase in the earth's temperature causes icebergs to melt at the north and south poles, thus increasing sea level. Besides, overexploitation of groundwater is believed to be the cause of land subsidence, this condition predominantly occurs on the coast of Indonesia. Research reveals that the geographic conditions of the Indonesian archipelago coast are lowland, coupled with the impact of melting icebergs in the north and south poles, and land subsidence, which is considered as a factor of sea flooding which takes place every year, this condition threatens the sustainability of Indonesia's coastal life. This condition causes more than 100 regencies/cities on the coast of Indonesia to have the potential to sink, but not all people on the coast are aware of this threat (Amindoni 2020).

Climate change also has an impact on the health of the Indonesian people. In Indonesia, extreme and unpredictable weather often occurs due to climate change, parts of Indonesia experience continuous rain accompanied by strong winds that cause flooding so that the environment becomes dirty and mosquitoes are easy to breed under this condition. This condition causes cases of malaria and dengue fever to increase. Meanwhile, in other regions, there is a prolonged drought which causes extreme temperatures from the sun's energy such as skin burn, rice fields drought, water sources drought, and forest fires that produce vapor, which has an impact on respiratory health. Furthermore, as a country with a tropical climate, Indonesia often experiences prolonged drought, this condition is very helpful for the growth of bacteria and parasites, this condition can cause diseases related to bacteria and air such as skin infections (Knowledge Centre Climate Change 2017).

The problem of climate change is not only in Indonesia, all countries feel the impact and contribute to this problem that makes climate change a world problem. Therefore, the United Nations, through a large SDGs agenda, lists climate change as one of the 17 goals (Johnston 2016). Indonesia is a country committed to the SDGs, and we must act together to make SDGs 2030 a success. A sector that contributes to climate change prevention is education, education can give birth to generations with a caring attitude towards nature preservation. This is reinforced by the fact that to deal with climate change, one of which is stated in target 13.3 which states building education, raising awareness, and human and institutional capacities related to mitigation, adaptation, impact reduction, and early warning of climate change. With indicator 13.3.1, i.e the number of countries that have integrated climate change mitigation, adaptation, impact reduction, and early warning in the curricula of primary, secondary schools, and colleges. (Kementerian Perencanaan Pembangunan Nasional/BAPPENAS 2017). Regarding the preservation of environmental learning that has been taught since elementary school, it turns out that it is too simple, causing misunderstandings. This misunderstanding has transformed into an inaccurate policy basis. This confirms that the role of education is very essential for the fate of the nation in the future, so education must be present to contribute to the problems that have long been occurring on this earth, namely social and economic problems that are contrary to nature conservation which results in the problem of climate change (Sumarto 1992). Education plays an important role in preserving the environment as well as supporting the development of natural resources in a sustainable manner (Khan, Ali, Khan, Shah, & Shoukat 2014).

Social studies education as a compulsory subject in elementary and middle schools plays a role in overcoming the problem of climate change that is now occurring. Social studies education consists of two points, the first is education and the second is social studies. Education is defined as a conscious and planned effort that allows students develop intellectually and emotionally as they are expected to possess high quality of character and the ability to build beneficial generation for themselves, their families, the community, and the nature in which they live (Basit & Maryani 2020). Meanwhile, social studies have an important role in developing the intellectual, emotional, cultural, and social aspects of the students. Such as ways of thinking, behaving, and behaving responsibly as individuals and citizens of the world. Furthermore, social studies play a role in obtaining the knowledge, skills, and values needed by students from various disciplines, and developing students' potentials to be sensitive to social problems in society, have a positive mental attitude to correct all inequalities, and are experts in overcoming every life problems that befall them or public (Büyükalan filiz & Baysal 2019; Maryani & Syamsudin 2009).

The objectives of social studies education are: (1) developing intellectual abilities which are oriented towards intellectual development relating to students and the interests of social studies subjects; (2) development of responsibility as a member of society that is oriented towards the development of students and the interests of the community; (3) student development in self-interest, society or science (Hasan 1996).

However, it is unfortunate that the implementation of social studies education has not supported ecological intelligence in dealing with climate change issues. The social studies learning process in schools which is carried out by the teacher still emphasizes the cognitive aspects, namely the transfer of theories only, thus learning tends to be focused on memorizing and emphasizes the thinking ability of students at a low level (Maftuh 2010). In the social studies learning process, the teacher only delivers subject matter, while the development of the ability to think and behave as provisions to become good citizens, is not given enough attention (Al-Muchtar 2007). The ecological intelligence of students can realize attitudes and real actions to protect the environment by living hygienically (Hendriyanto, Mustofa, & Sutopo 2018).

Human ecological intelligence is measured through (1) Knowledge of the impact of activities including how the impacts caused, and how they are affected; (2) Improvement of caring attitudes, i.e. caring for the environment, including transmitting understanding to others; (3) Skills to save the environment; and (4) involved in environmental activities (Goleman 2010).

Thus, it is necessary to strengthen the role of social studies education to ensure social studies learning does not only focus on cognitive, but also explore critical thinking skills, problem-solving, and ecological intelligence, therefore it can provide students to live more productively without harming Earth. In this article, the author will explore how social studies education improves ecological intelligence, which includes knowledge, empathy, and skills. People with knowledge, awareness, and life skills in harmony with the preservation of nature are the characteristics of humans with ecological intelligence. Ecologically intelligent people are people with an understanding of their attitudes and actions, not only impacting themselves and others but also impacting nature. A person's ecological intelligence will describe his actions

related to nature conservation (Supriatna 2016). Given the previous, ecological intelligence is complex. This intelligence is supported by intellectual, affective, social, and emotional elements, and psychomotor.

2 DISCUSSION

Human's unwise lifestyle causes the Earth's temperature to continue to rise, now all parts of the planet are experiencing climate change. People's ways of life are wrong because human understanding is still limited regarding climate change, such as knowledge of the causes of climate change, the adverse effects of climate change, and ways of tackling climate change.

To allow students to contribute to improving the problem of climate change, of course, students must have good ecological intelligence, one of the importance of ecological intelligence is 21st-century skills, including critical thinking and problem-solving abilities. Critical thinking and problem-solving skills are considered the foundation of 21st-century learning. Limited knowledge regarding the damage to the natural environment creates no empathy for environmental sustainability, as well as skills related to overcoming natural damage that occurs or skills to protect and preserve nature (Trilling & Fadel 2010).

This climate change problem involves a longstanding society's wrong behavior, the impact of the wrong behavior produces conditions that are not desired by the Earth's inhabitants. Thus, this problem is included in the list of societies' social problems that must be solved. In education, social issues are classified as the realm of social studies education, so social studies education must be present to solve the problem of climate change.

2.1 The role of social studies education to increase ecological intelligence in tackling climate change problems

Social studies education is a subject that aims to educate students to become good citizens. This is reinforced in Permendiknas No. 22 of 2006 concerning Standard Content of Social Studies Subjects, i.e.: (a) recognizing concepts related to people's lives and their environment; (b) have the basic skills of logical and critical thinking, curiosity, inquiry, problem-solving, and skills in social life; (c) commit to social and humanly values, and (d) can communicate, cooperate and compete in a pluralistic society at the local, national, and global levels (Depdiknas 2006). Furthermore, the function of social studies education in elementary schools is to develop the knowledge, values, attitudes, and social skills of students so they can study daily social life and foster a sense of pride and love for the country (Fajar 2002).

Thus, social studies learning has a major role in defeating the problem of climate change. Starting from growing knowledge about climate change, such as the causes of climate change, the negative impacts of climate change, and how to overcome the problems of climate change that occur. From the knowledge that has been built, it is hoped that critical thinking skills will emerge which will then lead students to empathize with climate change, and lead to the growth of problem-solving abilities that allow students to learn skills to overcome climate change problems that occur.

Another effort to achieve the goals of social studies education in overcoming the problem of climate change is to integrate local wisdom values into social studies learning materials. The role of local wisdom in preventing deforestation is proven to be effective, that in the tradition of the Sudanese people there is the term *Leuweung Larangan* or forbidden forest. The community cannot enter the forbidden forest at all and even use it, the Sundanese people believe that whoever violates will get bad luck in life. Communities who still apply the *Leuweung Larangan* include Kasepuhan Ciptagelar, Baduy, and Kampung Naga. Furthermore, local wisdom in warding off the consumptive culture of the people, namely in the Sundanese culture there is a proverb *saeutik mahi, loba nyesa,* or a little number of matters will be enough, but a huge number of matters will leave remains. So, when you have something in a small amount, you have to do whatever it takes to make it enough, do not let it turn you into wanting more, as well as when you have something that is a lot, you have to spare it, so you can use the remaining left.

The following effort in tackling climate change through social studies learning is to integrate the social studies learning values so the character of students is formed according to these values. In line with this, education aims to form students who have a responsible character by instilling norms and values in them. To comply with the prevailing norms and values requires discipline and responsibility that is created in each student (Yuliyanto, Fadriyah, Yeli, & Wulandari 2018). The values that must be developed in tackling climate change are religion, discipline, hard work, creativity, love for the country, friendship, the love to read, social care, environmental care, and responsibility. (1) Religious, i.e., students who accept religious orders and abandon their prohibitions, will keep them away from actions that harm the nature that may lead to climate change problems; (2) Discipline, i.e., students with disciplined characters such as bringing their food will reduce waste production; (3) Hard work, i.e., the problem of climate change, is not a simple matter, of course in overcoming it requires a hard work attitude from every human being who cares about this problem; (4) Creative, that is, the problem of climate change has yet to be resolved, even the condition is getting worse; (5) Creativity is required in finding solutions to solve them. Love for the Motherland will raise awareness of the problems that befall this country; (6) Friendship, i.e., the problem of climate change, including social problems caused by many people that have an impact on other people and themselves. So, solving this problem cannot be done alone, but it is necessary to solve it together; (7) The willingness to read is important

to solve the problem of climate change. As to solve this problem, one must have sufficient knowledge to increase the public knowledge in tackling the problem of climate change; (8) Social care, i.e., climate change has caused disasters, such as drought, floods, and many more. Affected communities will need help. Humans with high social care are easy to help those affected by disasters; (9) Care for the environment, i.e., the problem of climate change is the result of environmental damage, so, caring for the environment is required to overcome the problem of climate change; (10) Responsibility, i.e., the problem of climate change due to human activities that are not responsible for damaging the environment, so the responsibility is required to tackle the problem of climate change.

Of course, learning strategies are required to make these values attached to students. In developing values, two theories can be used, i.e., the theory of socialization and constructivism. The use of these two theories is usually correlated with a childhood developmental period, by the views of morality expressed by Piaget, namely: (1) Heteronomic morality (4–7 years), justice and rules are understood as the property of the world that cannot be changed, removed from people's control. The moral values at this stage are influenced by others, for example, wearing masks outside the home during a pandemic for fear of police raids; (2) Autonomous morality (>10 years), students are conscious of the rules and laws created by people and in judging an action, one must consider its aims and consequences. The values at this stage are not influenced by others, but arise from self-awareness, for example wearing a mask when leaving the house even though there are no police (Bybee 1982).

The strategy that is considered suitable for low-grade students at the primary school level is socialization, i.e., the process of teaching values by the teacher, by providing examples, such as disposing of garbage in its place, picking and sorting garbage, watering and caring for plants; etc. In this strategy the teacher uses a direct approach when learning, telling students that this is right or wrong, this strategy is considered effective as it can make students easily understand the message conveyed by the teacher. This strategy is easier for teachers to apply, only in practice the teacher will be more active than the students.

Furthermore, the constructivism strategy is considered more suitable for high-grade students at the elementary school level. In its implementation, the teacher does not tell directly about right or wrong to students, but the teacher facilitates students to find out whether something is right or wrong for themselves, for example, the teacher invites students to a discussion with a topic about the importance of reducing and processing waste, maintaining plants, using minimal vehicles, pollution, reduce food from processed meat, etc. The strength of this strategy is that students are more active than the teacher, but this strategy takes a long time. Learning values using this strategy must be comprehensive because it is worried that misperceptions will arise.

2.2 The role of teachers to increase ecological intelligence in tackling climate change problems

Teachers are the key to a successful education, teacher professionalism as an educator is required. Now teachers are not only the main source for learning, but teachers are also required to be role models, facilitators, and motivators for students in the social studies learning process, as students need to have real examples of concern for the problem of climate change that is happening.

Some things that teachers must pay attention to improve ecological intelligence in overcoming climate change problems: (1) Teachers must understand the objectives of social studies education, that social studies education is not only a cognitive problem but social studies learning must have strengths. Several principles of social studies learning have strengths, namely: meaningful, integrated, values-based, challenging, and active (National Council for the Social Studies 2006). Social studies learning with this kind of strength does not only develop cognitive aspects but they develop the affective and psychomotor aspects; (2) The teacher acts as a role model for students, so the teacher's actions must reflect behavior by the principles of tackling climate change problems. Teachers must increase knowledge about the problem of climate change, whether it is the causes, impacts, and how to tackle climate change problems; (3) Teachers must integrate social studies material with climate change issues and include it in the lesson plan. Starting from determining fundamental competencies that can develop students' real concern for tackling climate change problems, then designing learning goals, and designing learning activities, and assessments; (4) In the learning process, teachers must carry out innovative learning. Integrated thematic learning, cooperative learning, and active learning are alternatives in designing student-oriented learning skills to cope with climate change problems; (5) Teachers can invite students to make a collective agreement to tackle climate change, such as school residents are required to bring their supplies, obliged to sort and process the waste by themselves, must have a pet, and take care of the tree until it grows, using a bicycle, walk, or use public transportation to school, decrease meat consumption and substitute traditional foods. Students who succeed in realizing collective agreements receive appreciation from the teacher.

3 CONCLUSION

Ecological intelligence is required by students to overcome the problem of climate change which is currently getting worse. Social studies education plays an important role in improving students' ecological intelligence in tackling the problem of climate change. In improving ecological intelligence, social studies learning must have strength in the process, so students' cognitive, affective, and psychomotor aspects can be developed. The role of teachers in increasing ecological intelligence in overcoming climate change problems through social studies learning is very important. Teachers need to understand the goals of social studies, be role models, integrate ecological intelligence in lesson plans, create innovative social studies learning, and make collective agreements related to real actions in tackling climate change between teachers and students.

REFERENCES

Al-Muchtar, S. 2007. *Strategi Pembelajaran Pendidikan IPS.* Bandung: UPI Press.

Amindoni, A. 2020. Perubahan iklim: Pesisir Indonesia terancam tenggelam, puluhan juta jiwa akan terdampak. *BBC News Indonesia.*

Arora, N. K. 2019. Impact of climate change on agriculture production and its sustainable solutions. *Environmental Sustainability*, 2(2): 95–96. https://doi.org/10.1007/s42398-019-00078-w

Ashalatha, K. V, Munisamy, G., & Bhat, A. R. S. 2012. Impact of climate change on rainfed agriculture in India: a case study of Dharwad. *International Journal of Environmental Science and Development*, 3(4): 368–371.

Basit, R. A., & Maryani, E. 2020. Model Pembelajaran Active Learning Tipe Snowball Throwing dan Tipe Iindex Card Match (ICM) terhadap Pemahaman Konsep Siswa pada Pembelajaran IPS di Sekolah Dasar. *Jurnal Pendidikan Dasar*, 11(1): 118–125. https://doi.org/10.21009/10.21009/JPD.081

Büyükalan filiz, S., & Baysal, S. B. 2019. Analysis of Social Studies Curriculum Objectives According to Revised Bloom Taxonomy. *Inonu University Journal of the Faculty of Education*, 20(1): 234–253. https://doi.org/10.17679/inuefd.435796

Bybee, R. W. 1982. *Piaget for Educators.* Columbus: Charles E. Merrill Publishing Company.

Climate Central. 2019. Report: Flooded Future: Global vulnerability to sea-level rise worse than previously understood. *Climate Central.*

Depdiknas. 2006. Permendiknas no. 22 tahun 2006 tentang Standar Isi Mata Pelajaran. In *Departemen Pendidikan Indonesia.*

Fajar, A. 2002. *Portofolio dalam Pembelajaran IPS.* Bandung: PT. Remaja Rosdakarya.

Fawzy, S., Osman, A. I., Doran, J., & Rooney, D. W. 2020. Strategies for mitigation of climate change: a review. *Environmental Chemistry Letters*, 18(6): 2069–2094. https://doi.org/10.1007/s10311-020-01059-w

Global Carbon Report. 2019. Summary from Global Carbon Report 2019 and other press releases by GCP Tsukuba office. *Global Carbon Project*, 2–5. https://doi.org/10.5194/essd-11-1783-2019

Goleman, D. 2010. *Ecological intelligence: The hidden impacts of what we buy.* Currency.

Hasan, S. H. 1996. *Pendidikan Ilmu Sosial.* Jakarta: Dirjendikti, Depdikbud RI.

Hendriyanto, A., Mustofa, A., & Sutopo, B. 2018. Building Ecological Intelligence through Indonesian Language Learning Based on Kethek Ogleng Dance. *International Journal of Education*, 11(1): 50. https://doi.org/10.17509/ije.v11i1.10902

Humane Society International. 2011. *An HSI Report: The Impact of Animal Agriculture on Global Warming and Climate Change.* 1–27.

Johnston, R. 2016. Transforming Our World: The 2030 Agenda For Sustainable Development. In *Arsenic Research and Global Sustainability - Proceedings of the 6th International Congress on Arsenic in the Environment, AS 2016* (pp. 12–14). https://doi.org/10.1201/b20466-7

Kementerian Lingkungan Hidup dan Kehutanan. 2019. *Saatnya Berubah Aksi Korektif Siti Nurbaya Mengelola Lingkungan Hidup dan Kehutanan* (E. W. Soegiri, S. Murniningtyas, & T. Yanuariadi, Eds.).

Kementerian Perencanaan Pembangunan Nasional/ BAPPENAS. 2017. *Pedoman Penyusunan Rencana Aksi TPB/SDGs.*

Khan, H., Ali, F., Khan, H., Shah, M., & Shoukat, S. 2014. Estimating willingness to pay for recreational services of two public parks in Peshawar, Pakistan. *Environmental Economics*, 5(1): 21–26.

Knowledge Centre Climate Change. 2017. Dampak Perubahan Iklim Terhadap Kesehatan Manusia.

LIPI. 2015. *Pemahaman Masyarakat Terhadap Perubahan Iklim Minim*. Jakarta.

Maftuh, B. 2010. *Memperkuat Peran IPS dalam Membelajarkan Keterampilan Sosial dan Resolusi Konflik* (pp. 1–32). pp. 1–32.

Maryani, E., & Syamsudin, H. 2009. Pengembangan Program Pembelajaran IPS Untuk Meningkatkan Kompetensi Keterampilan Sosial. *Jurnal Metodik Didaktik, 9*(1): 1–15.

Mikhaylov, A., Moiseev, N., Aleshin, K., & Burkhardt, T. 2020. Global climate change and the greenhouse effect. *Entrepreneurship and Sustainability Issues*, 7(4): 2897–2913. https://doi.org/http://doi.org/10.9770/jesi.2020.7.4(21)

National Council for the Social Studies. 2006. *National Standards for Social Studies Teachers*. Retrieved from http://www.socialstudies.org/standards/teacherstandards

Sumarto, O. 1992. *Indonesia dalam Kancah Isu LIingkungan Global*. Jakarta: PT Gramedia.

Supriatna, N. 2016. *Ecopedagogy: Membangun Kecerdasan Ekologis dalam Pembelajaran IPS* (1st ed.; Nita, Ed.). Remaja Rosdakarya.

Trilling, B., & Fadel, C. 2010. 21St Century Skills: Learning for Life in Our Times. *Choice Reviews Online*, 47(10), 47-5788-47–5788. https://doi.org/10.5860/choice.47-5788

World Bank Group. 2018. Laporan Sintesis Sampah Laut Indonesia. *Public Disclosure Authorized*, (April), 1–49.

World Health Organization. 2002. The world health report 2002: reducing risks, promoting healthy life. In the *World Health Organization*.

Yuliyanto, A., Fadriyah, A., Yeli, K. P., & Wulandari, H. 2018. Pendekatan Saintifik untuk Mengembangkan Karakter Disiplin dan Tanggung Jawab Siswa Sekolah Dasar. *Metodik Didaktik*, 13(2): 87–98. https://doi.org/10.17509/md.v13i2.9307

Zalasiewicz, J., & Williams, M. 2009. A Geological History of Climate Change. In *Climate Change* (1st ed.). https://doi.org/10.1016/B978-0-444-53301-2.00006-3 '

Educational Innovation in Society 5.0 Era: Challenges and Opportunities – Purnomo & Herwin (Eds)
© 2021 the authors, ISBN 978-1-032-05392-9

Teacher's quality of pedagogical influence on a student's character in the society 5.0 era

N.K. Suarni, G.N. Sudarsana & M.N.M.I.Y. Rosita
Universitas Pendidikan Ganesha, Singaraja, Bali, Indonesia

ABSTRACT: The purpose of this study was to determine the effect of teacher pedagogical quality on student character development in the society 5.0 era. This type of research is a quasi-experimental research design with a post-test only control group. Respondents in this study were 140 elementary school students in the XI Cluster, Buleleng District, who were determined using random sampling techniques. The data was collected using an instrument in the form of a rubric. Data analysis was performed by using difference test (t-test). Based on the results of the analysis, it was obtained that the value of $t = -1.828$ with $p < 0.05$. This proves that there is an influence of the quality of the teacher's pedagogical on the character development of students in Society 5.0 Era. The results of this study have implications for the development of student character to face a smart society.

Keywords: Quality of Pedagogical; Society 5.0

1 INTRODUCTION

1.1 *Background of study*

Teachers have mastered pedagogical knowledge while studying in tertiary institutions, and practical simulations have also been experienced during undergraduate teacher education in tertiary institutions. However, after applying pedagogic knowledge in schools, there are still many facts gathered from several studies, that the application of teacher pedagogy in learning activity is not in accordance with pedagogical theoretical procedures and is as well difficult to adapt to the demands of changes that occur such as demands in the Industrial Revolution 4.0 Era and or Society 5.0 Era (Ardana 2018, Harta 2019, Karlina et al. 2017, Putra et al. 2019, Yuniari et al. 2019). This fact can be used as evidence that the quality of pedagogical (QoP) of teachers in learning activity is still low.

Quality of Pedagogical (QoP) of teachers is important to improve because it relates to the teacher's capacity, namely methods and insight. The method is a learning method that will be seen in the learning design or in the learning activities of students. Meanwhile, insights are things that are not visible in the learning design, unpredictable, occur spontaneously in learning (Hendayana et al. 2019). If the teacher dominates learning such as giving many lectures, there will be no insight so that the teacher cannot make personal observations to students. So, it is difficult to identify students who have learning difficulties or low learning motivation. From this, it will raise several learning problems such as students who are not enthusiastic about coming to school, causing low character strengthening in students.

Many efforts have been made by the state in improving the quality of teacher competencies such as education and training, increasing educational qualifications, improving life welfare, preparing learning support facilities both manual and ICT-based, but there is still much that can be done in improving the quality of learning so that students have the readiness maximally as a citizen in facing challenges that are so rapidly changing (Fildzah, 2020; Ghavifekr & Rosdy, 2015; Kurniati, Arafat, & Mulyadi, 2019; Muralidharan & Singh, 2020). Quality of pedagogy (QoP) has a great influence on the development of student character, especially in the education system.

The Ministry of National Education explained that there are 18 character values that need to be instilled in students, including: 1) religiousness, 2) honesty, 3) tolerance, 4) discipline, 5) hard work, 6) creativity, 7) independence, 8) democracy, 9) curiosity, 10) spirit of nationality, 11) love for the country, 12) respecting achievement, 13) friendliness/communication, 14) peacefulness, 15) love of reading, 16) caring for the environment, 17) sociability, and 18) responsibility (Kemendiknas 2011). However, these character values have crystallized into five main character values consisting of religiousness, nationalism, independence, integrity and cooperation in accordance with Presidential Regulation Number 87 of 2017 with the motto "Happy Learning in a Second Home" (Presidential Regulation No. 87 2017).

These character values should be instilled in students from an early age because character planting does not automatically emerge from within students, everything requires a learning process that may occur through their daily activities. It is then deepened through the formal education process in schools. So that it is considered important to strengthen the pedagogical quality of teachers to help students develop their own character through the learning process. This is because in the society 5.0 era it is expected that students will have a broad-based thinking character to adapt in the future, such as thinking analytically, critically, and creatively. This way of thinking is called higher-order thinking (HOTS: Higher-Order Thinking Skills) (Santoso 2019). In the society 5.0 era, all people are required to be able to solve various challenges and social problems by utilizing innovations that were born in the revolutionary of industry 4.0 era. So based on these demands this research can be used as a design to strengthen the quality of pedagogical teachers in helping students understand and develop their own character. It is feared that if this does not get attention, the possibility of deviation and incorrect character formation in students will very easily occur. As it is known that children of certain ages have different levels of curiosity, so that the teacher becomes one of the introductory figures to develop good and virtuous character. The problem is formulated: is there any influence of the Quality of Pedagogy (QoP) of teachers on the development of student character in the society 5.0 era?

1.2 Research purposes

The purpose of this study was to determine the effect of Quality of Pedagogical (QoP) teachers on character development of elementary school students in the XI Cluster, Buleleng District in the society 5.0 era.

2 THEORETICAL BASIS

2.1 Quality of Pedagogical (QoP)

A professional teacher is not only described as a model teacher as a teacher, but a person who can educate, guide, direct, train, assess, and evaluate their students. One of the competencies that teachers must have is pedagogical, namely the ability to manage student learning so that they can actualize various potentials within themselves. This competence prioritizes teacher management in learning and student activity in learning.

Quality of pedagogy is the quality of learning management carried out by teachers to achieve learning objectives. Quality of pedagogy builds the spirit of reflective learning activities and is related to methods and insights.

The following method is a learning method that will be visualized in a learning activity. The learning method has three indicators, namely 1) Pedagogy Knowledge (PK), is an educational theory that discusses teaching activities with the aim of changing children's behavior (Kurniasih & Sani 2017). In addition, it transforms knowledge and develops children's personalities (Sadulloh 2014); 2) Content Knowledge (CK), is the teacher's ability to master knowledge in various fields of science, at least including mastery of subject matter and deepening important content in learning. In other words, content knowledge leads to the specificity of the discipline or subject matter (Rosyid, 2016); 3) Pedagogy Content Knowledge (PCK), is a teacher's specific knowledge in teaching content with a strategy to be able to direct understanding to students in the form of a combination of understanding teaching material (content knowledge) and understanding how to educate (pedagogy knowledge) (Sukadi et al. 2015).

Insight is knowledge and insight into the characteristics of students. Insights relate to things that are not visible in a learning design such as the results of observing facial expressions and non-cognitive behaviors that may be shown by students (Hendayana et al. 2019). In its application, insight must be carried out silently, calmly, and slowly. Every child has their own face. This difference depends on the level of difficulty experienced by students. Insight occurs spontaneously and as a teacher who has a good quality of pedagogy must be responsive to understanding the changes shown by students. At this time, the quality of a teacher is shown, and how well the teacher can take decisions based on the results of the observations made is determined.

2.2 Strengthening character education

Emphasis on character education in children is not new anymore, because the activists of education since time immemorial have tried to implement character development as shaping the personality of everyone, especially the nation's future generations. As a result of the changing times, strengthening character education needs even more attention. It takes a place that is truly consistent in helping to develop the strengthening of character education. The character's values developed in the strengthening character education program are based on the philosophy of Ki Hajar Dewantara's character education, namely the harmonization of heart exercise (ethics), feeling (aesthetics), thought (literacy), and sports (kinesthetics).

In essence, education can be seen as a process of empowering and cultivating individuals so that they are able to meet development needs and meet public, cultural and religious demands on their environment. Education as a process of fostering students who are social, cultured, in an order of life that has local, national and global dimensions.

Character is a person's behavior based on values in accordance with the norms prevailing in society. It is strengthened again by the definition that character is a set of values that underlies a person's thoughts, feelings, attitudes and behavior, which are very important to be instilled from self (Hasanah & Deiniatur 2018). Whereas character education is intended as the formation of the foundation of students through the cultivation of character values in the form of educational actions as the next generation of quality and capable of living independently (Dalyono 2017; Wulandari & Kristiawan 2017).

Character education is also said to be an effort made by several parties such as schools, districts, states by instilling ethical values, care, honesty, justice, responsibility, and respect for themselves and others (Singh, 2019). Various efforts made by stakeholders such as educators, education experts, and the government in developing character education do not stop there. There are always improvements in strengthening character education to be able to achieve the goals of national education to be made. This improvement is evidenced by the strengthening of the character education movement through the strengthening character education program. In the program, the five main values of these characters are formulated, namely (1) religiousness, (2) nationalism, (3) independence, (4) cooperation, and (5) integrity (Sulistyarini et al. 2019).

The five character values contain the following sub-values:

1) Religiousness. The sub-values contained in religious values are loving and maintaining the integrity of creation, peace, steadfastness, confidence in loving the environment, mutual protection, tolerance, respect for differences in religion and belief, cooperation between followers of religions and beliefs, not imposing their will, sincere, and protect the small and the marginalized.
2) Nationalism. The sub-values contained in the nationalist values are protecting the environment and obeying existing regulations in schools such as following the flag ceremony solemnly, appreciating the nation's own culture, being willing to sacrifice, loving the homeland, obeying the law, respecting cultural, ethnic and religious diversity, discipline, and excellence.
3) Independence. The sub-values contained in the value of being independent are having a good work ethic (working hard), being tough and resilient, having a fighting spirit, being professional, creative, brave, and being a lifelong learner.
4) Cooperation. The sub-values contained in the value of cooperation are accustoming oneself to cooperate with one another by voluntary work, promoting deliberation and mutual respect between friends, non-violence, helping hand, solidarity, voluntary attitude, cooperation, inclusion, commitment to joint decisions, empathy, anti-discrimination and violence.
5) Integrity. The sub-values contained in the value of integrity are being actively involved in social life, consistent in acting and speaking based on truth, honesty, loyalty, commitment to morals, fairness, responsibility, and respect for individual dignity.

The goals of the strengthening character education program are: 1) human-resource development as a foundation for national development, 2) 21st-century skills needed by students such as character quality, basic literacy, and the 4Cs in order to realize excellence in competing in the golden generation of 2045, and 3) to prevent the tendency of degradation of morality, ethics, and manners.

Strengthening character education into attitudes and habituation behavior through education, which makes the basic capital of everyone's nation to be able to adapt to various changes, especially changes to life's challenges in the future, such as the era of industrial revolution 4.0 and/or society 5.0 era. This is supported by research results Triyani et al. (2020) dan Annisa et al. (2020) that the character education strengthening ceremony is the right means of strengthening character education for students because it integrates main values that are aligned with student development.

Society 5.0 is a new breakthrough for digital transformation that is centered on human life. According to Fukuyama, looking back at human history, various stages of society can be defined. Society 1.0 is defined as a group of people who hunt and gather, living in harmony with nature; society 2.0 forms groups based on agricultural cultivation, improved organization and nation building; society 3.0 is a society that promotes industrialization through the industrial revolution, enabling mass production; and society 4.0 is an information society that realizes the increased added value by connecting intangible assets as information networks. Society 5.0 is an information society built on top of Society 4.0, which aims for a prosperous human-centered society. The combination of cyber space and the real world (physical space) to produce quality data is the key to the realization that can create values and solutions to the challenges faced (Fukuyama 2018).

The goal of society 5.0 is to create a human-centered society, where economic development and community problem solving can be achieved, and everyone can enjoy an active and comfortable quality of life. An example that can be applied by teachers to be able to help students face society 5.0 is teaching students to search for material or learning materials from the internet by using their gadgets. Students can be directed to search for various learning videos available on Amazon Education, Khan Academy, Teacher's Room, Wikipedia, and others.

3 METHOD

3.1 Research design

This type of research is a quasi-experimental research design with posttest only control group research.

3.2 Research sample

The samples in this study were elementary school students in Cluster XI, Buleleng District, Buleleng Regency. Sample reduction is used by random techniques. Primary schools included in the XI Cluster of Buleleng District are Undiksha Lab Elementary

School, Kampung Bugis 1 Elementary School, Kampung Anyar 1 State Elementary School, and Kampung Anyar 3 Elementary School. After the equivalence, test was carried out, a lottery was conducted to determine the experimental and control classes. Thus, the obtained classroom experiment that Undiksha Lab Elementary School with 70 students and Kampung Anyar 3 Elementary School with a total of 70 students group experiment consists of two classes to implement the quality of pedagogical teachers in developing student character while the control group consisted of two classes given conventional learning.

3.3 Research procedure

The procedures for implementing this experiment are 1) the development of learning designs. The development of this learning design will be better if it is done collaboratively with peers or colleagues than independently. At this stage, it also discusses Pedagogy Knowledge (PK), Content Knowledge (CK), Pedagogy Content Knowledge (PCK) as well as deepening insight with peers; 2) observation of learning. This is done online through Google Meet due to the Covid-19 conditions that hit Indonesia so that the learning process cannot be done directly (face to face), observation is equipped with an instrument for assessing student character in the form of a rubric; 3) post-learning reflection, not to judge teachers as "good" or "bad" but to get feedback for teachers and inspiration for observers; 4) data analysis, the results of the character assessment instrument in the form of a rubric were analyzed to see the effect of QoP on student character development in the society 5.0 era.

3.4 Data collection

The method of collecting student character data in this study used an instrument in the form of a rubric which was used to assess student character through observation in online learning (Google Meet).

3.5 Data analysis

The data analysis method used to see the effect of quality of pedagogical (QoP) on students' character development used difference test (t-test)

$$t = \frac{\overline{X}_1 - \overline{X}_2}{\sqrt{\left(\frac{s_1^2}{n_1} + \frac{s_2^2}{n_2}\right)}} \quad (1)$$

where $\overline{X}_1$ = average post-test score of the experimental group students; $\overline{X}_2$ = average post-test score of the control group students; s_1^2 = variance of the experimental group; s_2^2 = variance of the control group; n_1 = number of experimental groups; n_2 = number of control groups

4 RESULT AND DISCUSSION

The formulation of the research problem is that there is an effect of the quality of the teacher's pedagogical strengths on the character development of students in the Society 5.0 Era. The following shows descriptive statistics in table 1.

Table 1. Descriptive statistics

	Group	N	Mean*	Std. Deviation	Std. Error Mean
Character	Control	70	93.34	10.441	1.248
	Experimental	70	96.39	9.218	1.102

Data calculated using SPSS 16.0 for windows

Based on the descriptive statistical table above, it can be seen that the number of subjects in the control group is 70 people, and the experimental group is 70 people. The mean value of student character in the control group was 93.34 while the experimental group was 96.39. These results indicate that the experimental group has a higher character mean. Furthermore, the independent simple t-test analysis is shown in table 2.

Table 2. Independent samples test

	t	df	Sig. (2-tailed)*	Mean Difference	Std. Error Difference
Equal variances assumed	−1.828	138	.070	−3.043	1.665
Equal variances not assumed	−1.828	135.911	.070	−3.043	1.665

Data calculated using SPSS 16.0 for windows

From the analysis, results obtained the value of $t = -1.828$ with $p < 0.05$. Then H0 is rejected and Ha accepted. This proves that this hypothesis means that there are differences in the effect of quality of pedagogical teacher against student character development in the society 5.0 era. The mean difference shows the difference between the means of the control group and the experimental group. The result of the mean difference analysis was -3.043, thus the value of the control group was lower than the experimental group. In other words, the experimental group showed higher character development than the control group. The teacher's pedagogical quality also shows a good influence on the character of students in the society 5.0 era.

The quality of pedagogical teacher is applied in several stages. The first stage, sharing learning design development. Sharing is done in the form of collaboration with peers or colleagues. Things that were developed were 1) the topic of the curriculum guide study which was considered difficult to understand by students. This collaboration will create learning innovations through sharing experiences between peers. 2) Reviewing the literature regarding the material to be conveyed through several sources, it aims to anticipate if there are deep and broad student questions. 3) Assessing the applicable curriculum, in addition to hard skills, students are also equipped with soft skills, namely 21st-century skills, which include critical thinking, creativity, collaboration, communication.

The second stage of learning observation, this is done online (Google Meet) due to the Covid-19 condition which makes direct learning impossible. So that when online learning is carried out, researchers also join in the learning to observe the implementation of learning that is focused on developing student characters.

The third stage is post-learning reflection. This stage aims to obtain feedback for the teacher. Then it is based on the instant transcript. The fourth stage, data analysis. Based on the results of observations of student character development through character instruments in the form of a rubric.

Based on the results of observations, there was a fairly good increase in each meeting in learning. This is shown from the learning design made, the learning media, and the teacher's appearance or appearance in teaching even though from a distance. The communication shown was also very good, it seemed that there were no problems in implementing distance learning. Some of the findings of researchers through the application of teacher of pedagogical, namely the method in learning there are three aspects: the first aspect is pedagogy knowledge (PK) teachers carry out lectures, discussions, and independent work. In the second aspect, the content knowledge (CK) of the teacher shows a correct, broad, in-depth and applicable explanation. The third aspect of pedagogy contents knowledge (PCK) occurs interactive dialogue between teachers and students then students and students. This was followed by a small discussion, but it caught the students' attention well. Whereas in insight, the teacher shows that it is not uncommon to reprimand students if there are those who are not focused and feel bored following the lesson. Teachers always have their own way of building fun interactions and learning such as teaching students to sing a little, making picture guesses, and making a few questions to repeat previous lessons.

The implementation of the quality of teacher pedagogical influence matters greatly on the development of student character. Even though they are far from teacher supervision, students can still study accompanied by their parents and complete all assignments well. In addition to the teacher's focus on learning, the teacher does not forget to maintain good relationships with the parents to ask questions about student learning progress at home or there may be difficulties in learning. Everything is done online, namely the WhatsApp group.

Some of the character development shown by students in learning such as 1) being religious, which is shown by the attitude of students praying before starting lessons, having good relationships with friends and teachers, respecting the opinions of others and not discriminating against friends; 2) nationalism, in which students obey teacher orders and rules such as using uniforms when learning online, being on time in taking online learning; 3) independence, where students collect assignments on time, answer their own questions given by the teacher, if there is a lesson that students do not understand, ask the teacher directly; 4) cooperation, if there are friends who find learning difficulties, students do not hesitate to invite them to study together, like to help friends who are in trouble, and collectively complete group assignments; 5) integrity, students listen to the teacher when explaining learning, do assignments as a form of responsibility as students, and are ready to accept sanctions if they violate rules such as the late submitting of assignments or even not doing assignments.

5 CONCLUSION

Descriptive statistics show that the number of subjects in the control group is 70 people and the experimental group is 70 people. The mean value of student character in the control group was 93.34 while the experimental group was 96.39. These results indicate that the experimental group has a higher character mean.

From the analysis, results obtained the value of $t = -1.828$ with $p < 0.05$. Then H0 is rejected, and Ha is accepted. This proves that the hypothesis in this study means that there is an influence of the quality of the teacher's pedagogical on the character development of students in the society 5.0 era. The mean difference shows the difference between the means of the control group and the experimental group. The result of the mean difference analysis was -3.043, thus the value of the control group was lower than that of the experimental group. In other words, the experiment group showed higher character development than control group. The teacher's pedagogical quality also shows a good influence on the character of students in the society 5.0 era.

The results of this study have implications for the character development of students to face a smart society. Teachers can develop pedagogical abilities in the education system according to the times, and all people can solve various challenges by taking advantage of innovations that were born in the revolutionary industry 4.0 era.

REFERENCES

Annisa, N., Hasibuan, P. H., & Siregar, E. F. S. 2020. Menyanyikan Lagu Indonesia Raya Sebagai Bentuk Impelementasi Penguatan Pendidikan Karakter di SDS Asuhan Jaya Kota Medan. *Jurnal Benderang (Pendidikan Guru Sekolah Dasar)* 1(1): 1–5.

Ardana, I. K. 2018. Pengaruh Model Discovery Learning Berbantuan Media Audio Visual dalam Setting Lesson

Study Terhadap Hasil Belajar IPA Mahasiswa PGSD Undiksha UPP Denpasar Tahun 2017. *Jurnal Ilmiah Sekolah Dasar* 2(1): 52–58.

Dalyono, B. 2017. Implementasi Penguatan Pendidikan Karakter di Sekolah. *Bangun Rekaprima* 3(3): 33–42.

Fildzah, Y. 2020. Comparative Study of Competency and Certification of Special Education Teachers in Indonesia and Another Various Country. *IJDS Indonesian Journal of Disability Studies* 7(1): 40–49.

Fukuyama, M. 2018. Society 5.0: Aiming for a New Human-centered Society. *Japan SPOTLIGHT* 27(August): 47–50.

Ghavifekr, S., & Rosdy, W. A. W. 2015. Teaching and learning with technology: Effectiveness of ICT integration in schools. *International Journal of Research in Education and Science* 1(2): 175–191.

Harta, J., Dharsana, I. K., & Renda, N. T. 2019. Pengaruh Model TSTS Melalui Lesson Study Terhadap Hasil Belajar IPA. *Jurnal Mimbar Ilmu* 24(1): 95–104.

Hasanah, U., & Deiniatur, M. 2018. Character Education in Early Childhood Based on Family. *Early Childhood Research Journal (ECRJ)* 1(1): 50–62.

Hendayana, S., & et. al. 2019. *DRAFT Panduan Pelaksanaan Kegiatan Riset Kompetitif Nasional Skema Penelitian Dasar 2019.*

Karlina, C. F., Dharsana, I. K., & Kusmariyatni, N. 2017. Pembelajaran Kooperatif Tipe (TSTS) Berbantuan Peta Pikiran Untuk Meningkatkan Hasil Belajar IPA Melalui Lesson Study. *MIMBAR PGSD Undiksha* 5(2): 1–12.

Kemendiknas, B. 2011. Pedoman Pelaksanaan Pendidikan Karakter. *Pusat Kurikulum dan Perbukuan*. Jakarta: Pusat Kurikulum dan Perbukuan.

Kurniasih, I., & Sani, B. 2017. *Pendidikan Karakter Internalisasi dan Metode Pembelajaran di Sekolah*. Jakarta: Kata Pena.

Kurniati, M., Arafat, Y., & Mulyadi, M. 2019. International Journal of Educational International Journal of Educational Review. *International Journal of Educational Review* 1(2): 1–8.

Muralidharan, K., & Singh, A. 2020. Improving Public Sector Management at Scale? Experimental Evidence on School Governance India. *NBER Working Paper* w28129.

Presidential Decree No. 87. 2017. *Penguatan Pendidikan Karakter* (p. 14).

Putra, P. G. N., Margunayasa, I. G., & Wibawa, I. M. C. 2019. Pengaruh Model Pembelajaran Kooperatif Tipe Group Investigation (GI) Berbasis Lesson Study Terhadap Penguasaan Konsep IPA. *Jurnal Pedagogi dan Pembelajaran* 1(2): 84–93.

Rosyid, A. 2016. Technological Pedagogical Content Knowledge: Sebuah Kerangka Pengetahuan Bagi Guru Indonesia di Era Mea. *Prosiding Seminar Nasional Inovasi Pendidikan* pp. 446–454.

Sadulloh, U. 2014. *Pengantar Filsafat Pendidikan (Edisi Kesembilan)*. Bandung: Alfabeta.

Santoso, K. A. 2019. Pendidikan untuk Menyamput Masyarakat 5.0. Retrieved from alinea. id Fakta, Data, Kata website: https://www. alinea. id/kolom/pendidikan-untuk-menyambut-masyarakat-5-0-b1XcI9ijL.

Singh, B. 2019. Character Education in the 21st Century. *Journal of Social Studies* 15(1): 1–8.

Sukadi, E., Cari, & Sarwanto. 2015. Implementasi Pedagogical Content Knowledge Pada Materi Listrik Dinamis Untuk Meningkatkan Kompetensi Calon Guru Fisika. *Jurnal Inkuiri* 4(1): 37–46.

Sulistyarini, S., Utami, T., & Hasmika, H. 2019. Project Citizen Model as Character Education Strengthening. *JETL (Journal Of Education, Teaching and Learning)* 4(1): 233–237.

Triyani, E., Busyairi, A., & Ansori, I. 2020. Penanaman Sikap Tanggung Jawab Melalui Pembiasaan Apel Penguatan Pendidikan Karakter Siswa Kelas III. *Jurnal Kreatif?: Jurnal Kependidikan Dasar* 10(2): 150–154.

Wulandari, Y., & Kristiawan, M. 2017. Strategi Sekolah Dalam Penguatan Pendidikan Karakter Bagi Siswa Dengan Memaksimalkan Peran Orang Tua. *JMKSP (Jurnal Manajemen, Kepemimpinan, dan Supervisi Pendidikan)* 2(2): 290–303.

Yuniari, K. M., Suarni, N. K., & Parmiti, D. P. 2019. Pengaruh Model Kooperatif Tipe Snowball Throwing Berbasis Penilaian Portofolio Terhadap Hasil Belajar PKn. *Jurnal Pedagogi dan Pembelajaran* 2(2): 223–232.

Educational Innovation in Society 5.0 Era: Challenges and Opportunities – Purnomo & Herwin (Eds)
© 2021 the authors, ISBN 978-1-032-05392-9

The effect of a simplified integrated learning environment on plagiarism behavior

G.N. Sudarsana, N.K. Suarni & I.K. Dharsana
Universitas Pendidikan Ganesha, Singaraja, Bali, Indonesia

ABSTRACT: This study aims to determine the effect of implementing a simplified integrated learning environment on plagiarism behavior in second semester students in the Guidance and Counseling Study Program. This type of research is a quasi-experimental design with a pre-posttest nonequivalent control group. The sample was second semester students, and the sample was taken using purposive sampling. The data were collected using the midterm exams as pre-test data, and final semester exams as post-test data and analyzed using the t-test. Based on the results of the analysis, it was found that the difference in the effect of implementing a simplified integrated learning environment between the experimental class with a mean value of -1.14 and the control class of -18.76. Furthermore, the significance value is $0.027 < 0.05$, meaning that there is a significant effect of the application of a simplified integrated learning environment on plagiarism behavior among students.

1 INTRODUCTION

1.1 *Background of study*

The tridharma of higher education is the basic task of a lecturer, in educating competent students starting from teaching. After that, the theory obtained by students is put into practice in real skills according to their field of knowledge, such as writing scientific papers in the form of essays, papers or theses. In writing, students must use their abilities in shaping ideas, as well as their analysis into the writing, and of course it must be based on relevant reviews. In using references made by other people's work, the academic community follows an academic ethic, namely citation, not just copy-pasting but also paraphrasing and linking it into writing ideas into skills that students must master. In fact, this process is running poorly and the skills that are taught to human beings in Indonesia are not what was intended. The indulgence offered in the industrial era 4.0 is the main reason for this.

This problem is a contemporary issue faced by all universities in Indonesia. The results of observations made by researchers during the 2019/2020 odd semester lecture process and cross-check with colleagues, researchers found several issues that were a factor in the emergence of plagiarism behavior, including: 1) low endurance in attending lectures, not focusing in class; 2) often delaying doing assignments; 3) being too lazy to find additional knowledge on their own beyond what is given in class; 4) preferring to copy and paste their friend's work; 5) being less intelligent in choosing references (not utilizing articles, only using outdated sources). This is an issue with

frequency and urgency that arises a lot among teachers and colleagues.

The student's point of view, there are several issues as follows: 1) low endurance in attending lectures, not focusing in class; 2) often delaying doing assignments; 3) being too lazy to find additional knowledge on their own beyond what is given in class; 4) prefering to imitate/wait for a friend's work; 5) finding lectures boring; 6) the material provided being difficult to understand; 7) it being easier to find material from Blogspot than to search from books/articles and scientific journals.

From several similarities in issues presented from the point of view of lecturers and students, it is necessary to identify factors using the Pareto Chart and Fishbone/Ishikawa Diagram. Pareto chart is a method invented by Pareto where if in a product, there is a 20% input and 80% output, then it becomes the law of 20/80 (Hossen et al. 2017, Powell & Sammut-bonnici 2014, Stojčetović et al. 2016)

Problem identification begins with distributing questionnaires to lecturers in the Guidance and Counseling Study Program. The researchers obtained 14 respondents with the following results:

1) The main problems that become obstacles for students in lectures are: a) low endurance in attending lectures, not focusing in class by 35.7%; 2) often delay doing assignments by 21.4%; 3) being too lazy to find additional knowledge on their own beyond what is given in the classroom by 21.4%; 4) students preferring to copy and paste the work of their friends by 14.3%; 5) students being less intelligent in choosing references (not utilizing articles, only using outdated sources) by 7.1%.

238

DOI 10.1201/9781003206019-45

2) Secondary problems that become obstacles for students in lectures were: a) being too lazy to find additional knowledge themselves beyond what is given in class by 57.1%; b) preferring to imitate/wait for a friend's job by 21.4%; c) having less qualified lecturers, 7.1%; d) the material provided is difficult to understand 7.1%; e) often delaying doing assignments by 7.1%.
3) The third-order problems that hinders students in lectures are: a) often delaying doing assignments by 42.9%; b) students prefer to copy and paste a friend's work by 35.7%; c) students who are less intelligent to choose references by 21.4%.
4) The fourth order problem that hinders students in lectures, obtained the following results: a) students are less intelligent in choosing references by 71.4%; b) students prefer to copy and paste a friend's work by 21.4%; c) students are too lazy to find additional knowledge on their own beyond what is given in the class by 7.1%.
5) The fifth order problems that hinder students in lectures are: a) poor attendance in lectures, not focusing in class by 30.8%; b) often delay doing assignments by 23.1%; c) lazy to find additional knowledge outside of what is given in the class by 15.4%; d) students prefer to copy and paste the work of their friends by 15.4%; e) students who are less intelligent in choosing references are 7.1%; f) low learning independence of 7.1%.

Based on the results of the analysis using the Pareto method and the results of the analysis are obtained in Figure 1, as follows:

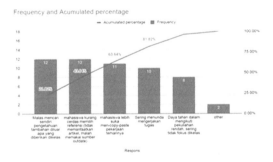

Figure 1. The results of the analysis of the Pareto method of lecturer respondents.

Based on Figure 1 above, there is a line in the range of 80%, so there are four (4) issues that are referred to be as the vital few and the rests are trivial many. These four (4) issues include: 1) being too lazy to find additional knowledge on their own beyond what is given in class; 2) Students are less intelligent in choosing references (not using articles, instead using outdate sources); 3) students prefer to copy and paste their friends' work; 4) students often delay doing assignments.

However, in this calculation, the distribution of data is evenly distributed with input 80% and output 20%, this will be cross-checked again with student responses. There were 44 student respondents, among others

1) The main factors that became the biggest overlay in the lecture process, the results were: a) low endurance in attending lectures, not focusing in class by 22.7%; b) often delaying doing assignments by 13.6%; c) being too lazy to find additional knowledge on their own outside of what is given in the classroom by 9.1%; d) the teaching methods of the lecturers were less attractive by 18.2%; e) the material given is difficult to understand by 13.6%; f) it is easier to find material from Blogspot by 22.7%.
2) The second factor that became an obstacle in the lecture process, obtained results: a) low endurance in attending lectures, not being focused in class at 13.6%; b) often delaying doing assignments by 11.4%; c) being lazy to look for themselves by 15.9%; d) prefering to imitate or wait for a friend's job by 2%; e) the teaching methods of the lecturers were less attractive by 20.5%; f) the material provided is difficult to understand by 5%; g) it is easier to find material from Blogspot by 27.3%; h) and others 4%.
3) The third factor that became an obstacle in the lecture process, obtained results: a) low endurance in attending lectures, not focused in class at 13.6%; b) often delay doing assignments by 18.2%; c) lazy to look for themselves by 13.6%; d) prefer to imitate or wait for a friend's job by 6.8%; e) the teaching methods of the lecturers were less attractive, 9.1%; f) the material given is difficult to understand by 20.5%; g) it is easier to find material from Blogspot by 11.4%; h) and others, 6.8%.
4) The fourth factor was the obstacle in the lecture process, obtained results: a) low endurance in attending lectures, not focusing in class at 13.6%; b) often delaying doing assignments by 22.7%; c) being too lazy to look for themselves by 15.9%; d) prefering to imitate or wait for a friend's job by 6%; e) the teaching methods of the lecturers were less attractive by 15.9%; f) the material provided is difficult to understand by 5%; g) it is easier to find material from Blogspot by 11.4%; h) and others at 9.5%.
5) The fifth factor that became an obstacle in the lecture process, obtained results: a) low endurance in attending lectures, not being focused in class by 15.9%; b) often delaying doing assignments by 9.1%; c) being too lazy to look for themselves by 27.3%; d) prefering to imitate or wait for a friend's work by 18.2%; e) the teaching methods of the lecturers were less attractive by 6%; f) the material given is difficult to understand by 9.1%; g) it is easier to find material from Blogspot by 9.1%; h) and others 5.3%.

Based on the results of the analysis using the Pareto method and the results of the analysis are obtained in Figure 2, as follows:

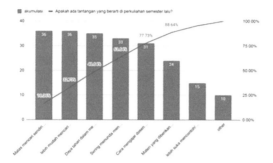

Figure 2. The results of the analysis of the student respondents' Pareto method.

If the results of the lecturer and student responses are combined, the Pareto graph results are obtained as follows:

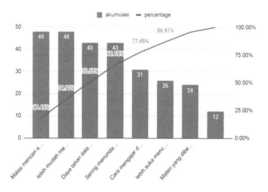

Figure 3. The results of the analysis of the Pareto method of lecturers and student respondents.

From the combined results of student responses and lecturer responses, there are five issues, namely: 1) laziness to seek additional knowledge on their own beyond what is given in class; 2) it being easier to find material from Blogspot than to search from books/articles and scientific journals; 3) low endurance in attending lectures, not being focused in class; 4) often delaying doing assignments; 5) finding lecturers boring.

These five issues are vital factors that are the root of the problem of plagiarism behavior among students. To see the dynamics of these five factors the researcher uses Ishikawa diagram/Fishbone diagram. The formulation of the problem in this study is whether there is an effect of the application of the simplified learning environment on plagiarism behavior among students of the Guidance and Counseling Study Program.

1.2 Research purposes

The purpose of this study was to determine the effect of implementing a simplified learning environment on plagiarism behavior among students of the Guidance and Counseling Study Program.

1.3 Research urgency

The urgency of this research includes several aspects: 1) through a simplified learning environment, it can reduce the tendency of problems to students, namely not being able to optimize their potential; 2) To become a guideline for the use of online tools for searching and processing references on the Internet; 3) For higher-education institutions, this research information becomes a reference for alleviating the problem of plagiarism that often occurs among students.

2 THEORETICAL BASIS

2.1 Plagiarism

Plagiarism in the large Indonesian dictionary plagiarism is defined as plagiarism that violates copyright, for more details it refers to Law number 28 of 2014 concerning copyright, as stated in article 1, which states "Copyright is the exclusive right of a creator who arises automatically based on the declarative principle after a creation is manifested in a tangible form without reducing restrictions in accordance with the provisions of the legislation." According to Harris (2017) in Using Sources Effectively, plagiarism that often occurs is planned plagiarism, this must be avoided because it will damage the image of one's character. This is reinforced by the opinion of Dhusia (2017) that someone who practices plagiarism actually already knows and understands if what they are doing is wrong, they assume that educators will not find this cheating.

Piracy or plagiarism is more likely to fall into ethical violations, but with a legal basis, the creator/copyright holder can take a last resort if the mediation fails to file a claim for compensation based on civil law or criminal law grounds.

Plagiarism among academics is also regulated in article 44. Therefore, the academic community is obliged to follow the provisions of article 44 in the context of point (a) to include or mention the source completely; this is a way to give recognition to the creator if a person uses his work in a paper/in research (Ambarawati & Purwanto 2019).

Aasheim, et al. (2012) stated that educators must continuously convey what can be done and what is not allowed to students, especially with regard to the development of academic ethics. Meanwhile, according to Awasthi (2019), libraries play an important role in creating awareness in students through organizing training.

The problem of plagiarism behavior by students who researchers have mapped through the Fishbone/Ishikawa diagram, there are two core factors that greatly influence this plagiarism behavior, namely: (1) being too lazy to find additional knowledge outside of what is given in class, and (2) it being easier to find materials from Blogspot rather than searching from books/articles and scientific journals (students are less intelligent in looking for references). The dynamics of these two factors are interrelated, as exemplified by a student who (1) is too lazy to find their own additional knowledge/skills (2) prefers to look for unreliable materials/references, and copy-paste, whereas most students now do this which in turn will affect other students who mimic this behavior.

2.2 Simplified Integrated Learning Environment (sILE)

Various methods are used to anticipate plagiarism, especially in the world of education. One of the ways to be applied in this research is the habituation method or habituation to students from an early age using a simplified integrated learning environment, hereinafter abbreviated as (sILE) in lectures. ILE is a system adopted from the United States Navy training system, is a combined group of automated systems that use information technology to streamline the learning process, automate learning management functions and provide learning to Navy personnel at home, at school or on the battlefield (Zeidan 2018). Uniform skills must be possessed by the American navy, in order to achieve their learning goals (in this case skills) they must use the techniques listed in the ILE direction, and during their education, they continue to use this ILE until the US Navy automatically has the same skill protocol/standard. Subsequently, applied to the learning system at several universities abroad, ILE at a glance is similar to online learning but the quality standard factor is the difference between these two systems.

Meanwhile, a simplified integrated learning environment (sILE) is a simplified version of ILE, taking the basic concepts of ILE without the use of sophisticated equipment/expensive hardware such as dedicated servers, special programs for ILE, and other IoT systems. sILE provides an environment that consists of tools to train students' skills, in this context the use of references, whether searching, quoting, or checking plagiarism. It is then integrated into lectures that are tailored to the course syllabus, teaching techniques to accommodate this sILE needs to be adjusted, using face-to-face teaching techniques, by giving network-integrated assignments (online).

3 METHOD

3.1 Research design

This type of research is a quasi-experimental research design with the pre-post-test nonequivalent control group.

3.2 Research sample

The subjects of this research were the second semester students of the Guidance and Counseling Study Program. This study uses nonprobability sampling method, which is categorized as purposive sampling, from problem identification and analysis, as well as the application design of sILE. The sample category is students in the first semester (1-2) of Bachelor at Undiksha Guidance and Counseling Study Program. This sampling is closely related to the application of sILE from an early age. It is hoped that students in the first semester will form a standard of skills in finding and using references, so that plagiarism behavior can be minimized from the start.

3.3 Research procedure

The procedure for implementing sILE in this study consists of: 1) granting a lecture contract with information about sILE listed in it; 2) the first meeting as well as the initial test of students' ability in writing (pretest); 3) the second meeting was providing modules and lectures using sILE (treatment); 4) the third meeting was providing modules and lectures using sILE (treatment); 5) the fourth meeting is providing modules and lectures using sILE, as well as a post-test by giving a paper writing assignment with instructions that have been listed in the sILE.

3.4 Methods of data collection and analysis

The data collection method in this study is to use the evaluation of learning outcomes from the results of students' midterm and final semester exams. Midterm exams as pre-test data, and final semester exams as post-test data in accordance with the research design. The goal is to take into account unexpected factors such as the adaptation of the subject when given treatment, supporting facilities such as tissue stability, equipment owned, and other factors outside that affect the course of treatment. Therefore, the time interval from the midterm exam to the final exam is the ideal interval to see the effect of this sILE. The data analysis method used in this study is the t-test.

4 RESULT AND DISCUSSION

Based on the results of the responses given by students and lecturers, there are five items that occur in semester II students of the Guidance and Counseling Study Program, namely: 1) being too lazy to seek additional knowledge themselves beyond what is given in class; 2) it being easier to find material from Blogspot than to search from books/articles and scientific journals; 3) low endurance in attending lectures, not being focused in class; 4) often delaying doing assignments; and 5) how to teach lecturers is less interesting/ boring.

The five issues above are vital factors that are the root of the problem of plagiarism behavior in students, to see the dynamics of these five factors the researcher uses Ishikawa diagram/Fishbone diagram. The analysis results are obtained as below:

Figure 4. Ishikawa diagram.

Based on the description of the Ishikawa/fishbone diagram above, it can be seen that the tendency of the problem to students is that they cannot optimize their potential, both from within themselves and their external/supporting potential. Likewise, the management of lecturers who assume that millennial students will automatically be able to use gadgets, take advantage of the internet, and use academic support software, but the fact is that students are confused and need guidance in these areas. It can be deduced that millennial students can use online facilities but cannot make optimal use of related facilities. Not all references on the internet are accountable and reliable, students who do not like to be bothered and confused in choosing this reference will then choose the one that is easiest for them to understand or that is in the form of the task the lecturer wants such as essay or paper and directly copies and paste without paraphrase. This will lead to habits that violate academic ethics.

Therefore, to overcome this problem, a simplified integrated learning environment was implemented in minimizing plagiarism behavior in the second semester students of the Guidance and Counseling Study Program. This simplified integrated learning environment is implemented through the following steps: 1) awarding a lecture contract with information about sILE contained therein; 2) the first meeting as well as the initial test of students' ability in writing (pretest); 3) the second meeting was providing modules and lectures using sILE (treatment); 4) the third meeting was providing modules and lectures using sILE (treatment); 5) for the fourth meeting, giving modules and lectures using sILE, as well as a post-test by giving a paper writing assignment with instructions that have been listed in the sILE.

The following shows descriptive statistics in table 1.

Table 1. Descriptive statistics

	Class	N	Mean*	Std. Deviation	Std. Error Mean
NGain_Percent	Experiment	21	−1.140	162.806	35.527
	Control	21	−18.759	97.491	21.274

*Data was calculated using SPSS 16.0 for Windows

Based on the descriptive statistical table above, it can be seen that the number of subjects in the experimental class was 21 people, and the control class was 21 people. The experimental class that was applied in a simplified integrated learning environment obtained a mean value of −1.140, while the control class that was not implemented by a simplified integrated learning environment had a mean value of −18,759. These results indicate that there are differences in the implementation of the simplified integrated learning environment on the plagiarism behavior of the second semester students of the Guidance and Counseling Study Program. Next, an independent simple t-test analysis was carried out, which is shown in table 2.

Table 2. Independent samples test

	Homogeneity test		t-test		
	F	Sig.	t	df	Sig. (2-tailed)*
Equal variances assumed	0.218	0.643	−2.300	40	.027

*Data was calculated using SPSS 16.0 for Windows

Based on the significance value in the Levene's Test for Equality of Variances of 0.643 > 0.05, the data variance for the experimental and contraction classes is homogeneous. Furthermore, for the independent sample t-test, n-gain was used. If seen from table 2, it is found a significance value of 0.027 < 0.05, it is concluded that there is a significant difference in the effect of the application of the simplified integrated learning environment (sILE) on plagiarism behavior in semester II students of the Guidance and Counseling Study Program.

These results are also supported by the behavior shown by students when doing assignments. The habit of looking for material easily and efficiently can be minimized. Students can already search for material based on various cutting-edge sources, such as national and international articles and electronic books. Although at first it felt difficult, students could feel the benefits directly. Now, students have a material vocabulary that is quite extensive and contains pretty good reading material. In addition, students can also know more deeply about how to quote properly and correctly to avoid plagiarism that might be done accidentally due to ignorance of how to quote. Gradually, students try to comply with academic ethics, which will greatly affect the progress of education in the future.

5 CONCLUSION

Based on the significance value on the Levene's Test for Equality of Variances of 0.643 > 0.05, the data variants for the experimental and contraction classes are homogeneous. Furthermore, for the independent sample t-test, n-gain was used. If seen from table 2, it is found a significance value of $0.027 < 0.05$, it can be concluded that there is a significant difference in the effect of the application of the simplified integrated learning environment (sILE) on plagiarism behavior in semester II students of the Guidance and Counseling Study Program. This research has implications for the habituation or habituation of students to develop critical and innovative thinking skills in the education system.

REFERENCES

Aasheim, C. L., et al. 2019. Plagiarism and programming: A survey of student attitudes. *Journal of information systems education* 23(3): 297–313.

Ambarawati, P. E. Y., & Purwanto, I. W. N. 2019. Pengaturan pengambilan tulisan pada karya tulis skripsi dalam menghindari plagiarisme. *Kertha Semaya: Journal Ilmu Hukum* 8(1): 1–12.

Awasthi, S. 2019. Plagiarism and Academic Misconduct: A Systematic Review. *DESIDOC Journal of Library & Information Technology* 39(2): 94–100

Dhusia, D. K. 2017. Strategies for Preventing Plagiarism - A Case Study of Top Indian Universities. *Global Journal of Enterprise Information System* 9(2): 84.

Harris, R. A. 2017. *Using sources effectively: Strengthening your writing and avoiding plagiarism*. Routledge.

Hossen, J., et al. 2017. An application of Pareto analysis and cause-and-effect diagram (CED) to examine stoppage losses: a textile case from Bangladesh. *Journal of the Textile Institute* 108(11): 2013–2020.

Law of the Republic of Indonesia Number 28 of 2014. *Hak Cipta*. 16 Oktober 2014. Lembaran Negara Republik Indonesia Tahun 2014 Nomor 266. Jakarta. https://dgip.go.id/peraturan-perundang-undangan-terkait-hak-cipta

Powell, T., & Sammut-bonnici, T. 2014. Pareto Analysis. *Wiley Encyclopedia of Management*, 1–2.

Stojčetović, B., et al. 2016. Application of the Pareto Analysis in. *9th International Quality Conference* (June 2015): 655–658.

Zeidan, B. A. 2018. Integrated Learning Environment ILE Editor. Retrieved from https://www.researchgate.net / publication / 327871719 _ Integrated _ Learning_ Environment_ILE

Educational Innovation in Society 5.0 Era: Challenges and
Opportunities – Purnomo & Herwin (Eds)
© 2021 the authors, ISBN 978-1-032-05392-9

The profile of pre-service elementary school teachers in developing lesson plans for science instruction

W.S. Hastuti, P. Pujiastuti, Pujianto & Purwono
Yogyakarta State University, Indonesia

ABSTRACT: This study aims to identify the profile of Pre-Service Elementary School (ES) Teacher problem solving and critical thinking skills at the Department of Primary School Teacher Education, Yogyakarta State University. It focuses on developing lesson plans to elaborate problem skills through the teaching-learning process in science instruction. Through this study, students are taught to be creative to utilize their problem solving and critical thinking skills on creating a learning scenario which is encourage the pupil on observing and describing, questioning and analyzing, exploring and creating, showing & telling, and reflecting. A qualitative study is conducted to determine their profile in all process of developing lessons plan. Data were gathered through observation, in depth interview and documentation of the product. Many students perform that their skills on developing steps of problems-solving process are extended. More activities in the introductory section, main section, concluding section, and application indicate that all of activities involved problem-solving steps. It can be inferred that ES teacher students use their critical thinking to create those sections of lesson plan.

1 INTRODUCTION

The quality of the implementation of learning cannot be separated from the quality of learning planning (Dorovolomo, Phan, & Maebuta, 2010). The two of them are linearly related. Lesson plans are plans for decision-making before learning is implemented (Panasuk, Stone & Todd, 2002). Choosing the right way to teach content requires critical thinking. This is one of the proofs that the teaching profession requires creative skills (Misra, 2015). Before teaching, the teachers must make a good lesson plan so that in practice they can teach well. Lesson plans are an important part of a tool that guides teachers in teaching. Lesson plans contains the content, how to deliver it and classroom management strategies to achieve the learning goal. In school, before entering the new academic year, teachers prepare some unit plans for one year. In addition, the teacher must also break it down into a weekly and daily plan. This becomes a guide for teachers.

Teacher candidates (pre-service ES teachers) need to have the skills to develop good lesson plans (Abdul-Gaafor & UmerFarooque, 2010, Bin-Hady, 2018). Thus, in lectures, they must be trained to be able to become creators. In fact, the curriculum (Curriculum-2013) has provided a teacher's book and a student book as a guideline for teachers to teach. However, lesson plans must be developed by teachers by taking into notice the potential of the school environment, and the features of students and subjects. Creating the lesson plan requires critical thinking skills.

Right now, the format of the lesson plans by the Minister of Education and Culture is simplified from 13 components to 3 components. The new lesson plan format consists of objectives, learning steps and assessments (The Ministry of Education Decree No. 14 of 2019 about the Simplification of Lesson Plans). The three main components must be in the lesson plan. This is intended so that the teacher takes more quality time for creative teaching than just making administrative completeness. Teachers are allowed to develop the lesson plan format. Hence, the lesson plans that are designed must be able to develop students' higher-order thinking skills (HOTs), usage interesting and fun learning methods, train students to have adaptive attitudes as needed by society in this 21st century.

Moreover, the lesson plan is not only used by the teacher as a guide in teaching (Bin-Hady, 2018) but more importantly as a basis for self-reflection (Gutierez, 2015). Lesson plans are framework for teachers to facilitate students in achieving the learning goals (Bin-Hady, 2018). Reflection skills are important to make individuals (teachers) understand their strengths and weaknesses (teaching process). When this becomes a habit by the teacher, continuous improvement in learning by teachers will be developed (Timperley, Wilson, Barrar, Fung: 2007). Improved learning emphasizes the clarity of learning objectives, activity procedures, implementation, and learning production. Various studies have shown that one of the factors that support learning is the effectiveness of the learning environment. Unfortunately, this cannot just happen because it requires careful planning. Therefore, teacher candidates need to be taught to develop good lesson plans.

During lectures, students as teacher candidates are trained to create lesson plans in order to bring theory closer to practice (Timperley et al., 2007). Some of studies indicate that pre-service teachers experience difficulty in designing quality lesson plans. These pre-service teachers need more time in doing lesson plan that is able to develop students' higher-order thinking skills. In line with this, the results of Johnson's research (2000) also shows that pre-service teachers find the initial lesson planning steps cumbersome.

The 2013 curriculum that applies in Indonesia since 2013 until now, especially for the elementary school level, uses an integrative thematic and scientific approach in the learning process. In learning at the elementary school level, the teacher must involve scientific activities consisting of 5 main components or what is commonly referred to as the 5M, namely observing, questioning, associating, doing, and communicating. According to Permendikbud No. 22/2016 about Process Standards, the five components do not have to be sequential. The purpose of the integrative thematic approach in the 2013 In elementary school learning, teachers must combine subjects and integrate them into one theme. For instance, science subject is integrated with Bahasa and Social Studies. Therefore, in designing good lesson plans, the student teachers need critical thinking. Robert Ennis stated that critical thinking is reasoning and reflective thinking that aims to determine what to believe and what to do (Hunter, 2009).

The results of the initial diagnosis made at the beginning of the lecture regarding the ability of students in designing thematic lesson plans containing science subject were: (a) learning activities that are designed do not show scientific/scientific activities based on laboratory performance, and have not developed critical thinking, (b) science assessment instruments compiled by students is still limited to the adoption of the teacher's book (Tools in the 2013 Curriculum), (c) between the learning process and assessment that is designed, there is less synchronization and an inability to integrate aspects of knowledge, attitudes, and skills. In fact, it is crucial for students' teachers to have the ability to develop lesson plans that improve the students' activities and build their own knowledge. Therefore, knowing pre-service teachers' profile regarding their ability to design lesson plans after the end of lectures is important to make improvements to future lectures. The expectation is that education graduates have good competence in designing lesson plans. This capability is needed by the professional teacher (Meier, 2017).

2 METHOD

Descriptive survey research was used in this study. This study was concerned with how what exists about pre-service teachers' competence in developing lesson plans. A cross-sectional study was used in this study, in which this study did a snapshot of students' abilities in developing lesson studies and compares work between individuals (Cohen, Manion & Morrison: 2005). Data collection and analysis techniques and drawing conclusions were carried out in the study to obtain a description of the subject's profile. This study involved all of 39 pre-service Elementary School Teacher Education (ESTE, namely student teacher) of YSU who took the Science Education course. Assessment of critical thinking skills and problem solving was conducted through assessing lesson plans and teaching simulations of the student teachers. The instrument focuses on whether students can think critically and creatively, share ideas, use good judgment and make decisions in designing lesson plan. Quantitative and qualitative descriptive analyzes were carried out in this study. The data were analyzed by means of a scoring technique with a scale of 4. Then the interpretation of the data was conducted by calculating the percentage, namely the formula.

$$\text{final score (\%)} = \frac{\text{Xscore}}{\text{Xmax}} \times 100\%$$

Description:
X = score of lesson plan

3 RESULTS AND DISCUSSION

The appraisal focus on the best strategies that are indicated by students, especially in developing laboratory activities for elementary students. A summary of the results of the design and simulation assessment is converted in Table 1.

Creativity includes 2 main factors: novelty and effectiveness (Mishra, Mehta & Henriksen: 2015). Therefore, teachers who are creative in developing lesson plans, mean that they are able to design their own lessons without plagiarizing other people's work but are also useful and of high quality (Aglazor, 2017). One of the qualities of teachers can be seen from their ability to develop lesson plans and their performance in teaching simulations. To find out the quality of lesson plans, it is necessary to conduct teaching simulations by students (Shepherd, Hair & Brown, 2010). The results of this study indicate that most student teachers are able to compile the stages in the lesson plan. However, there is still evidence that some student teachers have not been able to design it appropriately. In accordance with the core components that must be present in the lesson plan, namely learning objectives, learning steps, and learning assessments (Minister of Education and Culture Circular No. 14 of the 2019 concerning Simplification of Learning Plans), the study is focused on knowing student profiles on their ability to design these things.

Learning objectives are the essential strengths of learning which contain content and how to achieve it. The formulation of learning objectives is carried

Table 1. The result of the analysis of student teacher lesson plans

Components of lesson plan (adaptation from anonim)	Indicators (adaptation from anonim)	Rating	
		appropriate	inappropriate
Specific Learning Goals	Learning objectives in accordance with curriculum expectations	70%	30% the way/steps determined to understand the material is still not quite right.
Introductory Section	How to review the previous material is accurate	100%	0
	How to relate the previous material to the new material is properly	100%	0
	How to motivate students and make learning relevant to students is suitable	100%	0
Main Section	Activities designed to encourage students to think critically in applying the knowledge gained to deal with daily life problems	70% The activity begins with presenting the phenomenon/object, followed by the stage of asking inquiry questions	30% The activity begins by asking inquiry questions
	Activities are designed to encourage students to do scientific processes according to the primary school level	68% Activities conform to planned process skills	32% Activities do not match planned process skills
	The activity encourages students to design simple science experiments	70% Design can be developed (by student teachers) appropriately and easily done by elementary students	30% Design can be developed (by student teachers) appropriately but elementary students find it difficult to do so
	The activity encourages students to perform communicating skills	76% The design provide student opportunity to comminicate the data, idea, and experience creatively	24% The design just provide student opportunity to communicate the resullt of experiment or observing with deterimined format
Concluding Section	How to provide students opportunity to consolidate, practice, or to apply the learning in a new situation is appropriate	85% explore student understanding by providing opportunities for students to convey the implementation of concepts in daily life	15% Just concludes the lesson
	How to provide support for students who are not ready to practice this learning individually is appropriate	67% Provide detailed alternative methods for students who have difficulty practicing lessons outside of school. For example, through providing guidance.	33% Asking students to repeat activities accompanied by parents without providing guidance.
Application	There are tasks for student to demonstrate their learning	78% Give further assignments in accordance with the context of the material and the availability of facilities in the student's neighborhood	22% Giving further assignments according to the availability of facilities in the student's neighborhood, but the scope is too far from the context of the material
	There are assessments and evaluation designs to see that students have understood the lesson	68% Learning assessment can distinguish participants who are able and unable to do activities to achieve learning goals	32% Learning assessment is general in nature and is unable to distinguish those who are able and unable to carry out activities to achieve learning goals

out through the analysis of Standard Competence (SC) from the 2013 Curriculum. Learning objectives are specific targets of the general objectives (SC) in the curriculum (Akdeniz, et al., 2016). Objectives interpret the goals and focus and prioritize curricular components. They are narrow and time-bound to achieve a specific task which can be measured (Khan, Singh, Hande and Khumar, 2014:46). In learning objectives, in addition to directing students to understand the material, it also contains the attitudes and skills to be achieved (Mckimm & Swanwick, 2009). In instance of student teachers did is through experiments, students can explain the evaporation process scientifically. This can be interpreted as that besides students understanding that water can change phase to evaporate, it also aims to develop students' scientific reasoning about the science phenomenon. However, not all the student teachers can formulate the learning objective approriatelly, especially for methods to find the content.

The expectation from the implementation of the 2013 Curriculum is elementary school students reach at least SC as stated in the curriculum. SC is derived into learning indicators which are then developed into learning objectives. This becomes an estuary for teachers in developing learning and assessment steps. The results of this study indicate that 70% of student's teachers are able to develop learning goals correctly. It can be interpreted that some of students have been able to interpret the learning outcomes intended in the curriculum, and the others are less capable.

The second important component of the lesson plan is the learning steps. This component consists of 3 stages, namely the introductory section, the main section, and the concluding section. In this case, the truth of the content and the process of learning it becomes the focus of the lesson plan analysis (Cai, Perry & Wong: 2007). Most students have been able to develop it, and have the freedom of thinking and creating about determining the appropriate learning method (according to student characteristics and school potential as well as parents' conditions) however there are still some students who are deemed not proficient enough as described in Table 1. The data shows that most students have been able to develop lesson plans that encourage elementary students to: (a) think critically and apply the knowledge gained to deal with problems in everyday life, (b) are able to implement science processes at the elementary level (such as children can distinguish between observations and inferences, identify the data needed to answer simple questions, and the best techniques in collecting data), and (c) have the ability to design simple science experiments, for example planning procedures that can be done to obtain data, and (d) be able to provide student opportunity to communicate the data, idea, and experience creatively. However, a small proportion of other students still experienced difficulties. Developing critical thinking skills and problem solving in students is not easy enough to do. Systematic learning steps will make it easier for teachers to implement plans in their teaching

implementation. In that way, teacher can self-reflect on the achievement of learning objectives. The teacher can identify which steps may not have been carried out, or even discover which stages were less successful in making students achieve learning goals. Therefore lesson plans can be material for teacher reflection and material in determining follow-up plans.

Assessment is the final part of learning (application). This serves to find out whether students have achieved the set learning objectives (Jabbarifar, 2009). In addition, the results of the assessment can be a self-reflection for teachers (Xhaferi, 2017, Ratminingsih, 2017). Reflection on learning is important for students and teachers. For students, it can be used to understand their condition, so that it gives an idea of ??what things should be maintained and what things should be improved and even abandoned. Likewise, the teacher can make the results of the assessment into material for reflection, what are the successes and failures. Through self reflection, teachers can better develop their next lesson plans. The data shows that most students have developed assessments capable of measuring students' authentic abilities, but a small proportion of students still compile assessments that can be done by almost all students even though they do not follow the learning process.

In fact, teachers need to be involved in developing the curriculum. The curriculum is developed in order to meet the needs of students. The teacher is the one who understands that the best. Through their knowledge, experience, and competence, teachers can contribute to curriculum development in order to obtain guidelines that really suit the needs of students (Alsubaie, 2016). Teacher profiles regarding lesson plans as part of the curriculum can be an insight for lecturers in training students in developing curriculum for the next lecture.

4 CONCLUSION

Pre-service school teachers have realized that lesson plans are an essential part of the success of the learning process. In designing lesson plans, students have considered the characteristics of scientific uniqueness, namely according to the level of thinking of elementary school students. In more detail, students in developing lesson plans have been able to formulate learning objectives in accordance with curriculum expectations. In the introductory section: how to review the previous material, how to integrate the previous material with the material to be studied, and how to motivate so that the material to be studied is relevant and useful for students who are appropriate. Furthermore, in designing the main section, most students were able to: (a) encourage students to think critically by presenting phenomena/objects and then ask questions, (b) designing activities that train students to develop process skills according to elementary level, (c) presenting designs which are easy for elementary students to develop and do, (d) providing students opportunity to communicate the data, idea, and experience creativity. However, a small proportion of students: (a) are encouraging students to think critically directly ask questions without encouraging students to enter into phenomena, (b) are less able to synchronize learning activities and the process skills they want to train students, (c) are developing designs for students to develop experiments, there are those in the design that are difficult for elementary school students to do, and (d) just provide student opportunity to communicate the result of experiment or observing with determined format. As for the concluding section, first, most students were able to provide students with insight into the application of concepts that were appropriate to the context of student life, and a small proportion of students were still unable to adapt to the context of student life. Second, most students are able to provide detailed alternative methods for students who have difficulty practicing their learning outcomes independently, namely by providing guidelines to do with their parents, and a small proportion of students only ask students to repeat activities carried out at school. In the final stage, namely application, most students have given follow-up assignments in accordance with the material context and the availability of facilities in the student's neighborhood, but a small proportion of them are still too far from the material context. The learning assessment design prepared by most students can distinguish participants who are able and unable to engage activities to achieve learning objectives, but a small proportion of students in the design are still general in nature.

REFERENCES

AbdulGaafor, K. & UmerFarooque, T.K. 2010 Ways to improve lesson planning: A student teacher perspective. *International Seminar cum Conference on Teacher Empowerment and Institutional Effectiveness* November 2010: (9pg)

Aglazor, G. 2017. The role of teaching practice in teacher education programmes. Designing framework for best practice. *Global Journal of Educational Research* 16: 101–110.

Akdeniz, et al. 2016. Learning And Teaching. Theories, *Approaches And Models*. 2Nd Edition. Turkey: Çözüm Egitim Yayincilik.

Alsubaie, M.A. 2016. Curriculum Development: Teacher Involvement in Curriculum Development Journal of Education and Practice 7(9): 106–107.

Anonim. *Anonim Format Lesson Plan*. Canada: Queen University. http://educ.queensu.ca

Bin-Hady, W.R. 2018. How can I prepare an ideal lesson-plan? *International Journal of English and Education* 7(4): 275–289.

Cai, J., Perry, R. & Wong, N. Y. 2007. What is effective Mathematics teaching? A dialogue between East and

Cohen, L., Manion, L. & Morrison, K. 2005. *Research Methods in Education.* 5th ed. London: Taylor & Francis Dorovolomo, J., Phan, H.P. & Maebuta, J. 2010. Quality lesson planning and quality delivery: Do they relate? *International Journal of Learning* 17(3): 447–455.

Hunter, D.A. 2009. *A Practical Guide Ti Critical Thinking: Deciding What To Do and Believe.* New Jersey: John Wiley & Sons

Jabbarifar, T. 2009. The Importance of Classroom Assessment and Evaluation In Educational System *International Conference Of Teaching And Learning* (ICTL 2009) Proceedings Of The 2nd: 1–9.

Johnson, A. P. 2000. It's Time for Madeline Hunter to Go: A New Look at Lesson Plan Design. *Action in Teacher Education* 22(3): 72–78.

Khan, T. Singh, T. Hande, S. and Khumar, V. 2014. learning objectives:perfect is the enemy of good. International Journal of User-Driven Healthcare. 2(3):45–63.

Meier, M. 2007. Teacher Professional Learning, Teaching Practice and Student Learning Outcomes: Important Issues. *Handbook of Teacher Education:* : 409–414. Netherlands: Springer

Misra, P.. Mehta, R., & Henriksen, D. 2015. Creativity, Digitality, and Teacher Professional Development: Unifying Theory, Research and Practice. *Handbook Of Research On Teacher Education In The Digital Age*: 691–721. USA: IGI Global

Panasuk, R., Stone, W. & Todd, J. 2002. Lesson planning strategy for effective mathematics teaching. *Education, 122*(4): 808–827.

Ratminingsih, N. M., Artini, L. P., & Padmadewi, N. N. 2017 Incorporating Self and Peer Assessment in Reflective Teaching Practices. *International Journal of Instruction,* 10(4): 165–184.

Surat edaran Mendikbud No 14 tahun 2019 tentang Penyederhanaan Rencana Pelaksanaan Pembelajaran

Timperley, H., Wilson, A., Barrar, H., & Fung, I. 2007. *Teacher professional Learning and Development: Best evidence Synthesis Iteration.* Wellington, New Zealand: Ministry of education http://educationcouns.edcentre. govt.nz/goto/BES

Xhaferi, B., & Xhaferi, G. 2017 Enhancing Learning Through Reflection– A Case Study Of Seeu *De Gruyter* 4 : 53–68.

Mckimm, J, &Swanwick, T. 2009. Setting Lerning Objectives. *British Journal of Hospital Medicine* 70(7): 406–409.

Shepherd, C.K., Hair, M. & Brown, L. 2010. Investigating the use of simulation of teachng strategy. *Nursing Standard: official newspaper of the Royal College of Nursing* 24(35): 42–48.

Educational Innovation in Society 5.0 Era: Challenges and Opportunities – Purnomo & Herwin (Eds)
© 2021 the authors, ISBN 978-1-032-05392-9

Documenting factors influencing children like learning English as a foreign language

R. Rintaningrum
Institut Teknologi Sepuluh Nopember (ITS), Surabaya, Indonesia

S. Iffat Rahmatullah
King Khalid University, Abha, Kingdom of Saudi Arabia

ABSTRACT: When English is not only spoken by native speakers of English, English places itself more confidently in all aspects of life, including in the field of Education, and even learned by people with a variety of ages, including children. In an Asian country, particularly, Indonesia, children or young learners have opportunities to learn English from formal classroom instruction to non-formal education. However, since English in Indonesia spreads very largely, this inspires educators to open non-formal education that is English classes conducted outside of school curriculum, and held outside of formal classroom instruction. There are some Indonesian children taking these opportunities by attending the classes and programs offered. The aim of the study is to identify reasons why children like learning English. This study is gathering evidence on what factors that influence children like learning English. Data is obtained from one on one interview. The results of study indicate that there are a variety of reasons on why children like learning English in this context. The results of the interview show that five factors, namely, student level factors, parent level factors, teacher level factors, classroom level factors, and school level factors have a contribution on why children like learning English. The results of the study can be used to provide better assistance for students in order that educational goals can be met. In the future, the results of the study can be used to develop a model of teaching and learning English for children.

1 INTRODUCTION

Although the term 'global change' is not new, this term remains a hot debate across academic disciplines, particularly, science, technology, politics, economics, sociology, culture, including education. Global change has influenced the field of education in terms of change in educational processes and in particular, English language teaching and learning (Rintaningrum 2015). The dynamic changes occurring in this global era greatly affect the teaching and learning of English as a Foreign Language (EFL) across the non-English speaking world. English is growing to be the most popular language learned across the hemisphere of non-English speaking countries. Undoubtedly people in general, and children in particular have a great demand to learn English as a foreign language. Children in Asian countries (Fang 2009, p. 1; Kachru 2005; McArthur 2003; Rintaningrum 2015) such as Indonesia, Vietnam, Cambodia, Japan, Korea, Malaysia, Singapore, Thailand, Philippines, and China develop themselves to be proficient English users. Children in such countries start learning English in early stages. The English language has gained in status across Asian countries.

Developing English language proficiency within the setting of English as a foreign language is not an easy task. Cummins (1984) argued that this was because "the acquisition of language proficiency is not a dichotomous process but rather one that develops along a continuum". Moreover, the teachers of English in this setting are non-native speakers of English, and the target language is not spoken in this setting. The emergence of language learning theory namely, the 'Grammar Translation Method' dominate the teaching of English in Asian countries, such as Japan (Gorsuch 1998; Matsuura, Chiba, & Hilderbrandt 2001; Suzuki 1999), Indonesia (Alwasilah 1993; Widiati & Cahyono 2009), Korea (Vasilopoulos 2008), Thailand (Hiranburana 2017), and China (Zhang 2009) have used this method for years although in recent years there have been some changes in the method of teaching. Even when the latest teaching methods, namely, Communicative Language Teaching (CLT), is used, some of the characteristics of CLT make it difficult for a non-native-speaking teacher who may not have a high level of English language proficiency in the use of English to teach effectively. Teaching grammar rules and drills (the forms of language) are seen to be much simpler and easier than the use of the target language for non-native-speaking teachers to teach. Therefore, it is not surprising that students are able to achieve higher scores in writing (structure and written expression) since this is what they have been doing in

DOI 10.1201/9781003206019-47

class for a long time when learning English as a foreign language.

However, in line with a great demand in learning English, and due to the rapid changes in the development of Information, Communication, and Technologies (ICT) (Anderson 2005), a variety of modern techniques to teach and learn English have been improved including the use of technology for young learners (Bull & Ma 2001; Gee 1996; Harmer 2007; Prensky 2001; Rintaningrum 2016, 2018; Scott & Ytreber 2008; Tomlinson 2010; Yang 2013). The availability of the internet enables young learners to learn English through technology (Cameron. Lynne 2002) since young learners can gain some benefits of using technology for learning English (Clements & Sarama 2003). In a setting where English is learned as a foreign language, teaching Grammar to students is still important (Rintaningrum 2015). However, in order that students are attracted to actively involve during the process of learning, a combination of different methods of teaching is necessary (Larsen-Freeman & Anderson 2011; Rintaningrum & Aldous 2016).

In Indonesia, where study is conducted, English is learned starting from Primary schools, even some children learn English in Kindergarten. Some private Kindergartens in Indonesia use English as a medium of instruction in the class. The government of Indonesia put a great attention on the development of English in Indonesia. English has been introduced in primary school in Indonesia since 1994 as a school subject. Although the nature of English in primary schools in Indonesia is not compulsory, most schools in Indonesia take some advantages by putting English as a subject to be learned. English is a compulsory subject to be learned at high school.

Moreover, the great demand of learning English in Indonesia can be seen from the establishment of many English courses run as programs separate from schools' management. The characteristics of these English courses are non-formal education that mostly belongs to private business. Many Indonesian children provide their time to learn English after school hours by attending English courses held outside of the school. This study takes a closer look on what children are saying about the English language. This study does not investigate why English is important, or why English is necessary to learn.

This study simply investigates some reasons obtained from the voice of children on why children like learning English in order that better assistance can be provided in the process of teaching and learning of English. From the children's voice, it is expected that there are some new insights relating to teaching and learning English for children, in order to help children learn and develop the skills of English easily. There are some studies concerning why children learn English for adult, and not about why they like learning English, in particular for children. Therefore, a study for identifying factors that influence why children like learning English is essential.

2 RESEARCH QUESTION

A research question is advanced:

What factors that influence children like learning English?

3 METHOD

This study employs a descriptive qualitative method in order to obtain information from the respondents. In this study, the data collection is conducted through interviews. Face to face interview is chosen in order that rich information can be documented from the participants (Keats 1997). Participants have opportunities to answer the questions based on what they know and they feel.

The participants of the study are children between 7 and 10 years old attending an English course program conducted after school hours. There are about 15 children who participate in the study. The number of participants involved in the study is identified by the number of children attended the English course. The respondents in this study are identified as targeted participants. Scott A. & Ytreber H., n.d. (2008) argued that young learners in the range of age are able to argue and tell us why they think what they think. There are 7 girls, and 8 boys involved in the study. This study does not explain or compare about gender differences. In the middle of the program, participants are asked to express free ideas and comments concerning the English class they were attending while participants were having relaxed conversation and consultation to their teacher. Again, before the program ends, after participants have a variety of activities, the participants are given a question concerning why they like learning English.

The questions are not given like real interviews with adults. Interview is conducted in a good time, when participants are still doing their activities so the situation does not show that it is an interview. The participants have opportunities to provide a variety of answers based on what ideas they want to express. The results of the interview are reported in the name of planets such as Saturn, Mars, and Neptune.

Since the participants are children, as an ethic in a research tradition, a researcher needs to have permission from their gatekeepers, parents of the children. A researcher needs to pass the parents' gate before a study is conducted (Alderson 1995). Even at the introduction of the program parents are given information that children will be asked to give their reasons on why they like learning English. The researcher observes the class during the programs conducted, and sometimes teaches as well in order to establish relationships with children.

Interview with children is not conducted in serious ways, otherwise fun ways, in a happy situation. Researchers need to be able to create circumstances that do not make children feel like they are being interviewed. It is conducted in a circumstance when children are encouraged to involve in an activity that enables researchers to pass some questions. There is a window that enables to open a dialogue.

4 FINDING AND DISCUSSION

4.1 *Factors that influence children like learning English*

From the data collected through the interviews, some factors that contribute to why children like learning English are when they have a good attitude towards learning English, and when they have family support as well as their own individual interest.

From the interviews with the fifteen children, there are 23 theoretical categories that contribute to why it is for a learner to learn the English language in situations where English is taught as a foreign language. The categories are based on the ideas given by the respondents from the results of the interviews that are transcribed. From the transcripts of the interviews, it is an investigator's responsibility to identify a keyword in what the respondent says. Each appropriate key word emerges as a category in this study.

These 23 theoretical categories are reported and discussed together with comments that support why particular indicators are classified into a particular category. In this section, the indicators signify what the respondents say about the reasons why they like learning of English. The 23 categories are (a) classroom activities, (b) new experience, (c) individual interest, (d) student positive attitude, (e) emotional bonding, (f) students impression, (g) teacher creativity, (h) teachers personality, (i) teacher dress up play, (j) teachers support, (k) teachers sense of belonging, (l) teacher passion, (m) parental support, (n) opportunity to speak in English, (o) playing quiz, (p) the cost of the program, (q) teachers support, (r) classroom facilities, (s) innovative classroom setting, (t) a variety of teaching methodology, (u) learning English through technology, (v) student motivation and engagement, and (v) new knowledge and information. The result of the interview is reported below.

The results of interviews conducted with the participants show that factors that influence children like learning English are complex. This means that there is no single factor that is able to explain whether or not students like learning English since the factors are many and complex. Table 1 records five factors that influence why children like learning English.

Table 1. Factors influencing children like learning English.

Factors	Categories
Student or Individual Level	Student motivation and engagement Obtaining new experiences Opportunity to speak in English Individual interest in English Student positive attitude towards English Emotional bonding with teacher Student impression to their teacher
Parent Level	Giving children permission to take a course Paying the fee of the course Dropping children to school Picking up children at school
Teacher Level	Teacher personality Teacher knowledge Teacher experiences Teacher dress up play Teacher attitude Teacher passion Teacher support Teacher sense of belonging Teacher additional support Teacher teaching methodology
Classroom Level	Types of Classroom Activities Educational sense Entertainment sense Combination between Education and entertainment Playing quizzes Playing games Material presentation Multitasking learning Classroom facilities Innovative Classroom Setting Learning English through technology
School Level	Course fee Instructional design quality Curriculum Curriculum delivery or presentation Teaching methodology presentation

Table 1 shows that there are complex and many factors that influence why children like learning English. The factors which have influence on why children like learning English are (1) student level factors; (2) parent level factors; (3) teacher level factors; (4) classroom level factors; and (5) school level factors. The factors are explained as follows.

4.2 *Student level factors*

Student level factors are identified by student motivation and engagement, obtaining new experiences, opportunity to speak in English, individual interest

251

in English, student positive attitude towards English, emotional bonding with teacher, student impression to their teacher. From student level factors in Table 1 it can be explained that at individual level each students bring their own characteristics on why they like learning English. Internal factors play an important role for children to like English because these are factors that come from their own without any other interventions. From these internal factors, students like learning English. Their perception is also different from student to student. This means that participants have different reasons on why they like learning English.

4.3 Parent level factors

Parent level factors include giving children permission to take a course, paying the fee of the course, dropping children to school, and picking up children at school. From parent level factors it can be explained in Table 1 that parent is one of the factors that influence children like learning English. This indicates that parent support is very important because parent support assists children to have a positive attitude towards English. Children expect their parents to support them in order that children like learning English in a variety of ways. Children are more likely to feel safe in studying if their parent, or caregivers are able to accompany them before and after school in terms of dropping and picking up them.

4.4 Teacher level factors

Teacher level factors can be identified by teacher personality, teacher knowledge, teacher experiences, teacher dress up play, teacher attitude, teacher passion, teacher support, teacher sense of belonging, teacher additional teaching support, and teacher methodology. It can be explained from Table 1 that teachers have a great influence on why children like learning English. The results of the interview show that children have a big expectation of their teacher. It cannot be denied that children are also able to assess that teacher is one of the factors that influence children like learning English. Acknowledging student perception it is clear that the teacher plays an important role to assist students to have a positive attitude towards English. It is because students show their big hope to their teachers. Understanding students' hope to their teachers, it is time for teachers to reflect about what they have done and they need to improve in order that teachers can better assist their students learn English.

4.5 Classroom level factors

Classroom level factors include types of classroom activities, Educational sense activities, entertainment sense activities, combination between education and entertainment, playing quizzes, playing games, material presentation, multitasking learning, classroom facilities, innovative classroom setting, and learning English through technology. Table 1 indicates that classroom activities are one of factors that influence children to like learning English. Classroom activities that are designed properly and accurately are more likely to contribute to why children have a positive attitude towards English. It is very essential to well prepare classroom activities since one of children's characteristics is children have short attention and concentration span (Scott A. & Ytreber H. 2008), but they will learn best when they enjoy their work.

4.6 School level factors

School level factors include course fee, instructional design quality, curriculum, curriculum delivery or presentation, teaching methodology presentation. In this study school level factor is the top level factor that influences children like learning English. Table 1 records that school level factors have a big role to influence children to have a positive attitude towards English. School level factors are important factors to assist students to like learning English since these factors relate to school policies and decision makers of the school. The results of the interview indicate that participants' perspectives towards the learning of English can help schools to make policies that benefit not only for students but also for teachers. This means that school can better assist students if their policies align with student aspiration. At the same time when student aspiration is established and recognized by school, teachers are given access to conduct professional development programs in order that teachers are able to provide better assistance to students.

5 CONCLUSION

The results of the interview presented in Table 1 shows that there are several reasons why children like learning English. Children or young learners have different point of views on why they like learning English. The results of the interview give a lot of insights in order to help children learn English better. The reasons why children like learning English can be grouped into five factors, namely, student level factors, parent level factors, teacher level factors, classroom level factors, and school level factors. Children or young learners like learning English due to student level factors such as new experience, individual interest, positive attitude, emotional bonding, students impression, teacher level factors such as emotional bonding, teachers personality, acting like a character, teachers support, teachers sense of belonging, teachers' attitude, parent level factors such as parental support, and curriculum factors in terms of types of classroom activities, opportunity to speak in English, playing quiz, the cost, teachers support, classroom facilities, innovative classroom setting, a variety of teaching methodology, and learning English through technology as well as teachers additional support.

It is surprising that children are able to explain that internal factors such as attitude and motivation influence why they like English. Attitude and motivation are intrinsic factors that come from students. This means that attitude and motivation are not adult domination; children in the certain age ranges also bring their own intrinsic factors that contribute to their learning.

Parental or home support is the best determinant of student success at school. To what extent family encourages learning at home and involves themselves in their children's education helps children to achieve their success at school (PTA 2018). The results of the interview indicate that parental support is needed in children's school life. Parent engagement in children's school life means that both parents and school share responsibilities in order to help children learn and meet educational objectives. Hill & Tyson (2009) noted there is a relationship between parent involvement at school and student achievement.

Since children have short attention and concentration spans, it is necessary to create a variety of activities. Scott A. & Ytreber H. (2008) suggested that a variety of activities is a must. Moreover, children like playing and movement; therefore teachers must be creative in establishing tasks and activities to perform. Consequently, teachers need to organize and plan their lesson.

Moreover, since children in this era are living surrounded by technology, it is much more beneficial if they experience learning English through technological tools. This implies that facilities as media of learning need to be improved, teachers need to be technological literacy. Teachers need to follow the demand of the era meaning that teachers need to be up to date. Teachers need to be given some training and some professional development programs in order that they are up to date.

The results of study also show that curriculum plays an important role in order that the process of learning runs smoothly and fits the students need. This implies that lesson plan must be designed carefully by considering the characteristics of the existing context. Moreover, considering the characteristics of children that lack focus, there is flexibility for teachers to adjust the plan suitable with the students' needs, for example by combining a variety of methods of teaching. Moreover, the results of study help curriculum planners, and teachers to design the further and better curriculum.

Identifying factors that influence why children like learning English helps teachers to provide more assistance to students. It is expected that teachers have better decisions which parts need to be established and be improved to meet educational goals that results in better student achievement. The results of the interview show that children are able to assess and explain what they have done with their teacher during the class session. It is evident that children with their characteristics are a group of students who need a teacher with good personalities, good attitude, and good abilities (Donaldson 1978).

REFERENCES

Alderson, P. 1995. *Listening to Children. Children, Ethics and Social Research, London: Barnardos.*

Alwasilah, A. C. 1993. *Pengantar Sosiologi Bahasa.*

Anderson, J. 2005. *Information and communication technologies in Poland.* (Jonathan Anderson, Ed.), *Telecommunications Policy.* Paris: UNESCO. https://doi.org/10.1016/0308-5961(94)90052-3

Bull, S., & Ma, Y. 2001. Raising Learner Awareness of Language Learning Strategies in Situations of Limited Resources. *Interactive Learning Environments*, 9(2), 171–200. https://doi.org/10.1076/ilee.9.2.171.7439

Cameron. Lynne. 2002. Teaching Languages to Young Learners. L. Cameron. *ELT Journal*, 56(2), 201–203. https://doi.org/10.1093/elt/56.2.201

Clements, D. H., & Sarama, J. 2003. Strip mining for gold: Research and policy in educational technology — A response to "Fool's Gold." *Association for the Advancement of Computing in Education (AACE)*, 11(1), 7–69. Retrieved from http://www.editlib.org.ezproxy.csu.edu.au/f/17793

Cummins, J. 1984. *Bilingual and special education: Issues in assessment and pedagogy.* TX: PRO-ED.: Austion.

Donaldson, M. 1978. *Children's Mind.* London: Collins.

Gee, J. 1996. Social Linguistics and Literacies. *Social Linguistics and Literacies.* https://doi.org/10.4324/9781315722511

Gorsuch, G. J. 1998. Yakudoku EFL instruction. *JALT Journal*, 20(1), 6–32.

Harmer, J. 2007. The Practice of English Languag Teaching, 442.

Hill, N. E., & Tyson, D. F. 2009. Parental involvement in middle school: a meta-analytic assessment of the strategies that promote achievement. *Developmental Psychology*, 45(3), 740–763.

Hiranburana, K. 2017. Use of English in the Thai workplace. *Kasetsart Journal of Social Sciences*, 38(1), 31–38. https://doi.org/10.1016/j.kjss.2015.10.002

Keats, D. 1997. Interviewing for clinical research. In J. P. Keeves (Ed.), Educational Research, Methodology, and Measurement: An International Handbook. Oxford: Pergamon.

Larsen-Freeman, D., & Anderson, M. 2011. *Techniques & Principles in Language Teaching. Book* (Vol. 53). Oxford University Press. https://doi.org/10.1017/CBO9781107415324.004

Matsuura, H., Chiba, R., & Hilderbrandt, P. 2001. Beliefs about Learning and Teaching Communicative English in Japan. *JALT Journal*, 23(1), 69–91. Retrieved from http://jalt-publications.org/jj/articles/2668-beliefs-about-learning-and-teaching - communicative-english-japan

Prensky, M. 2001. The Games Generations: How Learners Have Changed. *Computers in Entertainment*, 1(1), 1–26. https://doi.org/10.1145/950566.950596

PTA, N. 2018. *Building Successful Partnerships: A Guide for Developing Parent and Family Involvement Programs.* Bloomington, Indiana: National PTA, National Education Service.

Rintaningrum, R. 2015. *Teaching And Learning Of English As A Foreign Language In A Global Context.* Flinders University. Retrieved from https://flex.flinders.edu.au/file/a0853ab2-5aaf-477b-9dd1-f8bb6d670451/1/Thesis-Rintaningrum-2015.pdf

Rintaningrum, R. 2016. Maintaining English Speaking Skill in Their Homeland through Technology: Personal Experience. *Asian EFL Journal*, 2016(SpecialEdition), 107–119.

Rintaningrum, R. 2018. Investigating Reasons Why Listening in English is Difficult: Voice from Foreign Language Learners By Ratna Rintaningrum Institut Teknologi Sepuluh Nopember (ITS), Surabaya, 1–7.

Rintaningrum, R., & Aldous, C. 2016. I Find It Easy To Learn English When. In Jambi University (Ed.). Jambi: Jambi University Press.

Scott A., W., & Ytreber H., L. 2008. *Teaching English to Children*. (N. Grant, Ed.). London New York: Longman.

Suzuki, T. 1999. NNihon-jin wa Naze Eigo ga Dekinaika [Why cannot Japanese master English?]. Tokyo: Iwanami Shoten. *IJALT*, 1(2).

Tomlinson, B. 2010. Principles and procedures of materials development for language learning: 3 Proposals for principled approaches to the development of ELT materials. *Materials in ELT: Theory and Practice*, (1995).

Vasilopoulos, G. 2008. Adapting Communicative Language Instruction in Korean Universities. *The Internet TESL Journal, Vol. XIV, No. 8*, 2. Retrieved from http://iteslj.org/Techniques/Vasilopoulos-CLT.html

Widiati, U., & Cahyono, B. Y. 2009. the Teaching of Efl Listening in the Indonesian Context: the State of the Art. *TEFLIN Journal*, 20(2), 139–150. https://doi.org/10.15639/teflinjournal.v20i2/194–211

Yang, Y. F. 2013. Exploring students' language awareness through intercultural communication in computer-supported collaborative learning. *Educational Technology and Society*, 16(2), 325–342.

Zhang, L. J. & A. W. 2009. Chinese senior high school EFL students' metacognitive awareness and.pdf. *Chinese Senior High School EFL Students' Metacognitive Awareness and Reading-Strategy Use*, 21(1), 37–59.

Educational Innovation in Society 5.0 Era: Challenges and
Opportunities – Purnomo & Herwin (Eds)
© 2021 the authors, ISBN 978-1-032-05392-9

The content validity analysis of the elementary school students' tolerance character measurement instrument

H. Sujati, Haryani, B.S. Adi, Kurniawati & T. Aprilia
Universitas Negeri Yogyakarta, Yogyakarta, Indonesia

ABSTRACT: This research aims to prove the content validity of the character measurement instrument of students' tolerance in elementary school. The instrument of measuring the tolerance character in this study was developed into three aspects, namely (a) respect, (b) acceptance, and (c) appreciation. Then, it is formulated into eight behavioral indicators, and developed into 35 items. This research method uses a descriptive research method with quantitative descriptive analysis techniques. The results of proofing validity are carried out by experts and analyzed using Aiken's V formula. Content validity data is obtained from three experts in different fields consisted of character education experts, measurement experts, and inclusion education experts. The results showed that the validity of the contents coefficient of 35 items in the tolerance character measurement instrument contained 33 items that were declared valid with a value ≥ 0.83. Thus, the 33 items on the tolerance character measurement instrument have good content validity, and support the validity of the instrument's overall content. Therefore, this instrument can be used to identify the level of tolerance character of elementary school students.

1 INTRODUCTION

Indonesia is a country that has a variety of things, ranging from ethnicity, culture, language, religion, and many others. This diversity does not divide Indonesian society. This is because Indonesia upholds the principle of diversity (being different but remaining one) so that unity and integrity are maintained and kept with tolerance. However, recently, many cases of intolerance have occurred in Indonesia, especially among students. Intolerance itself comes from the belief that the community, belief system, or lifestyle is higher than others, resulting in several consequences ranging from lack of appreciation to institutionalized discrimination (Hasani 2009). A survey conducted by UIN Jakarta showed that as many as 48.95% of student or student respondents felt that religion education influenced them not to associate with followers of other religions (Muthahhari 2017). In addition, cases of intolerance have also occurred in a high school in Yogyakarta regarding the school principal's attitude who is intolerant of the timing of school activities that coincide with other minority religious holidays in the school (Kumparan 2018). Based on these cases, it shows that there is a need for diversity education that must be instilled from an early age in students, as well as strengthening tolerance character education in all school stakeholders.

Tolerance itself is one of the characters that exist in the main values of strengthening character education

(PPK) in Indonesia. Tolerance as attitudes and actions that respect differences in religion, ethnicity, opinions, attitudes, and actions of others who are different from themselves (Kemendiknas 2010). Tolerance in the context of education is also understood as an educational activity encouraging students to be open-minded, and even giving positive appreciation for what is different from themselves, namely religion and attitude to life by (Hansen 2011). UNESCO states tolerance as a positive recognition of human rights and civil liberties (Unesco 1995a). Other opinion thinks that tolerance is behaviour of respecting and accepting all differences that exist within groups, and individuals to create a peaceful, and prosperous life (Lickona 2013); (Saulius 2013); (Samani and Hariyanto 2013). Therefore, it is important to instill a character of tolerance in students from an early age starting from the elementary school level.

In measuring the character of student tolerance, an assessment instrument is needed. The results of research by Muthahhari (2017) explained that the instrument for assessing student character which have been made is still on a small scale, and has not focused on the dominant personal variable character development at each level of education, which is only carried out in the upper secondary education level, so that for primary school, the instruments have not been made. In addition, based on research findings of Hadiwinarto (2010) it is stated that character assessment has not been carried out professionally by teachers.

DOI 10.1201/9781003206019-48

Character assessment is only carried out through haphazard direct observation without using clear assessment instruments. The same opinion was expressed by Simarmata et al. (2019) in which it is stated that assessing the character of students in thematic learning in elementary schools without exception in the character of tolerance, the teacher made an assessment only as a formality to carry out without understanding the essence of the character being assessed. Therefore it is necessary to develop a standardized instrument. It is to make it easier to measure tolerance characters. In the 2013 curriculum that is applied in schools, in addition to cognitive aspects, affective or attitude aspects are also assessed by educators. However, there is still no measuring instrument or non-cognitive measurement instrument or attitude aspect that has generally been standardized. Thus, the instrument which is being developed must be easy to use, and easy to understand. Tolerance is one part of the affective aspect, so the instrument which is developed should be in the form of a non-test instrument in the term of measuring.

The theory of developing a psychological scale in this study uses an implementation procedure that starts with (1) conducting a need for assessment in the field (which has been done in previous studies); (2) determining he latent variables to be developed by the instrument, (3) identifying behavioral indicators, (4) constructing grids, (5) compiling instrument items, (6) testing the content validity qualitatively and quantitatively, (7) writing instrument items, (8) selecting items that are suitable for content validity testing, (9) constructing validity tests, (10) doing item selection, (11) testing reliability, and (12) Making final compilation in the form of tolerance character measurement instruments (Azwar 2012a).

Proving the validity of the instrument content is an important first step in the instrument development process. Content validity concerns the extent to which the scale items represent the latent constructs being measured (Azwar 2012b). Based on this description, it is important to prove the validity of the content of the development of a tolerance character measurement instrument for elementary school students. The purpose of this study was to prove the validity of the content of the instrument for measuring the tolerance character of students in elementary schools.

2 RESEARCH METHODS

The research method which is used is descriptive research method. This research is part of the research to develop an instrument for measuring the tolerance character of students in elementary schools. The object of this research is an instrument for measuring tolerance of elementary school students, the subjects which were involved in proving the validity of the content of

the tolerance character measurement instrument are three experts from various fields, namely character education, measurement, and inclusive education. The three experts were asked to review items from the dimensions of construct representation, clarity, and relevance. The instruments which were used in data collection in this study were a questionnaire that was easy to quantify and a qualitative instrument review sheet. The data analysis technique which was used in this study was a quantitative descriptive analysis technique. Where the results of proving the validity of the content carried out by the expert were analyzed using the Aiken's V formula. The results of calculations and analysis using the Aiken's V formula can later be interpreted into the instrument validity category. The categorization refers to Polit et al. (2007) that items that have an index of ≥ 0.78 are an indication that the validity of the item's content is considered good.

3 RESULTS AND DISCUSSION

3.1 Results

The findings from literature studies through the analysis of journals and scientific books are the main reference for determining the character of tolerance. In the UNESCO (1995b), it explains that tolerance is respect, acceptance, and appreciation of the rich diversity of our world's cultures, our forms of expression and the ways of being human. This of course can be nurtured by knowledge, openness, communication, and freedom of thought, conscience, and belief. Tolerance can be interpreted as harmony in differences. The findings from the literature study reveal aspects, and indicators of tolerance character. The findings are then translated into a tolerance character instrument grid as shown in Table 1.

The lattice of the character instrument for tolerance becomes a reference for writing items. Item writing is done by taking into account the predetermined writing conventions. Furthermore, the first review is carried out by the researcher himself to double-check whether the item is in accordance with the behavioral indicators to be measured. The second review was conducted by experts or experts in character education, educational measurement, and inclusive education. This second stage is to prove the content validity of the tolerance character instrument that has been developed. The results of the assessment are in the form of qualitative (descriptive), and quantitative. The qualitative assessment results can be seen in Table 2.

Based on Table 2, the experts have recommended several item improvements as follows: shortening some statements, changing some diction, and adjusting the contents of the statements with the latent constructs measured. This improvement aims to facilitate the respondents' understanding of the content of the items and increase convergent, and discriminant validity. Quantitatively, the content validity is indicated by the expert agreement index (CVI). An item

Table 1. Instrument lattice of tolerance character.

No	Aspect	Indicator	Numbers of Items	No Item	
				Favorable	Unfavorable
1	*Respect*	Showing sympathy for others	4	1, 9, 17	25
		Respecting the beliefs of others	5	2, 10, 18, 26	33
		Respecting the opinions of others	4	3, 11, 19	27
2	*Acceptance*	Accepting the rights of others	4	4, 12, 20	28
		Accepting the differences of others	5	5, 13, 21	29, 34
		Accepting self	4	6, 14, 22	30
3	*Appreciation*	Getting benefit from diversity	4	7, 15, 23	31
		Interacting with people of different religions, cultures, and abilities	5	8, 16, 24, 32	35
	The total number of all items		35	26	9

is said to be feasible if it has a CVI ≥ 0.78. The results of the content validity analysis are presented in Table 3.

Table 2. Summary of expert assessment results.

Expert 1 (Character education):

1. There is a statement that is so sensitive that it needs to be replaced

2. Negative statements need to be balanced

3. Long, and ambiguous sentences

Expert 2 (Educational measurement):

1. Avoiding negative words, and frequency

2. Making the sentences simple

Expert 3 (Inclusive education):

1. Some statements need to conform to the indicator formulations in the grid

2. There is a statement that contains social desirability because there is a word "namely"

3. The language which is used is already communicative enough for students in level of elementary school.

4. In the case of research ethics involving children, usually the child's identity is not included in the data collection tools.

Based on the calculation of the V value in Table 3, the results are related to proving the validity of the content using the Aiken's V formula. The results show that of the 35 items validated by experts, 32 items were declared valid and 3 items were declared invalid. Items that are declared invalid need to be removed or revised. Overall, the percentage of content validity is 80%. Valid instruments are an important part of the assessment process because valid instruments are a fundamental goal for instrument development. A valid instrument is able to measure what should be measured in this case measuring students' cognitive abilities and students' scientific literacy abilities. Developing reliable and valid instruments is a long process. However, the most basic of the instruments must be content valid. This is because the resulting question items will represent the construct to be measured. Therefore, content validity is the key to a quality instrument.

3.2 *Discussion*

Measuring the affective aspect is different from measuring the cognitive aspect. Suryabrata (2000) explains that measuring attributes other than cognitive requires a response type of sentiment expression, which means that the type of response cannot be declared true or false, in fact, it can be said that all responses are considered true for each respondent. This also applies to measuring the character of tolerance, because the character of tolerance is one of the characters that cannot be directly seen and judged by others.

Character and practice are inseparable things in the discussion of tolerance. In the UNESCO (1995b) it explains that tolerance is respect, acceptance, and appreciation of the rich diversity of our world cultures, our forms of expression and ways of being human. This of course can be nurtured by knowledge, openness, communication, and freedom of thought, conscience, and belief. Tolerance can be interpreted as harmony in differences. A similar opinion was expressed by Lickona (2013) that tolerance is a form of contemplation of respect. Tolerance can eventually dissolve into a neutral relativism used to avoid prejudices regarding ethics. Tolerance becomes an attitude to have equality, and goals for individuals who have different thoughts, races and beliefs. Therefore there is a need for strong collaboration, and cooperation between teachers, students, parents, and schools in helping to implement the character of tolerance among students (Hernández et al. 2018).

Table 3. The results of content validation quantitatively.

Aspect	Indicator	No	Item	V-Value	Notes
Respect	Showing sympathy for others	1	I help my friends who has difficulty	1	Valid
		9	I stand up for a friend who is being teased	0,92	Valid
		17	I invite friends who lack confidence to come forward.	1	Valid
		25	I love to bother friends	0,83	Valid
	Respecting the beliefs of others	2	I respect friends of different religions to celebrate their religious holidays	0,92	Valid
		10	I respect friends of different religions to worship	0,92	Valid
		18	I am best friends with friends of different religions.	1	Valid
		26	I support friends of different religions to become class leaders.	1	Valid
		33	I disturb other friends of different religions	0,75	Invalid
	Respecting the opinions of others	3	I listen to friends who share their opinions.	1	Valid
		11	I support the results of deliberation decisions.	0,83	Valid
		19	I am happy to receive suggestions from friends.	1	Valid
		27	I get angry when my opinion is rejected.	0,92	Valid
Acceptance	Accepting the rights of others	4	I am happy to see friends getting gifts	0,92	Valid
		12	I understand that every student has the right to have many friends.	0,92	Valid
		20	I understand that school facilities can be used by all students.	0,92	Valid
		28	I forbid my friends to borrow books from the library.	0,83	Valid
	Accepting the differences of others	5	I am willing to be in the same group with all my friends.	1	Valid
		13	I am willing to help a friend in preparing for his religious celebration	0,92	Valid
		21	I am passionate about learning other people's cultures.	0,83	Valid
		29	I prefer staying at home to attending activities of different cultures.	0,83	Valid
		34	I impose opinions on other people.	1	Valid
	Accepting self	6	I have the same abilities as other friends.	0,75	Invalid
		14	I realized that I have weakness.	1	Valid
		22	I am proud of myself.	1	Valid
		30	I doubt my own abilities.	1	Valid
Appreciation	Getting the benefit of variety	7	I understand other people more, because I have a variety of friends	0,92	Valid
		15	I gained new knowledge from various friends	0,92	Valid
		23	I learned to respect other people from a variety of friends	0,92	Valid
		31	I get into trouble easily because I have friends of different religions	0,83	Valid
	Interacting with different people with different religion, culture, and ability	8	I enjoy spending time with friends of different religions.	1	Valid
		16	I like to discuss about cultural diversity.	1	Valid
		24	I am willing to study with friends who are not smart enough.	1	Valid
		32	I can easily forgive the mistakes of friends from different cultures.	0,75	Invalid
		35	I get annoyed when I play with friends who are not really smart.	1	Valid

The instrument for measuring the tolerance character that has been developed needs to be proven reliable. One of the steps to go through in proving the reliability of the instrument is the content validity. Azwar (2012b) states that content validity is carried out to ascertain whether the contents of the instrument measure precisely the situation to be measured through rational analysis. Content validity is usually exercised during the early stages of instrument development. The use of expert panels, such as those which are used in the Delphi method, is a representative way of testing content validity. The aim is to reduce the risk of errors in instrument development, and increase the probability of obtaining a construct validity index. To achieve good content validity it is necessary to use qualitative, and quantitative approaches (Retnawati 2015). Qualitative assessment is also known as a theoretical study. The theoretical analysis is carried out by considering the material, construction and cultural aspects. This theoretical study results in semantic changes, such as replacing ambiguous words, using more precise diction, and shortening certain phrases so that these items become more rational, and easily understood by respondents.

Quantitative content validity assessment has also been carried out. (Aiken 1985) offers three types of quantitative content validity assessment coefficients, namely the V, R, and H coefficients. In this study, the V coefficient is used. In proving the content validity, the researcher can determine the number of rating categories which are desired. The smallest number of rating categories which formulated by Aiken is 2 and the highest is 7 (Aiken 1985). This study used 5 rating categories with three raters. In this case Aiken (1985) determines the number 0.75 as the standard value of V, but in this study using the criteria set by Polit et al. (2007) namely, determining the number 0.78.

The results of the content validity assessment quantitatively showed that of the 35 items compiled, there were three items with a V value below 0.78, so that these three items were not included in the next analysis. Meanwhile, the other 32 items, because they have a content validity index above the standard, are expected to increase the validity of the construct. Because in essence, content validity measures how many items on a scale have a strong correlation with the construct it measures. It should be noted that the content validity is not a guarantee in identifying the measurement concept, but assessing the content validity supports the construct validity of an instrument (Yaghmaie 2003).

4 CONCLUSION

Based on the results of the analysis and discussion above, it can be concluded that the instrument for measuring the character of tolerance has proven the validity of the content. Content validity was carried out both qualitatively, and quantitatively. Qualitatively, the scale item has been improved in terms of diction, and simplification of statement formulation so that it is easier for respondents to understand. Quantitatively, three items were considered invalid by the expert team, while 32 other items were considered eligible for inclusion in the next analysis step, such as assessing construct validity and reliability.

REFERENCES

Aiken, L. R. 1985. Three Coefficients for Analyzing the Reliability and Validity of Ratings. *Educational and Psychological Measurement* 45(1): 131–142.

Azwar, S. 2012a. *Penyusunan skala psikologi.* Yogyakarta: Pustaka Pelajar.

Azwar, S. 2012b. *Reliabilitas dan validitas.* Yogyakarta: Pustaka Pelajar.

Hadiwinarto 2010. *Penajaman Penilaian Karakter dan Budi Pekerti.* Solo: Bahana Media Wirayuda.

Hansen, O. H. B. 2011. Teaching Tolerance in Public Education: Organizing the Exposure to Religious and Life-Stance Diversity. *Religion & Education* 38(2): 111–127.

Hasani, I. 2009. *Siding and Acting Intolerantly: Intolerance by Society and Restriction by the State in Freedom of Religion, Belief in Indonesia: Report of Freedom of Religion, Belief in Indonesia.* Jakarta: Setara Institute.

Hernández, A. S., Maria, M. I. & Daniel, M. 2018. *Teaching Tolerance in a Globalized World.* Switzerland: Springer.

Kemendiknas. 2010. *Pengembangan pendidikan budaya dan karakter bangsa.* Jakarta: Puskur-Balitbang.

Kumparan, T. J. 2018. *Sikap Intoleransi Sekolah Viral, Disdikpora DIY Tidak Tegas* [Online]. Available: https://kumparan.com/tugujogja/sikap-intoleransi-sekolah-viral-disdikpora-diy-tindak-tegas [Accessed 9 Februari 2020].

Lickona, T. 2013. *Character matters.* Jakarta: Bumi Aksara.

Muthahhari, T. 2017. *Survei UIN Jakarta: Intoleransi Tumbuh di banyak Sekolah dan Kampus* [Online]. Available: https://tirto.id/survei-uin-jakarta-intoleransi-tumbuh-di-banyak-sekolah-dan-kampus-czQL [Accessed 9 Februari 2020].

Polit, D. F., Beck, C. T. & Owen, S. V. 2007. Is the CVI an acceptable indicator of content validity? Appraisal and recommendations. *Res Nurs Health* 30(4): 459–67.

Retnawati, H. 2015. *Validitas Reliabilitas dan Karakteristik Butir.* Yogyakarta: Parama Publishing.

Samani, M. & Hariyanto. 2013. *Konsep dan model pendidikan karakter.* Bandung: Remaja Rosdakarya.

Saulius, T. 2013. What is "tolerance", and "tolerance education"? Philosophical perspectives. *Baltic Journal of Sport and Health Science* 89(2): 49–56.

Simarmata, N. N., Wardani, N. S. & Prasetyo, T. 2019. Pengembangan Instrumen Penilaian Sikap Toleransi dalam Pembelajaran Tematik Siswa Kelas IV SD. *Jurnal Basicedu* 3(1): 194–199.

Suryabrata, S. 2000. *Pengembangan Alat Ukur Psikologi.* Yogyakarta: Penerbit ANDI.

Unesco 1995a. *Declaration of principles on tolerance.* Paris: Unesco.

Unesco. 1995b. *Declaration of Principles on Tolerance. Proclaimed, and signed by the Member States of the UNESCO* [Online]. Available: http://www.un.org/en/events/toleranceday/pdf/tolerance.pdf [Accessed 22 September 2020].

Yaghmaie, F. 2003. Content Validity And Its Estimation. *J Med Edu* 3(1): 25–27.

Educational Innovation in Society 5.0 Era: Challenges and Opportunities – Purnomo & Herwin (Eds)
© 2021 the authors, ISBN 978-1-032-05392-9

Impoliteness language on social media: A descriptive review of PGSD UNY students

O.M. Sayekti, A. Mustadi, E. Zubaidah, S. Sugiarsih & E.N. Rochmah
Universitas Negeri Yogyakarta, Sleman, Yogyakarta, Indonesia

ABSTRACT: A student who is a prospective elementary school teacher must be able to provide examples of good and polite behavior, attitudes, and speech. A person's polite speech reflects direct or indirect communication through social media, which is currently the primary communication medium amid the COVID-19 pandemic. This study aimed to describe the existing impoliteness in social media PGSD FIP UNY students. The type of research used is the population's analysis is content with the active student in Elementary School Teacher Education Study Program Faculty of Education, State University of Yogyakarta. The sampling technique used purposive sampling. Data collection techniques divide into three techniques, documentation, observation, and literature study. This study's data were PGSD UNY students' utterances written on social media, Facebook, Instagram, WhatsApp, and Twitter. The method used to analyze the data used the content analysis schema technique, according to Krippendorf. This study indicates that students PGSD FIP UNY are still doing some impoliteness in their social media. Forms of impoliteness language include: (1) maxim of wisdom, (2) maxim of generosity, (3) variations maxims awards, (4) maxim of simplicity, (5) maxims of agreement, (6) maxim of generosity and agreement, (7) maxim of wisdom and consensus.

1 INTRODUCTION

Humans use language to communicate and interact with others. In the 4.0 revolution era, language is even more strategic because of technology that helps humans communicate. Computer Media Communication (CMC) is a new thing in the world of linguistics. CMC focuses on how language relates to computer communication media such as social media. In this case, language is used to communicate not only directly but also indirectly or through cyberspace. The existence of technology will make it easier to communicate even though we are far apart. Especially during the COVID-19 pandemic like today, the role of technology for communication can felt. That can imagine what would happen if the technology was not as advanced as it is now.

Communication via computer, one of which can do using social media. Humans today use social media a lot in their daily activities. They are starting from Facebook, Twitter, Instagram, WhatsApp, telegram, and many more. The wearer also from all walks of life. Both age and social status. If the first social status or age will limit the communication behavior, but with social media restriction, it becomes no longer exists. Currently, many critical national figures are active in social media. Many heads of regions, public officials, or the rector of a university active in social media. Thousands of followers do not see the social status.

They do this so that the people or students become closer to them.

Someone uses social media to communicate with other social users. In this case, social media works by utilizing internet networks (Hammod & Abdul-Rassul 2017). The term social media itself derived from the word "media" and "social." "Media" is defined as a means of communication (Laughey 2007 McQuail 2010). The word "social" can interpret that each individual's life can contribute to society. This statement is in line with Durkheim's statement in (Fuchs 2017) that media and all software are "social" or in the sense that both are products of social processes. From these various statements, it can conclude that social media is a digital-based communication tool used by the public in various forms of "social media." It is in line with the statement (Serwaa & Dadzie 2015) *"one cannot talk about technology without mentioning social communication media. Social media is an internet service which enables people to interact freely, share and discuss information about their lives"*.

Currently, social media presence has been able to change the world. The consequences also must be understood and skeptical by all parties. The presence of social media is increasingly an opportunity for individuals who are involved in it to free expression (Watie 2016). However, back to himself, each individual must have limits in his opinion so that what they convey does not hurt or harm the other party. In other words,

260

DOI 10.1201/9781003206019-49

individuals as communicators must apply language politeness. If a speaker adheres to language politeness, then the problems that often arise with freedom of communication on social media will be smaller. The opposite can happen if an individual is so free in a comment on social media, then the impoliteness cannot be avoided. Language impoliteness focuses on violating social norms or maxims in conversation (Oz et al. 2018). Papacharissi (2004) stated that one should uphold ethics, honesty, relevance, and giving if needed in communication. In line with Papacharissi, Locher (2010) stated that communication through social media is different from communication through direct face to face. Because of the interactions that occur on different social media with face-to-face interaction. Speakers must pay attention to the politeness language used when communicating.

Impoliteness seems to have occurred because of communication through different social media with direct communication. When communicating directly, people will pay more attention to the other person's face in front of them. However, if communicating through social media, face opponents said little attention. Because the interlocutor is not in front of the speaker, this will result in emotional communication. The speaker will write down what is on his mind, accompanied by the emotions that are in his head. As a result, impoliteness will appear (Liu 2017).

It also occurs in social media used by the Elementary School Teacher Education Faculty of the Education University of Yogyakarta. As the millennial generation, they tend to have freedom of speech. Their right of opinion has resulted in some deviation in language politeness. Impoliteness is a form of specific negative behaviors that occur in the context of speaking. Impoliteness arises because of the contradiction between desired and expected cause emotions speakers (Culpeper 2011). In Rahardi (2017), Mirriam also stated that language deviation is the behavior of someone who does not respect the interlocutor in a particular context. In this study, the impoliteness related to violations of politeness maxims. The maxims of language politeness include: (1) the maxim of generosity, (2) the maxim of wisdom, (3) the maxim of consensus, (4) the maxim of simplicity, (5) the maxim of appreciation, (6) the maxim of sympathy Leech in (Rahardi 2005) states that this research has a purpose of describing impoliteness of the language used by PGSD students on their social media accounts.

2 METHOD

This research includes content analysis research. Content analysis is a scientific technique for interpreting text or content. Krippendorff (2004) defines content analysis as a research technique to infer the meaning of the text or through procedures that can be trusted (reliable), can be replicated or applied in different contexts (replicable), and are valid. In general, there are three approaches to content analysis: description,

explanatory, and predictive. This study focused on a descriptive approach that describes the aspects or characters of a message or text (Eriyanto 2011). This study's data are PGSD UNY students' speeches written on social media, *Facebook, Instagram, WhatsApp, and Twitter*.

Meanwhile, this study's data sources were messages and stories written by PGSD UNY students on social media, Facebook, Instagram, WhatsApp, and Twitter. Selection of the research sample using a purposive sampling technique. That is, before sampling, the researcher determines the required criteria by being selected first. The method used to analyze the data used a content analysis scheme technique according to Krippendorff (2004) with the following stages: (1) data collection (unitizing), sampling, reducing, inferring, telling (narrating).

3 RESULT AND DISCUSSION

3.1 *Result*

The results of this research on Analysis of Impoliteness Language in Social Media: A Descriptive Review of PGSD UNY Students in the form of descriptions of language politeness and deviations that occur on *Twitter, WhatsApp, Instagram, and Facebook* social media used by PGSD UNY students. Data is taken from students' social media, both PGSD Central Campus, Mandala, and Wates. The data took approximately six months. The entire 502 cards wrote after going through data reduction to 435 speeches. Data in the form of impoliteness amounted to 80 utterances—the data took from three PGSD campuses, namely the Central Campus, Mandala Campus, and Wates Campus. The data distribution can see in Table 1.

Table 1. Impoliteness language on social media of PGSD UNY students.

No	Maxim Impoliteness	FB	Twit	IG	WA
1	Wisdom	–	5	2	24
2	Generosity	–	6	3	14
3	Simplicity	–	1	–	6
4	Sympathy	–	–	–	–
5	Consensus	–	–	–	4
6	Appreciation	–	–	–	8
7	Generosity and consensus	–	–	1	1
8	Wisdom and generosity	–	2	–	–
	Total	0	13	6	53

The data in Table 1 shows the distribution of language politeness maxims deviations on social media owned by PGSD FIP UNY students. Eighty-five speeches do not obey (deviate) from politeness. The maxims' most deviations are in the *WhatsApp* social media's wisdom maxims, 24 utterances. Meanwhile, there is no deviation at all on all social media belonging to PGSD FIP UNY students in the maxim of conclusion. On the Facebook social media, there were no deviations from civility in language because students are rarely active on the Facebook social media even though they have a Facebook account.

3.2 Discussion

3.2.1 Deviations of the maxim of wisdom

Deviation from the maxim of wisdom is a maxim that maximizes the benefits of others and minimizes the losses of others. In communication, there is often a deviation from the maxims of wisdom. Some of the characteristics of wisdom maxims deviation include: (1) Using harsh diction, (2) directing orders, (3) reprimanding with harsh diction, (4) giving direct suggestions (not using the word sorry.), (5) refusing with a high tone, (6) refuse with rough diction.

(165) "Ego vs. Keras Kepala. Mau sampe kapan? Terserah, dah muak aku/EFR." (Bahasa version)

(165) "Selfish vs. stubborn. How long? Whatever, I am fed up/ EFR." (English version)

Data number (165) is a speech that contains deviations from the maxims of wisdom. It refers to harsh diction, such as the word "disgusted" in his speech. Besides, speakers indirectly also provide suggestions without using the word sorry. The speech's context is that the speaker is annoyed with high egoism coupled with stubbornness. Finally, the speech number (165) appeared. Speech number 165 will avoid deviating from the maxims of wisdom if the speakers use subtle diction and provide suggestions accompanied by the word sorry.

(319) "Cewe, anjay ada yak cowo kek gitu brengsek banget. Cowo, anjy juga tuh ada cewek jahat bngt kek gitu, gak habis pikir (Foto kalimat motivasi tentang kesehatan hati)/P." (Bahasa version)

(319) "Girl, fuck, how could there is such a bastard boy? Guys, wtf, how could there is such a mean girl like her? it does not make any sense (Photo of a motivational sentence about liver health) / P." (English version)

Speech number (319) is also an aberration of the maxim of wisdom. In the speech, the speaker repeatedly mentions the harsh words "fuck and bastard." in Bahasa, "Anjay" is a slang language to express resentment. When a speaker utters this harsh word, it leads to the maxims of wisdom's deviant behavior. Because one of the maxims deviation characteristics is speaking harshly or using harsh diction, this is an immoral act given that the speaker is a prospective elementary school teacher. Elementary school teachers must be able to exemplify both attitudes, actions, and words.

3.2.2 Deviations of the maxim of generosity

The maxim of generosity is the maxim that maximizes self-sacrifice and minimizes self-gain. However, sometimes in communication, there is often a deviation from the maxim of generosity. Some of the characteristics of deviation from the maxim of generosity include (1) disrespect for the interlocutor (interrupting the conversation), (2) not allowing the interlocutor to argue, (3) prejudice against the interlocutor, (4) humiliating the other person.

(36) "Saku KKN 6 bulan Cuma 50 ribu itu ga wirth it. Jelas kurang yo. Subsidi kuota aja masih nggak nyampe. Itu mah tega banget/WH." (Bahasa version)

(36) "The six months KKN allowance of only IDR 50.000 is not worth it. Not enough. Quota subsidies have not been received yet. Very heartless/WH." (English version)

From the data number (36), it appears that there is a deviation from the maxim of generosity. The speech has a speaker context in which a student feels disadvantaged by the campus due to the KKN policy. The speaker then has a wrong prejudice against the speech opponent (campus). Speakers thought that the campus had the heart for students because they only gave them an allowance of IDR 50,000 for six months. However, behind this speech, there is a fact that is not yet known to the speakers. After a few days, the campus finally said that there would be an additional KKN allowance. Allowance of IDR 50,000 is a prefix, and allowance will give later. From the campus narrative, it can conclude that the speakers have prejudice before the speakers investigate further.

(175) "Ra ayem. Wes njajal iki kui yo tetep wae. Ramasyuk. Ono wae sek gagal ning h-1 pelepasan. Iki mah diculke bubar jalan bablas. Tur rarti dalan (KKN)/NM." (Javanese version).

(175) "Anxious. Having tried, but it is still the same. Disappointing. Some failed on H-1. That is what is called being released then dismissed but not knowing the way." (English version)

Data number (175) is still on the issue of KKN among students. Indeed, this issue became a hot topic at the time. Many students deviate from the maxim of generosity on the issue of KKN. As in data number (175) from this data, speakers feel disadvantaged by the campus. Bad speakers think by thinking that the campus does not release procedurally well. As a result, respondents felt that KKN was less conditioned and caused confusion among students when implementing it. It is a series of speakers' prejudices against the campus.

3.2.3 Award maxims deviation

Several characteristics, including characterize deviation from the maxims of appreciation (1) Giving criticism that puts others down, (2) talking that hurts others, (3) not saying thanks when receiving suggestions or criticism from others, (4) not respecting others' opinions. Some speech on social media conducted by PGSD students also leads to the characteristic deviation maxim awards.

(111) "Biasane jarang bahkan mungkin ratau ngomongke nyebahine dosen neng story. Tapi iki wis terlalu mumeti bagi mahasiswa. Intine ra gur pisan pindho ngenei, saben dina/NA." (Javanese version).

(111) "I seldom talked, even never talk about Lecturer in the story. However, it is too complicated for students. The point is that it happened not only once or twice, but it happens every day/NA." (English version)

Data number (111) has the characteristic that the speaker gives criticism that puts others down. The context of the speech is that the speaker gives criticism to one of the lecturers. The reason the criticism came out was that the assignment given by the Lecturer was too heavy. This task is not only once but many times and makes the speaker feel dizzy. In this case, the speaker does not appreciate the actions taken by the Lecturer.

(353) "Oke sebenere bahas SS yang kemarin saya kirimkan. Btw sebelum ini saya coba search di amazon dan beberapa toko online dunia yang fokus di sektor kesehatan. Hasilnya mengejutkan harga rapid tes nggak lebih dari ¡¢9 atau kalau dirupiahkan dengan kurs sekarang (14.429,45) HARGANYA 129.865,05. Nggak sampai 150.000 ya, ini untuk pembelian alat, btw saya ambil yang lumayan "mahal" untuk alat sebagai contoh. Oke ada yang komen, kan itu belum termasuk **bea cukai** Jawaban saya simpel, coba kamu cek peraturan (Terkait tarif rapid test)/FS." (Bahasa version)

(353) "Okay, discussing the SS that I sent yesterday. Before, I searched on Amazon and several online stores worldwide that focus on the health sector. The result is surprising that the rapid test price is not more than $9, or if it is converted into the current exchange rate (IDR 14,429.45), the price is IDR 129,865.05. Not up to IDR 150,000, yes, this is for purchasing equipment, by the way, I took a relatively "expensive" tool as an example. Okay, there are comments, right that does not include customs. My answer is simple, try to check the regulations (Regarding rapid test rates) / FS." (English version)

The data number (353) took from one of the students' WA statuses. The context of speech number (353) is a criticism that puts others down. The other person, in that case, is the government or the health service. The health office has the authority to set the highest price for a rapid test. He considered that the rapid tests in Indonesia were too expensive. Even though the price of the tool for rapid alone is only around IDR 15,000. Then why in Indonesia the cheapest rapid rate is IDR 150,000. This criticism has certainly brought down the government. Reading speech number (353) implies that the government is making a big profit from this rapid test price.

3.2.4 Perversion of the maxim of wisdom and generosity

There is often a deviation from the maxims of wisdom and generosity in communication on social media. These two maxims co-occur. The characteristics of the wisdom maxims deviation are: (1) Using harsh diction, (2) directing orders, (3) reprimanding with harsh diction, (4) giving direct suggestions (not using the word sorry), (5) refusing with a high tone, (6) refuse with harsh diction. Then the characteristics of deviation from the maxim of generosity include: (1) Disrespect for the interlocutor (interrupting the conversation), (2) not allowing the interlocutor to have an opinion, (3) prejudice against the interlocutor, (4) humiliating the interlocutor. The following are some of the utterances that indicate deviations from the simplicity maxim, including:

(202) "Asyuuu yak kerjanyaa ngoyo kuliah Kerja kuliah Kerja ninggalin rapat ninggalin tanggung jawab. Kok dituduh ambil uang/ RK." (Bahasa version)

(202) "Damn, I worked hard by going back and forth to study and work, leaving meetings, leaving responsibilities. How come you could be accused of taking money/ RK." (English version)

Deviations from the maxims of generosity and consensus have characteristics, including (1) disrespect for the interlocutor (interrupting the conversation), (2) not allowing the interlocutor to have an opinion, (3) prejudice against the interlocutor, (3) humiliating the interlocutor. The consensus maxims deviation characteristics include: (1) Giving criticism that puts others down, (2) speaking that hurts others, (3) not saying thanks when receiving suggestions or criticism from others, (4) not respecting other people's opinions.

(15) "Kalo kamu mau sambat soal kampus tapi takut DM atung Almira karena bisa aja di post difeed ig nya dan malah kena amukan buzzernya agagaga. Kalo kamu sambat soal kampus tapi kamu kesel mau sambat yang bagian apa, saking akehe sing arep disambati. Nganti bingung arep milih sing endi. Ya wis lur gpp. Cuman satu sambatan dari lubuk hatimu paling dalam soal kampus di link paling atas. Gasss ke aja." (Bahasa version)

(15) "If you want to hang around about the campus but are afraid of DM Atung Almira because you can post it on Instagram and instead

get the buzzer tantrum gaga. If you are hampered about the campus, you will get annoyed about what part of it you want to be obstructed because there are too many to complain about. Thus, it made it confused which one will have complained. It is okay, Guys. The Only splash from the bottom of your heart in the matter of campus is at the top link. go on." (English version)

Data number (15) is a form of speech where deviations from the maxims of generosity and consensus— deviation from the maxims of generosity indicated by speech that knocks down the interlocutor. The context of the speech is that speakers respond to complaints about campus problems. The speaker laughs at the interlocutor, who often gets tantrums from other students when the interlocutor complains about campus problems through the Chancellor's Instagram account. The deviation of the maxims of consensus is marked by the dropping of criticism that does not respect others. The other person in this speech is the campus. In the speech, the speakers dropped the campus's excellent name because they thought there were too many shortcomings provided by the campus. Such as speech:

"Kalo kamu sambat soal kampus tapi kamu kesel mau sambat yang bagian apa, saking akehe sing arep disambati. Nganti bingung arep milih sing endi".

If translated into excellent and correct Bahasa, it will say:
"Confused. What do you want to complain about because there are so many things currently being complained about."

(26) "Kalau berdasarkan kajian UNY bergerak kemarin kan dilampirkan ya terkait rencana pengeluaran dana dari setiap fakultas. Nah disitu banyak rencana anggaran yang tidak bisa dialokasikan selama pandemi. kebanyakan rencana anggaran tersebut bersumber dari UKT. Nah aku tuh ga paham dengan alasan mereka kekeh ngga mau nurunin UKT yang berlaku ke semua mahasiswa. Karena meskipun ada layanan penyesuaian UKT, mulai dari penyicilan, peringanan, penangguhan, namun permohonan tersebut belum tentu diterima semua dan kriteria kelolosan permohonannya pun tidak ada kejelasan/RW. " (Bahasa version)

(26) " Based on the previous UNY study, we attached the plan for spending funds from each faculty. So many budget plans cannot allocate during the pandemic. Most of these budget plans come from UKT. We do not understand why they do not want to reduce the UKT that applies to all students. Because even though there is a UKT adjustment service, starting from installments, mitigation, deferral, but all applications are not necessarily accepted and the criteria for passing the application are not clear / RW."

Data number (26) contains deviations from the maxims of generosity and consensus. The maxim of generosity deviates because, in the speech, the speaker prejudices against other people (campus). The wrong prejudice is in the form of speakers' dissatisfaction with the request to reduce UKT. Speakers considered that the criteria for passing the UKT reduction application were unclear. The deviation from the maxims of consensus is marked by giving criticism that puts others down (campus). Here, speakers write stories on social media, which indirectly bring down their campus. Without further researching the causes of the criteria for applying for UKT relief.

4 CONCLUSION

This study aims to describe the impoliteness of language in the social media of PGSD UNY students. This study's findings reveal that PGSD UNY students still practice impoliteness language on their social media accounts. The forms of language deviation include: (1) deviations from the maxims of wisdom, (2) deviations from the maxims of generosity, (3) deviations from the maxims of appreciation, (4) deviations from the maxims of simplicity, (5) deviations from the maxims of consensus, (6) deviations from the maxims of generosity and consensus, (7) deviations maxim of wisdom and consensus.

The weakness of this research is the limited number of research subjects. Because it only uses an active student population at PGSD UNY. Future research is expected to look at language deviations carried out on social media in terms of gender. Besides, the scope of research also needs to be expanded. Not only are language irregularities made by PGSD UNY students, but PGSD students outside YSU also need to be researched. This research is also beneficial for students because the results obtained can be used by students to learn language politeness.

From this conclusion, the suggestion is that students should learn about language politeness as a candidate for elementary school teachers. From now on, we have made it a habit to apply politeness in language. Because what they write on social media will be read by many people, even their students later. Students must be able to provide role models for their students or the wider community.

REFERENCES

Culpeper, J. 2011. Impoliteness: Using Language to Cause Offence.

Eriyanto 2011. Analisis Isi: Pengantar Metodologi untuk Penelitian Ilmu Komunikasi dan Ilmu-Ilmu Sosial Lainnya.

Fuchs, C. 2017. Social Media: A Critical Introduction.

Hammod, N. M. & Abdul-Rassul, A. 2017. Impoliteness Strategies in English and Arabic Facebook Comments. *International Journal of Linguistics* 9: 97.

Krippendorff 2004. Content Analysis: An Introduction to Its Methodology.

Laughey, D. 2007. Key Themes in Media Theory.

Liu, X. 2017. Impoliteness in Reader Comments on Japanese Online News Sites. *International Journal of Languages, Literature and Linguistics* 3: 62–68.

Locher, M. A. 2010. Introduction: Politeness and impoliteness in computer-mediated communication. *Journal of Politeness Research* 6: 1–5.

McQuail, D. 2010. Teori Komunikasi Massa. 392.

Oz, M., Zheng, P. & Chen, G. M. 2018. Twitter versus Facebook: Comparing incivility, impoliteness, and deliberative attributes. *New Media and Society* 20: 3400–3419.

Papacharissi, Z. 2004. Democracy online: Civility, politeness, and the democratic potential of online political discussion groups. *New Media and Society* 6: 259–283.

Rahardi, K. 2017. Linguistic Impoliteness in The Sociopragmatic Perspective. *Jurnal Humaniora* 29: 309.

Serwaa, N. A. & Dadzie, P. S. 2015. Social media use and its implications on child behaviour: a study of a basic school in Ghana. *International Journal of Social Media and Interactive Learning Environments* 3: 49.

Watie, E. D. S. 2016. Komunikasi dan Media Sosial (Communications and Social Media). *Jurnal The Messenger* 3: 69.

Educational Innovation in Society 5.0 Era: Challenges and
Opportunities – Purnomo & Herwin (Eds)
© 2021 the authors, ISBN 978-1-032-05392-9

Analysis of students' historical empathy in history education

S. Dahalan & A.R. Ahmad
Universiti Kebangsaan Malaysia, Malaysia

ABSTRACT: This study investigated students' historical empathy in the classroom. A total of 146 upper secondary school students were randomly selected to participate in a survey on historical empathy. The recorded Cronbach's alpha coefficient values ranged from 0.848 to 0.863, which reaffirmed the reliability of the measuring instrument in this study. The levels of historical empathy based on historical figures and events were analysed using IBM SPSS (version 24.0). Overall, the students' level of historical empathy was at a lower level. The results further revealed low levels of historical empathy based on historical figure ($M = 2.67$, $SD = 0.50$) and event ($M = 2.51$, $SD = 0.53$). Hence, it is essential for teachers to change their teaching strategies by focusing on historical empathy in history education.

1 INTRODUCTION

21st-century learning is defined as a form or process of learning that involves four types of capabilities that co-relate to the elements of collaboration, communication, and creativity in learning, namely (1) lifelong learning, (2) problem solving, (3) self-management, and (4) teamwork (Ester, Alexander, Jan, & Haan, 2017). With the goal of producing digital literate and responsible citizens in terms of life and career, the 21st-century learning requires students to holistically dominate the learning content through the application of communication skills, collaboration, creativity, critical thinking, and information literacy (Saucerman et al., 2017, Osler & Starkey, 2018, Dahalan et al., 2018, Sonley et al., 2007). As such, this study specifically discussed the emergence of historical empathy, the concept of historical empathy, and students' performance through historical empathy.

1.1 *Background of historical empathy*

There are various definitions of "empathy." For instance, Knight (1989) depicted empathy as an action and skill that can exist separately and together in individuals. Roberts (1972) defined empathy as a thinking process that involves the cognitive domain in order to understand individual actions and obtain information about situations that differ from the present situations. Boddington (1980) supported this definition of empathy and added a cognitive domain and the ability to support.

However, several researcher (Sutherland, 1986, Lee & Ashby, 2001, Wineburg, 2001, Abdullah, 2007, Perrotta & Bohan, 2017, Hoy, 2018, Stout & Stout,

2019) proposed empathy is a feeling to understand and experience life like historical figures and see the situation from their point of view. These studies linked empathy to the same feeling of understanding. The justification of historical figure with an empathy feature was also highlighted in relation to empathy, specifically the ability to know the events and actions of the historical figure, the ability to appreciate historical context, the ability to rationalize the action of the historical figure, and the ability to assume as a historical figure.

1.2 *Research objective*

The objectives of this research are:

a. To investigated students' level of Historical Empathy based on Historical Figures
b. To investigated students' level of Historical Empathy based on Historical Events

2 METHODOLOGY

A population is a group of individuals with similar characteristics and criteria for the purpose of a study (Creswell, 2014). Meanwhile, samples are sub-groups from the target population, with the aim of making generalisation over the population under study. For this study, the target population consisted of secondary school students in Hulu Langat, Selangor.

Wiersma and Jurs (2008) expressed that the steps in simple random sampling are to acquire random samples and produce samples that represent the overall characteristics of the population for more meaningful generalisation. Referring to the formula of sample size determination by Cohen et al. (2017), a sample size of 146 randomly sampled respondents was deemed adequate for the current study.

All survey responses were encoded accordingly. All information obtained were then processed and analysed using IBM SPSS (version 24.0). Analysing data using this software can produce accurate and error-free calculations (Pallant, 2007, Konting, 2000). The data analysis in this study specifically involved descriptive statistics.

Descriptive statistics refers to the analysis in the form of mean and standard deviation to describe the sample and address the highlighted research questions. Descriptive statistics discusses how to analyse the obtained data in order to ensure that all information accurately match the information of a sample; thus, confirming the hypothesised theory.

In particular, this study adapted the mean score performance by Ahmad (2002) for the interpretation of mean scores derived from the survey responses. The interpretation of quantitative data involved categorising the overall mean score into three categories to express historical empathy levels.

Table 1. Interpretations mean score

Mean Score	Level
1.00 – 1.99	Weak
2.00 – 2.99	Low
3.00 – 3.99	Medium
4.00 – 4.99	High

Source: Jamil Ahmad (2002)

For this study, the measuring instrument's reliability was analysed based on the Cronbach's alpha coefficient values, which ranged from 0.848 to 0.863. Meanwhile, the level of historical empathy based on historical figure recorded Cronbach's alpha coefficient value of 0.863. The level of historical empathy based on event recorded Cronbach's alpha coefficient value of 0.848. These recorded values showed that all 15 items used in this study had a high-reliability index and were appropriately used (Cronbach, 1951; Pallant, 2007).

3   DATA FINDINGS

3.1   *Students' level of historical empathy based on historical figures*

Table 2 presents the students' level of historical empathy based on historical figure. The overall level of historical empathy based on historical figure was found low based on the recorded mean score of 2.67 and standard deviation of 0.50. Item with the highest mean score was "*Saya dapat merasai penderitaan Bilal bin Rabah untuk berpegang teguh kepada ajaran Islam*" (M = 2.97, SD = 0.60). Meanwhile, the item with the lowest mean score was "*Saya dapat menjiwai sifat-sifat terpuji Nabi Muhammad SAW dalam konteks sifatnya yang pelbagai*" (M = 2.04, SD = 0.61).

Table 2.   Mean scores of historical empathy based on historical figures

No.	Item	M	SD	Interpretation
1.	*Saya dapat menjiwai sifat-sifat terpuji Nabi Muhammad SAW dalam konteks sifatnya yang pelbagai*	2.04	0.61	Low
2.	*Saya dapat merasai penderitaan Bilal bin Rabah untuk berpegang teguh kepada ajaran Islam*	2.97	0.60	Low
3.	*Saya dapat merasai kehebatan kepimpinan Nabi Muhammad SAW dalam perjanjian Hudaibiyah*	2.81	0.56	Low
4.	*Saya dapat menjiwai kebijaksanaan Uthman bin Affan sebagai perunding dalam perjanjian Hudaibiyah*	2.94	0.70	Low
5.	*Saya dapat menyelami sikap toleransi dan kepimpinan Khalifah Ali bin Abu Talib dalam konflik sesama Islam*	2.94	0.64	Low
6.	*Saya dapat merasai kehebatan Khalifah Umar ibn Abdul Aziz (Bani Umaiyah) dalam usaha beliau menjaga kebajikan rakyat*	2.93	0.73	Low
7.	*Saya dapat menjiwai semangat yang dimiliki Khalifah Harun al-Rasyid (Bani Abbasiyah) dalam usaha memajukan pendidikan negaranya*	2.09	0.71	Low
	Total	2.67	0.50	Low

Notes: M denotes mean score; SD denotes standard deviation.

3.2   *Students' level of historical empathy based on events*

With respect to the research objective, the students' level of historical empathy based on event involved four chapters from the content of the history subject. As shown in Table III, the overall level of historical empathy based on event was found low (M = 2.51, SD = 0.53). In particular, item with the highest mean score was "*Saya dapat merasai hikmah Nabi Muhammad SAW dalam menyelesaikan masalah meletakkan Hajarul Aswad antara puak-puak Arab Quraisy di Kota Makkah*" (M = 2.95, SD = 0.68). Meanwhile, item with the lowest mean score was "*Saya dapat menjiwai betapa pentingnya peristiwa hijrah dalam perubahan diri*" (M = 2.02, SD = 0.60).

Table 3. Mean scores of historical empathy based on events

No.	Item	M	SD	Interpretation
1.	*Saya dapat merasai bidang pertanian mampu mengangkat martabat ekonomi*	2.30	0.94	Low
2.	*Saya dapat menghayati kepentingan kemajuan kerajaan maritim dalam pembangunan ekonomi di peringkat antarabangsa*	2.31	1.07	Low
3.	*Saya dapat mendalami dan memahami pelbagai warisan agama, adat dan budaya terhadap masyarakat Malaysia hari ini*	2.40	1.08	Low
4.	*Saya dapat menyelami akhlak dan tingkah laku positif masyarakat Arab Jahiliah yang perlu dicontohi.*	2.36	1.10	Low
5.	*Saya dapat merasai hikmah Nabi Muhammad SAW dalam menyelesaikan masalah meletakkan Hajarul Aswad antara puak-puak Arab Quraisy di Kota Makkah*	2.95	0.68	Low
6.	*Saya dapat merasai keikhlasan dan kejujuran orang Ansar memilih Nabi Muhammad SAW sebagai pemimpin mereka*	2.87	0.62	Low
7.	*Saya dapat menjiwai betapa pentingnya peristiwa hijrah dalam perubahan diri*	2.02	0.60	Low
8.	*Saya dapat menghayati Piagam Madinah sebagai prinsip dalam politik Islam*	2.91	0.59	Low
	Total	2.51	0.53	Low

Notes: M denotes mean score; SD denotes standard deviation.

4 CONCLUSION

This study was conducted to determine historical empathy among students. Using IBM SPSS (version 24.0), the form of analysis considered in this study was descriptive statistics, specifically mean and standard deviation. There is a wide range of historical empathy. However, students in this study demonstrated low levels of historical empathy based on historical figures and events. In particular, the overall mean score for the historical empathy based on historical figure was higher than the recorded mean score for the historical empathy based on event. Hence, it can be concluded that the understanding and imagination of historical content play a major role in students' high-level historical empathy.

A suitable pedagogy is indispensable in dealing with low mastery learning among students, especially when analytical and evaluation thinking are involved (Cowgill & Waring, 2017, Lammert, 2020). The existing pedagogies used by teachers can only enhance the mastery that involves the level of historical empathy (Dillenburg, 2017, Huijgen et al., 2018). Therefore, teachers need to emphasize imagination (Leur et al., 2017) of the teaching content as well as analytical and evaluation thinking skills (Rantala et al., 2016) to ensure the thorough development of skills among students. Furthermore, pedagogical training for historical teachers should be carried out from time to time to ensure that they are well-informed about history education's current needs. In teaching history, the ability to master historical content imagination very well would contribute to the enhancement of students' empathy skills in history education.

REFERENCES

Abdullah, S. H. 2007. Empati sejarah dalam pengajaran dan pembelajaran Sejarah. *Jurnal Pendidik dan Pendidikan* 22: 61–74.

Ahmad, J. 2002. Pemupukan Budaya Penyelidikan Dikalangan Guru di Sekolah: Satu Penilaian. Thesis Dr. Fal. Fakulti Pendidikan, Universiti Kebangsaan Malaysia.

Boddington, T. 1980. British Journal of Educational Empathy and the teaching of history. *British Journal of Educational Studies* 28: 13–19.

Cohen, L., Lawrence, M. & Morrison, K. 2017. Research Methods in Education. 406–413.

Cowgill, D. A. & Waring, S. M. 2017. Historical Thinking: An Evaluation of Student and Teacher Ability to Analyze Sources. *Journal of Social Studies Education Research* 8: 115–145.

Creswell, J. W. 2014. Research Design: Qualitative, quantitative, and mixed methods approaches.

Dahalan, S. C., Ahmad, A. R. & Awang, M. M. 2018. Pembelajaran Abad ke-21, Mengapa dan Bagaimana Kemahiran Berfikir Aras Tinggi (KBAT) dalam Pendidikan Sejarah. *Transformasi dan pembangunan pendidikan di Malaysia.* Kementerian Pendidikan Malaysia.

Dillenburg, M. 2017. Understanding historical empathy in the classroom. Tesis Dr. Fal, School of Education, University of Boston.

Hoy, B. 2018. Teaching History With Custom-Built Board Games. *Simulation and Gaming* 49: 1–19.

Huijgen, T., van de Grift, W., van Boxtel, C. & Holthuis, P. 2018. Promoting historical contextualization: the development and testing of a pedagogy. *Journal of Curriculum Studies* 50: 410–434.

Knight, P. 1989. Empathy : concept , confusion and consequences in a national curriculum. *Oxford Review of Education*: 43.

Konting, M. M. 2000. Kaedah penyelidikan pendidikan. Cet. Ke5.

Lammert, C. 2020. Becoming inquirers: A review of research on inquiry methods in literacy preservice teacher preparation. *Literacy Research and Instruction* 8071.

Lee, P. & Ashby, R. 2001. Empathy, perspective taking, and rational understanding. *In:* Davis, O. L., Yeager, E. A. & Foster, S. J. (eds.) *Historical empathy and perspective taking in the social studies.* USA: Rowman and Littlefield.

Leur, T. D., Boxtel, C. V. & Wilschut, A. 2017. 'I Saw Angry People and Broken Statues': Historical Empathy in Secondary History Education. *British Journal of Educational Studies* 65: 331–352.

Osler, A. & Starkey, H. 2018. Extending the theory and practice of education for cosmopolitan citizenship. *Educational Review* 70: 31–40.

Pallant, J. 2007. SPSS Survival Manual: A step by step guide to data analysis using SPSS for Windows (Version 10). 350.

Perrotta, K. A. & Bohan, C. H. 2017. More than a feeling : Tracing the progressive era origins of historical empathy in the social studies curriculum , 1890 – 1940s. *The Journal of Social Studies Research*: 1–11.

Rantala, J., Manninen, M. & van den Berg, M. 2016. Stepping into other people's shoes proves to be a difficult task for high school students: assessing historical empathy through simulation exercise. *Journal of Curriculum Studies* 48: 323–345.

Roberts, M. 1972. Educational Objectives for the Study of History. *Teaching History* 2: 347–350.

Saucerman, J., Ruis, A. & Shaffer, D. 2017. Automating the Detection of Reflection-on-Action. *Journal of Learning Analytics* 4: 212–239.

Sonley, V., Turner, D., Myer, S. & Cotton, Y. 2007. Information literacy assessment by portfolio: A case study. *Reference Services Review* 35: 41–70.

Stout, R. & Stout, R. 2019. Empathy , Vulnerability and Anxiety Empathy , Vulnerability and Anxiety. *International Journal of Philosophical Studies* 2559.

Sutherland, M. B. 1986. Education and Empathy. *British Journal of Educational Studies* 34: 142–151.

Wiersma, W. & Jurs, S. G. 2008. Research mothods in education. An introduction. 9th Ed.

Wineburg, S. 2001. Historical Thinking and Other Unnatural Acts. *Phi Delta Kappan* 7: 81–94.

Educational Innovation in Society 5.0 Era: Challenges and Opportunities – Purnomo & Herwin (Eds)
© 2021 the authors, ISBN 978-1-032-05392-9

Analysis of the implementation of primary school teachers' professional duties during the Covid-19 pandemic in Sleman Yogyakarta

A. Hastomo, B. Saptono, S.D. Kusrahmadi, F.M. Firdaus & A.R. Ardiansyah
Universitas Negeri Yogyakarta, Yogyakarta, Indonesia

ABSTRACT: This research aims to describe the implementation of elementary school teachers' professional duties during the Covid-19 pandemic. This research type is quantitative with data collection methods using a scale of task implementation with data sources for teachers and school principals, peers. The research population was teachers in elementary schools in Ngaglik sub-district, Sleman Yogyakarta, with 43 schools. The data analysis technique in this study was descriptive statistical data analysis. The data obtained is quantitative in the form of a percentage of elementary school teachers' professional duties in Ngaglik sub-district, Sleman Yogyakarta. The results showed that the achievement of the implementation of the professional duties of teachers during the Covid 19 pandemic, which was carried out by using an online system of achievements with the data source of the teacher concerned in educational activities was 89.18%, teaching: 78.72%, guiding: 72.90%, directing: 78.72%, training: 72.67%, assessing: 77.79%, evaluating: 82.32%. The lowest aspect of task implementation, which was performed by the teacher, is training. Implementation of teacher professional duties with the principal's data source achieved by educating: 93.4375%, teaching: 84.21875%, guiding: 82.34%, directing: 85.47%, training: 81.40%, assessing: 85.46%, evaluating: 88.43% and the lowest aspect that the teacher does is training.

1 INTRODUCTION

The Covid-19 pandemic is a global health crisis in the form of outbreaks that attack many victims simultaneously in various countries. On March 11, 2020, the World Health Organization (WHO) has declared Covid-19 a global pandemic that potentially infects all world citizens. As reported from Kompas.com, the number of positive Covid-19 in Indonesia of May 22, 2020, reached 20,796 people. The pandemic has many impacts on various sectors of people's lives, including education. As of April 1, 2020, alone, UNESCO has recorded at least 1.5 billion school-age children affected by Covid-19 in 188 countries (Azoulay 2020). To reduce the spread of the virus, the government implemented a physical distancing policy, which resulted in all forms of formal and non-formal education having to be at home. It gives rise to the policy of the entire learning process carried out remotely. In Indonesia, distance learning during the pandemic carries out from the pre-school, elementary, junior high, high school to college levels.

Based on SE No. 15 of 2020, the Ministry of Education issued a circular on guidelines for learning from home in the emergency period of the spread of Covid-19. These guidelines contain guidance on online and offline distance learning. The guidelines aim to protect children's right to education services, prevent the spread and transmission of Covid-19 in the Education

unit, and ensure the fulfillment of psychosocial support for educators, students, and parents/guardians. Although the government has issued guidelines for learning from home, distance learning did not separate from some obstacles. Many schools are poorly prepared, especially in terms of minimal facilities and infrastructure, unstable internet access, the unavailability of adequate learning tools from the parents' side, such as laptops and gadgets (Aliyyah et al. 2020, Alea et al. 2020). Students feel compelled to study remotely without adequate facilities and infrastructure at home; there is no distance learning culture among students, students are used to being in school to interact with their friends and teachers; the school has been closed for too long, making students saturated, crying easily and losing their spirit in learning (Purwanto et al. 2020).

The constraints and urgency of learning during the pandemic require the teacher's responsibility to implement teaching and learning activities (van der Spoel et al. 2020). Teacher competence becomes the capital that determines the success of online learning implementation (Sudrajat 2020). One of the professional teachers is to have the readiness to act in a particular task, job, or situation (Carlsson 2016). Pandemic situations result in changes in various facets, such as learning more utilizing technology, teachers and students cannot face to face, the motivation of students learning decreases, and so on. The situation

270

DOI 10.1201/9781003206019-51

requires teachers to act professionally so that learners still get proper learning rights during the pandemic. As professionals, teachers need to improve services, improve knowledge, and give direction and encouragement to their students (Kosasi 2009). Under the Law. No. 14 of 2005, teachers are professional educators with the main task of educating, teaching, guiding, training, assessing, and evaluating learners. We can see the teacher's professionalism through his responsibility in carrying out all his devotion (Kunandar 2010). In distance learning during the pandemic, teachers must carry out their primary tasks by developing their competencies, adjusting education to students' needs, and learning materials (Wulandari 2018).

The implementation of teacher professional duties during the pandemic still has many shortcomings. Teachers only assess students' products at online learning time without prioritizing the process (Anugrahana 2020). Not all teachers can use direct applications and smart with laptop devices or headphones, lack of knowledge of teacher skills in designing effective distance learning, unstable internet access, and the implementation of distance learning (Simanjuntak & Kismartini 2020; Alea et al. 2020). Besides, teachers cannot face learning challenges during pandemics and struggle with maintaining caring relations with students (González et al. 2020, Jones & Kessler 2020).

The results of a preliminary study from May 18–30, 2020, found that the implementation of professional duties of grade V elementary school teachers in Ngaglik sub-district of Sleman Regency, Special Region of Yogyakarta, experienced obstacles due to pandemic Covid19, which demanded that learning should be done remotely. Teachers are shocked and have not been in a climate of distance learning. Many teachers only distribute books to students to be studied independently at home, and there are not many teachers who carry out professional tasks of teachers with distance learning modes. Therefore, the research analyzes elementary school teachers' professional duties during the Covid19 pandemics to obtains an overview of the percentage of teachers' professional duties.

According to the interview results with one of the elementary school teachers in Ngaglik sub-district, Sleman, Yogyakarta, the teachers usually focus more on learning activities in the classroom, but due to the Covid-19 pandemic, so that learning cannot be optimal. Based on observations and documentation studies of learning devices in elementary schools in the Ngaglik subdistrict in 2020, there are still no learning devices designed by distance learning modes. Based on some of these sources, most of the implementation of teachers' main task is more dominant in learning implementation. Teachers are perceived to lack in planning learning, assessing learning outcomes, guiding, and training learners. Based on some of these conditions, researchers are interested in further deepening basic teacher tasks in Ngaglik sub-district, Sleman, Yogyakarta during the Covid19 pandemics.

Based on the background above, in this study, research questions are formulated as follows: (1) How is the analysis of the implementation of elementary school teachers' professional duties in Ngaglik subdistrict, Sleman, Yogyakarta during the Covid19 pandemic? (2) What is the percentage of elementary school teachers' professional duties in Ngaglik sub-district, Sleman, Yogyakarta, during the Covid19 pandemics? (3) Based on the indicators of elementary school teachers' professional duties, which is the highest level of implementation in Ngaglik sub-district, Sleman, Yogyakarta during the Covid19 pandemics?

This study has the following objectives: (1) describing the results of the analysis of the implementation of professional tasks of elementary school teachers in Ngaglik sub-district, Sleman, Yogyakarta during the Covid-19 pandemic, (2) presenting the percentage of professional duties of elementary school teachers in Ngaglik subdistrict, Sleman, Yogyakarta, during the Covid19 pandemic, (3) knowing which indicators of professional duties of elementary school teachers are the highest level of implementation in Ngaglik subdistrict, Sleman, Yogyakarta.

Practical benefits expected to describe elementary school teachers' professional duties in Ngaglik sub-district, Sleman, Yogyakarta during the Covid-19 pandemics refer to other elementary school teachers' implementation policyholders will be useful information in decision making. Theoretical benefits: This research can be used as research materials in subsequent studies to be more scientifically perfect

2 RESEARCH METHODS

This study uses a quantitative descriptive research method to explain the phenomenon of implementing elementary school teachers' professional duties in the Covid-19 pandemic era by using numbers that describe the subject's characteristics under study. Descriptive research is research conducted to determine the value of independent variables, either one or more (independent) variables, without making comparisons or linking them with other variables (Sugiyono 2011).

The research subjects were all fifth-grade teachers in Ngaglik sub-district, Sleman Yogyakarta, totaling 43 Schools. This study's data collection technique uses a questionnaire instrument in the form of a scale for implementing professional teacher assignments through Google Form for all 43 teachers and principals in Ngaglik Sleman Yogyakarta district with a total number of 43 schools.

The data analysis technique in this study was descriptive statistical data analysis. In this technique, statistics are used to analyze data by describing the data that has been collected as it is without intending to make general conclusions or generalizations (Sugiyono 2011). This data analysis technique is carried out by data selection, data tabulation, and hypothesis testing.

3 RESULTS AND DISCUSSION

3.1 *Result*

The study results obtained from teacher respondents in the form of self-assessment of the implementation of professional teacher duties are presented in Table 1.

Table 1. Achievements of teacher duties.

No	Aspect	Achievements (%)
1	Educating	89,18
2	Teaching	78,72
3	Guiding	72,90
4	Directing	78,72
5	Training	72,67
6	Assessing	77,79
7	Evaluating	82,32
	Average	85,83/78,90

Based on Table 1 above, on the implementation of teacher professional duties with the data source being the teacher concerned, it can be seen that the aspect of the task that is highest achieved is to educate. Educating in the instrument to provide motivation, support children's achievement, maintain normative behavior, and help each other. The lowest aspect of implementing professional tasks is training activities to encourage students' physical activity, develop talents, and other sports activities.

The results of the research were obtained from the principal respondents as teacher supervisors in the form of supervisors assessing the implementation of professional teacher duties as follows

Table 2. Achievements of teacher duties.

No	Aspect	Achievements (%)
1	Educating	93,44
2	Teaching	84,22
3	Guiding	82,34
4	Directing	85,47
5	Training	81,40
6	Assessing	85,46
7	Evaluating	88,43
	Average	78,9/85,83

Based on Table 2, on the implementation of teacher professional duties with the data source being the teacher concerned, it can be seen that the aspect of the task that is highest achieved is to educate. Educating in the instrument to provide motivation, support children's achievement, maintain normative behavior, and help each other. The lowest aspect of implementing professional tasks is training activities to encourage students' physical activity, develop talents, and other sports activities.

When it is viewed from the two data sources, there was the same pattern. The implementation of teachers' high professional duties is an aspect of educating, and the lowest aspect is training.

3.2 *Discussion*

The research results prove that teacher performance can be better if it has abilities derived from each teacher's talents, interests, and personal attitudes. Keith Davis (Mangkunegara & Prabu 2000) suggested that the factor influencing performance is the motivation factor from the person's attitude in dealing with work situations. Motivation is a condition that moves employees to achieve the objectives of the organization. A mental attitude is a mental condition that encourages employees to strive to achieve maximum work achievement. Employees will be able to achieve maximum performance if he has high motivation.

The teacher must have the mental attitudes facing various situations, including during the Covid-19 pandemics. Teachers must be able to adapt to the circumstances and conditions in carrying out their professional duties. As (Carlsson 2016) stated, professional competence is a potential readiness of an individual to act in a task, situation, or job. Teachers must have professional competence in acting on their duties in every situation, such as the Covid-19 pandemic situation.

Teachers must provide excellent service to students from various situations and conditions. (Kosasi 2009) As professionals, teachers need to improve their services, improve their knowledge, and give direction and encouragement to their students. Therefore, teachers must improve education quality. The Covid-19 pandemic requires teachers to be more creative in arranging to learn and carrying out various professional tasks even though they cannot meet directly with students. Based on the self-assessment result in this study that explained the implementation of teaching tasks teachers, 78.72% proved that teachers are not 100% ready to face the Covid-19 pandemics. Therefore, there needs to be thorough preparation in carrying out teaching tasks during the Covid-19 pandemics.

The assessment results of the implementation of teaching tasks that only reached 78.72% will be improved. Simultaneously, teachers can teach using digital technology in carrying out distance learning in this Covid-19 pandemic era. (Goh & Wong 2014) reinforced this, which showed that teachers are obliged to improve skills through several professional development programs using technology in teaching and learning. Besides, in carrying out learning, teachers are working to move content and teaching materials into online spaces and be proficient in navigating the software required in teaching (Allen et al. 2020). Teachers must have competency mastery of literacy and science and technology, classroom management skills, communication, and social competencies. The development of these competencies can minimize online learning problems during this pandemic so that the learning process is better (Sudrajat 2020).

The work of education professionals in the current Covid-19 pandemics through Information and Communication Technology (ICT) with a neuroeducation contribution approach in managing emotions and motivational processes contributes to meaningful learning in students. ICT symbiosis and neuroeducation can contribute significantly to the current paradigm shift (Espino-Díaz et al. 2020). In addition to teaching, teachers also need to carry out other professional tasks such as educating, directing, training, guiding, assessing, and evaluating by Regulation of the Minister of National Education No. 16 of 2007. The research result showing the stability of elementary teachers' professional tasks in the Covid-19 pandemic era based on self-assessment that only reached 78.90% need to be improved again so that teachers can carry out professional tasks in various situations. In the work's professionalism, each individual has autonomy in increasing discretion based on collectively determined and monitored expertise (Stone-Johnson & Weiner 2020; Anugrahana 2020).

In addition to teachers' autonomy in carrying out professional work tasks, the implementation of these professional tasks must also be carried out collaboratively with the principal. As in this study, principals were allowed to assess teachers' professional duties who earned an average of 85.83%. Therefore, teachers and principals must collaborate intensively to improve the performance of professional tasks. (Purwanto et al., 2020) showed that the distance learning implementation would be useful if teachers and schools run it responsibly. Both sides need to understand the conditions while delivering the best performance, even if they work in different places. Besides, the headmaster's role is a vital relation to the competencies that must be possessed by a leader in the school environment during the Covid-19 pandemic and in ordinary situations. In these routine activities, the principal's role and competence must provide a positive space for teachers to improve learning competencies (Elfrianto et al. 2020).

The research conducted by (Wahyono et al. 2020) also showed that teachers' competencies and skills must continue to be enriched, supported by school policies that encourage teachers to continue learning. Related parties also need to evaluate online learning so that learning objectives can achieve optimally. The learning burden of learners must take into account, measured, both materially and timely. Teachers should not merely give assignments but must take them into account carefully. Teachers should not forget to appreciate the achievements of the students. A flexible and pandemic-ready curriculum is also needed.

Implementing the teacher's professional duties in teaching remotely in elementary school certainly does not require support from the principal alone, the parents, the committees, and the community.

The research result (Duraku & Hoxha 2020) shows online learning progress, the need for support from teachers, parents, and families, combined with practical advice for parties involved in education.

In the implementation of professional tasks, teachers need more attention in implementing pertussis learning, such as learning design; more attention needs to the pedagogical problems of teaching and learning (Flores & Swennen 2020).

4 CONCLUSION

Implementing professional duties for elementary school teachers in Ngaglik District, Sleman Regency used aspects of educating, teaching, guiding, directing, training, assessing, and evaluating. In principle, all aspects were carried out; it is just that there were several obstacles along with the implementation of learning with an online system. Teachers and students did not meet face-to-face in a free time so that the teacher's professional duties have obstacles in their implementation. The percentage of the implementation of elementary school teachers' professional duties in Ngaglik Sleman District is 82.37%. The primary school teachers' professional duties with the highest level of implementation are the educational aspect.

REFERENCES

Alea, L. A., Fabrea, M. F., Roldan, R. D. A. & Farooqi, A. Z. 2020. Teachers' Covid-19 awareness, distance learning education experiences and perceptions towards institutional readiness and challenges. *International Journal of Learning, Teaching and Educational Research* 19(6): 127–144.

Aliyyah, R. R., Rachmadtullah, R., Samsudin, A., Syaodih, E., Nurtanto, M. & Tambunan, A. R. S. 2020. The perceptions of primary school teachers of online learning during the covid-19 pandemic period: a case study in indonesia. *Journal of Ethnic and Cultural Studies* 7(2): 90–109.

Allen, J., Rowan, L. & Singh, P. 2020. Teaching and teacher education in the time of COVID-19. Taylor & Francis.

Anugrahana, A. 2020. Hambatan, solusi dan harapan: pembelajaran daring selama masa pandemi covid-19 oleh guru sekolah dasar. *Scholaria: Jurnal Pendidikan Dan Kebudayaan* 10(3): 282–289.

Azoulay, A. 2020. UNESCO. Dirjen UNESCO Director-General of the United Nations Educational, Scientific and Cultural Organization. (UNESCO) seen on.

Carlsson, M. 2016. Conceptualizations of professional competencies in school health promotion. *Health Education.*

Duraku, Z. & Hoxha, L. 2020. The impact of COVID-19 on education and on the well-being of teachers, parents, and students: Challenges related to remote (online) learning and opportunities for advancing the quality of education. *Retrieved online from* https://www. researchgate. net/publication/341297812.

Elfrianto, E., Dahnial, I. & Tanjung, B. N. 2020. The competency analysis of principal against teachers in conducting distance learning in covid-19 pandemic. *Jurnal tarbiyah* 27(1).

Espino-Díaz, L., Fernandez-Caminero, G., Hernandez-Lloret, C.-M., Gonzalez-Gonzalez, H. & Alvarez-Castillo, J.-L. 2020. Analyzing the impact of COVID-19 on education professionals. toward a paradigm shift: ICT and neuroeducation as a binomial of action. *Sustainability* 12(14): 5646.

Flores, M. A. & Swennen, A. 2020. The COVID-19 pandemic and its effects on teacher education. Taylor & Francis.

Goh, P. S. C. & Wong, K. T. 2014. Beginning teachers' conceptions of competency: Implications to educational policy and teacher education in Malaysia. *Educational Research for Policy and Practice* 13(1): 65–79.

González, Á., Fernández, M. B., Pino-Yancovic, M. & Madrid, R. 2020. Teaching in the pandemic: reconceptualizing Chilean educators' professionalism now and for the future. *Journal of Professional Capital and Community*.

Jones, A. & Kessler, M. 2020. Teachers' Emotion and Identity Work During a Pandemic. *Front. Educ* 5: 583775.

Kosasi, S. 2009. *Profesi keguruan*. Jakarta: Rineka Cipta.

Kunandar. 2010. *Guru profesional implementasi kurikulum tingkat satuan pendidikan (KTSP) dan sukses dalam sertifikasi guru*. Jakarta: Rajagrafindo Persada.

Mangkunegara, A. P. & Prabu, A. 2000. Manajemen sumber daya perusahaan. Remaja Rosdakarya.

Purwanto, A., Pramono, R., Asbari, M., Hyun, C. C., Wijayanti, L. M. & Putri, R. S. 2020. Studi eksploratif dampak pandemi covid-19 terhadap proses pembelajaran online di sekolah dasar. *EduPsyCouns: Journal of Education, Psychology and Counseling* 2(1): 1–12.

Simanjuntak, S. Y. & Kismartini, K. 2020. Respon pendidikan dasar terhadap kebijakan pembelajaran jarak jauh selama pandemi Covid-19 di Jawa Tengah. *Jurnal Ilmiah Wahana Pendidikan* 6(3): 308–316.

Stone-Johnson, C. & Weiner, J. M. 2020. Principal professionalism in the time of COVID-19. *Journal of Professional Capital and Community*.

Sudrajat, J. 2020. Kompetensi guru di masa pandemi covid-19. *Jurnal Riset Ekonomi Dan Bisnis* 13(2): 100–110.

Sugiyono. 2011. *Metode Penelitian Kuantitatif, Kualitatif dan R&D*. Bandung: Alfabeta.

Van Der Spoel, I., Noroozi, O., Schuurink, E. & Van Ginkel, S. 2020. Teachers' online teaching expectations and experiences during the Covid19-pandemic in the Netherlands. *European journal of teacher education* 43(4): 623–638.

Wahyono, P., Husamah, H. & Budi, A. S. 2020. Guru profesional di masa pandemi COVID-19: Review implementasi, tantangan, dan solusi pembelajaran daring. *Jurnal pendidikan profesi guru* 1(1): 51–65.

Wulandari, S. S. 2018. Peningkatan kompetensi profesional guru kewirausahaan melalui lesson study berbasis pantai dan laut. *Jurnal Pendidikan Edutama* 5(2): 69–78.

Educational Innovation in Society 5.0 Era: Challenges and
Opportunities – Purnomo & Herwin (Eds)
© 2021 the authors, ISBN 978-1-032-05392-9

Fun and interesting learning to improve students' creativity and self-confidence in the 5.0 social era

M. Susanti
Dehasen University of Bengkulu, Bengkulu, Indonesia

Y.P. Sari
Universitas Nahdlatul Ulama, Sidoarjo, Indonesia

K. Karim
Sekolah Tinggi Ilmu Ekonomi SAKTI ALAM Kerinci, Jambi, Indonesia

Sabri
Sekolah Tinggi Ilmu Ekonomi Haji Agus Salim Bukittinggi, Sumatera Barat, Indonesia

ABSTRACT: The national education goals listed in the Law of the Republic of Indonesia No. 20 of 2003 article 3 is to develop the potential of students to become human beings who believe and fear God Almighty, have a noble character, are healthy, knowledgeable, capable, creative, independent, and become democratic and responsible citizens (Kemdikbud, 2013). This study aims to develop teaching materials to improve students' abilities and self-confidence. The method used is a collaborative Quasi Experiment between students and teachers. Quasi-experimental research was carried out to determine the effectiveness of the learning model that had been designed, in the discussion of teaching materials on students' abilities and self-confidence. The subjects of this study were teachers and students from SMKN 01, SMKN 02, and SMKN 03 in Bengkulu City. Two classes from each school were chosen as the experimental class and the control class. The results of research with the project based learning model can improve students' abilities and confidence. Students are more creative, independent, confident and fun because students can complete learning projects in groups.

1 INTRODUCTION

The national education goal listed in the Republic of Indonesia Law No. 20 of 2003 article 3 is to develop the potential of students to become human beings who believe and fear God Almighty, have a noble character, are healthy, knowledgeable, capable, creative, independent, and become democratic and responsible citizens (Kemdikbud, 2013). These national education goals are achieved through the eight educational standards in the curriculum to improve Indonesia's quality of education. Efforts to improve education quality also develop along with changes and developments in science and technology information, especially in the twentieth century.

Efforts to improve student mastery of accounting lessons at the school level really need to be improved. This can provide sufficient provisions in life and the world of work so as to solve problems related to accounting. One of these efforts is the implementation of the 2013 curriculum which emphasizes students to think with the aim of developing Indonesian individuals who are productive, creative, innovative, and effective through strengthening attitudes.

Accounting lessons are very important in equipping students in real life. This is in accordance with the Ministry of National Education which states that the function of accounting subjects is to develop knowledge, skills, rational, thorough, honest, and responsible attitudes through recording procedures, grouping, summarizing financial transactions, preparing financial reports and interpreting companies based on Financial Accounting Standards (SAK). The importance of accounting lessons requires all parties to make improvements and improvements, especially those that are directly related to learning activities.

In addition, accounting subjects listed in economic subjects are used as one of the benchmarks for graduation at SMKN. Furthermore, in the selection of tertiary economic institutions, one of the subjects is a prerequisite for determining graduation in the SOSHUM choice. This shows that the important accounting lessons are controlled by students. However, the reality shows that student learning outcomes, one of which is in accounting subjects, need to be improved. Data from the Ministry of Education and Culture (2019) shows that the average value of the Computer-Based National Examination (UNBK) for the 2018/2019 SMKN level

DOI 10.1201/9781003206019-52

is 1.5% on a scale of 0-100. This shows that the subject tested, namely accounting, is classified as very low.

In accounting learning, there are many aspects that affect the achievement of learning outcomes. One of them is the aspect of student self-confidence in learning. According to Eggen & Kauchak (2010), self-confidence is a statement that describes a belief, a cognitive idea is accepted if it is true without the need to consider other things that support it.

In the accounting learning process in Vocational High Schools in Bengkulu city, student confidence still needs to be improved. The results of preliminary observations on accounting learning at one of the SMKN Kota Bengkulu show that it needs to be improved, especially in the attitude of self-confidence in students learning accounting. The results of distributing questionnaires to measure students' confidence in learning accounting are as in table 1 below;

Table 1. Results of the questionnaire for confidence of XI IPS students at SMKN Kota Bengkulu

Range	Number of Students	Percentage
Very High	1	3.45%
High	12	41.4%
Moderate	11	37.9%
Low	5	17.2%
Very Low		0%

This data indicates that only a portion of the students have self-confidence in the high category and one student in the very high category. This shows that it is necessary to make efforts to improve students' confidence in learning.

One of the efforts that can be made to improve the quality of learning is by designing learning so that it can facilitate students in developing abilities and confidence in learning. These efforts can be using learning media to teach materials that are specifically designed to develop students' abilities in understanding the concept of the material. However, schools' reality shows that it is still rare to find learning tools that teachers can use directly for learning, especially in developing students' abilities and confidence.

In designing learning, the selection of a learning model or approach is the primary key to implementing learning. One learning model that can be used is that learning can be done by involving students directly to develop their abilities, one of which is the project based learning (PjBL) model. The PjBL model facilitates students to make products in order to solve real-life problems. Product manufacturing projects can be carried out individually or in groups.

The PjBL model requires students to be able to produce products that can be used to solve real-life problems. In project implementation, students are required to be able to understand the concept well and produce products related to the concept. This is suitable with accounting learning where students

can directly practice in the field in finding material concepts. The results of an empirical study that states the effectiveness of the Project Based Learning model of learning outcomes are the results of research by Filcik, Bosch, Pederson, & Haugen (2012) which show that the learning model is effective in terms of the conceptual knowledge aspect. This study aims to use the PjBL model to provide interesting lessons that lead to student confidence.

2 LITERATURE REVIEW

2.1 Teaching materials

Teaching materials are an important component in the learning process in the classroom. According to Mudlofar (2012), teaching materials are all forms of materials that are used to assist teachers/instructors in carrying out teaching and learning activities in the classroom. The material in question can be written or unwritten. Meanwhile, according to Prastowo (2012), teaching materials are divided into four types, namely printed materials, listening teaching materials, listening point of view teaching materials, and interactive teaching materials.

Based on nature, teaching materials can be divided into four types: (1) print-based teaching materials, such as books, pamphlets, student study guides, tutorial materials, student workbooks, maps, charts photos of material from magazines and newspapers, and so on. (2) Technology-based teaching materials, for example, audio cassettes, radio broadcasts, slides, film strips, films, video cassettes, television broadcasts, interactive videos, computer-based tutorials, and multi-media. (3) Teaching materials used for practice or projects, such as science kits, observation sheets, interview sheets, etc. (4) Teaching materials needed for human interactive purposes (especially for distance education purposes), such as telephones, mobile phones, video conferencing, etc. In this study, the teaching materials' focus was print-based teaching materials, namely student workbooks.

Student workbooks refer to student worksheets based on project-based learning. Guidelines for preparing student workbooks refer to the preparation of student worksheets (LKS). According to McArdle (2010), LKS is a way of organizing learning activities which are an important part of the module and learning design (RPP). Arends & Kilcher (2010) provides guidelines for making worksheets as follows:

a) Give worksheets that are interesting and fun. Limit the use of standard worksheets.
b) Give a form of worksheet that can make students show success.
c) Adjust the length of time working on the worksheets with the age of the students.
d) Make continuous worksheets as guided practice, not an extension or continuation of learning.
e) The procedure must be clear, namely about what students do if they experience obstacles in doing

it and the next procedure for students who have finished working first or late.

f) Monitor student progress with worksheets, provide needed assistance and provide immediate feedback.

2.2 *Project-based learning model*

According to Patton (2012), project-based learning refers to student activities in designing, planning, and implementing projects that produce output in the form of products, publications, or presentations. The project carried out is adjusted to the characteristics of students and the concepts to be mastered. This shows that the project based learning model is a learning model that facilitates students to construct their own understanding of a concept as well as planning projects to produce products that can solve problems in real life.

Guo & Yang (2012) stated that project-based learning could be used as an effective approach to link teachers' professional development and student learning achievement. Based on a theoretical study of the steps for its application, it can be concluded that another advantage of the project-based learning model is that it can improve cooperation. The importance of group work in projects causes students to be able to develop and practice their communication skills and scientific performance. Based on this explanation, it can be concluded that group work is useful for training students' social attitudes. Students can help each other to complete projects, are good at helping those who are less clever, and remind each other to do their respective assignments well.

2.3 *The concept of self-confidence*

Eggen & Kauchak (2010) state that self-confidence is a statement that describes a belief, a cognitive idea is accepted if it is true without the need to consider other things that support it. In this case we see different ways in each of our self-beliefs that affect our motivation to learn, such as: (a) confidence in things to come, (b) confidence in intelligent thinking, (c) confidence in terms of skills, (d) confidence in terms of content (context), and (e) confidence in matters of achievement.

Furthermore, Willis (Gufron, 2010) argues that self-confidence is the belief that someone is able to cope with a problem with the best situation and can provide something fun for others. With self-confidence, when someone faces a problem, it can be resolved properly if they have self-confidence and can provide something of value to others. Lauster (Gufron, 2010) states that the aspects of self-confidence are as follows:

a. Self-confidence

A person's positive attitude about himself is a belief in one's ability.

b. Optimistism

It is about always having a positive outlook in dealing with all things about his abilities.

c. Objective

It is seeing things not according to himself but according to what should actually be

d. Responsible

Everything that a person bears that has become a consequence is someone's responsibility for something

e. Rational and realistic

Rational and realistic are thinking styles that are used to analyze something, an event, and a problems where thinking can be accepted by reason and in accordance with reality.

Thereby, the aspects of self-confidence adopted by the researcher are: belief in self-efficacy, optimism, responsibility, not being influenced by others, and being able to solve problems. These five aspects are used as references in measuring self-confidence in this study.

3 RESEARCH METHODS

The research method used was the Quasi-Experiment. The first thing to do is the development stage of the development of basic accounting teaching materials based on a project-based learning model, in addition to testing the effectiveness of the teaching materials being developed. The stages in this research are as follows.

3.1 *Preparation phase*

At this stage, coordination with the target schools for research and analysis of the learning curriculum in schools is carried out. At this stage, teachers also observe the use of teaching materials in schools.

3.2 *Collaborative experiment stage*

At this stage a quasi-experiment was carried out. In the implementation of experimental research at each research target school, two classes were selected by random sampling to be selected as a class given learning using project-based learning (experimental class) and conventional learning (control class).

Experimental research was conducted to determine the learning model's effectiveness that had been designed in teaching materials on students' creativity and confidence. This experimental research design used a post-test only control design (Sugiyono, 2012). The procedure in research is as follows:

3.3 *Data analysis*

Testing the validity of the main instrument used in this study is a list of questions distributed to respondents. Instruments made before distribution to respondents who are the sample of the study must be tested for validity and reliability through factor analysis. The list of questions created is actually able to reveal data so that it can answer problems until the research objectives are achieved. This validity test is intended to ascertain how well an instrument measures the concept that should be measured. By using a research instrument that has high validity, the results of the study are able to explain the research problem according to the actual situation or event with a significance below 0.05 and a Kaiser-Meyer Olkin (KMO) and a Measure of Sampling Adequacy (MSA) of at least 0.5 stated valid and the sample can be analyzed further. The data obtained were analyzed using Confirmatory Factor Analysis (CFA). This factor analysis is carried out to determine which indicators are relevant to the research variables. The component matrix is the loading factor value of the factor component variables. The required loading value, which is greater or equal to 0.5, is declared relevant. Santoso (2002: 104). This validity test is assisted by the IBM SPSS Statistics 23 tool.

3.4 Hypothesis testing

At the collaborative experimental stage, data analysis was carried out to test the following hypotheses.

$$H_0 : \mu_1 = \mu_2$$

$$H_1 : \mu_1 \neq \mu_2$$

μ_1 = Average Learning Outcomes by Teaching Using Teaching Materials
μ_2 = Average Learning Outcomes without Teaching Materials

The statistical hypothesis was tested using the t-test with the formula:

$$t = \frac{\bar{x}_1 - \bar{x}_2}{\sqrt{s^2 gab \left\{ \left(\frac{1}{n_1}\right) + \left(\frac{1}{n_2}\right) \right\}}} \qquad (1)$$

with $s^2 gab = \frac{(n_1-1)s_1^2 + (n_2-1)s_1^2}{n_1+n_2-2}$

4 RESULTS

4.1 Results of the validity of teaching materials

Experts assess Project-based learning on accounting teaching materials that have been compiled with the aim of seeing the quality of the LKS product in terms of content. The test results show that the teaching materials in the form of Student Activity Sheets meet the valid criteria. The results of the validity test of the Student Activity Sheets teaching materials are as follows.

Table 2. Validation results of content and material conformity

No	Rated aspect	Aiken Index	Criteria
1	Suitability of Content and Material	0,76	Valid
2	Construction	0,74	Valid
3	Language Accuracy	0,77	Valid
4	Practicality By Educators	0,75	Valid
5	Practicality by Students	0,78	Valid

The results of the validator's assessment of the teaching materials above indicate the valid category. This shows that in theory the accounting teaching materials developed with a project-based learning model have met the validity criteria and are suitable for use as teaching materials by Bengkulu City vocational high School or SMKN teachers.

4.2 Results of collaborative experiments

Description of student learning outcomes
The schools selected in the implementation of the collaborative experiment consisted of three schools, namely: (1) SMKN 01 Bengkulu City, and (2) SMKN 02 Bengkulu City, (3) SMKN 03 Bengkulu City. In each school, one class is selected as the experimental class, namely class X. Description of student learning outcomes after being given learning using project-based learning teaching materials is shown in the following table:

Table 3. Student learning outcomes data

Names of School	Total Student Score	Number of Students	Student's Average Score	Percentage KKM (%)
SMKN 01	2324	31	74.97	83,9
SMKN 02	2493	33	75,54	84.85
SMKN 03	2693	36	74.80	80.56

Based on the table above, it can be seen that the percentage of classical student learning completeness that reaches the KKM is more than 65%. In addition, the average score of the two trial classes has reached the KKM score. This shows that the learning tools developed have met the criteria for being effective.

4.3 Hypothesis testing results

The analysis results showed that there were differences in the average student learning outcomes on basic accounting material before and after the use of teaching materials. To analyze these differences statistically performed by t-test analysis. The hypothesis tested is as follows.

Table 4. T-test result data

Sample	t	df	Sig. (2-tailed)	Mean Difference	95% Confidence Interval of the Difference	
					Lower	Upper
SMKN1	70.859	30	.000	74.968	72.807	77.1284
SMKN2	75.184	32	.000	75.545	73.498	77.5922
SMKN3	70.976	35	.000	74.806	72.665	76.9452

H_0: There is a significant difference between student learning outcomes and KKM

H_1: There is no significant difference between student learning outcomes and KKM

With the testing criteria, namely if $t_{count} > t_{table}$ and the significant level < 0.05 then H_0 is accepted. if $t_{count} < t_{table}$ and significant level > 0.05 then H_0 is rejected. The results of the t-test for students' creativity after being given teaching materials are shown in the Table 5.

The t-test table shows the knowledge of students of SMKN 01, SMKN 02, and SMKN 03 of Bengkulu City after the use of teaching materials with a significant < 0.005. These results indicate a significant difference between student creativity and KKM at SMKN 01, SMKN 02, and SMKN 03 Kota Bengkulu before using project based learning and after using the project based learning model.

5 DISCUSSION

Experts have assessed project-based accounting and finance teaching materials with the criteria for assessing the suitability of content and material, learning activities, language accuracy, educator practicality, and student practicality testing. by using accounting and finance teaching materials can increase student confidence.

Evidence for improvement is in the pre-test and post-test results. The KKM score of students who used teaching materials was higher than students who did not use teaching materials (table 4).

The purpose of assessing the validity of teaching materials is to see the product's quality in terms of content. The test results show that the teaching materials in the form of Student Activity Sheets (LKS) meet the valid criteria (Table 3). There is a significant difference in learning outcomes using PjBL (Table 5) learning with this model can please students and can spur student enthusiasm in learning.

6 CONCLUSION

A project based learning model can improve students' abilities and self-confidence. Students are more creative, independent, confident and fun because students can complete learning projects in groups, besides that it can increase the creation of basic accounting teaching materials that meet valid and practical criteria and there is an effect of project-based learning teaching materials on student creativity and confidence.

After applying the project based learning model, the result is that the learning creativity and self-confidence of the students of SMKN 01, SMKN 02, and SMKN 03 Bengkulu City can be improved. This can be seen in the t-test results, where the t value results are greater than t-table. The t-test table shows students' knowledge after the use of teaching materials with t-count > t-table and

significant < 0.005. So, the results of this study are to accept HO and reject Ha, meaning that after applying the project-based learning model there has been an improvement in student creativity and self-confidence.

The effect of the project based learning model used in the learning method makes students feel happy, can improve student creativity and students are more confident in completing group assignments. This study's results are in line with the research of Susanti et al. (2020), which shows that the learning model with the project-based learning model can increase students' knowledge in receiving monitored lessons from the aspect of conceptual knowledge.

REFERENCES

Arends, R., & Kilcher, A. 2010. *Teaching for student learning: Becoming an accomplished teacher.* Routledge. New York.

Ausín, V., Abella, V., Delgado, V., & Hortigüela, D. 2016. Aprendizaje basado en proyectos a través de las TIC. Una experiencia de innovación docente desde las aulas universitarias. *Formacion Universitaria*, 9(3): 31–38. https://doi.org/10.4067/S0718-50062016000300005

Basilotta Gómez-Pablos, V., Martín del Pozo, M., & García-Valcárcel Muñoz-Repiso, A. 2017. Project-based learning (PBL) through the incorporation of digital technologies: An evaluation based on the experience of serving teachers. *Computers in Human Behavior*, 68: 501. https://doi.org/10.1016/j.chb.2016.11.056

Biasutti, M. 2015. Interdisciplinary project-based learning: an online wiki experience in teacher education. *Technology, Pedagogy and Education*, 24(3): 339–355. https://doi.org/10.1080/1475939X.2014.899510

Chang, S. 2018. Impacts of an augmented reality-based flipped learning guiding approach on students' scientific project performance and perceptions. *Computers and Education*, 125: 226–239. https://doi.org/10.1016/j.compedu.2018.06.007

Chu, S. K. W., Zhang, Y., Chen, K., Chan, C. K., Lee, C. W. Y., Zou, E., & Lau, W. 2017. The effectiveness of wikis for project-based learning in different disciplines in higher education. *Internet and Higher Education*, 33: 49–60. https://doi.org/10.1016/j.iheduc.2017.01.005

Dabae Lee, Yeol Huh, C. M. R. 2015. Collaboration, intra-group conflict, and social skills in project-based learning. *Instructional Science*, 43(5): 561–590. https://doi.org/10.1007/s11251-015-9348-7

Daniel Spikol, Emanuele Ruffaldi2 GiacomoDabisias, M. C. (2018). Supervised machine learning in multimodal learning analytics for estimating success in project-based learning. *Journal of Computer Assisted Learning*, 34(4): 366–377. https://doi.org/10.1111/jcal.12263

Depdikbud. 2013. Permendikbud Nomor 65 tahun 2013 tentang Standart Proses Pendidikan Dasar dan Menengah. Jakarta: Depdikbud.

Filcik, A. et al. 2012. The effects of project-based learning (PjBL) approach on the achievement and efficacy of high school mathematics students: a longitudinal study investigating the effects of PjBL approach in mathematics education. *Proceedings of the National Conference in Undergraduate Research (NCUR)*, 29–31 Maret 2012. Odgen Utah: Weber State University, Utah.

Guo, S., & Yang, Y. 2012. Project-based learning: an effective approach to link teacher professional development and student learning. *Journal of Educational Technology Development and Exchange,* Desember 2012, 5(2): 41–56.

Genc, M. 2015. The project-based learning approach in environmental education. *International Research in Geographical and Environmental Education,* 24(2), 105–117. https://doi.org/10.1080/10382046.2014.993169

Han, S. 2015. In-service teachers' implementation and understanding of STEM project based learning. *Eurasia Journal of Mathematics, Science and Technology Education,* 11(1): 63–76. https://doi.org/10.12973/eurasia.2015.1306a

Han, Sunyoung, Capraro, R., & Capraro, M. M. 2015. How Science, Technology, Engineering, and Mathematics (Stem) Project-Based Learning (PBL) Affects High, Middle, and Low Achievers Differently: the Impact of Student Factors on Achievement. *International Journal of Science and Mathematics Education,* 13(5): 1089–1113. https://doi.org/10.1007/s10763-014-9526-0

Ibrahim Bilgin, Yunus Karakuyu, Y. A. 2015. The effects of project based learning on undergraduate students' achievement and self-efficacy beliefs towards science teaching. *Eurasia Journal of Mathematics, Science and Technology Education,* 11(3): 469–477. https://doi.org/10.12973/eurasia.2014.1015a

Jian-Wei Lin a, C.-W. T. 2016. The impact of an online project-based learning environment with group awareness support on students with different self-regulation levels: An extended-period experiment. *Computers and Education,* 99: 28–38. https://doi.org/10.1016/j.compedu.2016.04.005

Kaj U. Koskinen. 2012. Organizational Learning in Project-Based Companies: A Process Thinking Approach. *Project Management Journal,* 39: 28–42. https://doi.org/10.1002/pmj

Kokotsaki, D., Menzies, V., & Wiggins, A. 2016. Project-based learning: A review of the literature. *Improving Schools,* 19(3): 267–277. https://doi.org/10.1177/1365480216659733

Mudlofar, A. 2012. *Aplikasi Pengembangan Kurikulum Tingkat Satuan Pendidikan dan Bahan Ajar dalam Pendidikan Islam.* Rajawali Pers: Jakarta

Plomp, T. 2010. An introductional to educational design research. *Proceedings of the seminar conducted at the east china normal University, Beijing*

Prastowo, A. 2012. Panduan Kreatif Membuat Bahan Ajar Inovatif. Diva Press. Yogyakarta

Patton, A. 2012. *Work that matters: the teacher's guide for project based learning.* The Paul Hamlyn Foundation. California

McArdle, G. 2010. *Instructional design for action learning.* Amacom. New York.

Rodríguez, J., Laverón-Simavilla, A., Del Cura, J. M., Ezquerro, J. M., Lapuerta, V., & Cordero-Gracia, M. 2015. Project Based Learning experiences in the space engineering education at Technical University of Madrid. *Advances in Space Research,* 56(7): 1319–1330. https://doi.org/10.1016/j.asr.2015.07.003

Scott Wurdinger, M. Q. 2014. Enhancing College Students' Life Skills through Project Based Learning. *InnovHighEduc.* https://doi.org/10.1007/s10755-014-9314-3

Zhang, Z., Hansen, C. T., & Andersen, M. A. E. 2015. Teaching Power Electronics with a Design-Oriented, Project-Based Learning Method at the Technical University of Denmark. *IEEE Transactions on Education,* 1–7.

Educational Innovation in Society 5.0 Era: Challenges and Opportunities – Purnomo & Herwin (Eds)
© 2021 the authors, ISBN 978-1-032-05392-9

Goal-orientation measurement model: A study of psychometric properties using a ranking scale

F.A. Setiawati & T. Widyastuti
Department of Psychology, Faculty of Education, Universitas Negeri Yogyakarta, Indonesia

ABSTRACT: Goal orientation, a patterned value which leads individual into various ways of working, can predict student's performance and learning success in education. Thus, assessing student's goal orientation is important. Unfortunately, even though there is lots of research in developing goal orientation scales in others countries, it is a bit difficult to find one in Indonesia. The aim of research was to develop goal orientation scale using a ranking type. This study involved 346 students of Yogyakarta State University, consisting of 63 male students and 283 female students. The data collection used of multistage random sampling. The goal-orientation scale in the study was developed based on the results of the previous studies which had translated the goal-orientation scale from Was into Indonesian. Goal orientation is divided into four dimensions, namely mastery orientation, performance-approach with focus on comparing ability, performance-approach with focus on proving ability, and avoidant goal orientation. The instrument was made by using a ranking type, and each item consisted of three statements showing three dimensions of goal orientation. Exploratory factor analysis supported that the items were grouped according to the theoretical concept referred, so the validity of the construct was met. The resulting scale has quite good psychometric properties with the reliability for each dimension ranging from 0.622–0.895.

1 INTRODUCTION

Success in achieving the desired goals is related to one's orientation. Orientation is a view that underlies thoughts, concerns, or tendencies. This orientation will determine the attitude, namely the right direction or place (Departemen Pendidikan Nasional, 2012). This orientation is associated with a goal so it is called a goal orientation. Goal orientation is patterned values in a person that can lead to various ways of working in areas to get various answers (Ames, 1992).

Research on goal orientation is important because it can predict the quality of learning outcomes in education. This is reinforced by Steinmayr, Bipp, & Spinath (2011) that goal orientation can predict academic performance better than intelligence and personality. Goal orientation also can predict work performance (Payne et al., 2007). In addition, goal orientation have a role in the student's focus and engagement in doing a task (Anderman et al., 2002); academic self-efficacy and self-regulated learning (Middleton & Midgley, 1997); psychological well-being (Kaplan & Maehr, 1999) as well as in life regulations (Becker et al., 2019).

In the earlier development, many researchers focused on two main approaches of goal orientation, namely mastery goal orientation and performance goal orientation (Middleton & Midgley, 1997). Individuals with mastery orientation focus to finish

and master a task. Whereas individuals with performance orientation focus on demonstrating their ability and comparing it to others (Was, 2006). Then, Elliot and Harackiewicz (1996) demonstrate that performance goal orientation can be divided into performance-approach and performance avoid. Performance-approach focus on demonstrating an ability to others. Meanwhile, performance-avoid focus on avoiding undesirable circumstances that make people incapable of a task. Pintrich (2000) then suggested that mastery goal orientation also can be divided into mastery-approach and mastery avoid. In mastery approach, individuals focus on mastering a task, studying, and understanding. In mastery-avoid, individuals focus on avoiding weakness in learning. The four dimensions of goal orientation was strengthened by Muis & Winne (2012; Erdem-Keklik & Keklik, 2013; Manrique-Abril et al., 2020).

However, the four dimensions of goal orientation are not adopted by many researchers (Elliot & McGregor, 2001). The prominent goal orientation construct divided goal orientation into three dimension, namely mastery goal orientation, performance-approach goal orientation, and performance-avoidant goal orientation (Elliot & Church, 1997; Middleton & Midgley, 1997; Midgley et al., 1998; Payne et al., 2007; Was, 2006). Those three dimensions of goal orientation are the most cited and replicated in other researchers.

DOI 10.1201/9781003206019-53

Since research on goal orientation is important, a scale in assessing goal orientation is needed. Goal-orientation measurement had been established in many studies. Of the several measuring instruments, each uses a different construct of goal-orientation. Such as the Achievement Goal Orientation Scale from Midgley et al. (1998, 2000) which is a part of the Patterns of Adaptive Learning Scales uses three components of goal orientation consisting of mastery goal orientation, performance-approach goal orientation, and performance-avoid goal orientation. The response format used five choices from (1) not at all true, to (3) somewhat true, to (5) very true. All three orientation subscales have good internal consistency.

Was (2006) developed a goal-orientation measuring instrument with 34 items describing four-goal orientations (mastery, performance approach, avoidant performance, and work avoidant). The response format in this study is Likert type with six choices of (1) very untrue, (2) mostly untrue, (3) somewhat untrue, (4) somewhat true, (5) mostly true, and (6) very true. The reliability on these measuring instruments ranged from .64 to .81.

The 2x2 Achievement Goal Orientation Scale (AGOS) was developed based on the theory of achieving goals for high school students in Turkey (Erdem-Keklik and Keklik, 2013). This scale was developed with four factors of the goal-orientation model, namely learning approach, learning avoidance, performance approach, and avoidance performance. Internal consistency with Cronbach's Alpha ranged from .72 to .82.

Even though research on goal orientation is important and proved by the 185.295 study on goal orientation on the website www.sciencedirect.com from 2003-2019. The research's interest on goal orientation in Indonesia is inadequate. In Indonesian Journal, Garuda Rujukan Digital (Garuda), from 2003 until 2019, there are only 43 studies relating to goal orientation in Indonesia. Some of them are correlational research on goal orientation (see research from Nurcahyanti et al., 2014; Variansyah & Listiara, 2017; Uyun, 2018) and comparison research (see Susetyo & Kumara, 2012). There is no article which focus on developing a goal orientation scale even though a scale is the main tool in research.

The goal orientation scale in previous researches tends to use a Likert scale. This type has weaknesses on faking (Honkaniemi et al., 2011), social desirability, and response bias (Xiao et al., 2017). The other type of measurement uses a force choice scale, one of which in the form of ranking scale can be an alternative. This type of instrument has advantages in avoiding social desirability and faking (Chernyshenko et al., 2009; Setiawati et al., 2013; Zwick & McDonald, 2000). On the other hand, goal-orientation measurement is more about measuring typical than maximum characteristics. Thus, the use of measuring instruments with ranking characteristics is appropriately carried out. This research aimed to develop a goal-orientation scale using a ranking type.

2 RESEARCH METHOD

This study used a quantitative approach with a focus on developing a goal orientation scale. The research subjects were the students of Yogyakarta State University. There were 346 Yogyakarta State University students in this study, consisting of 63 male students (18.21%) and 283 female students (81.79%) with age 17 until 23 years old (mean = 19.26, SD = .899). The data collection was done by multistage random sampling. Random at stage one was based on the faculties at Yogyakarta State University, while the next stage was carried out randomly at the study program level or department and classes in the study program. A total of 6 faculties was involved in this study including Faculty of Language and Art, Faculty of Economy, Faculty of Sport Education, Faculty of Education, Faculty of Social Education, and Faculty of Mathematics and Science. The delivery of the measuring instruments to all subjects of one class was done randomly.

The instrument was developed with a ranking type based on the goal orientation theory from Midgley et al. (1998) which divided the goal orientation into three dimensions covering mastery goal orientation, performance-approach goal orientation, and avoidant goal orientation. This theory was also proven by a previous study which adapted the goal orientation scale from Was (2006) into Indonesian from Romadhani, Setiawati, Noventira, Farida, & Adiframbudhi (2019).

Each item in the scale consists of three statements reflecting the three dimensions of goal orientation. The subjects were asked to give a response by ranking the three statements based on the compatibility to themselves. The most appropriate response was ranked 1, and the most inappropriate response was ranked 3. The evidence of validity based on the internal structure was carried out by exploratory factor analysis. Since goal orientation is a multidimensional construct, item analysis was done separately according to each dimension. Item selection was carried out using the corrected item-total correlation for each dimension. We also estimated reliability using Cronbach's Alpha. All analysis was performed using the SPSS program.

3 RESULT AND DISCUSSION

The development of the goal-orientation instrument was carried out based on Azwar (2016; Cohen & Swerdlik, 2010; Furr, 2011) by using five general steps of constructing a scale, i.e. (1) selecting psychological attributes, (2) preparing the specifications of instruments, (3) writing an item, (4) item selection, (5) validating the instruments.

In the first step, the researcher determined the goal orientation constructs by referring to Midgley et al.'s (1998) concept. The goal orientation was defined as a patterned value that can lead to various ways of working, in areas to get various answers. It consists of 3 dimensions: mastery goal orientation, performance-approach goal orientation, and avoidant orientation. That construct was proven by the previous research from Romadhani and Setiawati (2019; Elliot & Church, 1997; Van de Walle, 1997).

The second step was establishing specifications of the instrument covering the purpose of the test, characteristics of the target, type of instrument, and construct dimensions. The measurement objective is to understand one's goal orientation. The subject target were college students. The instrument was developed using the ranking scale by comparing the three statements for each number. The components of instruments consist of mastery, performance approach, and avoidance.

The third step was drafting the instrument. Items were made by presenting a situation that usually occurred in college life. This situation was responded based on the usual daily activities. A number of options (a, b, c) represented behaviors of the three dimensions (mastery, performance, and avoidant). The sample of items can be seen in Table 1. According to item selection in the fourth step, researchers conducted a field test via online considering the COVID-19 outbreak in 2020. The researchers looked for the contact of the lecturer of class. Students who had the opportunity were asked to fill informed consent. A total of 50 participants were selected randomly to get 25,000 rupiahs of internet credit as the reward. The result analysis was carried out to obtain the psychometric properties of the items.

Table 1. Sample item from each dimension

Statement			Ranking
I go to college for	M	Increasing my skill in my field	
	P	Proving to my friends and family that I can be the best	
	A	Avoiding bad judgment from my parents and friend of my lack	

Note: M (Mastery), P (Performance), A (Avoidant)

The last step was to validating the goal orientation construct. To validate the internal structure of goal orientation construct which was assessed by the scale, Exploratory Factor Analysis (EFA) with Orthogonal rotation namely Varimax was used to reduce the instrument items into dimensions. Factor reduction was carried out using the Principal Component Analysis (PCA). The number of dimensions in this factor analysis were set manually into three dimensions according to the construct theory used in this research. After reduction, the items were grouped into dimensions according to its loading factor. In this research we used a minimum 0.2 factor loading in the interpretation of data. Exploratory factor analysis found that three-factor explained 35% of the total variance.

Based on the loading factors, the items were grouped into three factors. Dimension 1 was related to mastering the material and achieving higher abilities than before. Item in dimension 1 represents the mastery goal orientation. Dimension 2 consists of items associated with comparing oneself with others. The item in dimension 2 reflected performance-approach goal orientation related to an effort to compare oneself to other people. The last, dimension 3 consists of items associated with an effort to prove that an individual is better than others and/or regarded as competent. Thus, dimension 3 represents the performance-approach related to demonstrate ability or competence. Items containing avoidance behavior were not grouped in one dimension. All avoidant items have negative coefficients. These items have a high coefficient all over the three factors. The high and negative coefficients indicated that the items can join in the same factor with the maximum coefficient, but has an opposite meaning. This means that if it belongs to the mastery dimension, the higher the item score, the lower the mastery dimension or it can be said as the opposite of mastery approach or mastery avoidance. The items of avoidance in all three dimensions mean that high avoidance items impact the low mastery score and approach performance. Thus, these items are still said to be avoidant items which are the opposite of the three mentioned dimensions.

Table 2. Reliability with Cronbach's alpha of goal orientation dimensions

Dimension	N	Reliability	Eliminated items	Reliability after item selection
Mastery	24	0.887	2(M8,M17)	0.895
Performance 1	24	0.603	4(P7, P8, P10, P24)	0.622
Performance 2		0.590	3 (P4, P17, P19)	0.670
Avoidant	24	0.874	1(A20)	0.876

There are two unique finding in this research. First, performance-approach goal orientation was divided into different two factors, dimension 2 and dimension 3. Item assessed performance-approach which focus on comparing ability to others were grouped in the dimension 2 whereas the item assessed performance approach with a focus on proving ability were grouped in dimension 3. This study may be a bit different with the previous which assessed performance-approach focus on comparing ability and proving ability in the same factor with the name performance-approach goal orientation (Romadhani and Setiawati, 2019) (Romadhani and Setiawati, 2019). The other finding is item assessed avoidant goal orientation were spread in the three dimensions with negative coefficient. However, this instrument still represents the same dimensions covering mastery goal orientation, performance approach, and avoidant like other studies even though with a note that item in performance goal orientation divided into two factors and item in avoidant goal orientation was spread in three

283

dimensions (see Elliot & Church, 1997; Middleton & Midgley, 1997; Midgley et al., 1998, 2000; Payne, Youngcourt & Beaubien, 2007).

Based on the loading factors on each item and component, item with loading factor less than 0.2 needs to be eliminated. Factor loading can be an indicator to analyze the characteristics of the items. There are two items in Mastery goal orientation which need to be eliminated (M8, M17). On Performance-approach, focusing on comparing ability, four items needed to be eliminated (P7, P8, P10, P24). Meanwhile, on the Performance-approach, there are three items with focus on proving ability (P4, P17, P19). The last, on avoidant there is only one item needed to be eliminated (A20). Thus there are 10 items needed to be eliminated. After deleting the item with a low factor loading, reliability of each dimensions were estimated. Cronbach's alpha reliability of four dimensions were promising. The reliability showed very high results on the Mastery Goal Orientation (.895) and Avoidant goal orientation (.876). The reliability in two other dimensions showed a moderate result (Performance 1 was .622 and Performance 2 was .670). The reliability can be seen in Table 2.

4 CONCLUSION AND SUGGESTIONS

This research has succeeded in constructing a goal-orientation measuring instrument by using ranking data. This measuring instrument has been tested for validity by Exploratory Factor Analysis. Exploratory Factor Analysis results showed that almost all mastery items are grouped in one dimension, whereas performance items are grouped in 2 dimensions. Avoidance items are in three dimensions with negative coefficients. Some items need to be corrected or dropped in the use of this instrument in order to get more accurate measurement results.

ACKNOWLEDGEMENT

This research was financially supported by Lembaga Penelitian dan Pengabdian kepada Masyarakat Universitas Negeri Yogyakarta [Institute of Research and Community Services of Universitas Negeri Yogyakarta].

REFERENCES

Ames, C. 1992. Classrooms: Goals , structures, and student motivation, *Journal of Educational Psychology* 84(3): 261–271.

Anderman, E. M., Austin, C. C. & Johnson, D. M. 2002. The development of goal orientation. *Development of Achievement Motivation*, 197–220. doi: 10.1016/b978-012750053-9/50010-3.

Azwar, S. 2016. *Penyusunan skala psikologi [Constructing a psychological scale]*. Pustaka Pelajar.

Becker, S. *et al.* 2019. Relations between life-goal regulation, goal orientation, and education-related parenting- A

person-centered perspective. *Learning and Individual Differences.* 76(September): 101786. doi: 10.1016/j.lindif.2019.101786.

Chernyshenko, O. S. *et al.* 2009. Normative scoring of multidimensional pairwise preference personality scales using IRT: Empirical comparisons with other formats. *Human Performance* 22(2): 105–127. doi: 10.1080/08959280902743303.

Cohen, R. J. & Swerdlik, M. E. 2010. *Psychological testing and assessment: An introduction to test and measurement 7th edition.* McGraw-Hill.

Departemen Pendidikan Nasional. 2012. *Kamus besar Bahasa Indonesia [Indonesian Dictionary].* Gramedia Pustaka Utama.

Elliot, A. J. & Harackiewicz, J. 1996. Approach and avoidance achievement goals and intrinsic motivation: A mediational analysis. *Journal of Personality and Social Psychology 70:* 968-980.

Elliot, A. J. & Church, M. A. 1997. A hierarchical model of approach and avoidance achievement motivation. *Journal of Personality and Social Psychology* 72(1): 218–232. doi: 10.1037/0022-3514.72.1.218.

Elliot, A. J. & McGregor, H. A. (2001). A 2 × 2 achievement goal framework. *Journal of Personality and Social Psychology,* 80(3): 501–519. doi: 10.1037/0022-3514.80.3.501.

Erdem-Keklik, D. & Keklik, İ. (2013). Exploring the factor structure of the 2x2 achievement goal orientation scale with high school students. *Procedia - Social and Behavioral Sciences* 84: 646–651. doi: 10.1016/j.sbspro.2013.06.619.

Furr, R. M. (2011) *Scale construction and psychometrics for social and personality psychology.* SAGE Publications Ltd.

Honkaniemi, L., Tolvanen, A. & Feldt, T. (2011). Personality and Social Psychology Applicant reactions and faking in real-life personnel selection. 376–381. doi: 10.1111/j.1467-9450.2011.00892.x.

Indonesia, *Garba Rujukan Digital (Garuda).* 2020. Available at: http://garuda.ristekbrin.go.id/documents?page=5&q=orientasi tujuan.

Kaplan, A. & Maehr, M. L. 1999. Achievement goals and student well-being. *Contemporary Educational Psychology* 24(4): 330–358. doi: 10.1006/ceps.1999.0993.

Midgley, C. *et al.* 1998. The development and validation of scales assessing students' achievement goal orientations. *Contemporary Educational Psychology,* 23(2): 113–131. doi: 10.1006/ceps.1998.0965.

Midgley, C. *et al.* 2000. Manual for the patterns of adaptive learning scales', *University of Michigan,* pp. 734–763.

Nurcahyanti, A. & Setyawan, I. 2014. Hubungan antara iklim sekolah dengan orientasi tujuan performa pada siswa sekolah menengah pertama', *Empati: Jurnal Karya Ilmiah S1 Undip,* 3(4): 62–73.

Payne, S. C., Youngcourt, S. S. & Beaubien, J. M. 2007. A meta-analytic examination of the goal orientation nomological net. *Journal of Applied Psychology* 92(1): 128–150. doi: 10.1037/0021-9010.92.1.128.

Romadhani, R. K. & Setiawati, F. A. 2019. *Goal orientation, well-being, dan kesuksesan akademik mahasiswa UNY semester awal.*

Setiawati, F. A., Mardapi, D. & Azwar, S. 2013. Penskalaan teori klasik instrumen Multiple Intelligences Tipe Thurstone dan Likert. *Jurnal Penelitian dan Evaluasi Pendidikan,* 17(2): 259–274. doi: 10.21831/pep.v17i2.1699.

Steinmayr, R., Bipp, T. & Spinath, B. 2011. Goal orientations predict academic performance beyond intelligence and personality. *Learning and Individual Differences* 21(2): 196–200. doi: 10.1016/j.lindif.2010.11.026.

Susetyo, Y. & Kumara, A. 2012. Orientasi tujuan, atribusi penyebab, dan belajar berdasar regulasi diri. *Jurnal Psikologi*, 39(1): 95–111.

Uyun, M. 2018. Orientasi Tujuan Dan Efikasi Akademik Terhadap Kecurangan Akademik Pada Mahasiswa Fakultas Psikologi Uin Raden Fatah Palembang. *Psikis: Jurnal Psikologi Islami* 4(1): 45–51. doi: 10.19109/psikis.v4i1.1938.

VandeWalle, D. 1997. Development and validation of a work domain goal orientation instrument. *Educational and Psychological Measurement* 57(6): 995–1015. doi: 10.1177/0013164497057006009.

Variansyah, V. & Listiara, A. 2017. Hubungan orientasi tujuan performa dengan kecemasan akademik pada siswa kelas X di SMA Negeri "A" Semarang. *Empati* 6(1): 419–424.

Was, C. 2006. Academic achievement goal orientation: Taking another look. *Electronic Journal of Research in Educational Psychology*, 4(10): 529–550.

Xiao, Y., Liu, H. & Li, H. 2017. Integration of the forced-choice questionnaire and the likert scale: A simulation study, *Frontiers in Psychology*, 8(MAY). doi: 10.3389/fpsyg.2017.00806.

Zwick, R. & McDonald, R. P. 2000. *Test theory: A unified treatment, Journal of the American Statistical Association.* doi: 10.2307/2669496.

Educational Innovation in Society 5.0 Era: Challenges and
Opportunities – Purnomo & Herwin (Eds)
© 2021 the authors, ISBN 978-1-032-05392-9

Multiliteracy education in 5.0 era on learning entrepreneurship with the project based learning model

B. Afriansyah
Raflesia Polytechnic of Bengkulu, Indonesia

N. V. Yustanti
Dehasen University of Bengkulu, Indonesia

ABSTRACT: The implementation of the 2013 curriculum emphasizes the improvement and balance of soft and hard skills. Students who are in the next generation must have individual abilities in combining and coordinating learning activities so that they can be active students in the learning process. The purpose of this study is to determine the level of effectiveness of the literature used by teachers in improving student achievement. Multiliteracy learning needs to be applied so that students can have it after they finish their studies in school. The sample used in this study were students of SMAN 01 and SMAN 02 Rejang Lebong Bengkulu Province in Indonesia. The model used in this study is project-based learning (PjBL) model in order to improve student creativity. The PjBL model facilitates students to make products to solve real-life problems. Product manufacturing projects can be tried individually or in groups. With a multiliterate approach in distributing education to students with the project-based learning (PjBL) model, the results can increase students' self-confidence.

1 INTRODUCTION

The development of the world today has come to the era of a 5.0 society. The form of human life has been in the form of information. Therefore, in a successful process that is of high quality and is able to compete globally, technological developments are important for all human beings and for the future of the nation (Ahlah & Melianah, 2020). Here, digital literacy needs to be developed in the world of education to build a better national character and to be better prepared for facing the era of 5.0 public education. The rapid development of information communication; and mutual cooperation can be instilled through; encouraging the progress of national integrity; matching students in preparing themselves; being able to position someone as an important part for the independence of nationalism and religiosity. In community 5.0, students must prepare themselves for a new world – who will be the individuals that determine how they do this (Rajab Agustini, 2020)? Students who are the next generation must have individual abilities in combining and coordinating learning activities so that they can be active students in the learning process.

National education targets are listed in the Indonesian Republic Law no. 20 of 2003 article 3 develops the potential of students, namely to become human beings who believe and fear God Almighty, have a noble character, are healthy, knowledgeable, capable, creative, independent, and become democratic and responsible citizens. The national education target is achieved

through 8 educational standards in the curriculum to improve the quality of education in Indonesia (Fadlillah, 2014). The rapid development of information and communication technology with all its impacts is a reality that cannot be avoided at this time (Garton, 2014). Teachers as educators as well as agents of change, driving the nation's progress, must be able to position themselves as an important part of the changes that are currently taking place. Teachers must be able to accompany students to preparing themselves in welcoming a new era of an advanced society, the society of 5.0. Students will be the individuals who determine how the Republic of Indonesia in the future must have a reliable character (Rajab Agustini, 2020). Efforts to improve student mastery of entrepreneurship at the school level really need to be improved. This can provide sufficient provisions in life and the world of work so as to be able to solve problems related to entrepreneurship. One of these efforts is the implementation of 2013 curriculum which emphasizes students to think with the aim of producing productive, creative, innovative, effective Indonesian individuals through strengthening attitudes.

Entrepreneurs are economic actors who will make changes (Kourilsky, 1998; Baum & Locke, 2004; Man et al., 2002; Ren, 2011; Falck, 2012). According to the theory of balance (equilibrium theory), achieving balance requires actions and decisions of economic actors (actors) that must be re-peated in "the same way" until it reaches a balance (Larson, 1994; Bhave, 1994; Lu & Beamish, 2001; Wennekers & Thurik, 1999;

DOI 10.1201/9781003206019-54

Sarasvathy, 2008). So, the keyword "repeats the same way," which is called "static situation," and the situation will not bring change. This means that people who are static or act like most people won't make a difference.

Multiliteracy learning is the process of learning that is developed based on scientific work (Untari, 2017). This learning aims to form students who are ready from various aspects to living life both in school, the workplace, and society. Multiliteracy learning needs to be applied to students so that students have the skills to be able to carry out learning in implementing the 2013 curriculum properly. Multiliteracy learning focuses on multicompetence. In multiliteracy learning, students acquire not only one competency, but also a variety of character and attitude competencies (Shamim, 2008; Gilmore, 2019; Ludwig, 2015; Joutsenlahti & Kulju, 2017; Martín-Peña, 2018; Rendón-Galvis, 2020). To establish a multiliteracy and writing center as a forum that supports 21st century communication skills, one effort that can be made to increase student creativity in running small businesses is to design learning so that it can facilitate students in developing learning abilities and self-confidence. Creating an intelligence assessment system that is desired for teaching materials (Chen & Tzeng, 2011). Critical Perspectives on Language Teaching Materials can unite a collection of critical voices (Gray, 2013). This effort can be done by using learning media such as teaching materials specifically designed to develop students' abilities. despite the fact that it is still rare to find learning tools that can be used by teachers directly for learning, especially in developing students' abilities and confidence. There is a significant influence in the use of learning models on economic learning achievement Sipahutar, 2018; Ritonga, 2018; Fitria et al., 2019; Susanti et al., 2020; Putra et al., 2020; Wijaya et al., 2020).

The approach used in learning Curriculum 2013 is an integrative scientific and thematic approach. Learning with a scientific approach is carried out by a scientific process. In this case, what students get is done with their own senses and minds so that they experience directly in the process of gaining knowledge. Learning using a scientific approach is carried out through the process of observing, questioning, trying, reasoning, and communicating. Learning like this is intended to improve and shape the attitudes, skills, and knowledge of students maximally. The PjBL approach is a term in education science that defines a concept that when talking about teaching and learning about knowledge content focuses on the activities to be carried out, (Hernández et al., 2015). Using a learning model with modules can increase students' self-confidence, reasoning can interact with one another to produce delusions in individuals who are vulnerable to self-confidence, (Andreou, 2014). Multimedia Teaching Materials provided by instructors can improve education technology, (Armenteros et al., 2013). Awayed-Bishara, (2015) analyzed the cultural content of the materials used to teach English

to Arabic speakers in high schools in Israel. Classroom experiments and teaching materials on OLED with semiconductor polymers can contribute to experimental and conceptual approaches for curricular integration of semiconductor polymers, (Banerji et al., 2005). Cakmak & Takayama, (2014) have taught people how to teach robots to give learning system effectiveness and capabilities. Assessment of student acceptance and satisfaction were done with video-based teaching materials to teach practical skills from a distance (Donkor, 2011). With a multiliteracy approach in giving lessons to students with the project-based learning (PjBL) model, it is hoped that students' self-confidence in developing their creative ideas can increase.

Project-based learning (PBL) is a form of student-centered active learning characterized by student independence, constructive questions, goal setting, collaboration, communication, and reflection in real-world practice and involving teachers in various projects. The application of the project-based learning model in entrepreneurship learning is assumed to facilitate students in developing students' abilities and self-confidence. Several empirical studies show that PjBL has a role in improving learning outcomes in schools. One of the results of research by Susanti et al., (2020) shows that the application of the PjBL learning model can increase student creativity. Project-based learning involves the active role of students to produce products or projects that are able to encourage students' ability to understand knowledge through a systematic syntax.

As for the purpose of this study is to determine the level of effectiveness of literature used by teachers in improving student achievement.

2 METHOD

The research design used was quasi-experimental research. The purpose of this study was to test the practical level of literacy used by teachers, namely project-based learning models. With this PBL model, teachers can improve the quality of learning for high school students in the Department of Social Sciences (IPS) in Bengkulu Rejang Lebong district. The sample used in this study were students of SMAN 01 and SMAN 02 Rejang Lebong Bengkulu Province in Indonesia.

The data collection technique is divided into two stages, namely development research carried out by observation and distributing validity and practicality assessment sheets. Validity data collection was done online by contacting experts, namely economics lecturers. Meanwhile, the clinical test was carried out in small groups by visiting high school students and teachers in Bengkulu city by observing health protocols. In the experimental stage, data collection was carried out by giving tests to students after the treatment was given. In addition, observation was carried out to observe the implementation of the learning stages according to the *project based learning* model.

2.1 Data analysis

2.1.1 Validity analysis

The formula determines the validation of the teaching materials used in reference to what Aiken has stated, namely; $V = \frac{\sum s}{n(c-1)}$, with $s = r - I_0$

Information:
$V =$ item validity index
$s =$ the number assigned to each appraiser which has been reduced by the lowest score
$r =$ rater's score
$I_0 =$ lowest score
$c =$ the number of categories
$n =$ the number of rater

2.1.2 Practicality analysis

The qualitative data generated from the sum of experimental data with a scale of 5 were taken from Widoyoko's research (2009) as in the Table 1.

Table 1. Learning tool practicality

Score Interval	Category
$X > \overline{X}_i + 1,8sb_i$	Very Practical
$\overline{X}_i + 0,6sb_i < X \leq \overline{X}_i + 1,8sb_i$	Practical
$\overline{X}_i - 0,6sb_i < X \leq \overline{X}_i + 0,6sb_i$	Moderate
$\overline{X}_i - 1,8sb_i < X \leq \overline{X}_i - 0,6sb_i$	Less Practical
$X \leq \overline{X}_i - 1,8sb_i$	Not Practical

3 RESULTS AND DISCUSSION

3.1 Development research results on validity analysis

This entrepreneurship teaching material is a project-based learning system, where students can learn to practice the types of small and medium enterprises. This book has been valid and has been used in high school students to assist teachers in improving the quality of students. The results of the validity test of teaching materials are as follows.

Table 2. Validation results of teaching materials

No	Rated Aspect	Aiken Index	Criteria
1	Formulation and Purpose	0,70	Valid
2	Conformity of Content and Material	0,75	Valid
3	Learning Activities	0,73	Valid
4	Language accuracy	0,74	Valid
5	Learning Resources	0,71	Valid
6	Application of PjBL	0,72	Valid

It was found that the results of all assessment categories were valid and suitable for use as teaching materials. All assessment categories indicate valid and workable projects.

3.2 Development research results on practicality analysis

The schools that were selected in the implementation of this collaborative experiment consisted of two schools, namely: SMAN 01 and SMAN 02 Rejang Lebong Bengkulu Province. In each school, one class was selected as the experimental class, namely class XI, majoring in social studies. Description of student learning outcomes after being given learning using literacy in the form of textbooks as follows:

Table 3. Score interval using LKS teaching materials

Names of Schools	t	df	Sig. 2-tailed	Mean Difference	95% Confidence Interval of the Difference Lower	Upper
SMAN 1	81.90	28	.000	75.83	73.93	77.72
SMAN 2	80.72	31	.000	76.23	74.30	78.16

Test Value $= 82, 83, 81$

The t-test table shows the students' knowledge of SMAN 01 and SMAN 02 Rejang Lebong Bengkulu Province there is an increase in scores from before using the LKS and after using the LKS. There is a difference 3.79 in SMA 1 (77.72 - 73.93) and there is a difference 3.86 at SMA 2 (78.16 - 74.30).

Literacy used in this research is as illustrated by the book in the Figure 1.

Figure 1. Literacy on learning entrepreneurship with the project based learning model

This book contains ways to develop talents and interests through entrepreneurial activities. Students are asked to practice directly, make works they like. For example, students who like to make handicrafts, the teacher teaches how to make products from waste materials that have economic value. The first part contains the use of oil palm frond waste, the second part contains the use of plastic waste and the third part contains ways to manage vegetable and animal materials into an international food.

Then students are given an assessment by the teacher. The results of student assessments can be seen in the Table 4.

Table 4. Score cumulative student

Schools	Total Student Score	Total Student / class	Average Score Student	% Cumulative Student/KKM
SMAN1	2199	29	75.82	89
SMAN2	2363	31	76.22	87.09

Based on the table above, it can be seen that the percentage of classical student learning completeness who reaches the KKM is more than 65%. In addition, the mean scores of the two trial classes have reached cumulative scores. This shows that there is an increase in student creativity in SMAN 01 and SMAN 02 Rejang Lebong Bengkulu Province.

4 CONCLUSION

This learning model is in line with the equilibrium theory put forward by Larson (1994). Getting into balance requires action. This action is in the form of student activities in entrepreneurship, through the learning media of entrepreneurship books students can increase their achievement. This is where there is a balance between literacy and the quality of education. If education and student creativity can run in a balanced manner, there will be an increase in human resources.

Project-based learning (PBL) is a form of student-centered active learning characterized by student independence, constructive questions, goal setting, collaboration, communication, and reflection in real-world practice and involving teachers in various projects. From the results of our research, it shows that there is an increase in student scores, before using entrepreneurship books and after using entrepreneurship textbooks. After applying the model in learning with a project-based learning approach it was found that the learning abilities and self-confidence of the students of SMAN 01 and SMAN 02 Rejang Lebong Bengkulu Province could improve.

So, this research is in line with the research of Susanti et al., (2020) which shows that the learning model with a project-based learning model can increase students' knowledge in receiving lessons in terms of the conceptual knowledge aspect. In theory, the learning method using the PjBL mode can increase student creativity. It can provide a positive impact on the world of education in the future.

This Project Based Learning model can be sustainable by using other literacies outside of entrepreneurship worksheets, for example by using teaching aids in entrepreneurship so that students can pursue their hobbies more.

REFERENCES

Ahlah, S., & Melianah. 2020. Membangun Karakter Siswa Melalui Literasi Digital Dalam Menghadapi Pendidikan Abad 21 Era Society 5.0. *Prosiding Seminar Nasional Pendidikan Program Pascasarjana Universitas PGRI Palembang 10 Januari 2020*: 805–814.

Andreou, C. 2014. Dopaminergic modulation of probabilistic reasoning and overconfidence in errors: A double-blind study. *Schizophrenia Bulletin*, 40(3): 558–565. https://doi.org/10.1093/schbul/sbt064

Armenteros, M., Liaw, S. S., Fernández, M., Díaz, R. F., & Sánchez, R. A. 2013. Surveying FIFA instructors' behavioral intention toward the Multimedia Teaching Materials. *Computers and Education*, 61(1): 91–104. https://doi.org/10.1016/j.compedu.2012.09.010

Awayed-Bishara, M. 2015. Analyzing the cultural content of materials used for teaching English to high school speakers of Arabic in Israel. *Discourse and Society*, 26(5): 517–542. https://doi.org/10.1177/0957926515581154

Banerji, K., Gundersen, D. E., & Behara, R. S. 2005. Quality management practices in Indian service firms. *Total Quality Management and Business Excellence*, 16(3): 321–330. https://doi.org/10.1080/14783360500053881

Bhave, M. P. 1994. A process model of entrepreneurial venture creation. *Journal of Business Venturing*, 9(3): 223–242. https://doi.org/10.1016/0883-9026(94)90031-0

Cakmak, M., & Takayama, L. 2014. Teaching people how to teach robots: The effect of instructional materials and dialog design. In *ACM/IEEE International Conference on Human-Robot Interaction* (pp. 431–438). https://doi.org/10.1145/2559636.2559675

Chen, C. H., & Tzeng, G. H. 2011. Creating the aspired intelligent assessment systems for teaching materials. *Expert Systems with Applications*, 38(10): 12168–12179. https://doi.org/10.1016/j.eswa.2011.03.050

Desi Fitria, Melly Susanti, M. D. I. 2019. Project Based Learning Model in Improving The Ability and Trust. *International Journal of Science, Technology & Management*, 1(3): 237–243. https://ijstm.inarah.co.id/index.php/ijstm/issue/view/10.46729

Donkor, F. 2011. Assessment of learner acceptance and satisfaction with video-based instructional materials for teaching practical skills at a distance. *International Review of Research in Open and Distance Learning*, 12(5): 71–88. https://doi.org/10.19173/irrodl.v12i5.953

Ermy Wijaya, Noprianysah, M. S. 2020. Model Pembelajaran Berbasis Proyek dalam Meningkatkan Kemampuan dan Kepercayaan Siswa. *Pedagogia Jurnal Ilmu Pendidikan*, 18(02): 136–147.

Fadlillah, M. 2014. *Implementasi Kurikulum 2013 Dalam Pembelajaran SD/MI. SMP/ Mts, &SMA/MA*. Ar-Ruzz Media.

Garton, S. 2014. Identifying a research agenda for language teaching materials. *Modern Language Journal*, 98(2): 654–657. https://doi.org/10.1111/modl.12094

Gilmore, G. 2019. Intercultural praxis: Australian diploma of early childhood education preservice teacher visual literacy comparisons between Australia and Vietnam. *Intercultural Education*, 30(6): 634–657. https://doi.org/10.1080/14675986.2019.1627112

Gray, J. (Ed.). 2013. *Critical perspectives on language teaching materials*. Switzerland: Springer Nature. https://doi.org/10.1057/9781137384263

Hernández, M. I., Couso, D., & Pintó, R. 2015. Analyzing Students' Learning Progressions Throughout a Teaching Sequence on Acoustic Properties of Materials with a Model-Based Inquiry Approach. *Journal of Science Education and Technology*, 24(2–3): 356–377. https://doi.org/10.1007/s10956-014-9503-y

Joutsenlahti, J., & Kulju, P. 2017. Multimodal Languaging as a Pedagogical Model—A Case Study of the Concept of Division in School Mathematics. *Education Sciences*, 7(1): 9. https://doi.org/10.3390/educsci7010009

Larson, B. A. 1994. Changing the economics of environmental degradation in Madagascar: Lessons from the national environmental action plan process. *World Development*, 22(5): 671–689. https://doi.org/10.1016/0305-750X(94)90043-4

Lu, J. W., & Beamish, P. W. (2001). The internationalization and performance of SMEs. *Strategic Management Journal*, 22(6–7): 565–586. https://doi.org/10.1002/smj.184

Ludwig, C. 2015. Narrating the "Truth": Using autographics in the EFL classroom. In *Learning with Literature in the EFL Classroom* (pp. 299–320). https://doi.org/10.3726/978-3-653-04297-9

Martín-Peña, M. L. 2018. The digitalization and servitization of manufacturing: A review on digital business models. *Strategic Change*, 27(2): 91–99. https://doi.org/10.1002/jsc.2184

Putra, I. U., Susanti, M., & Bengkulu, U. D. 2020. Development of Entrepreneurial Teaching Materials by Using Project-Based Learning Model in Improving Students' Creativity and Self-Confidence. *Jurnal Keilmuan Manajemen Pendidikan*, 6(02): 153–162. https://doi.org/10.32678/tarbawi.v6i02.2948

Rajab Agustini, M. S. 2020. Penguatan Pendidikan Karakter Melalui Literasi Digital Sebagai Strategi Menuju Era Society 5.0. In *PROSIDING SEMINAR NASIONAL PENDIDIKAN PROGRAM PASCASARJANA* (pp. 624–633). UNIVERSITAS PGRI PALEMBANG.

Rendón-Galvis, S. C. 2020. The use of ICT to promote reading in public libraries with the intervention of librarians. *Investigacion Bibliotecologica*, 34(83): 129–144. https:// doi.org / 10.22201 / iibi.24488321xe.2020.83. 58095

Ritonga, L. 2018. Pengaruh Penggunaan Model Pembelajaran Jigsaw Terhadap Hasil Belajar Ekonomi Pada Materi Penawaran Di Kelas X SMA Negeri 1 Padangbolak Julu. *Jurnal MISI Institut Pendidikan Tapanuli Selatan*, 1(1): 430–439.

Sarasvathy, S. D. 2008. Effectuation: Elements of entrepreneurial expertise. In *Effectuation: Elements of Entrepreneurial Expertise*. https://doi.org/10.4337/97818 48440197

Shamim, F. 2008. Trends, issues and challenges in English language education in Pakistan. *Asia Pacific Journal of Education*, 28(3): 235–249. https://doi.org/10.1080/ 02188790802267324

Sipahutar, J. 2018. Pengaruh Penggunaan Model Pembelajaran Picture and Picture Terhadap Hasil Belajar Ekonomi Siswa Pada Materi Pengangguran Kelas Xi IPS SMA Negeri 1 Sibabangun. *Jurnal Misi*, 1(1): 104–104.

Susanti, M., Herfianti, M., Damarsiwi, E. P. M., Perdim, F. E., & Joniswan. 2020. Project-based learning model to improve students' ability. *International Journal of Psychosocial Rehabilitation*, 24(2): 1378–1387. https://doi.org/10.37200/IJPR/V24I2/PR200437

Untari, E. (2017). Pentingnya pembelajaran multiliterasi untuk mahasiswa pendidikan guru sekolah dasar dalam mempersiapkan diri menghadapi kurikulum 2013. *Wahana Sekolah Dasar*, 25(1), 16-22.

Wennekers, S., & Thurik, R. 1999. Linking Entrepreneurship and Economic Growth. *Small Business Economics*, 13(1): 27–56. https://doi.org/10.1023/A:1008063200484

Author index

Abidin, Z. 96
Adi, B.S. 255
Afriansyah, B. 286
Ahmad, A.R. 18, 266
Aman 36
Aprilia, T. 255
Ardiansyah, A.R. 270
Arifin 151
Arifin, Z. 96
Awang, M.M. 18
Azim, M. 169
Azizah, N. 204

Basit, R.A. 226
Budiman, J. 36
Budimansyah, D. 142
Bulan, I. 138
Bunyanuddin, E. 204

Chen, M.-Y. 80

Dahalan, S. 266
Damayanto, A. 50
Datuk, A. 151
Desmaiyanti 54
Dharsana, I.K. 238
Dwiningrum, S.I.A. 11

Fadhlia, H.N. 101
Fajriatin, K. 27
Fathurrohman 175
Febriani, Y. 169
Firdaus, F.M. 270
Firmansyah 40
Fleer, M. 7
Fragkiadaki, G. 7

Gafur, A. 27

Handoyo, R.R. 50
Hapsari, W.P. 132, 163
Hardiani, W. 64
Hariadi, P. 169
Haritani, H. 169
Haryani 255
Haryanto 132, 163, 220
Hastomo, A. 270
Hastuti, W.S. 244
Hermanto 64

Hidayah, L.R. 199
Hidayat, M. 142
Hidayati 40

Iffat Rahmatullah, S. 249
Ishartiwi 50
Ismail, M.H. 91
Ismiyasari, F.N. 96
Isvandari, S.N. 182

Jabar, C.S.A. 182
Jamilah 178
Jaswanti, B.D. 209
Juantara, B. 138
Jusuf, R. 147

Karim, K. 275
Kartono 196
Kasiyan 119
Kawuryan, S.P. 40, 199
Kurniawan, W. 96
Kurniawati 255
Kusrahmadi, S.D. 270

Labib, U.A. 132
Liasari, I.W. 125

Maftuh, B. 226
Mahendra, I.K. 45
Mahendra, Y. 31
Manullang, B. 196
Maryani, Sc. 196
Masami, H. 1
Minarsih, N.M.M. 45
Mohamad, N.A. 18
Mujinem 40
Mulyatiningsih, B.E. 107
Mustadi, A. 80, 186, 214, 260

Nasiwan 31
Nio, T.H. 196
Nopriansah 157
Nugraha, T. 69
Nuryani, H. 220

Oktresia, E.E. 169

Prabawanto, S. 69
Prasetia, H. 175

Pujianto 244
Pujiastuti, P. 192, 244
Purba, D. 138
Purbani, W. 101, 113
Purwandari, E. 50
Purwanta, E. 209
Purwono 244
Putri, H.E. 226

Rafsanjani, A. 169
Rahayu, S. 74
Rahmawati, D. 192
Rahmawati, E. 96
Rahmia, S.H. 31
Rai, P. 7
Ridwan, M. 178
Rintaningrum, R. 249
Rochmah, E.N. 260
Rosita, M.N.M.I.Y. 232

Sabri 275
Sadjim, U.M. 147
Saptono, B. 270
Sari, Y.P. 275
Sartono, E.K.E. 175
Sayekti, O.M. 260
Senen, A. 40
Setiawati, F.A. 281
Shodiq, S.F. 142
Sholihah, U. 113
Suarni, N.K. 232, 238
Sudarsana, G.N. 232, 238
Sugiarsih, S. 214, 260
Sugiman 54
Sujati, H. 60, 255
Sujati, K.I. 163
Sukitman, T. 178
Sulistyo, A. 119
Supardi 74
Suparlan 175
Suprayitno, E. 36
Suresman, E. 142
Surya, P. 80
Susanti, M. 157, 275
Sutama 96
Suyitno, H. 196
Syahrul 151
Syamsudin, A. 125, 163

Verrawati, A.J. 186

Wangid, M.N. 199
Widyasari, C. 96
Widyastuti, T. 281
Wijaya, E. 157

Wuryandani, W. 175, 186

Yatini, A. 107
Yuliana, T.P. 169

Yuliyanto, A. 226
Yustanti, N.V. 286

Zubaidah, E. 214, 260